Aircraft
Basic Science

Seventh Edition

Aviation Technology Series

Aircraft Powerplants, *Kroes/Wild/Bent/McKinley*
Aircraft Maintenance and Repair, *Kroes/Watkins/Delp*
Aircraft Basic Science, *Kroes/Rardon*
Aircraft Electricity and Electronics, *Eismin/Bent/McKinley*

GLENCOE Aviation Technology Series

Aircraft
Basic Science

Seventh Edition

Kroes • Rardon

GLENCOE

Macmillan/McGraw-Hill

New York, New York   Columbus, Ohio   Mission Hills, California   Peoria, Illinois

AIRCRAFT BASIC SCIENCE, Seventh Edition
International Editions 1993

Exclusive rights by McGraw-Hill Book Co. - Singapore for manufacture and export.
This book cannot be re-exported from the country to which it is consigned by McGraw-Hill.

1 2 3 4 5 6 7 8 9 0 KHL PMP 9 8 7 6 5 4 3

Library of Congress Cataloging-in-Publication Data

Kroes, Michael J.
    Aircraft basic science / Michael J. Kroes, James R. Rardon. - 7th ed.
        p.   cm. - (Aviation technology series)
    Rev. ed. of: Aircraft basic science / Michael J. Kroes . . . [et al.].   6th ed.   1988
    Includes index.
    ISBN 0-02-801814-1
    1. Airplanes - Design and construction.   I. Rardon, James R.   II. Title.   III. Series.
TL671:2.K74   1993
629.13-dc20   92-35355
                                                                        CIP

When ordering this title, use ISBN 0-07-112517-5

Printed in Singapore

Contents

Preface

Aircraft Basic Science is one of five textbooks in the Aviation Technology Series. Its purpose is to provide the general technical information needed as a foundation for work as a technician in the field of aviation maintenance. The subjects covered in this text are those that are usually applicable to both airframes and powerplants and their associated systems. The material is of a general nature since, in addition to applying to the maintenance technician, it is also applicable to many related fields of study in aviation.

This edition has been updated to include many recent advances in aviation technology and, with the exception of electricity, has been revised to meet the requirements for general subject material specified in FAR Part 147. Electricity and electronics are not covered but are treated in detail in another text in this series.

The current "aging-aircraft" situation requires that the maintenance technician have more knowledge of materials, processes, and inspection procedures. The chapter involving fabrication processes has been expanded to provide more information in this area.

Throughout the text an increased emphasis has been placed on relating the material to both general-aviation and transport-category aircraft. In addition, the chapters on aerodynamics have been reorganized and expanded to provide a solid foundation in the theory of flight.

Due to the increased emphasis on understanding the Federal Aviation Regulations and on the ability to use technical maintenance publications, this material has been expanded and divided into two new chapters.

This book, when used along with the other books in the series and in conjunction with classroom and shop instruction, will provide the student with the technical information needed to qualify for certification as an airframe and powerplant technician.

This book is intended as a text and general reference book for those individuals involved in aircraft maintenance and operation. The material presented is for use in understanding aircraft materials, processes, practices, and operations. Technical information contained in this book should not be substituted for that provided by manufacturers.

Michael J. Kroes
James R. Rardon

Acknowledgments

The authors wish to express appreciation to the following organizations for their generous assistance in providing illustrations and technical information for this text: Aircraft Spruce and Specialty, Fullerton, California; Air France, New York, New York; Aeroquip Corporation, Jackson, Michigan; Atlas Copco, Inc., Orange, California; Aluminum Association, Inc., Washington, D.C.; Aluminum Company of America, Los Angeles, California; Avtec Corp., San Pablo, California; Beechcraft Corp., Wichita, Kansas; Bell Helicopter Textron, Division of Textron, Inc., Fort Worth, Texas; Boeing Company, Seattle, Washington; Boeing Vertol, Division of Boeing Company, Philadelphia, Pennsylvania; Cessna Aircraft Co., Wichita, Kansas; Cherry Fasteners, Townsend Division, Textron, Inc., Santa Ana, California; Cleveland Twist Drill, an Acme Cleveland Company, Cleveland, Ohio; Deutsch Metal Components Division, Los Angeles, California; Douglas Aircraft Company, Division of McDonnell Douglas Corporation, Long Beach, California; Evergreen Weigh, Inc., Lynwood, Washington; Federal Learjet Corp., Tucson, Arizona; General Electric Corp., Lynn, Massachusetts; Grumman Aircraft Corp., Bethpage, New York; Hi-Shear Corporation, Torrance, California; Hobart Brothers Co., Troy, Ohio; International Aviation Publishers, Riverton, Wyoming; International Nickel Corporation, New York, New York; Lockheed Corp., Burbank, California; Lufkin Division, Cooper Group, Apex, North Carolina; National Aeronautics and Space Administration, Washington, D.C.; Nicholson File Division, Cooper Group, Apex, North Carolina; Northrop Corporation, Hawthorne, California; Northrop University, Inglewood, California; Peterson Publishing Co., Los Angeles, California; Piper Aircraft Corporation, Vero Beach, Florida; Proto Tool Company, Los Angeles, California; Purdue University, West Lafayette, Indiana; Resistoflex Corporation, Roseland, New Jersey; Rocketdyne Division, Rockwell International Corp., Canoga Park, California; Sikorsky Aircraft Division, United Technologies, Stratford, Connecticut; Snap-on Tools, Kenosha, Wisconsin; Stanley Tools, New Britain, Connecticut; Stellite Division of the Cabot Corp., Kokomo, Indiana; Stratoflex, Inc., Fort Worth, Texas; VOI-SHAN Division, VSI Corporation, Culver City, California.

In addition, the authors wish to thank the many aviation schools and instructors who provided valuable suggestions, recommendations, and technical information for the revision of this text.

Fundamentals of Mathematics 1

INTRODUCTION

The science of mathematics, so important to the modern age of technology, had its beginnings in the dim ages of the past. It is probable that prehistoric people recognized the differences in quantities at an early age and therefore devised methods for keeping track of numbers and quantities. In the earliest efforts at trade it was necessary for the traders to figure quantities. For example, someone might have traded ten sheep for two cows. To do this the trader had to understand the numbers involved.

As time progressed, the ancient Babylonians and Egyptians developed the use of mathematics to the extent that they could perform marvelous engineering feats. Later the Greeks developed some of the fundamental laws which are still in use today. One of the great Greek mathematicians was a philosopher named Euclid, who prepared a work called *Elements of Geometry*. This text was used by students of mathematics for almost 2000 years. Another Greek mathematician was Archimedes, who is considered one of the greatest mathematicians of all time. One of his most important discoveries was the value of π (pi), which is obtained by dividing the circumference of a circle by its diameter. Archimedes discovered many other important mathematical relationships and also developed the early study of calculus. Modern differential and integral calculus were discovered by Sir Isaac Newton in the seventeenth century. These discoveries are considered some of the most important in the history of mathematics.

Today's modern technology, including aircraft maintenance, is greatly dependent upon mathematics. Computing the weight and balance of an aircraft, designing a structural repair, or determining the serviceability of an engine part are but a few examples of an aviation maintenance technician's need for mathematics. Electronic calculators and computers have made mathematical calculations more rapid and usually more accurate. However, these devices are only as good as the information put into them and do not excuse the technician from learning the fundamentals of mathematics.

It is expected that you, the aviation technician/student, have taken or are taking mathematics courses that go beyond the material in this chapter. The purpose of this chapter is to refresh your understanding of fundamental mathematical processes. Emphasis is placed on those mathematical terms or problems that you will encounter in portions of your technical studies or employment.

ARITHMETIC

Numbers

The ten single-number characters, or **numerals**—1, 2, 3, 4, 5, 6, 7, 8, 9, and 0—are called **digits**. Any number may be expressed by using various combinations of these digits. The arrangement of the digits and the number of digits used determine the value of the number being expressed.

Our number system is called a **decimal** system, the name being derived from the Latin word *decem*, meaning "ten." In the decimal system the digits are arranged in columns, which are powers of 10. The column in which a certain digit is placed determines its expressed value. When we examine the number 3 235 467, we indicate the column positions as follows:

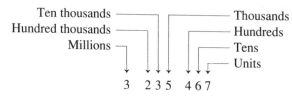

We may analyze the total number by considering the values expressed by each column, thus:

Units	7	7
Tens	6	60
Hundreds	4	400
Thousands	5	5 000
Ten thousands	3	30 000
Hundred thousands	2	200 000
Millions	3	3 000 000

We may now observe that the total number consists of 3 millions, 2 hundred thousands, 3 ten thousands, 5 thousands, 4 hundreds, 6 tens, and 7 units. The total number is read "three million, two hundred thirty-five thousand, four hundred sixty-seven."

There are several classes of numbers. **Whole numbers**, also called **integers**, are those which contain no fractions. Examples of such numbers are 3, 10, 250, and 435. A **fraction** is a part of a unit. A **mixed number** contains a whole number and a fraction. An **even number** is one which is divisible by 2. The numbers 2, 4, 6, 8, 10, 48, and 62 are even. **Odd numbers** are those which are not divisible by 2. The numbers 3, 5, 11, 13, 53, and 61 are odd.

Addition and Subtraction

Addition and subtraction may be considered the simplest of mathematical operations; however, these operations require practice to do quickly and accurately.

Addition.   Addition is the process of combining the values of two or more numbers into a single value. The combined value is called the **sum** of the values (numbers). The sign for addition is the **plus sign** (+). This sign placed between numbers indicates that they are to be added. Numbers to be added may be arranged horizontally or vertically in columns, as shown here:

$$324 + 25 + 78 = 427$$

$$\begin{array}{r} 324 \\ 25 \\ + \ 78 \\ \hline 427 \end{array}$$

Numbers to be added are usually arranged in columns for more speed and convenience in performing the addition.

$$\begin{array}{r} 7 \\ 6 \\ 3 \\ 8 \\ +5 \\ \hline 29 \end{array} \qquad \begin{array}{r} 32 \\ 420 \\ 8 \\ 19 \\ 26 \\ +248 \\ \hline 753 \end{array} \qquad \begin{array}{r} 4382 \\ 276 \\ 1820 \\ 2753 \\ 47 \\ + \ 238 \\ \hline 9516 \end{array}$$

Practice is one of the surest ways to learn to add accurately and rapidly. If you want to attain proficiency, you should take time to make up problems or find problems already prepared and then practice solving the problems until you feel comfortable.

It is recommended that you practice adding by sight. It is quite easy to learn to add by sight when the numbers to be added contain only one digit. With a little practice, the sight of any two digits will immediately bring the sum to mind. Thus when seeing the digits 6 and 5, for example, you should immediately think *11*, or upon seeing 9 and 7, you should instantly think *16*.

When we want to add two-digit numbers by sight, it is merely necessary to add the units and then the tens. Suppose that the numbers 45 and 23 are presented for addition. The units are 5 and 3, so we immediately think *8 units*. The tens are 4 and 2, so we think *6 tens*. The sum of 6 tens and 8 units is 68. If the units in an addition add to a sum greater than 9, we must remember to add the ten or tens to the sum of the tens. If we wish to add 36 and 57, for example, we see that the units add to 13, or 1 ten and 3 units. We record the 3 units and **carry** the ten, adding it to the 3 tens and 5 tens. The result is 9 tens and 3 units, or 93.

Subtraction.   Subtraction is the reverse of addition. The sign for subtraction is the minus sign (−). In ordinary arithmetic a smaller number is always subtracted from a larger number.

In subtraction the number from which another is to be subtracted is called the **minuend**, the number being subtracted from the other is called the **subtrahend**, and the result is called the **difference**.

$$\begin{array}{rl} 675 & \text{minuend} \\ -342 & \text{subtrahend} \\ \hline 333 & \text{difference} \end{array}$$

In subtraction it is important to remember the components of a number, that is, the units, tens, hundreds, and so on. This will make it easier to perform the necessary operations. In the preceding example, the numbers in the subtrahend are smaller than the corresponding numbers in the minuend, and the operation is simple. If a number in the minuend is smaller than the corresponding number in the subtrahend, it is necessary to **borrow** from the next column. For example,

$$\begin{array}{r} 853 \\ -675 \\ \hline 178 \end{array}$$

In the first column we find the 3 smaller than the 5, and therefore we must borrow 1 ten from the next column. We then subtract 5 from 13 to obtain 8. We must remember that there are only 4 tens left in the second column, and we have to borrow 1 hundred from the next column to make 140. We subtract 70 from 140 and obtain 70, and so we place a 7 in the tens column of the answer. Since we have borrowed 1 hundred from the 8 hundreds of the third column, only 7 hundreds are left. We subtract 6 hundreds from 7 hundreds, thus leaving 1 hundred. We therefore place a 1 in the hundreds column of the answer.

Multiplication

The act of **multiplication** may be considered multiple addition. If we add $2 + 2$ to obtain 4, we have multiplied 2 by 2, because we have taken 2 two times. Likewise, if we add $2 + 2 + 2 + 2$ to obtain 8, we have multiplied 2 by 4, because we have taken 2 four times.

In multiplication the number to be multiplied is called the **multiplicand**, and the number of times the multiplicand is to be taken is called the **multiplier**. The answer obtained from a multiplication is the **product**. The following example illustrates these terms:

$$\begin{array}{rl} 425 & \text{multiplicand} \\ \times \quad 62 & \text{multiplier} \\ \hline 850 & \\ 25 \ 50 & \\ \hline 26 \ 350 & \text{product} \end{array}$$

Note that the terms *multiplicand* and *multiplier* may be interchanged. For example, 2 × 4 is the same as 4 × 2.

When we use multiplication to solve a specific problem, the names of the terms have more significance. For example, if we wish to find the total weight of 12 bags of apples and each bag weighs 25 pounds (lb), then the multiplicand is 25 and the multiplier is 12. We then say 12 times 25 lb is 300 lb, or 12 × 25 = 300.

We can understand the multiplication process by analyzing a typical but simple problem, such as multiplying 328 by 6.

$$
\begin{array}{r}
328 \\
\times \quad 6 \\
\hline
48 \quad = 6 \times 8 \\
120 \quad = 6 \times 20 \\
1800 \quad = 6 \times 300 \\
\hline
1968
\end{array}
$$

In actual practice we do not write down each separate operation of the multiplication as shown in the foregoing problem, but we shorten the process by carrying figures to the next column. In the problem shown we can see that 6 × 8 is 48 and that the 4 goes into the tens column. Therefore, when we multiply, we merely carry the 4 over and add it to the next multiplication, which is in the tens column. When we use this method, the operation is as follows:

$$
\begin{array}{r}
14 \\
328 \\
\times \quad 6 \\
\hline
1968
\end{array}
$$

The first step in this operation is to multiply 8 by 6.

$$6 \times 8 = 48$$

Record the 8 (units) and carry the 4 (tens), then multiply 2 by 6.

$$6 \times 2 = 12$$

Add the 4 to obtain 16. Record the 6 (tens) and carry the 1 (hundreds). Then multiply 3 by 6.

$$6 \times 3 = 18$$

Add the 1 and obtain 19. Record the 19.

When there is more than one digit in the multiplier, we repeat the process for each digit, but we must shift one column to the left for each digit. This is because the right-hand digit of the multiplier is units, the next digit to the left is tens, the next is hundreds, and so on. If we multiply 328 by 246, we proceed as follows:

$$
\begin{array}{r}
328 \\
\times \quad 246 \\
\hline
1\ 968 \\
13\ 120 \\
65\ 600 \\
\hline
80\ 688
\end{array}
\qquad
\begin{array}{r}
328 \\
\times \quad 246 \\
\hline
1\ 968 \\
13\ 12 \\
65\ 6 \\
\hline
80\ 688
\end{array}
$$

Zeros were placed at the end of the second and third multiplications in the first example to show that we were multiplying by 40 and 200, respectively. In actual practice the zeros are not usually recorded. In the preceding multiplication we multiplied 328 first by 6, then by 40, and finally by 200. When we added these products, we obtained the answer, 80 688.

Accurate multiplication requires great care. First, it is important to know the multiplication tables. Second, care must be taken to record products in the correct column. Third, the addition must be made carefully and accurately. In order to acquire proficiency in multiplication, practice is essential.

In any mathematical problem it is smart to check the answer for accuracy. There are a number of methods for checking multiplication, and the most obvious is to divide the product by either the multiplicand or the multiplier. If the product is divided by the multiplicand, the quotient (answer) should be the multiplier.

Another method for checking multiplication is to repeat the problem, reversing the multiplicand and multiplier. If the product is the same in each case, the answer is probably correct.

Division

The act of **division** may be considered the reverse of multiplication; that is, division is the separating or dividing of a number into a certain number of equal parts. The symbol for division is the division sign (÷), and it is read "divided by." For example, 98 ÷ 4 is read "98 divided by 4." In arithmetic there are two commonly used methods for the division of whole numbers. These are **short division** and **long division**. The terms used to describe the elements of a division problem are **dividend**, which is the number to be divided; **divisor**, the number of times the dividend is to be divided; and **quotient**, the number of times the divisor goes into the dividend. In the problem 235 ÷ 5 = 47, the number 235 is the dividend, 5 is the divisor, and 47 is the quotient.

The process of short division is often used to divide a number by a divisor having only one digit. This is accomplished as follows:

$$
\begin{array}{r}
3 \\
7\overline{)3857} \\
\hline
551
\end{array}
$$

The first step is to divide 38 by 7. Since 7 × 5 = 35, it is obvious that after the division of 38 by 7 there will be a remainder of 3. This 3 is held over in the hundreds column and becomes the first digit of the next number to be divided. This number is 35, and 7 goes into 35 five times without leaving a remainder. The only number left to divide is the 7, into which the divisor goes once. The quotient is thus 551. The process of division as just explained may be understood more thoroughly if we analyze the numbers involved. The dividend 3857 may be expressed as 3500 + 350 + 7. These numbers divided separately by 7 produce the quotients 500, 50, and 1. Adding these together gives 551, which is the quotient obtained from the short division.

Long division is employed most often when the dividend and the divisor both contain more than one digit. The process is somewhat more complex than that of short division, but with a little practice, long division may be accomplished easily and accurately.

To solve the problem 18 116 ÷ 28, we arrange the terms of the problem as shown here:

$$\begin{array}{r} 647 \\ 28{\overline{\smash{\big)}\,18\ 116}} \\ \underline{16\ 8} \\ 1\ 31 \\ \underline{1\ 12} \\ 196 \\ \underline{196} \\ 0 \end{array}$$

The first step in solving the problem is to divide 181 by 28, because 181 is the smallest part of the dividend into which 28 can go. It is found that 28 will go into 181 six times, with a remainder of 13. The number 168 (6 × 28) is placed under the digits 181 and is subtracted. The number 13, which is the difference between 168 and 181, is placed directly below the 6 and 8 as shown, and then the number 1 is brought down from the dividend to make the number 131. The divisor 28 will go into 131 four times, with a remainder of 19. The final digit 6 of the dividend is brought down to make the number 196. The divisor 28 will go into 196 exactly seven times. The quotient of the entire division is thus 647.

If we study the division shown in the foregoing example, we will find that the dividend is composed of 28 × 600 = 16 800, 28 × 40 = 1120, and 28 × 7 = 196. Then by adding 16 800 + 1120 + 196, we find the sum, which is 18 116, the original dividend. We could divide each part of the dividend by 28 separately to obtain 600, 40, and 7 and then add these quotients together; however, it is usually quicker and simpler to perform the divisions as shown.

If a divisor does not go into a dividend an even number of times, there will be a remainder. This remainder may be expressed as a whole number, a fraction, or a decimal. Fractions and decimals are discussed later in this chapter.

In the following example the divisor will not go into the dividend an even number of times, so it is necessary to indicate a remainder:

$$\begin{array}{r} 223\frac{10}{16} \\ 16{\overline{\smash{\big)}\,3578}} \\ \underline{32} \\ 37 \\ \underline{32} \\ 58 \\ \underline{48} \\ 10 \end{array}$$

Fractions

A **fraction** may be defined as a part of a quantity, unit, or object. For example, if a number is divided into four equal parts, each part is one-fourth ($\frac{1}{4}$) of the whole number. The parts of a fraction are the **numerator** and the **denominator**, separated by a line indicating division.

In Figure 1–1 a rectangular block is cut into four equal parts; each single part is $\frac{1}{4}$ of the total. Two of the parts make $\frac{1}{2}$ the total, and three of the parts make the fraction $\frac{3}{4}$ of the total.

A fraction may be considered an indication of a division. For example, the fraction $\frac{3}{4}$ indicates that the numerator 3 is to be divided by the denominator 4. One may wonder how a smaller number, such as 3, can be divided by a larger number, such as 4. It is actually a relatively simple matter to accomplish such a division when we apply it to a practical problem. Suppose we wish to divide 3 gallons (gal) of water into four equal parts. Since there are 4 quarts (qt) in a gallon, we know that there are 12 qt in 3 gal. We can then divide the 12 qt into 4 equal parts of 3 qt each. Three quarts is $\frac{3}{4}$ gal; thus we see that 3 divided by 4 is equal to $\frac{3}{4}$. The principal fact to remember concerning fractions is that a fraction indicates a division. The fraction $\frac{1}{2}$ means that 1 is to be divided by 2, or that the whole is to be cut in half.

A fraction whose numerator is less than its denominator is called a **proper fraction**. Its value is less than 1. If the numerator is greater than the denominator, the fraction is called an **improper fraction**.

A **mixed number** is a combination of a whole number and a fraction, such as $32\frac{2}{5}$ and $325\frac{23}{35}$, which mean $32 + \frac{2}{5}$ and $325 + \frac{23}{35}$.

Fractions may be changed in form without changing their values. If the numerator and the denominator of a fraction are both multiplied by the same number, the value of the fraction remains unchanged, as shown in the following example:

$$\frac{3 \times 3}{4 \times 3} = \frac{9}{12}$$

The value of $\frac{9}{12}$ is the same as $\frac{3}{4}$. In a similar manner, the value of a fraction is not changed if both the numerator and the denominator are divided by the same number.

$$\frac{24 \div 12}{36 \div 12} = \frac{2}{3}$$

Thus we see that a large fraction may be simplified in some cases. This process is called **reducing the fraction**. To reduce a fraction to its lowest terms, we divide both the numerator and the denominator by the largest number that will

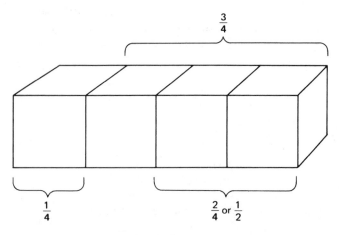

FIGURE 1–1 Fractions of a whole.

go into each without leaving a remainder. This is accomplished as follows:

$$\frac{36 \div 4}{40 \div 4} = \frac{9}{10} \quad \text{and} \quad \frac{525 \div 25}{650 \div 25} = \frac{21}{26}$$

Addition and Subtraction of Fractions. In order to add or subtract fractions, the denominators of the fractions must have equal values. For example, it is not possible to add $\frac{1}{3}$ to $\frac{2}{5}$ until the denominators of the fractions have been changed to equal values. Since 3 and 5 will both go evenly into 15, we can change $\frac{1}{3}$ to $\frac{5}{15}$, and $\frac{2}{5}$ to $\frac{6}{15}$. In this case, 15 is called the **lowest common denominator** (LCD) of the fractions being considered. It is now a simple matter to add the fractions.

$$\frac{5}{15} + \frac{6}{15} = \frac{11}{15}$$

We can see that this addition makes sense because $5 + 6 = 11$. Since both the 5 and the 6 denote a specific number of fifteenths, we add them to obtain the total number of fifteenths.

The foregoing principle may be understood more easily if we apply it to a practical problem. Suppose we wish to add 3 gal and 5 qt and 1 pint (pt) of gasoline. The most logical method is to convert all quantities to pints. In 3 gal of gasoline there are 3×8 or 24 pt; in 5 qt there are 5×2 or 10 pt. Then we add 24 pt + 10 pt + 1 pt. The answer is 35 pt. If we wish to convert this quantity to gallons, we must divide the 35 by 8. We find that we have 4 gal and 3 pt, or $4\frac{3}{8}$ gal.

To prepare fractions for adding or subtracting we proceed as follows:

1. Find the LCD.
2. Divide the LCD by each denominator.
3. Multiply the numerator and denominator of each fraction by the quotient obtained when the LCD was divided by the denominator.

To practice these steps, perform the following addition:

$$\frac{3}{4} + \frac{7}{8} + \frac{5}{6}$$

The LCD is 24. Divide the LCD by the first denominator, and then multiply the fraction by this quotient:

$$24 \div 4 = 6$$

$$\frac{3 \times 6}{4 \times 6} = \frac{18}{24}$$

Do the same for the second fraction:

$$24 \div 8 = 3$$

$$\frac{7 \times 3}{8 \times 3} = \frac{21}{24}$$

And for the third fraction:

$$24 \div 6 = 4$$

$$\frac{5 \times 4}{6 \times 4} = \frac{20}{24}$$

Then add all the fractions:

$$\frac{18}{24} + \frac{21}{24} + \frac{20}{24} = \frac{59}{24} = 2\frac{11}{24}$$

Adding and Subtracting Mixed Numbers. When adding and subtracting mixed numbers, we must consider both the whole numbers and the fractions. To add $5\frac{3}{8} + 7\frac{2}{3}$, we should first add 5 and 7 to obtain 12, and then we must add the fractions. We find that $\frac{3}{8} = \frac{9}{24}$, $\frac{2}{3} = \frac{16}{24}$, and thus $\frac{9}{24} + \frac{16}{24} = \frac{25}{24}$, or $1\frac{1}{24}$. Then $12 + 1\frac{1}{24} = 13\frac{1}{24}$, the total sum of the mixed numbers.

Subtraction of mixed numbers is accomplished by subtracting the whole numbers and then the fractions. For example, subtract $8\frac{2}{3}$ from $12\frac{3}{4}$.

$$\begin{array}{r} 12\frac{3}{4} = 12\frac{9}{12} \\ - 8\frac{2}{3} = - 8\frac{8}{12} \\ \hline 4\frac{1}{12} \end{array}$$

If the fraction of the subtrahend is greater than the fraction of the minuend, it is necessary to borrow 1 from the whole number in the minuend to increase the fraction of the minuend. If we wish to subtract $5\frac{7}{8}$ from $9\frac{1}{3}$, we must increase the $\frac{1}{3}$ to a value greater than $\frac{7}{8}$. The LCD of the fractions is 24, and so $5\frac{7}{8}$ becomes $5\frac{21}{24}$ and $9\frac{1}{3}$ becomes $9\frac{8}{24}$. We must then borrow 1 from 9 and add the 1 to $\frac{8}{24}$. The minuend then becomes $8\frac{32}{24}$. The final form of the problem is then

$$\begin{array}{r} 8\frac{32}{24} \\ -5\frac{21}{24} \\ \hline 3\frac{11}{24} \end{array}$$

Multiplication of Fractions. Multiplication of fractions is accomplished by placing the product of the numerators over the product of the denominators. This result is then reduced to lowest terms. For example,

$$\frac{2}{5} \times \frac{1}{2} \times \frac{3}{4} = \frac{6}{40} = \frac{3}{20}$$

Where possible in the multiplication of fractions, cancellation is employed to simplify the fractions before final multiplication takes place.

$$\frac{\cancel{5}}{\cancel{8}} \times \frac{\cancel{1}}{\cancel{5}} \times \frac{\cancel{9}^{3}}{10} \times \frac{\cancel{4}}{\cancel{5}} = \frac{3}{10}$$

In the preceding problem we have canceled all values except the 3 in the numerator and the 10 in the denominator. First we canceled the 5s, and next we divided the 8 in the denominator by the 2 in the numerator. The 4 that was left in the denominator was then canceled by the 4 in the numerator. The 9 in the numerator was divided by the 3 in the denominator to leave a 3 in the numerator. The product, $\frac{3}{10}$, contains the product of the numerators and the product of the denominators reduced to their lowest terms. This may be proved by multiplying the numerators and denominators without canceling any values, as follows:

$$5 \times 2 \times 9 \times 4 = 360$$

$$8 \times 3 \times 10 \times 5 = 1200$$

Then,

$$\frac{360}{1200} = \frac{3}{10}$$

The reduction of the fraction was accomplished by dividing both the numerator and the denominator by 120.

A problem involving a few more operations than that given previously is

$$\frac{\cancel{25}^{5}}{\cancel{6}} \times \frac{\cancel{36}^{3}}{7} \times \frac{\cancel{11}}{\cancel{20}^{4}} \times \frac{9}{\cancel{44}_{4}} = \frac{135}{56} = 2\frac{23}{56}$$

In this problem note that the 25 in the numerator and the 20 in the denominator were both divided by 5 to obtain a 5 in the numerator and a 4 in the denominator.

Dividing Fractions. The division of fractions is simply accomplished by inverting the divisor and multiplying. Inverting a fraction means to turn it over; for example, if we invert $\frac{3}{4}$, it becomes $\frac{4}{3}$. It is also of interest to note that $\frac{4}{3}$ is the reciprocal of $\frac{3}{4}$. If we invert a whole number, we merely place a 1 above it. Hence, 3 becomes $\frac{1}{3}$ when it is inverted. To practice, divide $\frac{5}{8}$ by $\frac{7}{15}$.

$$\frac{5}{8} \div \frac{7}{15} = \frac{5}{8} \times \frac{15}{7} = \frac{75}{56} = 1\frac{19}{36}$$

Multiplying and Dividing Mixed Numbers. Mixed numbers may be multiplied or divided by changing the mixed numbers to improper fractions and then proceeding as with fractions. For example, try multiplying $5\frac{7}{8}$ by $3\frac{2}{3}$.

$$5\frac{7}{8} = \frac{47}{8} \text{ and } 3\frac{2}{3} = \frac{11}{3}$$

$$\frac{47}{8} \times \frac{11}{3} = \frac{517}{24} = 21\frac{13}{24}$$

For another example, divide $9\frac{3}{4}$ by $4\frac{2}{3}$.

$$9\frac{3}{4} = \frac{39}{4} \text{ and } 4\frac{2}{3} = \frac{14}{3}$$

$$\frac{39}{4} \div \frac{14}{3} = \frac{39}{4} \times \frac{3}{14} = \frac{117}{56} = 2\frac{5}{56}$$

The procedures explained in the preceding sections are not the only possible methods. Other methods will provide the same results, but those given here are commonly used.

Decimals

Decimal fractions, or **decimals**, provide a means of performing mathematical operations without using the time-consuming and complex methods of common fractions. A decimal fraction is a common fraction converted to tenths, hundredths, thousandths, or other power of ten. For example, if we convert the common fraction $\frac{3}{4}$ to a decimal, we find that it becomes 0.75. This is accomplished by dividing the numerator by the denominator:

$$\begin{array}{r} 0.75 \\ 4\overline{)3.00} \\ 2\ 8 \\ \hline 20 \\ 20 \\ \hline \end{array}$$

Any fraction may be converted to a decimal by this same process. Let us assume that we wish to convert the fraction $\frac{28}{35}$ to a decimal.

$$\begin{array}{r} 0.80 \\ 35\overline{)28.00} \\ 28.00 \\ \hline \end{array}$$

The decimal 0.80 is the same as 0.8 and may be read "eighty hundredths" or "eight tenths."

If we want to convert a fraction to a decimal when the denominator will not go evenly into the numerator, the decimal will be carried to the nearest tenth, hundredth, thousandth, or ten-thousandth according to the degree of accuracy required for the problem. For example, we may wish to convert the fraction $\frac{25}{33}$ to a decimal carried to the nearest ten-thousandth. We proceed as follows:

$$\begin{array}{r} 0.7576 \\ 33\overline{)25.0000} \\ 23\ 1 \\ \hline 1\ 90 \\ 1\ 65 \\ \hline 250 \\ 231 \\ \hline 190 \\ 198 \\ \hline \end{array}$$

Rounding Off Decimals. In the preceding problem, the answer would be alternately 7 and 5 indefinitely if we continued to carry the division onward. Instead we **round off** the answer to the degree of accuracy required. The accuracy required will be a function of the equipment being used. To calculate a sheet metal layout to one-thousandth (0.001) of an inch is not necessary if the scale being used can only measure in tenths (0.1) of an inch. To round off to tenths of an inch, the calculation would be carried to two decimal places. If the last, or second, digit is less than five, it is dropped. If it is five or more, one is added to the preceding number, and then the second number is dropped. To round off the above example to one decimal place, the calculation would be 0.75. The second digit is 5, so one is added to the first digit, resulting in a rounded-off answer of 0.8. If an accuracy of one-hundredth (.01), or two decimal places, was required, the calculation is out to three places (0.757). Since the third digit (7) is greater than five, the rounded-off answer is 0.76. If the calculation had resulted in a figure of 0.6336, the third digit, being less than 5, would have been dropped. The answer, accurate to two decimal places, would have been 0.63.

Multiplication of Decimals. The multiplication of decimals is performed in the same manner as the multiplication of whole numbers except that we must use care in placing the decimal point in the product. Let us assume that we wish to multiply 37.5 by 24.2.

$$
\begin{array}{r}
37.5 \\
\times\, 24.2 \\
\hline
750 \\
1500 \\
750 \\
\hline
90750 \\
\end{array}
$$

Having completed the multiplication of numbers containing decimals, we count the number of decimal places in the multiplicand and the multiplier and point off this many places in the product. In the foregoing example there is one decimal place in the multiplicand and one in the multiplier. We therefore point off two places in the answer. After the answer is obtained, we may drop any zeros at the right-hand end of the answer. The answer of the problem would then be 907.5 and be read "nine hundred seven and five tenths." If the zero were left in the answer, the decimal portion would be read "fifty hundredths."

It is often necessary to multiply decimals in which there are no whole numbers. For example,

$$
\begin{array}{rl}
0.056 & \text{(fifty-six thousandths)} \\
\times\, 0.325 & \text{(three hundred twenty-five thousandths)} \\
\hline
0280 \\
0112 \\
0168 \\
0\,000 \\
\hline
0.018200 & \text{(one hundred eighty-two ten-thousandths)}
\end{array}
$$

Since there is a total of six decimal places in the multiplicand and multiplier, we must point off six places in the product. This makes the answer 0.018200, which would be read "eighteen thousand two hundred millionths." In order to simplify the answer, we drop the two zeros at the right and read the answer "one hundred eighty-two ten-thousandths."

Note: It is customary with some writers to omit the zero ahead of a decimal point not preceded by a whole number. Thus 0.04 would be written .04. In either case the value is the same and the decimal is read "four hundredths." It is not necessary to use the zero before the decimal point, but it may aid in preventing mistakes.

Addition of Decimals. The addition of decimals is a simple matter provided the decimals are properly placed. In adding a column of numbers with decimals, the decimal points should be kept in line in a column, as shown here:

$$
\begin{array}{r}
23.065 \\
2.5 \\
354.2 \\
+\quad 0.637 \\
\hline
380.402 \\
\end{array}
$$

In the foregoing problem observe that the digits in the first column to the right of the decimal points add up to more than 10. When this occurs, we carry the 1 over into the units column. If the column should add up to 20, we would carry the 2 over to the units column.

Subtraction of Decimals. Subtraction of decimals is almost as easy as subtracting whole numbers. It is necessary, however, that we use care to avoid mistakes in the placing of decimal points. This is illustrated in the following problems:

$$
\begin{array}{r}
652.25 \\
-\quad 28.64 \\
\hline
623.61 \\
\end{array}
\qquad
\begin{array}{r}
2568.2300 \\
-\quad 376.4532 \\
\hline
2191.7768 \\
\end{array}
\qquad
\begin{array}{r}
320.000 \\
-215.365 \\
\hline
104.635 \\
\end{array}
$$

Observe in the foregoing problems that where there are fewer decimal places in the minuend than in the subtrahend, we add zeros to fill the spaces. This aids in avoiding mistakes which could otherwise occur. The addition of the zeros does not affect the value of the decimals.

Division of Decimals. The division of decimals requires much more care than the addition, subtraction, or multiplication of these numbers. This is because it is easy to misplace the decimal point in the quotient. The principal rule to remember in dividing decimals is to place the decimal point of the quotient directly above the decimal point of the dividend. This is illustrated in the following problems:

$$
\begin{array}{r}
3.32 \\
28\overline{)92.96} \\
84 \\
\hline
8\,9 \\
8\,4 \\
\hline
56 \\
56 \\
\hline
\end{array}
\qquad
\begin{array}{r}
0.476 \\
34\overline{)16.184} \\
13\,6 \\
\hline
2\,58 \\
2\,38 \\
\hline
204 \\
204 \\
\hline
\end{array}
\qquad
\begin{array}{r}
0.0002268 \\
435\overline{)0.0986500} \\
870 \\
\hline
1165 \\
870 \\
\hline
2950 \\
2610 \\
\hline
3400 \\
3480 \\
\end{array}
$$

In the third problem just illustrated the division does not come out evenly, and so the answer is rounded off with an 8 to provide an accuracy to the nearest ten-millionth.

When the divisor contains decimals, we move the decimal point to the right until the divisor is a whole number. We then move the decimal in the dividend the same number of points to the right. This is equivalent to multiplying both the dividend and the divisor by the same number, and so the quotient remains the same.

To illustrate this point, let's divide 34.026 by 4.538.

$$
4.538\overline{)34.026}
$$

$$
\begin{array}{r}
7.49\ + \\
4538\overline{)34026.00} \\
31766 \\
\hline
22600 \\
18152 \\
\hline
44480 \\
40842 \\
\hline
3538 \\
\end{array}
$$

To further illustrate, let's divide 20.583 by 3.06.

$$
3.06\overline{)20.583}
$$

$$\begin{array}{r} 6.726\ + \\ 306\overline{)2058.300} \\ \underline{1836} \\ 222\ 3 \\ \underline{214\ 2} \\ 8\ 10 \\ \underline{6\ 12} \\ 1\ 980 \\ \underline{1\ 836} \\ 144 \end{array}$$

Now divide 23.42 by 4.3867.

$$4.3867\overline{)23.4200}$$

$$\begin{array}{r} 5.34\ - \\ 43867\overline{)234200.00} \\ \underline{219335} \\ 14865\ 0 \\ \underline{13160\ 1} \\ 1704\ 90 \\ 1754\ 68 \end{array}$$

The small + and − signs placed after the quotients in the preceding examples indicate that a small amount is to be added or subtracted if the number is to be made exact; that is, the exact answer is a little more or a little less than the answer shown.

Converting Decimals to Common Fractions. It has been stated that a decimal is a fraction, and of course this is true. A decimal fraction is a fraction that has 10, 100, 1000, or another power of ten, for the denominator. The decimal 0.34 is read "34 hundredths" and may be shown as $\frac{34}{100}$. Also, the decimal fraction 0.005 may be written as $\frac{5}{1000}$. To convert a decimal fraction to a common fraction, we merely write it in the fraction form and then reduce it to its lowest terms by dividing the numerator and denominator by the same number. To convert 0.325 to a common fraction, we write $\frac{325}{1000}$ and then divide the numerator and denominator by 25:

$$\frac{325 \div 25}{1000 \div 25} = \frac{13}{40}$$

To convert 0.625 to a fraction, we divide as follows:

$$\frac{625 \div 125}{1000 \div 125} = \frac{5}{8}$$

It is obvious that many decimals cannot be converted to small common fractions because the numerator and denominator may not have common factors. However, it may be possible to arrive at an approximate fraction which is within the accuracy limits required. For example, 0.3342 may be converted to approximately $\frac{1}{3}$.

Percentage

The term *percentage* is used to indicate a certain number of hundredths of a whole. The expression 5% means $\frac{5}{100}$, or 0.05. To find a certain percentage of a number we multiply the number by the number of the percentage and then move

the decimal point two places to the left. For example, to find 6% of 325 we multiply 325 by 6 to obtain 1950, and then we move the decimal two places to the left and find the answer 19.50, or 19.5. We could just as easily multiply by 0.06 to obtain the same answer.

Certain percentages are equal to commonly used fractions, and it is well to be familiar with these: 25% = $\frac{1}{4}$, 50% = $\frac{1}{2}$, 75% = $\frac{3}{4}$, $12\frac{1}{2}$% = $\frac{1}{8}$, and 33% = $\frac{1}{3}$. Familiarity with these fractions and their equivalent percentages is helpful in many computations.

If we wish to find what percentage one number is of another, we divide the first number by the second. For example, 26 is what percentage of 65?

$$\begin{array}{r} 0.40 \\ 65\overline{)26.00} \\ \underline{26\ 0} \end{array}$$

Since we change a decimal to a percentage by moving the decimal point two places to the right, 0.40 becomes 40%. Thus 26 is 40% of 65.

Ratio and Proportion

A **ratio** is the numerical relation between two quantities. If one man has two airplanes and another has three airplanes, the ratio of their airplane ownership is 2 to 3. This may also be expressed as $\frac{2}{3}$ or 2:3. Thus we see that a ratio is actually a fraction, and it may also be used mathematically as a fraction.

A ratio may be reduced to lowest terms in the same manner as a fraction. For example, the ratio 24:36 may be reduced to 2:3 by dividing each term of the ratio by 12. If a certain store has 60 customers on Friday and 80 on Saturday, the ratio is 60:80, or 3:4.

A **proportion** expresses equality between two ratios. For example, 4:5::12:15. This may also be expressed 4:5 = 12:15 or $\frac{4}{5} = \frac{12}{15}$.

In a proportion problem the outer numbers (such as 4 and 15 in the example just given) are called the **extremes**, and the two inside numbers (5 and 12) are called the **means**. In a proportion, the product of the means is equal to the product of the extremes. We may demonstrate this rule by using the preceding example.

$$5 \times 12 = 4 \times 15 = 60$$

We may use the rule to find an unknown term in a proportion.

$$6:16 = 9:?$$

Using x to denote the unknown quantity, we can say

$$6 \times x = 16 \times 9 \text{ or } 6x = 144$$

Then,

$$\frac{6x}{6} = \frac{144}{6} \text{ or } x = 24$$

We can prove the foregoing answer by using it in the original proportion.

$$6:16 = 9:24 \text{ or } \frac{6}{16} = \frac{9}{24} = \frac{3}{8}$$

Powers and Roots

A **power** of a number represents the number multiplied by itself a certain number of times. For example, $5 \times 5 = 25$; hence the **second power** of 5 is 25. If we multiply $5 \times 5 \times 5$ to obtain 125, we have found the third power of 5. The third power of 5 is indicated thus: 5^3. It is read "5 cubed" or "5 to the third power." The second power of a number is called the **square** of a number. This terminology is derived from the fact that the area of a square is equal to the length of one side multiplied by itself. The term **cube** is derived in a similar manner because the volume of a cube is equal to the length of one edge raised to the third power. Any power of any number may be found merely by continuing to multiply it by itself the indicated number of times. For example, 2^6 is equal to $2 \times 2 \times 2 \times 2 \times 2 \times 2 = 64$.

The small index number placed above and to the right of a number to indicate the power of the number is called an **exponent**. The number to be raised to a power is called the **base**. In the expression 25^4, the small number 4 is the exponent and the number 25 is the base. If we multiply 25 by itself the number of times indicated by the exponent 4, we find that $25^4 = 390\,625$.

A **factor** of a number is another number which will divide evenly into the first number. For example, 3 is a factor of 12. Other factors of 12 are 2, 4, and 6, because each of these numbers will divide evenly into 12. A **root** of a number is a factor which when multiplied by itself a certain number of times will produce the number. For example, 2 is a root of 4 because it will give a product of 4 when multiplied by itself. It is also a root of 8 because $2 \times 2 \times 2 = 8$. A **square root** is the root of a number which when multiplied by itself once will produce the number. For example, 3 is the square root of 9 because $3 \times 3 = 9$. It is the **cube root** of 27 because $3 \times 3 \times 3 = 27$. A root which must be multiplied by itself four times to produce a certain number is the fourth root of that number. Hence 3 is the fourth root of 81 because $3 \times 3 \times 3 \times 3 = 81$.

When the square root of a number is indicated, we place the number under the radical sign: $\sqrt{64}$. If a larger root is to be extracted, we place the index of the root in the radical sign: $\sqrt[3]{27}$. This indicates that the value expressed is the cube root of 27, or 3.

Many formulas in technical work require extraction of square roots. The development of the electronic calculator has greatly simplified this task. The speed and accuracy of the calculator has made learning the procedure for manually extracting a square root unnecessary. The procedure is shown in Figure 1–2 for those that are interested. Many tables containing various powers and roots, such as Table 1–1, are available. Tables of this type can be very useful when performing manual calculations.

Powers and Roots of Fractions. When a fraction is to be raised to a certain power, the numerator is multiplied by the numerator and the denominator is multiplied by the denominator. For example, if we wish to find the third power (cube) of $\frac{3}{4}$, we multiply $3 \times 3 \times 3$ for the new numerator and $4 \times 4 \times 4$ for the new denominator. Since $3 \times 3 \times 3 = 27$

Problem: *Extract the square root of 104 976.*

1. Place the number under the radical sign and separate it into **periods** of two digits each starting from the right of the number.

$$\sqrt{10'49'76}$$

2. Determine the nearest perfect square smaller than the first period on the left, and subtract this square from the first period. Place the root of the square above the first period. Bring down the next period to form the new dividend 149.

$$
\begin{array}{r}
3 \\
\sqrt{10'49'76} \\
9 \\
\hline
60)\,1\;49
\end{array}
$$

3. Multiply the root 3 by 20 and place the product to the left of the new dividend. The product 60 is the **trial divisor**. Determine how many times the trial divisor will go into the dividend 149. In this case 60 will go into 149 two times. Add 2 to the trial divisor to make 62, which is the complete divisor. Place 2 above the second period and then multiply the complete divisor by 2. Place the product 124 under the dividend 149 and subtract. Bring down the next period to make the new dividend.

$$
\begin{array}{r}
3\;2 \\
\sqrt{10'49'76} \\
9 \\
\hline
60)\,1\;49 \\
2 \\
\hline
62 \quad 1\;24 \\
25\;76
\end{array}
$$

4. Multiply the partial answer 32 by 20 to obtain the new trial divisor 640. Determine how many times 640 will go into 2576. Inspection indicates that it will go into 640 four times. Add the 4 to 640 to obtain 644, which is the complete divisor. Place the 4 above the third period. Then multiply the complete divisor 644 by 4 to obtain 2576. This product is equal to the dividend; hence the computation is complete.

$$
\begin{array}{r}
3\;2\;\;4 \\
\sqrt{10'49'76} \\
9 \\
\hline
60)\,1\;49 \\
2 \\
\hline
62 \quad 1\;24 \\
640)\;\;\;25\;76 \\
4 \\
\hline
644 \\
25\;76
\end{array}
$$

FIGURE 1–2 Procedure for extracting a square root.

and $4 \times 4 \times 4 = 64$, the cube of $\frac{3}{4}$ is $\frac{27}{64}$. To extract a particular root of a fraction, we must extract the roots of both the numerator and the denominator. For example, the square root of $\frac{4}{9} = \frac{2}{3}$. This is because the square root of 4 is 2 and the square root of 9 is 3.

Scientific Notation

Scientific notation is the process of using powers of ten to simplify mathematical expressions and computations. Fig-

TABLE 1-1 Tables of Squares, Cubes, Square Roots, and Cube Roots

No. n	Sq. n^2	Cube n^3	Square root $\sqrt{n}$	Cube root $\sqrt[3]{n}$	No. n	n^2	n^3	$\sqrt{n}$	$\sqrt[3]{n}$
1	1	1	1.000	1.000	51	2 601	132 651	7.141	3.708
2	4	8	1.414	1.259	52	2 704	140 608	7.211	3.732
3	9	27	1.732	1.442	53	2 809	148 877	7.280	3.756
4	16	64	2.000	1.587	54	2 916	157 464	7.348	3.779
5	25	125	2.236	1.710	55	3 025	166 375	7.416	3.803
6	36	216	2.449	1.817	56	3 136	175 616	7.483	3.825
7	49	343	2.645	1.913	57	3 249	185 193	7.549	3.848
8	64	512	2.828	2.000	58	3 364	195 112	7.615	3.870
9	81	729	3.000	2.080	59	3 481	205 379	7.681	3.893
10	100	1000	3.162	2.154	60	3 600	216 000	7.746	3.914
11	121	1331	3.316	2.224	61	3 721	226 981	7.810	3.936
12	144	1728	3.464	2.289	62	3 844	238 328	7.874	3.957
13	169	2197	3.605	2.351	63	3 969	250 047	7.937	3.979
14	196	2744	3.741	2.410	64	4 096	262 144	8.000	4.000
15	225	3375	3.873	2.466	65	4 225	274 625	8.062	4.020
16	256	4096	4.000	2.519	66	4 356	287 496	8.124	4.041
17	289	4913	4.123	2.571	67	4 489	300 763	8.185	4.061
18	324	5832	4.242	2.620	68	4 624	314 432	8.246	4.081
19	361	6859	4.358	2.668	69	4 761	328 509	8.306	4.101
20	400	8000	4.472	2.714	70	4 900	343 000	8.366	4.121
21	441	9261	4.582	2.758	71	5 041	357 911	8.426	4.140
22	484	1 0648	4.690	2.802	72	5 184	373 248	8.485	4.160
23	529	1 2167	4.795	2.843	73	5 329	389 017	8.544	4.179
24	576	1 3824	4.899	2.884	74	5 476	405 224	8.602	4.198
25	625	1 5625	5.000	2.924	75	5 625	421 875	8.660	4.217
26	676	1 7576	5.099	2.962	76	5 776	438 976	8.717	4.235
27	729	1 9683	5.196	3.000	77	5 929	456 533	8.775	4.254
28	784	2 1952	5.291	3.036	78	6 084	474 552	8.831	4.272
29	841	2 4389	5.385	3.072	79	6 241	493 039	8.888	4.290
30	900	2 7000	5.477	3.107	80	6 400	512 000	8.944	4.308
31	961	2 9791	5.567	3.141	81	6 561	531 441	9.000	4.326
32	1024	3 2768	5.656	3.174	82	6 724	551 368	9.055	4.344
33	1089	3 5937	5.744	3.207	83	6 889	571 787	9.110	4.362
34	1156	3 9304	5.831	3.239	84	7 056	592 704	9.165	4.379
35	1225	4 2875	5.916	3.271	85	7 225	614 125	9.219	4.396
36	1296	4 6656	6.000	3.301	86	7 396	636 056	9.273	4.414
37	1369	5 0653	6.082	3.332	87	7 569	658 503	9.327	4.431
38	1444	5 4872	6.164	3.362	88	7 744	681 472	9.380	4.448
39	1521	5 9319	6.245	3.391	89	7 921	704 969	9.434	4.464
40	1600	6 4000	6.324	3.420	90	8 100	729 000	9.486	4.481
41	1681	6 8921	6.403	3.448	91	8 281	753 571	9.539	4.497
42	1764	7 4088	6.480	3.476	92	8 464	778 688	9.591	4.514
43	1849	7 9507	6.557	3.503	93	8 649	804 357	9.643	4.530
44	1936	8 5184	6.633	3.530	94	8 836	830 584	9.695	4.546
45	2025	9 1125	6.708	3.556	95	9 025	857 375	9.746	4.562
46	2116	9 7336	6.782	3.583	96	9 216	884 736	9.798	4.578
47	2209	10 3823	6.855	3.608	97	9 409	912 673	9.848	4.594
48	2304	11 0592	6.928	3.634	98	9 604	941 192	9.899	4.610
49	2401	11 7649	7.000	3.659	99	9 801	970 299	9.949	4.626
50	2500	12 5000	7.071	3.684	100	10 000	1 000 000	10.000	4.641

ure 1–3 shows the values of ten for various powers. By using powers of ten to express very large numbers or very long decimals, the amount of computation necessary for multiplication, division, and extracting roots is reduced. Many calculators and computer programs use scientific notation to display large numbers or long decimals.

With scientific notation, long numbers can be simplified. The number to be simplified is divided by a power of ten. For example, let's express 2 600 000 in scientific notation with one digit to the left of the decimal point. (*Note:* $10^6 = 1\ 000\ 000$.)

$$\frac{2\ 600\ 000}{1\ 000\ 000} = 2.6$$

Thus,

$$2\ 600\ 000 = 2.6 \times 1\ 000\ 000$$

or

$10^0 = 1$	
$10^1 = 10$	$10^{-1} = 0.1$
$10^2 = 100$	$10^{-2} = 0.01$
$10^3 = 1000$	$10^{-3} = 0.001$
$10^4 = 10000$	$10^{-4} = 0.0001$
$10^5 = 100000$	$10^{-5} = 0.00001$
$10^6 = 1000000$	$10^{-6} = 0.000001$
$10^7 = 10000000$	$10^{-7} = 0.0000001$

FIGURE 1–3 Powers of 10.

$$2\,600\,000 = 2.6 \times 10^6$$

Other examples are

$$37\,542\,000 = 3.7542 \times 10^7$$
$$123\,000 = 1.23 \times 10^5$$

The exponent of 10 can easily be determined for scientific notation by counting the number of places that the decimal point is moved to the left. In the first example (37 542 000) the decimal has been moved seven places to the left. In the second example it has been moved five places, giving 10 the exponent of 5.

For numbers less than 1.0 the exponent will be negative. It can be determined by counting the number of places the exponent has moved to the right.

$$.0372 = 3.72 \times 10^{-2}$$
$$.000\,045\,67 = 4.567 \times 10^{-5}$$

To change from scientific notation to the actual number, move the decimal point the number of places indicated by the exponent. Move it to the right if the exponent is positive and to the left if it is negative.

Multiplication and division are simplified with scientific notation. In multiplying two numbers, as shown in Figure 1–4, the two base numbers are multiplied. The exponent for the answer is determined by adding the two exponents.

Two numbers stated in scientific notation may be divided as shown in Figure 1–5. After dividing the base numbers, the exponent is assigned a value obtained by subtracting the exponent of the divisor from that of the dividend.

The square root of a number in scientific notation can be found by finding the square root of the base number and dividing the exponent by 2 (see Figure 1–6).

ALGEBRA

Introduction

We may define **algebra** as the branch of mathematics which uses positive and negative quantities, letters, and other symbols to express and analyze relationships among units of quantitative data. The process of algebra enables us to make

---

Problem: *Multiply 12 000 000 by 173 000.*

1. Convert to scientific notation.

$$(1.2 \times 10^7) \times (1.73 \times 10^5)$$

2. Multiply both base numbers.

$$1.2 \times 1.73 = 2.076$$

3. Add the exponents.

$$7 + 5 = 12$$

4. Answer.

$$2.076 \times 10^{12} \text{ or } 2\,076\,000\,000\,000$$

FIGURE 1–4 Multiplication with scientific notation.

Problem: *Divide 12 000 000 by 173 000.*

1. Convert to scientific notation.

$$\frac{1.2 \times 10^7}{1.73 \times 10^5}$$

2. Divide both base numbers.

$$\frac{1.2}{1.73} = 0.694$$

3. Subtract the exponents.

$$7 - 5 = 2$$

4. Answer.

$$0.694 \times 10^2 \text{ or } 69.4$$

FIGURE 1–5 Division with scientific notation.

Problem: *Find the square root of 4.0 × 10⁶.*

$$\sqrt{4.0 \times 10^6}$$

1. Square root of 4 = 2
2. Square root of $10^6 = 10^3$
3. $\sqrt{4.0 \times 10^6} = 2.0 \times 10^3$

FIGURE 1–6 Finding a square root with scientific notation.

calculations and arrive at solutions which would be difficult or impossible through normal arithmetic methods. All mathematical systems beyond arithmetic employ the methods of algebra for computation.

Aviation maintenance technicians use many algebraic formulas and expressions on a daily basis. In many cases these operations have become so routine that many do not realize that algebra is being used. The formula for the area of a rectangle, $A = l \times w$, or $A = lw$, is an algebraic expression (A = Area, l = length, and w = width). The formula for finding the force on a hydraulic piston may be expressed $F = P/A$, where F is force in pounds, A is area of the piston in square inches, and P is the pressure of the fluid in pounds per square inch (psi). In computing the weight and balance of an aircraft, the technician is not only working with algebraic formulas but with positive and negative quantities. Most computations involving these operations have been simplified to where the problem can be solved by placing the proper numbers into the formula. However, a knowledge of algebra is essential for the technician to understand what is happening in the procedure or system.

Equations

An equation is a mathematical expression of equality. For example, $2 + 6 = 8$ is a simple equation. In the general terms of algebra this equation would be $a + b = c$. If the value of any two of the symbols is known, the other one can be determined. If $a = 2$ and $b = 6$ in the equation, then we know that $c = 8$ because $2 + 6 = 8$.

If we ask the question "What number added to 6 will produce 10?" we can express the question in algebraic terms thus: $6 + x = 10$. To find the value of x, we must subtract 6 from 10, and so we rearrange the equation to $x = 10 - 6$. Note that we changed the sign of the 6 when we **transposed** it (moved it to the opposite side of the equals sign). We complete the solution and state the simplified equation as $x = 4$.

Positive and Negative Numbers

In algebra we use the same signs that are used in arithmetic; however, in algebra the signs sometimes have a greater significance than they do in arithmetic. All terms in algebra must have either a positive or a negative value. Terms having a positive value are preceded by a plus sign $(+)$ or by no sign at all. Negative terms are preceded by a minus sign $(-)$. A positive number or expression has a value greater than zero, and a negative number has a value less than zero. This may be understood by considering temperature. If we were told that the temperature was $10°$, we would not know for sure what was meant unless we knew whether it was $10°$ above zero or $10°$ below zero. If the temperature was above zero, it could be shown as $+10°$, and if below zero, it could be shown as $-10°$.

Algebraic Addition

Algebraic addition is the process of combining terms to find the actual value of the terms. The sum $5 + 6 + 8 = 19$ is algebraic addition as well as arithmetic addition. The sum $-5 - 6 + 4 = -7$ is also algebraic addition, but we do not use this method in arithmetic. *To add the terms in an algebraic expression when there are both negative and positive quantities, we combine the terms with the same sign, subtract the smaller value from the larger, and then give the answer the sign of the larger.* To add $8 -9 -4 +6 +7 -3$, we combine the 8, 6, and 7 to obtain $+21$ and then combine -9, -4, and -3 to obtain -16. We then subtract the 16 from 21 to obtain $+5$. If the negative quantity were greater than the positive quantity, the answer would be negative.

When we are combining numbers or terms containing letters or other symbols, we cannot add those terms having different letters or symbols. For example, we cannot add $3b$ and $5c$. The indicated addition of these terms would merely be $3b + 5c$. We can add $3a$ and $5a$ to obtain $8a$, in which case we would show the expression as $3a + 5a = 8a$.

In the term $3a$ the figure 3 is called the **coefficient** of a. Thus we see that a coefficient is a multiplier. Remember also that a number placed above and to the right of another number or symbol to show a power is called an **exponent**. For example, in the term x^2 the figure 2 is the exponent of x, and the term is read "x square."

In algebra the letters used in place of numbers are called **literal** numbers. Thus the equation $x + y = z$ contains all literal numbers. When we wish to add terms containing different literal numbers, we combine those terms having the same letter or symbol. To solve or simplify the expression $4a + 5b - 2a - 6c + 9b - 3a + 8c - 3c - 4b + 3c$, we may proceed as follows:

$$
\begin{array}{rrr}
+4a & +5b & -6c \\
-2a & +9b & +8c \\
-3a & -4b & -3c \\
   &   & +3c \\
\hline
-a & +10b & +2c
\end{array}
$$

Then,

$$4a + 5b - 2a - 6c + 9b - 3a + 8c - 3c - 4b + 3c = -a + 10b + 2c$$

Algebraic Subtraction

The rule for subtraction in algebra is the following: *change the sign of the subtrahend and add.* If we wish to subtract 4 from 10, we change the sign of the 4 to minus and add. We then have $10 - 4 = 6$. If we want to subtract $-2x + 3y + 4z$ from $6x + 5y - 8z$, we proceed as shown here:

$$
\begin{array}{lll}
6x + 5y - 8z & \textit{Change signs of} & 6x + 5y - 8z \\
-2x + 3y + 4z & \textit{subtrahend:} & \underline{2x - 3y - 4z} \\
\overline{} & & 8x + 2y - 12z
\end{array}
$$

Use of Parentheses

Parentheses are used in algebra to indicate that two or more terms are to be considered as a single term. For example, $3 \times (5 + 2)$ means that the 5 and the 2 are both to be multiplied by 3. Alternately, the 5 and 2 can be added and then multiplied by 3, and the answer will be the same.

$$3 \times (5 + 2) = 3 \times 5 + 3 \times 2 = 21$$
$$3 \times (5 + 2) = 3 \times 7 = 21$$

If the parentheses were not used in the above example, the solution would be

$$3 \times 5 + 2 = 15 + 2 = 17$$

From this we observe that parentheses cannot be ignored in the solving of an algebraic problem.

In an expression where no multiplication or division is involved and parentheses are used, careful attention must be paid to the signs of the various terms. In the expression $(3a + 7b - 6c) + (4a - 3b - 2c)$, the parentheses actually have no effect and the expression could be written $3a + 7b - 6c + 4a - 3b - 2c$. If, however, a minus sign precedes the term enclosed by parentheses, then the signs of the quantities in the parentheses must be changed when the parentheses are removed.

$$(3a + 7b - 6c) - (4a - 3b - 2c)$$
$$= 3a + 7b - 6c - 4a + 3b + 2c$$

The foregoing expression means that the quantity $4a - 3b - 2c$ is to be subtracted from the quantity $3a + 7b - 6c$. From the rule for subtraction we know that the sign of the subtrahend must be changed and then the terms added.

$$
\begin{array}{r}
3a + 7b - 6c \\
\underline{-4a + 3b + 2c} \\
-a + 10b - 4c
\end{array}
$$

Sometimes **brackets** [ ] are also used to group terms which are to be considered as one term. Usually the brackets are used only when parentheses have already been used inside the bracketed expression.

The following expression illustrates the use of brackets:

$$9 + [7a - (3b + 8x) - 2y + 4z] - 2c$$

The most common use of parentheses is to indicate multiplication of terms. For example, $(2x + 3y)(4x - 7y)$ indicates that the quantity $2x + 3y$ is to be multiplied by the quantity $4x - 7y$. The expression $5a(x + y)$ means that the quantity $x + y$ is to be multiplied by $5a$.

Multiplication

In order to explain multiplication clearly, certain arrangements of algebraic terms not previously defined must be discussed. These are **monomials**, **binomials**, and **polynomials**. A monomial is an expression containing only one term, such as x, ab, $2z$, xy^2m, $2x3y$, and a^2b^3y. A binomial is an expression containing two terms connected by a minus $(-)$ or plus $(+)$ sign, such as $a + b$, $2x + 3y$, $abc + xyz$, and $4y^2 - 3z$. A polynomial is any expression in general containing two or more algebraic terms.

In the multiplication of algebraic terms, monomials, binomials, and other polynomials can be multiplied by any other expression regardless of whether it is a monomial, binomial, or other polynomial. Fractional terms and expressions can be multiplied by any other term or expression.

In the algebraic multiplication of terms and expressions, the signs of each term or expression must be carefully noted and properly handled. The following rules apply:

1. When two terms of *like signs* are multiplied, the sign of the product is positive.
2. When two terms of *unlike signs* are multiplied, the product is negative.

These rules may be demonstrated as follows:

$$2 \times 3 = 6$$
$$-2 \times 3 = -6$$
$$-2 \times -3 = 6$$
$$2x \times -3y = -6xy$$
$$-4x \times -6y = 24xy$$
$$-5(a + b) = -5a - 5b$$
$$-3a(2x - 4y) = -6ax + 12ay$$

To multiply purely literal terms which are unlike, the terms are merely gathered together as a unit.

$$a \times b = ab \quad ab \times cd = abcd \quad aby \times cdx = abcdxy$$

To multiply literal terms by like terms, the power of the term is raised.

$$a \times a = a^2 \qquad ab \times ab = a^2b^2$$
$$abx \times aby = a^2b^2xy \quad bc \times bc \times bc = b^3c^3$$
$$abc \times bcx \times cxy = ab^2c^3x^2y$$

Multiplication in algebra can be indicated in four ways, as shown in the following examples:

$$ab = a \times b = a \cdot b = (a)(b)$$
$$xyz = x \times y \times z = x \cdot y \cdot z = (x)(y)(z)$$

When performing multiplication in algebra you must be alert to observe how multiplication is indicated and to perform the computation accordingly. Note whether there is no sign, a regular times sign, a dot, or parentheses placed between the terms to be multiplied.

To multiply a binomial by a monomial, multiply each term of the binomial separately by the monomial.

$$a(b + c) = ab + ac$$
$$a(ab + xy) = a^2b + axy$$
$$-4a(2b - 3c) = -8ab + 12ac$$
$$2b(a + 3c) = 2ab + 6bc$$
$$3x^2(4xy - 2z) = 12x^3y - 6x^2z$$

Since the purpose of this chapter is to serve as a review or refresher, multiplication and division of binomials and polynomials are not covered. Those needing more information on algebra topics should consult an algebra textbook.

Division

In algebra, **division** may be considered the reverse of multiplication, just as in arithmetic. The division sign $(\div)$ is not usually employed, and division is indicated by making the dividend the numerator of a fraction while the divisor becomes the denominator of the fraction. For example,

$$a \div b \text{ is usually written } a/b$$

and

$$(2a + 5b) \div (x + y) \text{ is written } \frac{2a + 5b}{x + y}$$

A simple division may be performed as follows:

$$\frac{4a + 6ab}{2a} = 2 + 3b$$

Note that the monomial divisor $2a$ (the denominator) was divided into both terms of the binomial dividend (the numerator). In this example the divisor divided evenly into both terms of the dividend. If, however, the divisor will not divide evenly into both terms, a part of the quotient will have to be fractional.

$$\frac{3x + 2y}{x} = 3 + \frac{2y}{x}$$

Order of Operations

In solving an algebraic expression or equation, certain operations must be performed in proper sequence. Indicated multiplications and divisions must be completed before additions are made. This is demonstrated in the following equation:

$$7x + 3(x - y) - \frac{8x - 4}{2} = 2(x + 5) - 3y$$

The terms enclosed in parentheses are to be multiplied by the coefficients 3 and 2, respectively. Thus $3(x - y)$ becomes $3x - 3y$, and $2(x + 5)$ becomes $2x + 10$. As a result of the division indicated, $-(8x - 4)/2$ becomes $-(4x - 2)$. The original equation is then

$$7x + 3x - 3y - 4x + 2 = 2x + 10 - 3y$$

To solve for x, all x terms are transposed to the left side of the equation. Note that when a term is moved from one side of the equation to the other, the sign must be changed to maintain the equality.

$$7x + 3x - 4x - 2x = 10 - 2 - 3y + 3y$$

Then all the terms are combined.

$$4x = 8 \qquad x = 2$$

In the foregoing equation, the term

$$\frac{-8x - 4}{2}$$

must be treated as a single quantity. Therefore, when the division is made, the term is placed in parentheses to indicate that the negative sign applies to the complete term: $-(4x - 2)$. When the parentheses are removed, the $4x$ takes a negative sign and the 2 becomes positive. When the value of 2 is substituted for x, $-(4x - 2)$ becomes $-(8 - 2)$, or -6. Remember that whenever a mathematical expression is enclosed in parentheses or brackets, it is treated as a single quantity. If it is preceded by a minus sign, all the terms within the parentheses or brackets must have their signs changed when the parentheses are removed. Note the following examples:

$$-(a + b + c) = -a - b - c$$
$$-(x - y + z) = -x + y - z$$

Solution of Problems

When an algebraic expression contains only one unknown quantity, expressed by a letter, it is comparatively simple to find the value of the unknown quantity. In the equation $5x + 2 - 3x = 14 - 4x$, we can easily find the value of x by transposing and combining. A rule to be remembered at this point is that when a term or quantity is moved from one side of an equation to the opposite side, the sign of the term or quantity must be changed. The solution of the equation just mentioned is as follows:

$$5x + 2 - 3x = 14 - 4x$$
$$5x - 3x + 4x = 14 - 2$$
$$6x = 12$$
$$x = 2$$

Note that in this operation the sign of $-4x$ was changed and the sign of the $+2$ was changed. This was done because the $-4x$ and the $+2$ were transposed, or moved from one side of the equation to the other. When the quantities were combined, $6x$ was found equal to 12. It is quite apparent then that x is equal to 2. This is also shown by dividing both sides of the equation by 6.

Algebra is particularly useful in solving certain problems which are more difficult to solve by arithmetic or which may not be solved by arithmetic. The following examples show how some of the less difficult types may be solved by algebraic methods.

In the first example, let's say that one number is 3 times another number. The sum of the numbers is 48. If this is true, what are the numbers?

$$\text{Let } x = \text{the smaller number}$$
$$\text{Then } 3x = \text{the larger number}$$
$$x + 3x = 48$$
$$4x = 48$$
$$x = 12, \text{the smaller number}$$
$$3x = 36, \text{the larger number}$$

In the next example, one number, increased by 5, is equal to one-half another number. The sum of the numbers is 55. What are the numbers?

$$\text{Let } x = \text{the smaller number}$$
$$\text{Then } 2(x + 5) = \text{the larger number}$$
$$x + 2(x + 5) = 55$$
$$x + 2x + 10 = 55$$
$$3x = 55 - 10 = 45$$
$$x = 15, \text{the smaller number}$$
$$2(x + 5) = 2(15 + 5) = 40, \text{the larger number}$$
$$15 + 40 = 55$$

In the final example, a man has five times as many dimes as he has quarters. The total value of his dimes and quarters is $5.25. What number of each does he have?

$$\text{Let } x = \text{the number of quarters}$$
$$\text{Then } 5x = \text{the number of dimes}$$
$$25x + 10(5x) = 525$$
$$25x + 50x = 525$$
$$75x = 525$$
$$x = 7, \text{the number of quarters}$$
$$5x = 35, \text{the number of dimes}$$

Note that x must be multiplied by 25 to find the total number of cents represented by the quarters. Since $x = 7$, we find that the money represented by quarters is 175 cents, or $1.75. Also, we find that $5x \times 10 = 35 \times 10$, or 350 cents. Then $1.75 + $3.50 = $5.25.

There are many types of problems which may be solved with methods similar to those shown for the foregoing problems. You can gain skill in solving such problems through practice. The ability to interpret word problems and reduce them to equation form is the most important requirement.

GEOMETRY

The word **geometry** is derived from *geo*, a Greek word meaning "earth," and *metria*, meaning "measurement." Geometry can be said to literally mean the measurement of earth or land. In actuality, geometry deals with the measurement of areas, volumes, and distances.

The proof of geometrical propositions by means of axioms, postulates, or corollaries constitutes the major portion of most geometry courses. It is expected that you have previously had or are taking additional math courses, including geometry. In this section, definitions and applications of geometry will be emphasized. Theory will be introduced only to the extent necessary to support application.

Definitions

The following terms are essential to the understanding of the application of geometrical principles. Figure 1–7 provides graphic examples of the terms.

Point. A point has no length, breadth, or thickness but has only **position**.

Line. A line has no breadth or thickness but has length.

Surface. A surface has no thickness but has length and breadth.

Plane, or **plane surface**. A plane, or plane surface, may be defined in several ways, as follows:

1. A surface such that a straight line that joins any two of its points lies wholly in that surface.
2. A two-dimensional extent of zero curvature.
3. A surface any intersection of which by a like surface is a straight line.

Solid. A solid, in the geometric sense, has three dimensions, that is, length, breadth, and thickness.

Lines

The following terms describe different types of lines:

Straight line. A line having the same direction throughout its length. If a portion of a straight line is placed so that both ends fall within the ends of the other part, the portion must lie wholly within the line.

Equal lines. Two lines are equal if when placed one upon the other, their ends can be made to coincide.

Curved line. A line which continuously changes direction.

Broken line. A line consisting of a number of different straight lines.

Parallel lines. Lines in the same plane which can never intersect no matter how far they are extended.

Angles

The following terms are used to define and describe angles. Figure 1–8 provides illustrations to assist with understanding the definitions.

Angle. An angle is the opening between two straight lines drawn in different directions from the same point.

Acute angle. An angle which is less than a right angle.

Right angle. An angle which is one-fourth of a circle, that is, 90°.

Obtuse angle. An angle of more than 90°.

Straight angle. An angle whose sides form a straight line, that is, an angle of 180°.

Bisector. A bisector is a point, line, or surface which divides an angle into two equal parts.

Vertex of an angle. The common point from which the two sides of an angle proceed.

Adjacent angles. Two angles having a common side and the same vertex.

Vertical angles. Two angles with the same vertex and with sides that are prolongations of the sides of each other.

Perpendicular line. A straight line which makes a 90° angle with another straight line.

A common practice is to identify angles by upper-case letters. Most of the angles shown in Figure 1–8 can be identified by a three-letter combination as either angle *AOB* or *BOA*. The center letter of the combination will be the one located at the vertex. Three different angles can be identified in the illustration of the adjacent angles: *AOB*, *BOC*, and *AOC*. By using the three-letter combination, the specific angle will be clearly identified.

Shapes

Circles. A **circle** is a closed curve, all portions of which are in the same plane and equidistant from the same point (see Figure 1–9). The **diameter** of the circle is the length of a straight line passing through the center of a circle and limited at each end by the circle. The **radius** of the circle is a straight line from the center of the circle to the circle perimeter. The radius is equal to one-half the diameter. An **arc** is any portion of the circle. A major arc is one of more than 180°, while a minor arc is less than 180°. A **semicircle** is an arc of 180°. A **sector** is the area within a circle bounded by two radii and the arc connecting the two radii. A **quadrant** is a sector with an arc of 90°. A **chord** is any straight line connecting two points on a circle. A **secant** is a straight line which intersects a circle. A **tangent** is a straight line of unlimited length which only has one point in common with a circle. An **inscribed angle** in a circle is an angle whose vertex is on the circle. A **central angle** is an angle whose vertex is at the center of the angle.

The **circumference** of a circle is the length of the perimeter. The circumference can be computed by multiplying

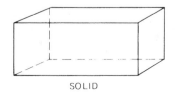

SOLID

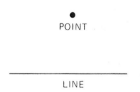

•
POINT

LINE

SURFACE

FIGURE 1–7 Geometric terms.

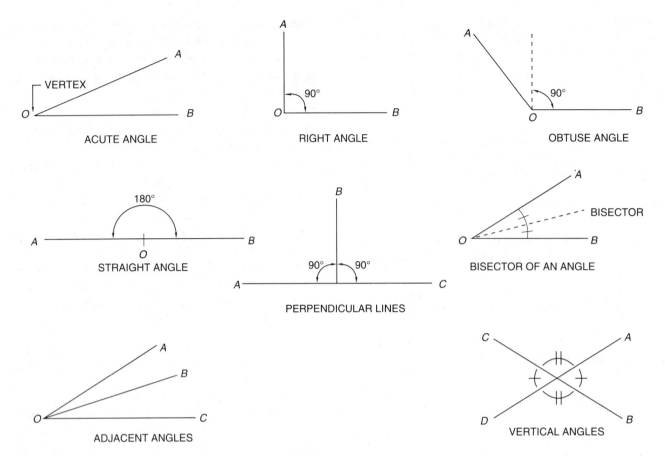

FIGURE 1–8 Angles.

the diameter times **pi**. Pi (Greek letter π) is a constant equal to 22 divided by 7. Pi is usually rounded off to 3.14 or 3.1416. The formula for circumference is

circumference = pi × diameter

or $C = \pi D$

Since the diameter is 2 times the radius, an alternate formula is

circumference = 2 × pi × radius

or $C = 2\pi r$

Polygons. A **polygon** is a plane, closed figure bounded by straight lines joined end to end. Polygons may have any number of sides from three upward. A **regular polygon** has all sides and angles equal. Some common polygons include the following:

3 sides—triangle	6 sides—hexagon
4 sides—quadrilateral	8 sides—octagon
5 sides—pentagon	

All polygons can be considered as being made up of a number of triangles, as shown by the pentagon in Figure 1–10a. The number of triangles will be equal to the number of sides minus 2. The sum of the interior angles of a triangle is 180°. Thus the sum of the interior angles of any polygon is equal to the number of sides minus 2 times 180°. A triangle has a total of 180° [(3 − 2) × 180°], a quadrilateral, 360° [(4 − 2) × 180°], a pentagon, 540° [(5 − 2) × 180°], and so on. By definition a regular polygon has equal angles; therefore a regular hexagon would have included angles of 120°.

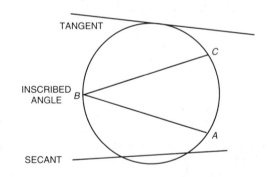

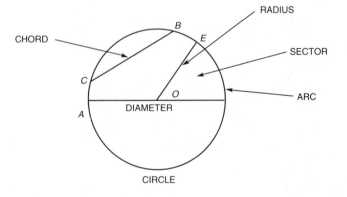

FIGURE 1–9 Parts of a circle.

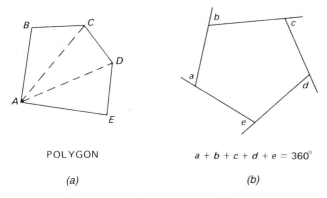

POLYGON

(a)

$a + b + c + d + e = 360°$

(b)

FIGURE 1–10 Angles of polygons.

If the sides of any polygon, such as the pentagon in Figure 1–10*b*, are extended consecutively in the same direction, an angle will be formed that will be the supplement of the internal angle. The sum of the supplementary angles will always equal 360°.

The aviation maintenance technician can expect to encounter a number of different polygon shapes in his or her work. The majority of the shapes encountered will be in the categories of triangles, quadrilaterals, and hexagons.

Triangles.   A **triangle** is a plane bounded by three sides, or a three-sided polygon with a total included angle of 180°. A number of various types of triangles exist within this definition. Variations include the sizes of the angles and the length of the legs.

An **acute triangle** is one in which all angles are less than 90° (see Figure 1–11). An **obtuse triangle** has one angle greater than 90°. A **right triangle** has one 90°, or right, angle. Remember that the sum of the three angles must be 180°.

An **equilateral triangle** has all sides of equal length. An equilateral triangle is also a regular polygon. The length of the sides and the included angles are all equal. The angles are each 60°. An **isosceles triangle** has two sides equal in length and two equal angles. A **scalene triangle** has no equal sides or angles.

A drawing of a triangle will normally have the angles identified by an upper-case letter and the sides by a lower-case letter. Each side will have the same letter as the angle it is opposite. This allows each angle and/or side to be clearly identified.

The side opposite the 90° angle in a right triangle is called the **hypotenuse**, side *b* in Figure 1–12. The **Pythagorean theorem** states that in a right triangle the square of the hypotenuse is equal to the sum of the squares of the other two sides. This can be written algebraically as $a^2 + b^2 = c^2$. With this theorem, if we know the length of two sides of a right triangle, we can easily calculate the third.

In addition to angles and length of sides, triangles are dimensioned for computation purposes by **base** and **height**. Any one of the sides may be picked as the base. The height of the triangle is the length of a line perpendicular to and extending from the base to the vertex of the opposite angle. Figure 1–13 shows the correct way to measure height for different-shaped triangles.

Quadrilaterals.   A **quadrilateral** is a four-sided polygon with the sum of the included angles equal to 360°. A regular quadrilateral with all sides of equal length and with equal angles is a **square**. A **rectangle** has four angles of 90° and two pairs of parallel sides. One pair is longer than the other as shown in Figure 1–14. A **parallelogram** is a four-sided figure whose opposite sides are equal and parallel. This definition would also include any rectangle. In common practice, however, a parallelogram does not have any right angles. A **trapezoid** is a four-sided plane that has two parallel sides and two that are not parallel.

Squares are dimensioned by the length of any two of the equal sides. Rectangles are dimensioned by the two different lengths referred to as length times width or length times height. In common practice the shorter side will be specified as the width or height. Parallelograms are dimensioned by base and height. The longest dimension is normally used as the base. The height is the length of a line perpendicular to, and extending from, the base to the opposite parallel line

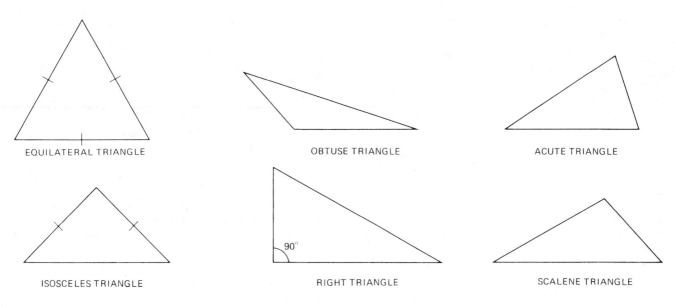

EQUILATERAL TRIANGLE   OBTUSE TRIANGLE   ACUTE TRIANGLE

ISOSCELES TRIANGLE   90°   RIGHT TRIANGLE   SCALENE TRIANGLE

FIGURE 1–11 Types of triangles.

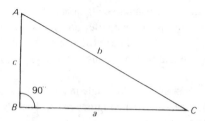

FIGURE 1–12 Right triangle.

(see Figure 1–14). Trapezoids are dimensioned for height the same as the parallelogram. Since the parallel sides are unequal in length, both are identified for dimensions, such as base 1 and base 2 ($b1$, $b2$).

Hexagon. The six-sided regular hexagon is a familiar shape for the aircraft maintenance technician. This is a standard shape used for aircraft nuts and bolts. When used for aircraft hardware, a hexagon shape is dimensioned "across the flats," as shown in Figure 1–15. For calculation purposes the length of each flat and the distance across corners is also used.

Formulas

The aviation maintenance technician needs to be able to calculate the amount of area in a plane, the volume of a solid, and the surface area of a solid. The need for this ability will be found when working in such areas as sheet metal layout and repair, powerplants, hydraulic systems, and fuel systems. While rote memorization of formulas may suffice for this task, it is recommended that you learn how the formula is derived. This will enable better retention of the information and a wider use of its application.

Area. Area is measured in units of **square inches (in²)** or square centimeters **(cm²)**. The **area of a circle** is a func-

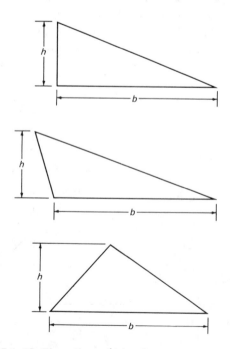

FIGURE 1–13 Dimensions of triangles.

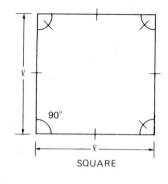

SQUARE

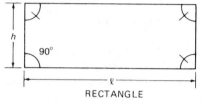

RECTANGLE

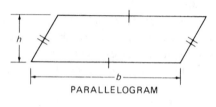

PARALLELOGRAM

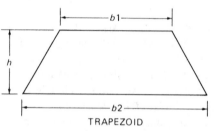

TRAPEZOID

FIGURE 1–14 Dimensions of quadrilaterals.

tion of the radius. Three different formulas are in common use. They are given in the examples that follow. For each example, find the area of a circle with a radius of 2 in (see Figure 1–16).

1. Area equals pi times the radius squared, $A = \pi r^2$.

$$A = 3.14 \times 2 \times 2 = 12.56 \text{ in}^2$$

2. Area equals pi times the diameter squared divided by four, $A = (\pi d^2)/4$, or $.7854d^2$. (*Note:* 3.1416/4 = 0.7854).

$$A = 0.7854 \times 4 \times 4 = 12.56 \text{ in}^2$$

3. Area equals one-half the radius times the circumference, $A = \frac{1}{2}rC$. (*Note:* $C = \pi d$.)

$$A = \frac{1}{2} \times 2 \times 3.14 \times 4 = 12.56 \text{ in}^2$$

All three formulas give the same answer. The one to use will depend on the data available and personal preference.

To find the **area of a sector**, first find the area of a full circle and divide the answer by $N/360$. N is the number of degrees included in the sector. To demonstrate, find the area of a 60° sector with a radius of 2 in (see Figure 1–16).

Area of circle with 2 in radius = 12.56 in²

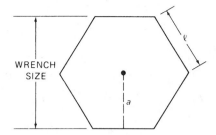

FIGURE 1–15 Dimensions of a hexagon.

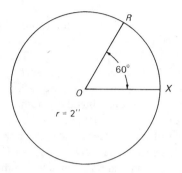

FIGURE 1–16 Calculating the area of a circle or a sector.

$$\text{Area of sector} = \frac{12.56}{\left(\frac{360}{60}\right)}$$

$$\frac{12.56}{6} = 2.09 \text{ in}^2$$

The **area of any regular polygon** can be found by using the formula $A = \frac{1}{2}ap$, with a being the perpendicular distance from a side to the center (apotherm) and p being the perimeter, or sum of the sides. The similarity should be noted between this formula and one of the formulas for the area of a circle ($A = \frac{1}{2}rC$). To demonstrate, find the area of the hexagon shown in Figure 1–15.

$$\text{Distance across flats} = \tfrac{5}{8} \text{ in}$$
$$a = \tfrac{5}{16} = 0.3125$$
$$\text{Length of each side} = 0.360 \text{ in}$$
$$p = 2.16 \text{ in}$$

$$A = \tfrac{1}{2}ap = \tfrac{1}{2} \times 0.3125 \times 2.16$$
$$A = 0.338 \text{ in}^2$$

The polygon formula will work with any regular polygon. However, in the case of squares and equilateral triangles, other formulas are usually more convenient. In Figure

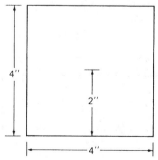

FIGURE 1–17 Calculating the area of a square.

1–17 is a square that is 4 in on each side. To use the polygon formula we determine that $a = 2$ and $p = 16$. Using the formula $A = \frac{1}{2}ap$ gives an area of $\frac{1}{2} \times 2 \times 16$, or 16 in². The same result can be found by simply multiplying the length by the height ($4 \times 4 = 16$). Since all four sides are equal in length, the **area of a square** may be stated as $A = l^2$. To find the **area of a rectangle** with two different lengths of sides, the formula will be $A = lh$, with l equal to length and h equal to height.

To find the **area of a triangle** look at the example in Figure 1–18a. Rectangle $ABCD$ is drawn with a diagonal from B to C. The diagonal creates two triangles, ABC and BCD. Each triangle has a height and a base equal to the height and base of the rectangle. The two triangles are equal and therefore the area of each triangle equals one-half the area of the rectangle. To find the area of one triangle the formula would be $A = \frac{1}{2}bh$, with b representing the length of the base and h representing the height.

To demonstrate this formula, find the area of a triangle with a base of 6 in and a height of 4 in (Figure 1–18b).

$$\text{Area} = \text{one-half times the base times the height}$$
$$A = \tfrac{1}{2}bh$$
$$= \tfrac{1}{2} \times 6 \times 4 = 12 \text{ in}^2$$

This formula is valid for any type of triangle.

The area of an equilateral triangle can be found by an alternate formula which states that the area is equal to one-quarter of the square of the base times the square root of 3.

$$A = \tfrac{1}{4}b^2\sqrt{3}$$

Figure 1–19 shows an equilateral triangle with 6 in on a side. The height of this triangle has been computed to be 5.2 in. Find the area of the triangle in Figure 1–19 using both formulas.

1.
$$A = \tfrac{1}{2}bh$$
$$= \tfrac{1}{2} \times 6 \times 5.2 = 15.6 \text{ in}^2$$

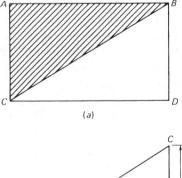

(a)

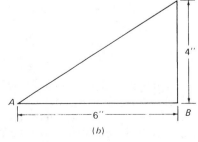

(b)

FIGURE 1–18 Calculating the area of a triangle.

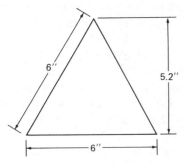

FIGURE 1–19 Calculating the area of an equilateral triangle.

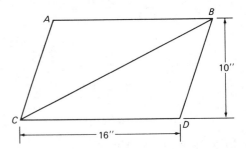

FIGURE 1–20 Calculating the area of a parallelogram.

2.
$$A = \tfrac{1}{4}b^2\sqrt{3}$$
$$= \tfrac{1}{4} \times 6 \times 6 \times 1.732 = 15.6 \text{ in}^2$$

Figure 1–20 illustrates the derivation of a formula for the **area of a parallelogram**. The parallelogram *ABCD* is divided with a diagonal from *B* to *C*. As with the rectangle in Figure 1–18a, we have created two equal triangles. Each triangle has a base equal to the base dimension of the parallelogram, and a height also equal to the height of the parallelogram. By visual inspection it can be seen that the area of parallelogram *ABCD* is equal to two times the area of triangle *BCD*. The formula for the area of the triangle is $A = \tfrac{1}{2}bh$. The area of a parallelogram can be found by multiplying the base times the height. To demonstrate this formula, find the area of the parallelogram in Figure 1–20.

$$A = bh$$
$$= 16 \times 10 = 160 \text{ in}^2$$

Care should be taken that the height is measured with a line that is perpendicular to the base and not by the length of an end.

Figure 1–21 illustrates a method for calculating the **area of a trapezoid**. Trapezoid *ABCD* has height *h*, base 1 equals line *AB*, and base 2 equals line *CD*. From point *B* extend line *AB* a length equal to line *CD* to point *B'*. From point *D* extend line *CD* a length equal to line *AB* to point *D'*. Connect point *B'* and point *D'*. Line *AB'* = *AB* + *CD*. Line *CD'* = *CD* + *AB*. Thus line *AB'* = *CD'*. A parallelogram has been created. Further visual inspection of the figure will show that *ABCD* and *BB'CC'* are equal in area. The area of *ABCD* would then be half of the area encompassed by *AB'CD'*. The formula for the **area of a trapezoid** is one-half of the sum of base 1 plus base 2 times the height. This can be expressed in a formula as $A = \tfrac{1}{2}(b1 + b2)h$.

To demonstrate this formula, find the area of the trapezoid in Figure 1–21.

$$b1 = 3 \quad b2 = 2 \quad h = 1.5$$
$$A = \tfrac{1}{2}(b1 + b2)h$$
$$= \tfrac{1}{2} \times (3 + 2) \times 1.5$$
$$= 3.75 \text{ in}^2$$

The area of an irregular polygon can be found by dividing it into shapes such as triangles or rectangles. Figure 1–10a on page 17 shows a pentagon divided into three triangles. Using trigonometry it is possible to calculate the area of each triangle. The sum of the areas would be the area of the pentagon.

Volume. Volume requires that an object have length, breadth, and depth. Volume is expressed in units of cubic inches (in^3) or cubic centimeters (cm^3). The volume of a rectangular solid is equal to the product of the height, length, and width: $V = hwl$ (see Figure 1–22). A solid cube has equal edge dimensions; thus the volume of a cube will equal the cube of one dimension, or $V = l^3$.

The volume of a cylinder is equal to the product of its cross-sectional area and its height ($V = Ah$ or $V = r^2h$). The volume of a cone or a pyramid is equal to one third the product of the area of the base and the altitude ($V = a\pi r^2/3$). The correct way to measure altitude *a* is shown in Figure 1–23a. The volume of a sphere is equal to the product of one-third times 4 pi and the cube of the radius [$V = (4\pi/3)r^3$].

Surface Area. It is occasionally necessary to calculate the surface area of an object. For an item which has surfaces made up of circles or polygons it is a simple matter of finding the sum of the individual surfaces. The surface area of a cylinder may be found by multiplying the circumference by the height. This would give the surface area of a cylinder that had no ends. If the ends are to be included, they can be calculated as the area of a circle and added to the other answer.

The surface area of a cone, called the **lateral area**, is equal to one-half the product of its **slant height** and the circumference of the base [$A = (sl \times c/2)$]. Figure 1–23b shows the correct way to measure slant height (*sl*). Lateral area

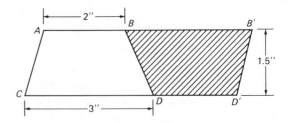

FIGURE 1–21 Calculating the area of a trapezoid.

$$V = hwl$$

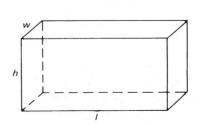

FIGURE 1–22 Volume of a rectangular solid.

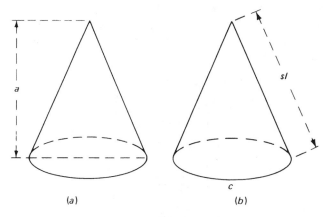

FIGURE 1–23 Volume and surface area of a cone.

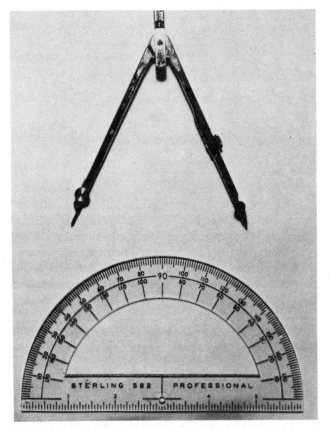

FIGURE 1–24 Compass and protractor.

does not include the base. If surface area of the base is required it can be calculated as the area of a circle.

The surface area of a sphere is equal to the product of 4π and the square of the radius ($A = \pi 4 r^2$).

Geometric Constructions

During the maintenance of aircraft it is often necessary to layout or transfer geometric shapes to new material. It is possible with a compass, a ruler, and a protractor to construct many geometrical figures which accurately fulfill their definitions or descriptions. In most cases the protractor is not needed. A compass and a protractor are shown in

Figure 1–24. The following list describes how to make some common geometric constructions.

1. *Bisect a straight line.* (See Figure 1–25a.) Adjust the compass so that it spans a greater distance than one-half the length of the line. Place the point on A and strike an arc CDE as shown. Using B as a center, strike an arc FGH. Connect the points J and K with a straight line. The line JK bisects the line AB.

2. *Draw a perpendicular from a point to a line.* (See Figure 1–25b.) From the point P use the compass to strike arcs at A and B, using the same radius in each case. Then from points A and B strike intersecting arcs at C. Connect the points P and C with a straight line. The line PC is perpendicular to AB.

3. *Bisect an angle.* (See Figure 1–25c.) Given the angle AOB, place the point of a compass at O and strike arcs at A and B so that $OA = OB$. From points A and B strike intersecting arcs at C using equal radii. Draw the line OC. OC is then the bisector of the angle.

4. *Duplicate a given angle.* (See Figure 1–25d.) Given the angle AOB, draw the line $O'D$. Strike an arc AB such that $OA = OB$. Draw the arc CD using the radius OA. Using the distance AB as a radius, strike an arc at C with D as a center. Draw the line OC. The angle $CO'D$ is equal to angle AOB.

5. *Duplicate a given triangle.* (See Figure 1–25e.) Given the triangle ABC, draw a horizontal line DX. Using AB as a radius and D as the center, draw an arc cutting DX at E. Using AC as a radius and D as a center, draw an arc in the vicinity of F. Using CB as a radius and E as a center, draw an arc to intersect the other arc at F. Draw the lines DF and EF. DEF is the duplicate of triangle ABC.

TRIGONOMETRY

The branch of mathematics which makes possible the solution of unknown parts of a triangle is called **trigonometry**. When the values of certain angles and sides of a triangle are known, it is possible to determine the values of all the parts through the use of trigonometric processes.

Trigonometric Functions

Trigonometric functions are based on the ratios of the sides of a right triangle to one another. In the diagram of Figure 1–26, the right triangle $AB'C'$ is superimposed on right triangle ABC with the angles at A coinciding. The lines $B'C'$ and BC are parallel; hence the triangles are similar. In similar triangles the ratios of corresponding sides are equal, and so $AB/AC = AB'/AC'$. In like manner, the ratios of the other sides are also equal. Furthermore, any right triangle which has an acute angle equal to A will have the same ratios as those for the triangles shown in Figure 1–26.

In trigonometry the ratios of the sides of a right triangle to one another are given particular names. These are **sine**, **cosine**, **tangent**, **cotangent**, **secant**, and **cosecant**. These ratios are called **trigonometric functions** and may be ex-

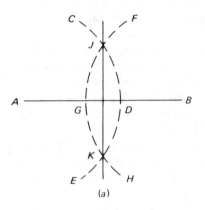

(a)

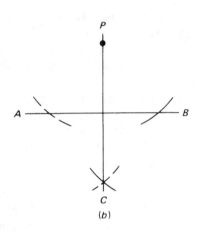

(b)

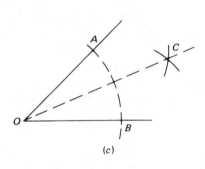

(c)

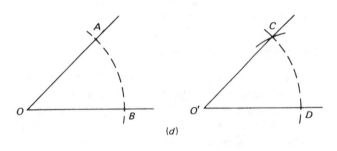

(d)

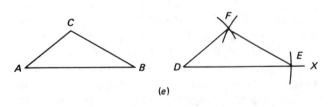

(e)

FIGURE 1–25 Geometric constructions.

plained by the use of the triangle in Figure 1–27.

In the triangle ABC, side c is the hypotenuse, side b is the side adjacent to angle A, and side a is the side opposite angle A. The functions of angle A are then as follows:

The sine of angle A, called *sin A*, is

$$\frac{\text{side opposite}}{\text{hypotenuse}} \quad \text{or} \quad \frac{a}{c}$$

The cosine of A, called *cos A*, is

$$\frac{\text{side adjacent}}{\text{hypotenuse}} \quad \text{or} \quad \frac{b}{c}$$

The tangent of A, called *tan A*, is

$$\frac{\text{side opposite}}{\text{side adjacent}} \quad \text{or} \quad \frac{a}{b}$$

The cotangent of A, called *cot A*, is

$$\frac{\text{side adjacent}}{\text{side opposite}} \quad \text{or} \quad \frac{b}{a}$$

The secant of A, called *sec A*, is

$$\frac{\text{hypotenuse}}{\text{side adjacent}} \quad \text{or} \quad \frac{c}{b}$$

The cosecant of A, called *csc A*, is

$$\frac{\text{hypotenuse}}{\text{side opposite}} \quad \text{or} \quad \frac{c}{a}$$

The importance of the foregoing functions lies in the fact that a particular function always has the same value for the same angle. For example, sin 50° is always equal to 0.7660. This means that in a right triangle which has an acute angle of 50°, the sine of 50° will always be 0.7660 regardless of the size of the triangle. The table "Trigonometric Functions" in the appendix of this book may be used to determine the values of the functions of any angle.

In the triangle shown in Figure 1–27, the functions of the angle B are the **cofunctions** of angle A. That is,

$$\sin B = \cos A \qquad \cos B = \sin A$$
$$\tan B = \cot A \qquad \cot B = \tan A$$
$$\sec B = \csc A \qquad \csc B = \sec A$$

These relationships can easily be shown by noting the sides adjacent to and opposite angle B.

Solution of Right Triangles

If any side and one of the acute angles of a right triangle are known, all the other values of the triangle may be determined. For example, if an acute angle of the right triangle in Figure 1–28 is 35° and the side adjacent to this angle is 6 in long, we may determine the other values as follows: From the table of functions we find cos 35° = 0.8192. Then,

$$\frac{b}{c} = 0.8192 \quad \text{or} \quad \frac{6}{c} = 0.8192$$

$$c = \frac{6}{0.8192} = 7.32$$

From the table of functions we find tan 35° = 0.7002. Then,

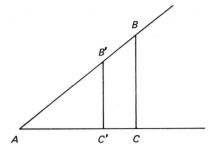

FIGURE 1–26 Similar right triangles superimposed.

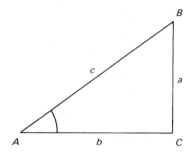

FIGURE 1–27 Triangle to show functions of an angle.

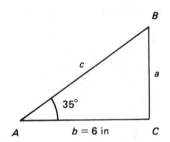

FIGURE 1–28 Solution of a right triangle.

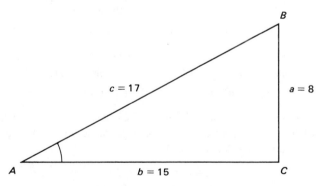

FIGURE 1–29 Solving for an angle.

$$\frac{a}{b} = 0.7002 \text{ or } \frac{a}{6} = 0.7002$$

$$a = 4.2012$$

Since the sum of the angles of a triangle is 180°, the other acute angle of the triangle is 55°.

The sides of the triangle are 4.2012, 6, and 7.32. We can verify these answers by the formula $a^2 + b^2 = c^2$, which shows that the square of the hypotenuse of a right triangle is equal to the sum of the squares of the other two sides.

If the sides of a right triangle are known, the angles can also be determined. This is shown in the problem of

Figure 1–29. In the triangle ABC, side $a = 8$, $b = 15$, and $c = 17$. Thus,

$$\sin A = \frac{a}{c} = \frac{8}{17} = 0.4706$$

From the table of functions, we know that

$$0.4706 = \sin 28° \, 4' \text{ (approximately)}$$

Then,

$$\text{Angle } A = 28° \, 4' \text{ (approximately)}$$

Changes in Values of Functions

An examination of Figure 1–30 reveals two right triangles with different angles but with the hypotenuse of each being equal in length. The denominator of both the sine and the cosine of an angle is the value of the hypotenuse. Assigning a value of 1 to each hypotenuse in Figure 1–30 results in the value of the sine being represented by the length of AB and $A'B'$. The cosine for each triangle will be represented by the length of OA and OA'. An examination of the diagram shows that as angle AOB increases the value of the sine will also increase. As the angle decreases the value of the sine will decrease. As the angle becomes smaller this value will continue to decrease until at an angle of 0° the sine equals 0. At 90° the side opposite, AB, will equal the length of the hypotenuse, and the value of the sine will be 1.0. The value of the sine will vary from 0.0 to 1.0 as the angle goes from 0° to 90°. The cosine will increase as the angle gets smaller. At 0° the length of the side adjacent, OA, is equal to the length of the hypotenuse and will equal 1.0. At 90° the side adjacent, OA, will have decreased to zero, and the value of the cosine will be 0.0.

Figure 1–31 shows a graph of the functions of a sine and a cosine through 360°. The graph for the sine is called a sine wave. The sine wave shows that the value of the sine increases from 0° to 90° and decreases in value from 90° to 180°. From 180° the value increases, but in a negative direction until it reaches 270°, at which time it begins decreasing to zero at 360°. At this point the sine value has completed a full cycle of 360° and starts over again. Figure 1–31 also

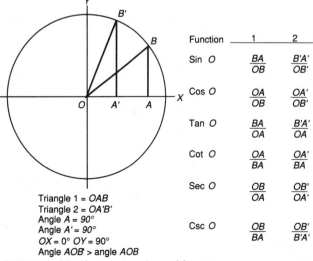

Function	1	2
Sin O	$\frac{BA}{OB}$	$\frac{B'A'}{OB'}$
Cos O	$\frac{OA}{OB}$	$\frac{OA'}{OB'}$
Tan O	$\frac{BA}{OA}$	$\frac{B'A'}{OA}$
Cot O	$\frac{OA}{BA}$	$\frac{OA'}{BA}$
Sec O	$\frac{OB}{OA}$	$\frac{OB'}{OA'}$
Csc O	$\frac{OB}{BA}$	$\frac{OB'}{B'A'}$

Triangle 1 = OAB
Triangle 2 = OA'B'
Angle A = 90°
Angle A' = 90°
OX = 0° OY = 90°
Angle AOB > angle AOB

FIGURE 1–30 Changes in values of functions.

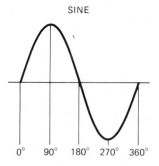
SINE

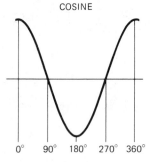
COSINE

FIGURE 1–31 Functions of the sine and the cosine.

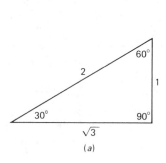
(a)

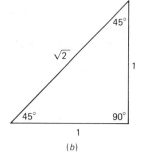
(b)

FIGURE 1–32 Functions of 30°, 45°, and 60° angles.

shows that the cosine has a similar wave form but is reaching maximum and minimum values 90° from the sine wave. The aviation technician/student will find the sine wave used extensively in other subjects, such as electricity.

The tangent in Figure 1–30 becomes larger as the angle increases. The side opposite becomes longer and the side adjacent becomes shorter. At 0° the value of AB would be 0, making the value of the tangent 0.0. As the angle approaches 90° the value of the side adjacent approaches 0. As a result the tangent value becomes very large and at 90° is said to be infinite (∞). Figure 1–30 can also be used to show the following ranges of values: The cotangent will have an infinite value at 0° and a value of 0.0 at 90°. The secant will have values from 1.0 to infinity in the 0° to 90° range. The value of the cosecant will vary from infinity to 1.0 over the same range.

The values, as the angle changes from 0° to 90°, can be summarized as follows:

$$\sin a, \text{ 0 to 1}$$
$$\cos a, \text{ 1 to 0}$$
$$\tan a, \text{ 0 to } \infty$$
$$\cot a, \infty \text{ to 0}$$
$$\sec a, \text{ 1 to } \infty$$
$$\csc a, \infty \text{ to 1}$$

It should also be noted that in a right triangle, such as Figure 1–30, the sine of one angle will equal the cosine of the second. For example,

$$\text{sine of angle } AOB = \frac{AB}{OB} = \text{cosine of angle } OBA$$

Many other relationships exist among the trigonometric functions, but they are beyond the scope of this text.

TABLE 1–2 Comparison of Number Systems

Decimal	Binary	Octal	Hexadecimal
0	0	0	0
1	1	1	1
2	10	2	2
3	11	3	3
4	100	4	4
5	101	5	5
6	110	6	6
7	111	7	7
8	1000	10	8
9	1001	11	9
10	1010	12	A
11	1011	13	B
12	1100	14	C
13	1101	15	D
14	1110	16	E
15	1111	17	F
16	10000	20	10
17	10001	21	11
.	.	.	.
.	.	.	.
.	.	.	.
100	1100100	144	64
1000	1111101000	1750	3E8

Functions of Particular Angles

The angles of 30°, 45°, and 60° are frequently used angles. Triangles containing these angles have relationships that are easy to remember and frequently will reduce the amount of calculation required for a problem. Figure 1–32a shows the relationship of length for the sides of a 30°-60°-90° triangle. The shortest side, opposite the 30° angle, is assigned the value of 1. The hypotenuse of this triangle will be two times the length of the shorter side. The length of the third side, adjacent to the 30° angle, will be equal to the length of the shortest side times the square root of three. Thus,

$$\text{sine } 30° = \tfrac{1}{2} = 0.5000 \text{ (also cosine } 60°)$$
$$\text{cosine } 30° = \tfrac{\sqrt{3}}{2} = 0.8660 \text{ (also sine } 60°)$$
$$\text{tangent } 30° = \tfrac{1}{\sqrt{3}} = 0.5773 \text{ (also cotangent } 60°)$$
$$\text{cotangent } 30° = \tfrac{\sqrt{3}}{1} = 1.7320 \text{ (also tangent } 60°)$$

The 45° triangle in Figure 1–32b has two equal sides, each assigned a value of 1. The hypotenuse has a value of the square root of 2, or 1.414.

$$\sin 45° = \tfrac{1}{\sqrt{2}} = .707$$
$$\cos 45° = \tfrac{1}{\sqrt{2}} = .707$$
$$\tan 45° = \tfrac{1}{1} = 1.000$$
$$\cot 45° = \tfrac{1}{1} = 1.000$$

You should also remember the Pythagorean theorem ($a^2 + b^2 = c^2$, where c is the hypotenuse of a right triangle). If any two sides of a right triangle are known, the third side can be calculated.

ALTERNATIVE NUMBER SYSTEMS

All of the mathematics in this chapter have used a system with a base of 10. This system, using 10 digits and known as the **decimal system**, was discussed in the first part of the chapter. The development of digital electronics introduced the need for other systems.

Every number system has three concepts in common: (1) a **base**, (2) **digit value**, and (3) **positional notation**. The base is the number of digits used in the system. Each digit of a specified system has a distinct value. Each number position carries a specific weight, depending upon the base of the system. For example, we can express the decimal number 546 as follows:

$$
\begin{aligned}
6 \times 10^0 &= 6 \times 1 &=&\ \ \ 6 \\
4 \times 10^1 &= 4 \times 10 &=&\ \ 40 \\
5 \times 10^2 &= 5 \times 100 &=&\ 500 \\
&&&\ \overline{546}
\end{aligned}
$$

Electronic devices work with a system of only two numbers, 0 and 1. This system is known as the **binary** system and uses powers of two. A binary number of 1101 is equal to a decimal number of 13. For example, we can express the binary number 1101 as follows:

$$
\begin{aligned}
1 \times 2^0 &= 1 \times 1 &=&\ \ 1 \\
0 \times 2^1 &= 0 \times 2 &=&\ \ 0 \\
1 \times 2^2 &= 1 \times 4 &=&\ \ 4 \\
1 \times 2^3 &= 1 \times 8 &=&\ \ \underline{8} \\
&&&\ \overline{13}\ \ \text{(decimal)}
\end{aligned}
$$

The **octal** system uses a base of 8 and digits from 0 through 7. The **hexadecimal** system uses 16 digits. These include the 10 digits from 0 to 9 and the first six letters of the alphabet, A through F. The hexadecimal positional value is based on powers of 16. Table 1–2 shows the relationship for some numbers in the four systems. An examination of Table 1–2 reveals that large quantities expressed in the binary system require a large number of positions and can become awkward to handle. The octal and hexadecimal systems can ex-

press large quantities in numbers with three or four positions. These two systems are easily converted to binary information.

CHARTS AND GRAPHS

Charts and graphs are extensively used in aircraft maintenance and operation manuals to present and to aid in the use of mathematical data. The time and effort required to make mathematical calculations or to understand an operation can often be greatly reduced by using such aids. Charts and graphs of many types are found in technical literature related to aircraft maintenance. You should know the different types and when they are used.

A chart may be used to present many types of information. The information may be presented in a variety of ways. Table 1–1 of this chapter is a chart presenting various functions of powers and roots. It is made up of numerical lists. Other charts may have combinations of numerical data and text, as in the case of a trouble-shooting chart. The service section of a maintenance manual may have a lubrication chart with pictorial diagrams, text, numbers, and symbols.

Graphs

Graphs are charts which provide numerical or mathematical information in graphical form, that is, with lines, scales, bars, sectors, and so forth. The graph usually shows the changes in the value of one or more variables as another variable changes.

A **broken-line graph** or a **bar graph** is used to show comparative quantitative data. The broken-line graph is useful to show trends in quantitative data over a period of time. This use is illustrated in Figure 1–33. In the illustra-

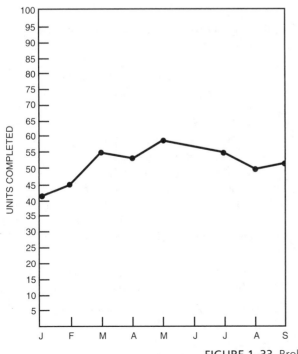
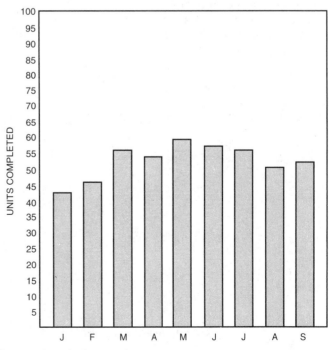

FIGURE 1–33 Broken line graph and a bar graph.

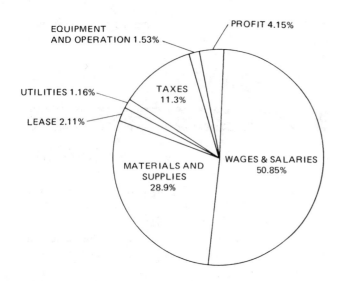

FIGURE 1–34 Circle graph.

tion, the broken-line graph and bar graph provide the same information.

The **circular**, or **pie**, chart is used to graphically represent the division or distribution of a whole. Figure 1–34 is a circle graph indicating how, and in what proportion, a company's operating expenses were distributed. If it was desired that this year's distribution be compared with last year's, a bar or broken line graph would be the better choice.

Continuous-Line Graphs. The broken-line graph is made up of a number of finite points connected with a line. The space between the points has no significance. The **continuous-line graph** has a line connecting points which have been measured or calculated. The line provides continuous information in that a reading could be taken at any point on it. Figure 1–35 is an example of a continuous-line graph. The two variables are stress and strain of a metal. The strain, plotted on the horizontal axis, produced by varying amounts

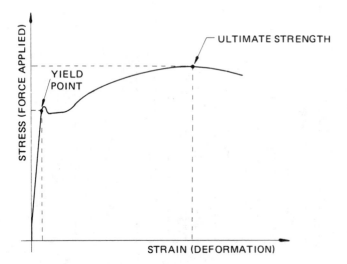

FIGURE 1–35 Continuous-line graph.

of stress, plotted on the vertical axis, is shown by this graph. If values were recorded on the graph, the amount of strain produced by a given amount of stress could be determined. In addition to the values that can be read, the shape of the line will have significance to those working with it. Figure 1–36 shows three continuous lines on one graph. All three of the lines represent an independent variable being compared to a common variable. The common variable is the angle of attack. The relationship to angle of attack is shown for each independent variable. By having several variables on one graph it is possible to determine what values or ranges will give the best combined results.

Graphs may be used to show limits. In Figure 1–37 the altitude and airspeed combinations that are safe for autorotation of a helicopter have been plotted. The portions of the graph which represent unsafe combinations have been shaded. The safe combinations can be readily determined. For example, 30 knots (kn) airspeed is safe only above 350 feet. Figure 1–38 shows two graphs used for aircraft loading. The first graph converts weights at specific locations (i.e., front seats) into index units. The total index units are then plotted against the weight of the aircraft on the second chart. If the plotted point is within the limits shown on the chart, the aircraft is properly loaded.

Figure 1–39 shows a graph of three variables: distance, time, and speed. If any two of the three variables are known, the approximate value of the other can be quickly determined. The dotted line is an example of a known time and speed giving a distance. In the example, a speed of 375 kn for 2.5 hours would result in a distance of approximately 940 miles. A **nomograph**, also called an alignment chart, is a chart used for calculations. It has scales showing the values of three or more variables. The distances between the scales and the values on each scale are placed in such a manner that the user may use a straightedge to line up two known values and obtain a third value.

Graphs of Mathematical Functions. Graphs can be used to aid in solving mathematical problems. Equations involving the values x and y can be plotted on a graph to provide a visual indication of the value of each variable as the other changes. Such a graph is shown in Figure 1–40. The first equation plotted on this graph is $x + y = 8$. When x is given a value of 0, $y = 8$ and is plotted on the y coordinate at +8. In the same manner, when y is given the value of 0, $x = 8$ and is plotted on the x coordinate (axis) at +8. The line drawn between the plotted points provides all the values of x for any value of y and vice versa. When the equation $x - y = 5$ is plotted on the same graph, the line for the equation intersects the first line at a point where $x = 6.5$ and $y = 1.5$. These values satisfy both equations.

It is apparent that the functions of x and y for two independent equations, also called simultaneous equations, can be solved graphically. For example, the following two equations,

Equation 1:

$$2x + 3y = 12$$

Equation 2:

$$x - 2y = -6$$

are solved graphically in Figure 1–41. In Equation 1, when $x = 0$, then $y = +4$, and when $y = 0$, then $x = +6$. In Equa-

tion 2, when $x = 0$, $y = +3$, and when $y = 0$, $x = -6$. When the lines are plotted on the graph, they intersect at a point where $x = \frac{6}{7}$, and $y = 3\frac{3}{7}$. These values satisfy both equations.

Graphs of algebraic equations may produce other than straight lines. Examples are shown in Figure 1–42.

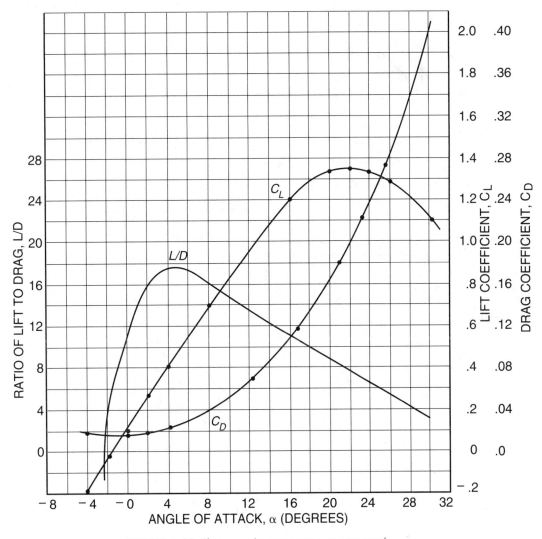

FIGURE 1–36 Three continuous curves on one graph.

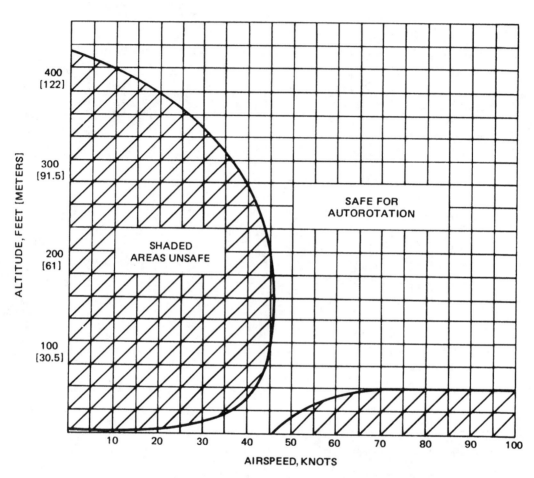

FIGURE 1–37 Continuous-line graph showing operating limits.

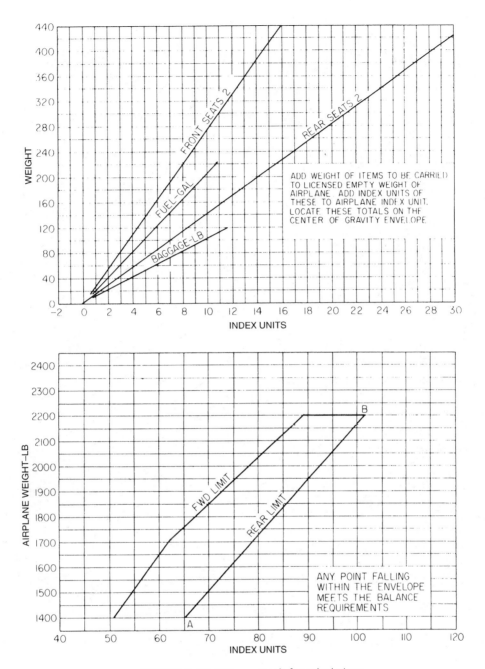

FIGURE 1–38 Using a graph for calculations.

DISTANCE, NAUTICAL MILES × 100

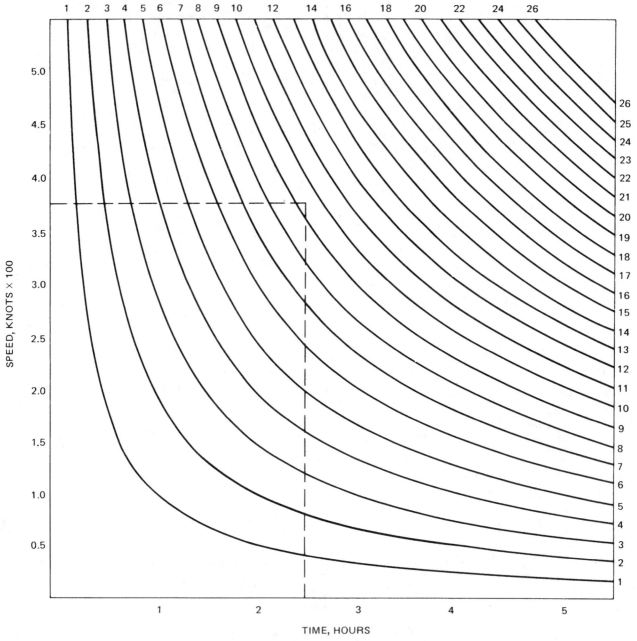

FIGURE 1–39 Graph relating three variables.

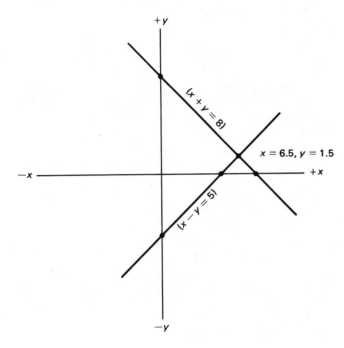

FIGURE 1–40 Graphical solution of equations.

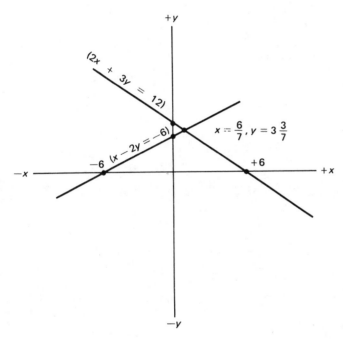

FIGURE 1–41 Graphical solutions.

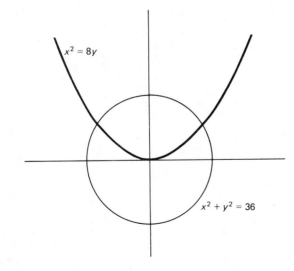

FIGURE 1–42 Graphs of algebraic equations.

1. What do we mean by *decimal system*?
2. What is a whole number?
3. Explain even and odd numbers.
4. What term is used to designate the answer for an addition problem?
5. What is the term used to identify a number being subtracted from another?
6. Which is listed first, the multiplier or the multiplicand?
7. A number to be divided is called the _____.
8. What is a proper fraction?
9. What is meant by *reducing* a fraction to its lowest terms?
10. How are two fractions multiplied?
11. Explain how one fraction may be divided by another.
12. When rounding-off a number, what determines if the last digit stays the same or is increased by 1?
13. What is meant by the *extremes* and *means* of a proportion?
14. Define *a power of a number*.
15. Define *the root of a number*.
16. How does scientific notation simplify the expression of large numbers?

17. How are two numbers expressed in scientific notation multiplied?
18. How are positive and negative quantities added?
19. Define *right angle*, *straight angle*, *acute angle*, and *obtuse angle*.
20. Define *diameter*, *radius*, *arc*, *chord*, and *sector*.
21. What is meant by *pi*, π?
22. What is the Pythagorean theorem?
23. Give formulas for the following items:

   Area of a rectangle
   Area of a triangle
   Area of a trapezoid
   Area of a circle
   Volume of a rectangular solid
   Surface area of a sphere
   Volume of a sphere
   Lateral area of a cone
   Volume of a cone

24. What digits are used by the hexadecimal number system?
25. Which chart would be best to show the distribution of a whole?

Science Fundamentals 2

INTRODUCTION

During the twentieth century, more progress has been made in technology and science than was made in all the previous centuries since the beginning of time. The space shuttle program has established the reality of civilians traveling in space (see Figure 2–1). Interplanetary communication and exploration has been accomplished. Although less spectacular than space travel, great advances have been made in air transportation and the automatic control of aircraft. The jumbo jets, air buses, and supersonic transport aircraft have brought air travel throughout the world to millions of people who could not have had this advantage a few years ago (see Figure 2–2). All these accomplishments are the result of the application of the laws of physics.

Today, the person concerned with the operation, maintenance, or design of any of the thousands of mechanical and electronic devices is constantly faced with the application of scientific law. A technician must have an understanding of the common laws of physics in such areas as motion, heat, light, and sound if he or she is to acquire the basic technical knowledge necessary to maintain and repair today's advanced aircraft and space vehicles.

MEASUREMENTS

In order to arrive at values of distance, weight, speed, volume, pressure, and so on, it is necessary to become familiar with the accepted methods for measuring these values and the units used to express them. Through the ages, human beings have devised many methods for measuring. In this text the **English system** and the **metric system**, both of which are used extensively throughout the world, will be examined.

Length and Distance

The system commonly used in the United States is the **English system**; it involves such units as the inch, foot, yard, mile, pint, gallon, pound, and ton.

Originally, the units *inch*, *foot*, *yard*, and *mile* were not exact multiples or factors of one another, but, for the sake of convenience, the foot was made equal to 12 inches (in), the yard was made 3 feet (ft), and the mile was made 5280 ft, or 1760 yards (yd). The **nautical mile (nmi)**, used internationally for navigation, is based on one sixtieth of 1° of the earth's circumference at the equator. It is approximately 6080 ft. A speed of 1 nmi/h is called 1 knot (kn). Most airplane airspeed indicators are calibrated in knots.

Most other countries, and scientists in this country, make use of a system called the **International System of Units**, or **SI**. This system is commonly referred to as the **metric system**, which has many advantages over the English system. In the metric system, prefixes are used to make the base units larger or smaller by multiples of 10. The prefixes most often used are shown in Table 2–1.

In the metric system all the measurements of length are

FIGURE 2–1 Space shuttle.   *(NASA)*

FIGURE 2–2 Boeing 767.   *(Boeing Commercial Airplane Co.)*

TABLE 2–1 Metric Prefixes

Prefix	Symbol	Multiplying Factor
kilo-	k	1000
deci-	d	0.1
centi-	c	0.01
mili-	m	0.001
micro	μ	0.000 001

either multiples or subdivisions of the meter, based on multiples of 10. The following table shows how the units of length are related:

$$10 \text{ millimeters} = 1 \text{ centimeter}$$
$$10 \text{ centimeters} = 1 \text{ decimeter}$$
$$10 \text{ decimeters} = 1 \text{ meter}$$
$$10 \text{ meters} = 1 \text{ decameter}$$
$$10 \text{ decameters} = 1 \text{ hectometer}$$
$$10 \text{ hectometers} = 1 \text{ kilometer}$$

One meter (m) is equal to 39.37 in, which is a little longer than the U.S. yard. Thus 1 decimeter (dm) is equal to 3.937 in. One centimeter (cm) is equal to 0.3937 in, and 1 millimeter (mm) equals 0.03937 in. In practice, the units of length most commonly used are the millimeter, the centimeter, the meter, and the kilometer.

Since today's aircraft are used all over the world, many manufacturers will provide both systems of measurement in their technical manuals. An example of this is illustrated in Figure 2–3.

Area

Measurements of area are usually indicated in units that are the squares of the units of length. In the English system the units of area are the **square inch (in²)**, the **square foot (ft²)**, the **square yard (yd²)**, and the **square mile (mi²)**.

Area in the metric system is indicated in square metric units. These are the **square centimeter (cm²)**, the **square meter (m²)**, and the **square kilometer (km²)**.

Volume and Capacity

The amount of space occupied by an object is the object's volume or capacity. The units used to measure volume are derived from units of length. Volume and capacity are indicated in three-dimensional units, or cubes, of the basic linear units. The **cubic inch (in³)**, **cubic foot (ft³)**, and **cubic yard (yd³)** are the common units of volume in the English system. An example of a cubic inch is shown in Figure 2–4.

Other units of volume and capacity are the **pint**, **quart**, and **gallon**. In general, when speaking of volume, cubed units are used, and when speaking of capacity, the pint, quart, or gallon measurements are used.

The metric system employs the cubed metric units as units of volume. The most commonly expressed units are the **cubic centimeter (cm³)** and the **cubic meter (m³)**. For capacity, the **liter** is generally employed. The liter (L) is equal to 1000 cm³. It is also equal to 1.056 U.S. liquid quarts. One m³ equals 1000 L.

Weight

Units of weight in the English system are no more standardized than other units of measure used in the English system. Among the units of weight in common use are the

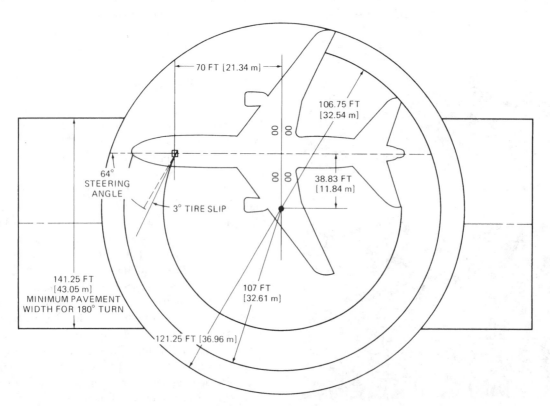

FIGURE 2–3 Turning radii for the L-1011 given in both English and metric measurements. *(Lockheed Corp.)*

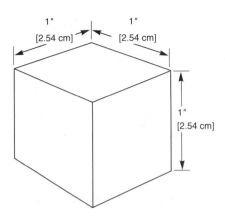

FIGURE 2–4 One-inch cube.

grain, **troy ounce**, **avoirdupois ounce**, **troy pound**, **avoirdupois (avdp) pound**, and **ton** (short ton). The following table shows the relationship among the units of weight just mentioned:

$$7000 \text{ grains} = 1 \text{ avdp pound}$$
$$5760 \text{ grains} = 1 \text{ troy pound}$$
$$12 \text{ troy ounces} = 1 \text{ troy pound}$$
$$16 \text{ avdp ounces} = 1 \text{ avdp pound}$$
$$2000 \text{ avdp pounds} = 1 \text{ ton (short)}$$

The **grain** is the smallest unit of weight in the English system and was derived from the weight of a grain of wheat. It is used principally for the measurement of medicinal components or drugs. The **troy** ounce and pound are used for weighing precious metals such as platinum, gold, and silver. **Avoirdupois** weights are used for almost all materials and objects in the English-speaking countries. These are the common ounce (oz), pound (lb), hundredweight (cwt), and ton.

The most uniform system of weights is the metric system. The following shows the relationship among the metric units of weight:

$$1000 \text{ grams} = 1 \text{ kilogram}$$
$$1000 \text{ kilograms} = 1 \text{ metric ton}$$

The **gram** (g) is the weight of 1 cm³ of pure water at a temperature of 4°C or 39.2°F, which is the point of greatest density for water. The gram is equal to about 15.432 grains, and the **kilogram** is 2.2046 lb (avdp).

The metric system of weights is much less complex than the English system because the weights can be converted from one unit to another merely by moving the decimal point to the right or left. For very small measurements of weight, the **milligram** (mg) is used in the metric system. The milligram is 0.001 g.

Other Units of Measurement

Units of measurement are required for many more purposes than those mentioned in the foregoing paragraphs. For example, items such as **force**, **density**, **electrical values**, **light intensity**, **sound intensity**, **velocity**, **energy**, and numerous other values must also be measured. In this section only the most commonly known units of measurement have been discussed. As other areas of physical laws and phenomena

are explored, the units of measurement required in each area will be discussed.

Conversion factors to show the relationships among various units are given in the appendix of this text.

GRAVITY, WEIGHT, AND MASS

Gravity, or **gravitation**, is the universal force that all bodies exert upon one another. It is defined by the **universal law of gravitation**, which states: The attraction between particles of matter is directly proportional to the product of their masses and inversely proportional to the square of the distance between them. This can be expressed by the equation

$$F = G \ \frac{m_1 m_2}{r^2}$$

where

$$F = \text{attractive force}$$
$$r = \text{distance between two bodies (particles)}$$
$$m_1 \text{ and } m_2 = \text{masses of bodies}$$
$$G = \text{the universal gravitation constant}$$
$$(G = 6.67 \times 10^{-11} \text{ newton} \cdot \frac{m^2}{kg^2})$$

The terms *weight* and *mass* are, many times, used interchangeably; however, *weight* and *mass* have different definitions.

The **weight** of a body is the pull exerted upon the body by the gravitation of the earth. The weight of a body may change depending upon its distance from the center of the earth. The farther away an object is from the center of the earth, the less it will weigh.

The **mass** of a body is a measure of the amount of material the body contains. Regardless of the location to the earth's center, the mass of a body will never change as long as no matter is added to or removed from it. Remember, however, that 1 lb of mass will not be exactly 1 lb of weight if the mass is not at the proper distance from the center of the earth.

Both mass and weight are measured in units of pounds or kilograms. Another unit of measuring mass that may be encountered is that of the slug. A **slug** is defined as being a unit of mass having a value of approximately 32.175 lb under standard atmospheric conditions.

Density

An important physical property of a substance is its **density** (d), which is its mass per unit volume. Table 2–2 is a list of the densities of various common substances. In metric units, density is properly expressed in kilograms per cubic meter. The density of water is 1000 kg/m³, since 1 m³ of water has a mass of 1000 kg.

In the English system, density should be expressed in slugs per cubic foot, since the slug is the unit of mass used in this system. In these units the density of water is 1.94 slugs/ft³.

Because weights rather than masses are usually specified in the English system, a quantity called **weight density** is

TABLE 2–2 Weights of Materials

Gas

| Gas | Density* | |
	lb/ft³	kg/m³
Air (at 59°F)	0.076 510	1.225 60
Air	0.080 710	1.293 00
Carbon dioxide	0.123 410	1.977 00
Carbon monoxide	0.078 070	1.250 70
Helium	0.011 140	0.178 46
Hydrogen	0.005 611	0.089 89
Nitrogen	0.078 070	1.250 70
Oxygen	0.089 212	1.429 20

Liquids

| Liquid | Specific Gravity | Density* | |
		lb/U.S. gal	kg/m³
Alcohol (methyl)	0.81	6.80	810
Benzine	0.69	5.80	690
Ethylene glycol	1.12	9.30	1 120
Gasoline	0.72	6.00	720
Glycerine	1.26	10.50	1 260
Jet fuel	0.80	6.70	800
Lubricating oil	0.89	7.40	890
Mercury	13.55	113.00	13 550
Sulfuric acid	1.84	15.40	1 840
Water	1.00	8.35	1 000

Metals

| Metal | Specific Gravity | Density* | |
		lb/ft³	kg/m³
Aluminum	2.700	168.56	2 700
Brass	8.400	524.40	8 400
Copper	8.920	556.88	8 920
Gold	19.300	1 204.90	19 300
Iron	7.860	490.70	7 860
Lead	11.344	708.21	11 344
Mercury	13.550	845.93	13 550
Platinum	21.450	1 339.12	21 450
Silver	10.500	655.52	10 500
Titanium	4.500	280.95	4 500
Zinc	7.140	445.75	7 140

*At standard sea-level pressure and 0°C.

commonly used. As the name suggests, weight density is weight per unit volume, and its units are pounds per cubic foot.

Changes in temperature will not change the mass of a substance but will change the volume of the substance by expansion or contraction, thus changing its density. To find the density of a substance, its mass and volume must be known. Its mass is then divided by its volume to find the mass per unit volume.

Specific Gravity

The **specific gravity** of a substance is the ratio of the density of the substance to the density of water. To determine the specific gravity of a substance when the density is known, divide the weight of a given volume of the substance by the

weight of an equal volume of water. For example, if the specific gravity of lead is unknown and the density in pounds per cubic foot is 708.21 lb, divide 708.21 by 62.4 (the density per cubic foot of water) and 11.34 is obtained, which is the specific gravity of lead. Table 2–2 lists the specific gravity for various substances.

A device called a hydrometer is used for measuring the specific gravity of liquids. This device consists of a tubular-shaped glass float contained in a larger glass tube (see Figure 2–5). The float is weighted and has a vertically graduated scale. To determine specific gravity, the scale is read at the surface of the liquid in which the float is immersed. A specific gravity of 1000 is read when the float is immersed in pure water. When immersed in a liquid of greater density, the float rises, indicating a greater specific gravity. For liquids of lesser density the float sinks, indicating a lower specific gravity.

A hydrometer is often used to determine the specific gravity of the electrolyte (battery liquid) in an aircraft battery. When a battery is discharged, the calibrated float immersed in the electrolyte will indicate a specific gravity of approximately 1150. The indication of a charged battery is between 1275 and 1310 (see Figure 2–5).

The specific gravity and density of solids and liquids are expressed as explained in the foregoing paragraphs; however, gases require a different treatment. The weight of a given

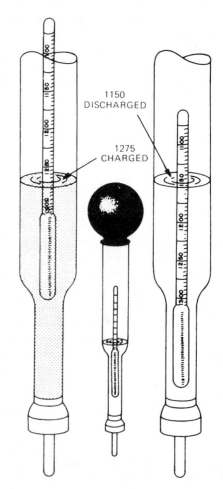

FIGURE 2–5 Hydrometer measuring the electrolyte in an aircraft battery.

volume of a gas will depend upon the pressure and the temperature. For this reason, the density of a gas is given according to standard pressure and temperature conditions, that is, a pressure of 76 cm of mercury and a temperature of 0°C. Under these conditions it is found that the density of dry air is 1.293 g/L, or 0.081 lb/ft³. Since the density of air is used as a standard, the specific gravity of air is given as 1. The specific gravity of any other gas is the ratio of the mass of a given volume of the gas to an equal volume of dry air, with both the gas and the air being under standard conditions.

Speed

An object's **speed** is a measure of how fast it is moving. Speed tells us how fast the object has moved during a certain time without respect to direction. Speed involves only the length of the path traveled by a body and the time required to travel the path. For example, the average speed of the aircraft in Figure 2–6 would be 350 mph because it required 1 h to travel a distance of 350 mi.

Speed is indicated as the ratio of the distance traveled to the time of travel. This definition is expressed in the equation

$$\text{speed} = \text{distance traveled/time required}$$

Velocity

The words **velocity** and **speed** are often used in the same sense, that is, to indicate how fast something is moving. The two words are similar in some respects, but there is an important difference. *Velocity* refers to both the speed and direction of the object. The velocity of an object is the rate at which its position changes over time and the direction of the change.

If an aircraft travels from point A to point B (in Figure 2–6) in 1 h following the irregular path, the velocity of the aircraft is 200 mph east-northeast.

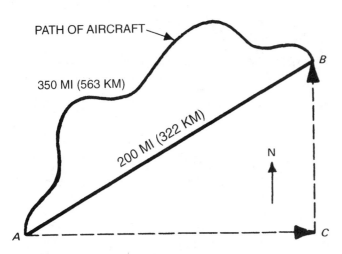

FIGURE 2–6 Diagram to illustrate speed and velocity.

Units of Force

Force may be defined as a push or a pull upon an object. In the English system the **pound** (lb) is used to express the value of a force. For example, we say that a force of 30 lb is acting upon a hydraulic piston.

A unit of force in the metric system is the **newton** (N). The newton is the force required to accelerate a mass of 1 kilogram (kg) 1 meter per second per second (m/s²).

The **dyne** (dyn) is also employed in the metric system as a unit of force. One dyne is the force required to accelerate a mass of 1 g 1 centimeter per second per second (cm/s²). One newton is equal to 100 000 dynes [0.225 lb].

Newton's First Law of Motion

Force is required to produce motion in a body that is at rest, and force is also required to stop the motion of a body. An **unbalanced force** on an object causes the object to accelerate in the direction of the force. The acceleration is proportional to the force and inversely proportional to the mass of the object. The concepts of force and motion are expressed by Newton's laws of motion. Newton's first law of motion states: A body at rest tends to remain at rest and a body in motion tends to remain in motion in a straight line unless forced to change its state by an external force.

The first law of motion concerns the property of matter called inertia. **Inertia** is defined as the tendency of matter to remain at rest if at rest or to continue in motion in a straight line if in motion. Newton's first law of motion defines inertia, stating: Objects at rest tend to stay at rest. It takes a net force to start or stop an object. The property of inertia is demonstrated by the fact that satellites will continue to orbit around the earth for months or years even though they have no means of propulsion. This is because there is no appreciable outside force in space to retard the speed of the satellite. The only substantial force acting on a satellite orbiting the earth is the force of gravity. Gravity causes the satellite to curve around the earth instead of shooting off in space.

Inertia can also be demonstrated by a simple experiment with a glass of water. If a glass of water is placed on a piece of paper on a smooth surface, the paper can be jerked from under the glass without disturbing the glass or its contents. The inertia of the glass of water causes it to remain at rest when the paper is moved.

Newton's Second Law of Motion

The term **acceleration** may be defined as the rate at which velocity changes. If the body increases in velocity, it has positive acceleration; if it decreases in velocity, the acceleration is negative. Newton's second law of motion explains acceleration. The law states: The acceleration of a body is directly proportional to the force causing it and inversely proportional to the mass of the body. This means that a given body will accelerate in proportion to the force applied to it. For example, if a 10-lb [4.536-kg] weight accelerates at 32.2 ft/s² [9.82 m/s²] when a 10-lb [44.5-N] force is applied to it, it will accelerate at 64.4 ft/s² when a 20-lb [89-N] force is applied to it.

Airplanes must have rapid acceleration in order to take off from relatively short runways. In the case of the airplane pictured in Figure 2–7, the acceleration is provided by two turbine engines each capable of producing 3700 lb [16 000 N] of thrust.

If the friction of air acting on a freely falling body is ignored, the rate of acceleration will be 32.2 ft/s² [9.82 m/s²]. This is called the **acceleration of gravity** and is indicated by the letter symbol g. From this it is known that when a force equal to the weight or mass of a body is applied to the body, the body will accelerate at 32.2 ft/s² [9.82 m/s²] if there is no friction or other force opposing the applied force. **Terminal velocity** is the highest velocity reached by a falling object. The object keeps falling at a constant velocity and no longer accelerates. Knowledge of these concepts is useful in developing an equation based upon Newton's second law.

$$F = \frac{Ma}{g}$$

where

$$F = \text{force, lb [N]}$$
$$M = \text{mass, lb [N]}$$
$$a = \text{acceleration, ft/s}^2 \text{ [m/s}^2]$$
$$g = 32.2 \text{ ft/s}^2 \text{ [9.82 m/s}^2]$$

With this formula, if the weight or mass is known, the force required to produce the acceleration can be determined.

In calculating the amount of force required to accelerate a 4000-lb [1814-kg] aircraft to a speed of 60 mph [26.8 m/s] in 10 s, it must first be determined what the rate of acceleration is. Sixty miles per hour is 88 ft/s; therefore if the aircraft is accelerated from 0 to 88 ft/s in 10 s, the rate of acceleration is 8.8 ft/s² [2.683 m/s²]. Then, by the formula,

$$F = \frac{(4000 \times 8.8)}{32.2} = 1093.1 \text{ lb [4862 N]}$$

From the above equation it is calculated that it requires a force (push) of 1093.1 lb [4862 N] to accelerate the aircraft from 0 to 60 mph in 10 s.

Friction

The force that opposes motion between two surfaces that are touching is called **friction**. The amount of friction depends on the types of surfaces and the force pressing them together. It does not depend on how fast the objects are moving or on how large an area is in contact.

Newton's Third Law of Motion

Action and reaction are explained by Newton's third law of motion: For every action there is an equal and opposite reaction. This law indicates that no force can exist with only one body but that existence of a force requires the presence of two bodies. One body applies the force, and the other body receives the force. In other words, one body is **acting** and the other is **acted upon**. This is clear in the operation of an automobile. The wheels of an automobile exert a force against the road, tending to force the road to the rear. Since the road cannot move to the rear, the automobile is forced to move forward. Other examples of action and reaction are the recoil of a gun, the backward force of a fire hose when the water is turned on, the thrust of a propeller, and the thrust of a turbine engine, as is shown in Figure 2–8.

Thrust

The thrust of a propeller, rocket, or gas-turbine engine depends upon the acceleration of a mass (weight) in accordance with Newton's second law. **Thrust** is defined as being a reaction force which is measured in pounds. A propeller accelerates a mass of air, a rocket accelerates the gases resulting from the burning of fuel, and a turbine engine accelerates both air and fuel gases. The quantity (mass) of air and gases accelerated and the amount of acceleration determine the thrust produced.

The basic formula for the thrust of a gas-turbine engine is

$$F = \frac{w}{g}(V_2 - V_1)$$

where

$$F = \text{force, lb}$$
$$w = \text{flow rate of air and fuel gases}$$
$$g = \text{acceleration of gravity (32.2 ft/s}^2) \text{ [9.82 m/s}^2]$$
$$V_2 = \text{final velocity of gases}$$
$$V_1 = \text{initial velocity of gases}$$

The space shuttle illustrated in Figure 2–9 can produce more than 5 000 000 lb of thrust [25 000 000 N] on lift-off.

FIGURE 2–7 Gates Model 55 Learjet. *(Gates Learjet Corp.)*

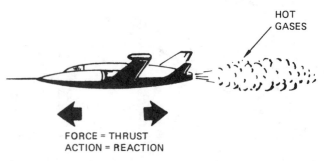

HOT GASES

FORCE = THRUST
ACTION = REACTION

FIGURE 2–8 Newton's third law of motion.

FIGURE 2–9 Space shuttle on lift-off. *(NASA)*

Powerful rocket engines are utilized to overcome the force of gravity in allowing an object to go into space. Objects sent into space must reach **escape velocity**, which is the velocity that an object must be given so that it does not fall back to Earth. The escape velocity varies for each planet, but for Earth it is about 24 850 mph [40 000 kph].

Momentum

Another quantity that often must be taken into account in situations that involve moving bodies is **momentum**. There are two kinds of momentum: **linear** and **angular**.

Linear momentum is a measure of the tendency of a moving body to continue in motion along a straight line. It is defined as the product of the mass of a body times its velocity:

$$\text{momentum} = \text{mass} \times \text{velocity}$$

Angular momentum is a measure of the tendency of a rotating body to continue to spin about an axis. It is the rotational quantity that is similar to linear momentum. A simple example of angular momentum is a top, which would spin indefinitely if not for the friction between its tip and the ground and between its body and the air. The precise definition of angular momentum is more complicated than linear momentum because it depends not only upon the mass of the body and speed that it is turning but also upon how the mass is distributed in the body. The farther away from the axis of rotation the mass is distributed, the more angular momentum it will have.

CENTRIFUGAL AND CENTRIPETAL FORCE

A weight attached to the end of a cord and twirled around as shown in Figure 2–10 will produce a force tending to cause the weight to fly outward from the center of the circle. This outward pull is called **centrifugal force**. There is an equal and opposite force pulling the weight inward and preventing it from flying outward; this is called **centripetal force**.

From Newton's first law of motion it is known that a body in motion tends to continue in motion in a straight line. Therefore, when a body is moved in a circular path, a continuous force must be applied to keep the body in the circular path. This is centripetal force.

Centripetal force is always directly proportional to the mass of the object in circular motion. Thus, if the mass of the object in Figure 2–10 is doubled, the pull on the string must be doubled to keep the object in its circular path, provided the speed of the object remains constant.

Centripetal force is inversely proportional to the radius of the circle in which an object travels. If the string in Figure 2–10 is shortened and the speed remains constant, the pull on the string must be increased, since the radius is decreased and the string must pull the object from its linear path more rapidly.

Two formulas are used for determining the magnitude of centripetal force, depending upon whether the metric system or the English system of measurements is being used. In the metric system the centripetal force is calculated in **dynes** by the formula

$$F = \frac{mv^2}{r}$$

where

$$F = \text{force, dyn or N}$$
$$m = \text{mass, g or kg}$$
$$v = \text{speed, cm/s or m/s}$$
$$r = \text{radius of circle, cm or m}$$

The result in dynes if divided by 100 000 will give the force in newtons.

To find the force in pounds, use the following formula:

$$F = \frac{Wv^2}{gr}$$

where

$$F = \text{force, lb [N]}$$
$$W = \text{weight, lb [N]}$$
$$v = \text{speed, ft [m/s]}$$
$$g = \text{acceleration of gravity, 32.2 ft/s}^2 \text{ [9.82 m/s}^2]$$
$$r = \text{radius, ft [m]}$$

The value of the centripetal force required to keep a 10-lb [4.5-kg] weight in a circle with a 20-ft [7-m] radius while moving at a speed of 60 ft/s [1.83 m/s] may be determined as follows:

$$F = \frac{(10 \times 60^2)}{(32.2 \times 20)} = \frac{36\ 000}{644} = 55.9 \text{ lb [248.7 N]}$$

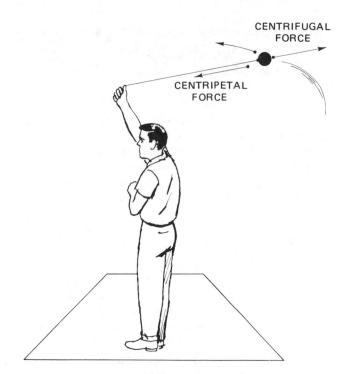

FIGURE 2-10 Centrifugal and centripetal forces.

COMPOSITION AND RESOLUTION OF FORCES

Vectors

A vector quantity is any quantity involving both magnitude and direction. A **vector** is represented by a straight arrow pointing in the direction in which the quantity is acting, and the length of the arrow represents the magnitude of the quantity. For example, the arrow **OY** in Figure 2-11 represents a force of 40 lb [18.1 kg] acting upward.

In Figure 2-12 a force of 10 lb [4.5 kg] is acting upward from a point O and a force of 5 lb [2.3 kg] is acting to the right from the point, as shown by the vectors. The **resultant** force, or **resultant**, can be determined by drawing the parallelogram (a rectangle in this case) *OXAY* and then measuring the diagonal vector **OA**. The vector **OA** is the resultant and by measurement is found to be approximately 11.2 lb [5 kg]. Since *OAY* is a right triangle, the formula used is as follows:

$$OA = \sqrt{(10^2 + 5^2)} = 11.18$$

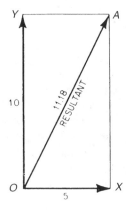

FIGURE 2-11 Vector representing 40 lb directed upward.

To solve a problem involving a right triangle such as that shown in Figure 2-12, it is convenient to use basic trigonometric formulas. These formulas are based upon trigonometric functions such as the **sine**, **cosine**, and **tangent** of a particular angle, as explained in the previous chapter of this text. That is, the values of the functions of a right triangle such as that shown in Figure 2-13 may be expressed as follows:

Sine = side opposite/hypotenuse   or $\sin A = \dfrac{a}{c}$

Cosine = side adjacent/hypotenuse   or $\cos A = \dfrac{b}{c}$

Tangent = side opposite/side adjacent   or $\tan A = \dfrac{a}{b}$

If the sine or tangent of the angle *OAX* in Figure 2-12 is located on a table of trigonometric functions, it is found that the angle is about 63°27′. The tangent of the angle is $\frac{10}{5} = 2$, which is the tangent of the angle 63°27′, as shown in the tangent table. Therefore, the direction of **OA** is 63°27′ counterclockwise from **OX**.

The determination of the resultant force is particularly

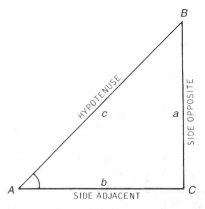

FIGURE 2-12 Resultant of two vectors acting from the same point.

FIGURE 2-13 Functions of a right triangle.

useful in calculating the effects of lift and drag forces that are imposed on airplanes.

It must be noted that the combination of vector forces, that is, forces having both magnitude and direction, requires something more than the mere addition of quantities. Vectorial addition is accomplished by means of parallelograms or triangles, as demonstrated in the foregoing paragraphs.

WORK, ENERGY, AND POWER

Work

In the scientific or technical sense, **work** refers to the application of force to a body and the displacement of the body in the direction of the force. Work is the product of the force applied to an object and the distance the object moves in the direction of the force.

$$\text{work (W)} = \text{force (F)} \times \text{distance (D)}$$

If the distance is zero, no work is done by the force no matter how great it is. Also, even if something moves through a distance, work is not done on it unless a force is acting on the object.

In the English system of units, the unit in which work is measured is called the **foot-pound (ft-lb)**. One foot-pound is the amount of work done by a force of 1 lb that acts through a distance of 1 ft.

In the metric system, work is measured in **joules**, where 1 joule (J) is the amount of work done by a force of 1 N that acts through a distance of 1 m. Therefore,

$$1\ J = 1\ N \cdot m$$

To convert from one system of units to another, it should be noted that

$$1\ J = 0.738\ \text{ft-lb}$$
$$1\ \text{ft-lb} = 1.36\ J$$

It is safe to say that work is performed whenever a force is applied and a movement occurs as the result of the force. If a lever is used to raise a weight, it can be seen how a machine performs work. In Figure 2–14 a lever is used to raise a box weighing 500 lb. The distance from the **fulcrum** (f) to the center of gravity of the box, B, is 2 ft, and the distance from the fulcrum to the applied force is 4 ft. By measurement it is found that to lift the box a distance of 6 in ($\frac{1}{2}$ ft), the opposite end, A, of the lever must be moved a distance of 1 ft. The work applied to the lever is then

$$250\ \text{(lb)} \times 1\ \text{(ft)} = 250\ \text{ft-lb [340 J]}$$

The work done on the box is expressed by the equation

$$W = FD = 500\ \text{(lb)} \times \tfrac{1}{2}\ \text{(ft)} = 250\ \text{ft-lb}$$

Thus it is shown that the work applied to the lever is equal to the work done by the lever, or **output = input.** This is true for any machine that is 100% efficient. In reality, however, there is always some loss in a machine because of friction, so the output cannot be quite as great as the input.

Energy

The term **energy** may be defined as the capacity for doing work. There are two forms of energy: potential energy and kinetic energy. **Potential energy** is the form of energy possessed by a body because of its position or configuration. For example, if a 10-lb weight was raised against the pull of gravity 2 ft higher than it was before, it is now capable of exerting a force of 10 lb (its weight) through a distance of 2 ft in returning to its original position. As long as the weight is elevated, the 20 ft-lb of work done on it is stored, ready to be released whenever the weight is lowered. It has, in other words, the potentiality of doing this amount of work, which is called its potential energy.

The notion of potential energy is not necessarily associated only with the force of gravity. A tightly wound spring or a gas compressed in a metal cylinder is also able to produce mechanical work that can be measured in the same units. Potential energy in chemical, rather than mechanical, form is stored in an automobile fuel tank that is filled with gasoline or in a charge of high explosive in an artillery shell. Potential energy in still another form lies in the nuclear energy of the fissionable fuel rods in a nuclear reactor.

Kinetic energy is the energy possessed by a body because of its motion. The amount of kinetic energy a moving

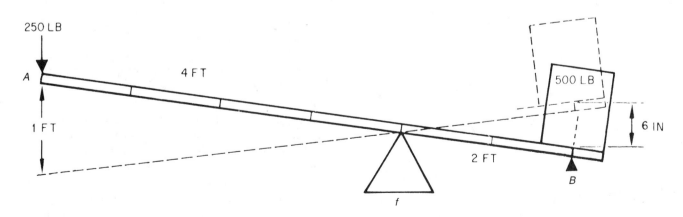

FIGURE 2–14 Work done by means of a lever.

object has depends on the velocity and mass of the object. The greater the velocity and the mass, the greater the kinetic energy.

When a hammer is used to drive a nail, the kinetic energy of the hammer does the work of driving the nail. In the case of a water-driven turbine, when the water is stored, it possesses potential energy; when it is released through the turbine, it has kinetic energy, and this energy is imparted to the turbine.

There are many different types of energy, but they are all in reality different forms of potential or kinetic energy. Each kind of energy can be used to exert a force to do work. Chemical energy is the energy stored within matter. Chemical energy stored in a battery can be used to produce the electrical energy needed to start an aircraft. When a match is lit, the chemical energy in the match is changed to light and thermal energy. One form of energy can be changed to another form.

The **law of conservation of energy** states that energy can be neither destroyed nor created; it can be changed only in form. The total amount of energy in the universe always remains constant. This means, of course, that the amount of energy imparted to a body is equal to the energy released by the body when the body is returned to its former state; that is, the condition of position, temperature, or configuration the body was in before energy was imparted to it.

One of the principal facts to remember concerning energy is that any change in the state of a body requires either that energy be given to the body or that energy be given up by the body. If an aircraft is flying straight and level at a constant speed and it is to be accelerated, the engine must be given more fuel energy, which in turn is transmitted to the propeller as mechanical energy. When landing, if the aircraft speed is to be decreased, the aircraft must give up energy to the air (air friction) and to the runway in the form of friction. For a more rapid decrease in speed, the brakes must be applied, thus converting the kinetic energy of the aircraft to heat energy at the brake discs.

Energy is expressed in the same units as those used for work. If a weight of 20 lb [9 kg] is raised 10 ft [3 m], the potential energy it acquires is 200 ft-lb [271 J] because

$$E = Fs \text{ or } E = 20 \times 10 = 200$$

where

$$E = \text{energy}$$
$$F = \text{force}$$
$$s = \text{distance}$$

In all cases, energy is the product of a force times a distance.

Kinetic energy can be expressed by the equation

$$E_k = \frac{Wv^2}{2g}$$

where

$$W = \text{weight}$$
$$v = \text{velocity}$$
$$g = \text{acceleration of gravity}$$

Potential energy is expressed as

$$E_p = Fs \text{ or } E_p = Wh$$

where

$$W = \text{weight of a body}$$
$$h = \text{height to which it has been raised}$$

Power

The rate of doing work is called **power**, and it is defined as the work done in unit time. As a formula, this would be

$$\text{power} = \text{work done/time taken to do the work}$$

Power is expressed in several different units, such as the **watt**, **ergs per second**, and **foot-pounds per second**. The most common unit of power in general use in the United States is the **horsepower**. One horsepower (hp) is equal to 550 ft-lb/s or 33 000 ft-lb/min. In the metric system the unit of power is the **watt** (W) or the **kilowatt** (kW). One watt is equal to $\frac{1}{746}$ hp; that is, 746 watts = 1 hp and 1 kW = 1.34 hp.

To compute the power necessary to raise an elevator containing 10 persons a distance of 100 ft in 5 s (assuming the loaded elevator weighs 2500 lb), proceed as follows:

$$\text{power} = \frac{(2500 \times 100)}{(5 \times 550)} = 90.9 \text{ hp } [67.81 \text{ kW}]$$

If there were no friction, the power required would be 90.9 hp; however, there is always a substantial amount of friction to overcome, and so the actual power required would be considerably greater than that calculated.

In the foregoing problem the amount of work to be done is 2500 × 100 ft-lb. This work is to be done in 5 s, or at a rate of 50 000 ft-lb/s. Since 1 hp = 550 ft-lb/s, divide 50 000 by 550 to find the horsepower.

To calculate the horsepower required to fly a light airplane at 100 mph [86.84 kn] when it is known that it requires 200 lb of thrust to overcome the drag at this speed, use the following steps: One hundred mph is equal to 146.67 ft/s, and 200 lb at 146.67 ft/s is equal to 29 334 ft-lb/s. 29 334 is then divided by 550 (ft-lb/s) to obtain the horsepower, which is about 53.1 hp [39.6 kW].

MACHINES

Nearly any mechanical device which aids in doing work can be called a machine. A **machine** can make a job easier by changing the size or direction of an applied force. Machines are devices which make use of the law of conservation of energy to change either the direction or the magnitude of a force.

No machine is capable of doing more work than the driving agent does on the machine. If more work could be done by the machine than was done upon the machine to make it run, the machine would be creating the ability to do work within itself. This, of course, it cannot do. Moreover, since any practical machine will have some **friction**, energy will be lost due to friction work within the machine. Therefore, more work has to be done on the machine than is required to do the work without the machine. Some machines make a job easier because they increase the amount of force that is available. The force applied to a machine is the **effort force**. The **resistance force** is the force produced by the machine. For example, using a screwdriver to open a can of paint allows a force to be exerted at the tip which is greater than the force exerted by your hand. This greater force is gained by moving your hand a greater distance than the screwdriver tip moves. The number of times a machine increases the effort force is called the **mechanical advantage (MA)**.

The mechanical advantage of a machine is not always greater than 1. Machines are sometimes used to increase the distance an object is moved or its speed, in which case the MA will be less than 1. Some machines only change the direction of the effort force, in which case the MA will be equal to 1.

An understanding of the principles of simple machines provides a necessary foundation for the study of compound machines, which are combinations of two or more simple machines. There are six types of simple machines. They are the lever, pulley, wheel and axle, inclined plane, screw, and wedge.

Lever

The **lever** is one of the simplest machines which enables a person to exert greater force than the person's direct effort can produce. The point at which a lever pivots is called the **fulcrum**. When an **effort force** is applied to a lever, a resistance can be overcome. Figure 2–15 shows a lever being used to raise one end of a heavy box. The distance from the fulcrum, *f*, to point *A* is 3 times the distance from *f* to *B*. Under these conditions a 100-lb force applied at *A* will lift a 300-lb weight at *B*.

Every lever has a fulcrum, an effort arm, and a resistance arm. The length of the effort arm is the distance from the fulcrum to the point where the effort force is applied. The length of the resistance arm is the distance from the fulcrum to the place where the resistance force acts. The resistance is the weight of the object or the frictional force to be overcome by using the lever. There are three classes, or different types, of levers, as shown in Figure 2–16. These classes are based on the location of the fulcrum, the resistance force, and the effort force. In a first-class lever, the fulcrum is between the effort and resistance forces. In a second-class lever, the resistance is between the effort and the fulcrum. In a third-class lever, the effort is between the fulcrum and the resistance.

Levers may provide mechanical advantages, since they can be applied in such manner that they can magnify an applied force. The mechanical advantage (*MA*) of a lever may be determined by dividing the weight (*W*) lifted by the force (*F*) applied.

$$MA = \frac{W}{F}$$

Mechanical advantage is usually expressed in terms of a ratio; for the lever in Figure 2–15, it would be 3:1.

Another principle of the lever is illustrated in Figure 2–17. When the lever is balanced, we find that

$$F_1D_1 = F_2D_2$$

In the illustration, $50 \times 40 = 200 \times 10$. The product of the force times the distance is called a **moment**. The distance from the reference point (in this case *f*) is called the **arm**. In an aircraft weight-and-balance problem, the force is usually a weight, and so the moment is equal to the weight times the arm. This principle is used in determining the weight-and-balance conditions of an airplane.

Pulleys

Several **pulleys** are often used to provide a mechanical advantage. In Figure 2–18*a*, a single pulley is shown with a rope to support a weight of 50 lb [22.7 kg]. In order to raise the weight, at least 50 lb must be applied downward on the end of the rope. This gives an *MA* of one. In Figure 2–18*b*, two pulleys with a rope provide a 2:1 advantage. Observe here that the weight is being supported by two sections of rope; therefore one section of rope only must support one-half the total weight. It is therefore possible to apply 50 lb at

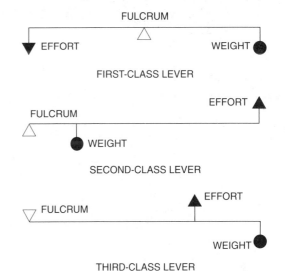

FIGURE 2–16 Three classes of levers.

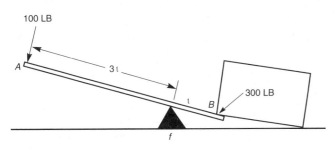

FIGURE 2–15 Mechanical advantage of a lever.

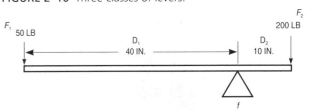

FIGURE 2–17 Balancing a lever.

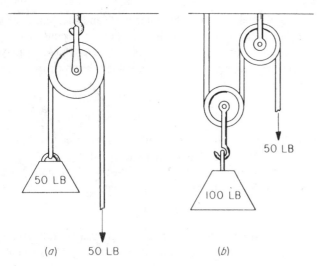

FIGURE 2–18 Mechanical advantage of pulleys.

the end of the rope and raise 100 lb [45.4 kg] with the pulley (the weight of the lifting pulley is not being considered). In any arrangement of pulleys (block and tackle), the number of ropes actually supporting the weight determines the mechanical advantage of pulley combination.

In Figure 2–19 a set of double pulleys is shown. It will be noted that there are four sections of rope supporting the weight of 80 lb [36.3 kg]. This means that each rope is required to support only 20 lb [9 kg], disregarding the weight of the pulleys. It is necessary to apply only slightly over 20 lb to the traction rope to raise the 80-lb weight. The mechanical advantage is therefore 4:1, since the weight is four times as great as the effort required to raise it.

Gears and Pulleys

Several **gears** also are employed to provide a mechanical advantage. If one gear having a diameter of 2 in is meshed with a gear having a diameter of 4 in, as shown in Figure 2–20, the mechanical advantage is 2:1. The smaller gear must turn two revolutions in order to rotate the larger gear one revolution. The turning force applied to the small gear need be only one-half the force applied to the large gear by the driven load. The turning force is called **torque**, or **twisting moment**. A train of many gears can be used to provide an extreme mechanical advantage. However, if the number of gears is too great, the friction will eventually become so large that the mechanical advantage is lost.

When a machine is driven by a belt, mechanical advantage can be obtained by driving a large pulley with a small pulley, as shown in Figure 2–21. The diameter of the large pulley is twice that of the smaller pulley; therefore, the mechanical advantage is 2:1. The large pulley will turn one-half the speed of the small pulley, but the torque required on the small pulley shaft will be only one-half the torque of the large pulley shaft.

Inclined Plane

The **inclined plane** offers a simple example of mechanical advantage that is used in many devices. Figure 2–22 illustrates the principle of the inclined plane. Assume that B is a 120-lb barrel of flour that must be raised 2 ft. The work to

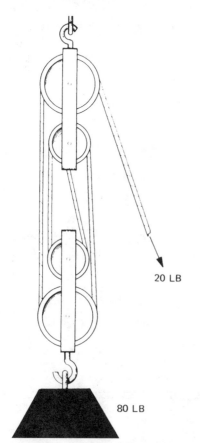

FIGURE 2–19 Multiplication of forces by means of pulleys.

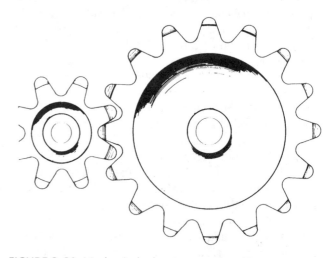

FIGURE 2–20 Mechanical advantage produced by gears.

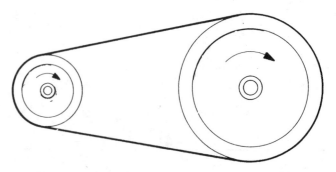

FIGURE 2–21 Mechanical advantage with a pulley drive.

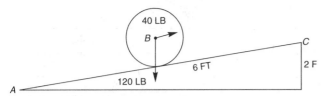

FIGURE 2–22 Principle of the inclined plane.

be done is 240 ft-lb. If the barrel is moved a distance of 6 ft to do the work, then the force need be only 40 lb, because 6 × 40 = 240. Therefore, the ratio of the length of the inclined plane to the vertical distance is the mechanical advantage, disregarding friction.

Screw

The **screw** is actually an adaptation of the inclined plane principle. A screw jack can be used to raise buildings through human power by providing a large multiplication of the human effort. There is considerable friction in a screw arrangement, but even with the friction, the screw makes possible a great multiplication of force.

Compound Machines

A **compound machine** is made by combining two or more simple machines. Most machines are actually compound machines. For example, a combination of a screw and a gear, called a **worm-gear arrangement**, is often used in machines to provide a large mechanical advantage. A worm-gear drive is shown in Figure 2–23. One revolution of the drive shaft will move the rim of the driven gear the distance of one tooth. The mechanical advantage is therefore equal to the number of teeth on the driven gear. If the gear has 20 teeth, the mechanical advantage is 20:1.

Combining several simple machines can make jobs easier; however, remember that it is impossible to get more work out of a machine than is put into it. Due to friction, the more moving parts in a machine, the more work will be lost. The ratio of work output to work input is the **efficiency** of a machine. *High efficiency* means much of the work input is changed to useful work done by the machine. *Low efficiency* means much of the work input is lost to friction and does not result in useful work. Efficiency is calculated by dividing

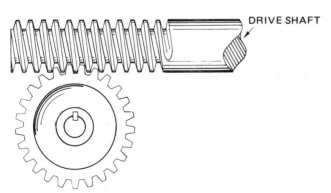

FIGURE 2–23 Worm-gear drive to develop mechanical advantage.

the work output by the work input. It is usually expressed as a percent. The equation is

$$\text{efficiency} = \frac{W^{out}}{W^{in}} \times 100\%$$

The efficiency of a machine can be increased by reducing friction. This is generally accomplished in aircraft systems and engines through the use of lubricants and oils.

HEAT

Another form of energy, **heat**, is manifested in matter by the motion of the matter's molecules. As heat is increased, the motion of the molecules increases. This adds to the internal energy of the material to which the heat is applied. If heat is applied to one end of a metal rod, it will be found that the other end of the rod gradually becomes warmer. This is because the molecules in the heated end of the rod increase their motion and strike other molecules along the rod with greater force, which increases the motion of the molecules progressively all along the rod. When this occurs, it is said that the rod is conducting heat.

When a heated object is in contact with a cold object, the heat transfers from the hot object to the cold object. This also is brought about by the motion of the molecules in the hot object striking the molecules of the cold object, thus increasing the motion of the molecules in the cold object.

Measuring Temperature

The degree of heat or cold (heat energy) measurable in a body is called **temperature**. Temperature is measured with a thermometer; its value is expressed in degrees.

In aviation, temperature can be expressed using a number of different scales. The scales used are Fahrenheit, Celsius, Rankine, and Kelvin. The Celsius temperature scale is based on the freezing and boiling points of water. The freezing point of water is given the value of 0°C. The boiling point of water is labeled as 100°C. In the International System (SI), temperature is measured using the Kelvin scale. In the Kelvin scale, zero is the same as **absolute zero**, the coldest possible temperature.

Absolute zero, one of the fundamental constants of physics, is commonly used in the study of gases. If the heat energy of a given gas sample could be progressively reduced, some temperature would be reached at which the motion of the molecules would cease entirely. If accurately determined, this temperature could then be taken as a natural reference, or as a true absolute zero value. Absolute temperatures are expressed in kelvins or degrees Rankine (°R). The various temperature scales are compared in Table 2–3.

Note from Table 2–3 that there is a difference of 180°F or °R between the point where water boils and ice melts. For this same range of temperature there is a difference of 100°C or 100 kelvins. From this we know that Fahrenheit and Rankine degrees have the same size and Celsius degrees and kelvins have the same size. Furthermore, we can see that 100° in the Celsius scale is equivalent to 180° in the

TABLE 2–3 Comparison of Temperature Scales

	°Fahrenheit	°Celsius	°Rankine	Kelvins
Water boils	212	100	672	373
Ice melts	32	0	492	273
Absolute zero	−460	−273	0	0

Fahrenheit scale. A detailed conversion table is given in the appendix.

To convert one type of scale to the other we can use the following formulas:

$$°F = \tfrac{9}{5}°C + 32 \quad °C = \tfrac{5}{9}(°F - 32)$$

Since the kelvin is the same size as the Celsius degree and begins counting 273° higher on the scale,

$$°C + 273 = \text{kelvins}$$

The Fahrenheit and Rankine scales are 460° apart, so to convert degrees Fahrenheit to degrees Rankine,

$$°F + 460 = °R$$

Effects of Heat

The effects of heat make possible many of the powerful machines in use today. Various fuels, such as gasoline, may be burned to cause a great expansion of the air and gases of combustion. The expanded gases are used to move the pistons in gasoline engines, thus causing the crankshaft to rotate and develop power for turning a propeller. In gas-turbine engines the burning of fuel with oxygen causes a great expansion of gases, which drives the turbine of the engine to compress the air, and the exhausted gases cause the jet thrust. Similarly, the burning of either liquid or solid fuels in a rocket causes a great expansion of gases, which produces the thrust due to the acceleration of the gases as they are ejected from the rocket nozzle.

The energy available from a fuel is determined by the amount of heat it produces when burned. In order to measure heat energy it is necessary to employ heat units. Heat units have been established on the basis of heating value. In the metric system the heat unit is called the **calorie (cal)**. One calorie is the amount of heat required to raise the temperature of 1 g of water 1°C and is equal to approximately 4.186 J. In the English system the unit of heat measurement is that amount of heat necessary to raise the temperature of 1 lb of water through 1°F. This quantity of heat is called the **British thermal unit (Btu).**

The amount of work that can be performed by a certain amount of heat has also been determined. For example, it has been found that 1 Btu can do 778 ft-lb of work. Also, 1 cal can produce 4.186 J of work, or about 3.09 ft-lb. From these calculations we may determine how much work can be obtained from a certain amount of fuel, provided that the heat value of the fuel is known. The heat values of a few common fuels are given in Table 2–4.

From the information in the table it can be determined how much power can be developed when a certain amount of gasoline is being burned in a given time. For example, if

TABLE 2–4 Heat Value

Fuel	Heat Value	
	Btu/lb	Cal/g
Wood	7 000–8 000	4 000–4 500
Gasoline	20 000–20 500	11 000–11 400
Coal	13 500–15 000	7 600–8 400
Gas	9 900–11 500	5 500–6 400

an engine is burning 40 lb/h of gasoline, how much power will the engine deliver if it is 35% efficient?

40 lb/h of gasoline would produce 800 000 Btu/h
800 000 Btu/h = 13 333.3 Btu/min
13 333.3 Btu/min = 10 373 307.4 ft-lb/min

Since 1 hp equals 33 000 ft-lb/min,

10 373 307.4 ft-lb/min = 314 hp
314 × 0.35 = 110 hp, approximately

From the foregoing example, approximately 110 hp [82.03 kW] can be obtained from an engine burning 40 lb/h [18.4 kg/h] of gasoline when the engine is 35% efficient.

Specific Heat

The **specific heat** of a substance is the number of calories required to raise the temperature of 1 g of the substance 1°C or the number of Btu's required to raise 1 lb of the substance 1°F. The value is the same for each. The specific heat of water is 1, and that of other substances is usually less than 1. Table 2–5 gives the specific heats of a variety of common substances.

Table 2–5 shows that only 0.22 Btu is required to raise the temperature of 1 lb of aluminum 1°F and that 0.11 Btu will raise the temperature of 1 lb of iron 1°F. The specific heat of different substances varies substantially.

TABLE 2–5 Specific Heat for Various Substances

Substance	Specific Heat
Water	1.000
Ice	0.500
Alcohol	0.590
Aluminum	0.220
Copper	0.093
Iron	0.110
Silver	0.056
Lead	0.031
Mercury	0.033
Platinum	0.032

Change of State

Another interesting heat phenomenon is noted when a substance melts or when it is converted to a vapor. For example, when 1 g of water changes to ice at 32°F [0°C], it gives up 80 cal. When 1 g of ice is melted, it absorbs 80 cal. This accounts for the fact that ice can be forced to melt by the application of salt, and this melting process absorbs heat and lowers the temperature of the water-salt mixture that we call brine. A home ice cream freezer utilizes this principle in freezing ice cream. The energy needed to change from a solid state to a liquid state or from a liquid to a solid is called the **heat of fusion**. Once all of the ice has melted, the temperature begins to rise. When the temperature reaches the boiling point (212°F), it stops rising. The water then changes to water vapor. The energy involved in changing from a liquid to a gas or from a gas to a liquid is called the **heat of vaporization**.

When a substance is warmed, such as during evaporation, and its temperature increases, it is because the added energy increases the kinetic energy of the particles. Condensation and freezing are the opposites of evaporation. Condensation and freezing are warming processes because energy is given off to the environment. Melting and evaporation are cooling processes because energy is absorbed from the environment.

Expansion

As heat flows into a material, the kinetic energy of its particles increases. As their energy increases, the particles move faster. As the particles move faster, they collide with each other more violently. These violent collisions push the particles farther apart. As the particles move farther apart in a material, the material's volume increases. Thus, the material expands.

The effect of heat on metals is particularly important in the design and operation of heat engines. Metals usually expand with an increase in temperature, and this expansion must be accounted for in the design of an engine. Materials expand at different rates. The increase in length of a metal per unit length per degree of rise in temperature is called the **coefficient of linear expansion**. As shown in Table 2–6, for iron the coefficient of linear expansion is 0.000 012 cm/°C. This means that 1 cm of iron will have a length of 1.000 012 cm after the temperature is increased 1°C. The coefficient of expansion for aluminum is twice that of iron. A bimetallic strip of metal can be made of iron and aluminum bonded together. The strip will bend when heated. Since aluminum expands faster than iron, the strip will bend toward the iron. As illustrated in Table 2–6, the modern alloys used in turbine engines often expand much more than ordinary iron or steel. For this reason a turbine engine must be designed to "grow" as its temperature increases. A large engine may increase in length more than an inch at operating temperature. This is one of the reasons the technician must be careful to allow correct clearances when assembling a turbine engine.

TABLE 2–6   Coefficient of Thermal Expansion
for Selected Materials

Material	Coefficient of Linear Expansion 10^{-6}/°C
Steel	11.7
Iron	12.0
Aluminum	23.6
Brass	20.0
Magnesium	26.0
Titanium	9.5
Rubber	162.0

This same precaution must be taken in assembling any device that is subject to large changes in temperature during operation.

Laws of Thermodynamics

The term **thermodynamics** is defined as the branch of the science of physics dealing with the mechanical actions and relations of heat. There are two principal laws of thermodynamics which are of particular interest to the aviation technician.

The first law of thermodynamics is similar to the law of the conservation of energy. That is, heat energy cannot be destroyed; it can only be changed in form.

The second law of thermodynamics states that heat cannot flow from a body of a given temperature to a body of a higher temperature. That is, heat will only flow from a warmer body to a cooler body. **Coldness** is the absence of heat.

HEAT TRANSFER

The ability of a substance to either retain or transfer heat plays an important role in selecting the materials from which aircraft components and engines will be constructed. There are three basic methods by which heat is transferred between locations and substances; they are **conduction**, **convection**, and **radiation**.

Conduction

The transfer of energy through a conductor by means of molecular activity and without any external motion is called **conduction**. As a rule, conduction is more effective in solids than in liquids or gases.

Materials that are poor conductors are used to prevent the transfer of heat and are called **heat insulators**. Certain materials, such as finely spun glass or asbestos, are particularly poor heat conductors. The heat conductivities of some familiar materials are shown in Table 2–7. These figures were determined by using a cube of the material 1 cm on a side, with one face of the cube kept just 1°C cooler than the

TABLE 2–7   Heat Conductivity of Different Materials*

Material	Heat Conductivity (at 18° C)
Silver	0.970 00
Copper	0.920 00
Aluminum	0.480 00
Iron, cast	0.110 00
Lead	0.080 00
Mercury	0.016 00
Glass	0.002 50
Brick	0.001 50
Water	0.001 30
Wood	0.000 30
Asbestos	0.000 20
Cotton wool	0.000 04
Air	0.000 06

*Expressed in calories per second flowing through 1 cm² when the temperature gradient is 1°C/cm

opposite face. Heat will flow from the warmer face to the cooler, and the number of calories per second flowing through the cube is the heat conductivity, or thermal conductibility, of the material.

Convection

The process by which heat is transferred through fluids by the movement of matter is called **convection**. Liquids and gases are called fluids because they flow. This flow has its basis in the fact that heated bodies increase their volume and therefore decrease their density. In a teakettle, the water near the bottom is heated by immediate contact with the hot metal. It becomes lighter than the rest of the water in the kettle and floats up; its place is taken by the cooler water from the upper layers. These convection currents carry the heat up "bodily" and mix the water in the kettle. A similar phenomenon takes place in the atmosphere when, on a hot summer day, air heated by contact with the ground streams up to be replaced by cooler air masses from above.

As the air rises to cooler layers of the atmosphere, the water vapor in the air condenses into a multitude of tiny water droplets and forms the cumulus clouds so characteristic of hot summer days.

Radiation

Conduction and convection involve the transfer of energy between particles of matter. A third way in which heat energy can be transferred from one body to another is by **radiation**. Radiation is a transfer of energy that does not require the presence of matter. Standing outdoors at an open fire on a winter day, the heat received does not come by conduction through the air or the ground, since both of these are cold. The heat is not transferred by convection, since the hot air over the fire rises into the sky, taking its heat away with it. Just as the bright flames and the glowing coals radiate light, they also send out an even greater amount of radiant heat that travels unimpeded through the air, to be absorbed by skin and clothing.

The term *radiation* refers to the continual emission of energy from the surface of all bodies. All the energy we receive from the sun has been radiated in this way across 93 million miles of vacuum. Only a small part of this energy is in the form of light; most of the rest is radiant heat. Conduction and convection usually take place very slowly, while radiation takes place at the speed of light.

A hot object, such as an aircraft engine, may transfer heat by conduction, convection, and radiation.

Effects of Heat on Aircraft Materials

In maintaining, servicing, repairing, and operating aircraft and power plants, aviation technicians must be constantly aware of the effects of heat on the different materials and structures upon which they are working. A seemingly simple operation, such as sharpening a drill or cutting a piece of steel, can generate sufficient heat to damage the drill or soften the teeth of the metal-cutting saw. Special lubricants or cutting oils are often used in these cases.

Aluminum alloys used in aircraft structures are usually heat-treated to provide the maximum strength, hardness, and toughness possible. If such an alloy should be subjected to temperatures beyond a prescribed level, the alloy would lose its strength and might fail during operation. For this reason, some aluminum alloys cannot be welded.

Carbon steels and alloy steels are damaged by excessive temperatures. Some steels will maintain adequate strength after welding, while others must be re-heat-treated after welding. The technician must, therefore, make sure to follow the prescribed procedure and use the correct materials.

Steels and other metals in an engine are often damaged when the engine is operated beyond safe temperature limits. Excessive heat in an engine causes such problems as feathered piston rings, scored cylinders, burned valves, burned piston heads, burned bearings and bushings, and a number of other damages. The operator of an aircraft engine must observe carefully all operating limitations specified by the manufacturer.

The very high temperatures experienced in the operation of gas-turbine engines make it necessary that certain parts of the engines be constructed of high-temperature alloys often referred to as **exotic metals**. Such alloys often contain large percentages of cobalt, nickel, columbium, and other metals which, when alloyed, provide great strength at high temperatures as well as high-temperature corrosion resistance. During the operation of gas-turbine engines, great care must be taken to observe the temperature limitations.

In servicing aircraft heating and air-conditioning systems, the technician must ascertain that ducting for hot air be of a type and material adequate to withstand the temperatures involved. The temperature of air in hot-air ducts is often at a level which will burn or otherwise damage materials that are not designed for high temperatures. In replacing ducts, the technician must replace parts with the correct part number as specified by the manufacturer.

Many types of plastic materials are used throughout modern aircraft. Some plastics can withstand relatively high temperatures, while others will melt or burn. **Thermoplastic plastics** will melt or soften when exposed to excessive temperatures, while **thermosetting plastics** will burn, char, shrink, or crack when overheated. The technician must assure that the type of plastic material being installed is correct for the temperatures to which it will be subjected.

As explained previously, metals expand and contract with changes in temperature. An increase in temperature usually means an expansion of metal. Since different metals have different coefficients of expansion, designers must allow clearances and other features which will permit the expansions and contractions to occur without warping or overstressing the materials involved. For example, an aluminum alloy fuselage expands more rapidly than steel control cables when temperature increases. It is necessary, therefore, to install automatic tension adjusters with control cables for large aircraft.

Properties of Liquids

A **liquid** is defined as a substance that flows readily and assumes the shape of its container but does not tend to expand indefinitely. Liquids retain their total volume; a gallon of water will remain a gallon whether it is poured into a flat dish or into a tall, narrow container. The molecules in a liquid are free to move throughout the confining space, but they are bound with a force, one with another, so that they tend to remain together. Liquids are virtually incompressible. The highest pressures obtainable with modern laboratory equipment are able to squeeze water only into about three-fourths of its original volume.

Viscosity

When a force is applied to a liquid, the fluid deforms permanently under the force, and we say that the liquid flows. Some liquids are more fluid, that is, flow more readily, than others. It is convenient to have a quantity which measures the resistance of the liquid to flow. Such a quantity is the **viscosity** of the liquid. Viscosity is commonly defined as the resistance of a fluid to flow. Viscosity refers to the "stiffness" of a fluid, or its internal friction.

The **viscosity index** is a measure of the change in the viscosity of a fluid with a change in its temperature. For most liquids, viscosity decreases with increasing temperature. This is a reflection of the fact that the molecules are less tightly bound together at higher temperatures, and, therefore, the friction between them is less. For water at 0, 50, and 100°C, the viscosity is 1.79, 0.55, and 0.28 centipoise (cP), respectively. The viscosity of several fluids is given in Table 2–8.

Archimedes' Principle

The buoyancy principle, first discovered by **Archimedes**, is as follows: A body placed in a liquid is buoyed up by a force equal to the weight of the liquid displaced. It therefore follows that a floating body will displace its own weight in liquid. This can be demonstrated by placing a block of wood in a container of liquid that is filled to the overflow point, as shown in Figure 2–24. When the block is placed in the liquid, an amount of the liquid will flow out that is equal in weight to the weight of the block. This can be proved by weighing the liquid that has flowed out of the container.

TABLE 2–8   Viscosities of Liquids at 30°C

Material	Viscosity, cP
Air	0.019
Acetone	0.295
Methanol	0.510
Water	0.801
Ethanol	1.000
SAE No. 10 oil	200.000

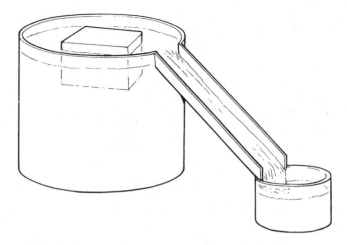

FIGURE 2–24 Displacement of a liquid by a floating solid.

Buoyancy is the effect of liquid force on a body immersed or submerged in the liquid. For example, if a particular object which has a volume of 3 ft³ [0.08 m³] is submerged in water, the buoyant force (BF) on it will be equal to the weight of the displaced water, namely, the weight of 3 ft³ of water. Since the weight density of water is 62.4 lb/ft³ [1000 kg/m³], the buoyant force will be 3 times 62.4, or 187 lb [85 kg].

Fluid Pressure

Many types of liquids are known, but the most common is water. Therefore, in considering the characteristics of a liquid, water will be used as an example. Consider water in a cylindrical glass. Because of its weight, the water exerts a force on the bottom of the glass, a force the same as that which would be produced by a cylindrical piece of ice if the water were frozen and the glass walls removed. This force is distributed over the entire bottom of the glass so that each square centimeter of the area of the bottom carries its own equal share of the load. This force per unit area is called the pressure, P, and is equal to the total force divided by the area over which it is exerted:

$$F = PA \text{ and } P = \frac{F}{A}$$

Consider the bowl of water shown in Figure 2–25. The arrows represent the direction of force acting on the sides and bottom of the bowl. The amount of force exerted at any particular point depends upon the vertical distance from the surface of the water to the point where the force is to be measured. The force exerted at point 1 is determined by the distance a. Likewise, the force at 2 is determined by the distance b, and the force on the bottom at 3 or any other point on the bottom of the bowl is determined by the distance c.

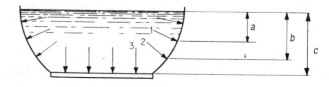

FIGURE 2–25 Force exerted by a liquid.

To compute total force, the force per unit area, or pressure, is used. Pressure is expressed in pounds per square inch, in grams per square centimeter, or in kilopascals (kPa). One pascal is a pressure of $1 \text{ N} \cdot \text{m}^2$.

Pressure due to fluid height (h) also depends on the density (d) of the fluid. Water, for example, weighs 62.4 lb/ft³, or 0.036 lb/in³, but a certain oil might weigh 55 lb/ft³, or 0.032 lb/in³. In Figure 2–25, if the distance c is 4 in [10.16 cm], the pressure, P, at the bottom of the bowl will be

$$P = hd = 4 \text{ in} \times 0.036 \text{ lb/in}^3$$
$$= 0.14444 \text{ psi } [0.996 \text{ kPa}]$$

Water, or any other fluid, since it does not have a rigid shape, will exert equal pressure in all directions. Thus the formula P = hd is equally useful in calculating the pressure against the wall of a container at any depth, no matter at what angle the wall happens to be. For this reason, the shape of the container makes no difference. In Figure 2–26, there are three containers of water. The areas of the bottoms of the containers are equal, and the depth of the water, h, is the same in each container. Therefore, the total force on the bottom of each container is the same as that on the other two containers.

In Figure 2–26, the containers are filled with a liquid of density d, the pressure at the bottom is hd for all, and at the points marked A, the pressure in all three is hd.

Pressure and Force in Fluid Power Systems

In a hydraulic system the fluid is confined, and so pressure applied to the fluid at any point is immediately transmitted to every other point touched by the fluid. Figure 2–27 illustrates **Pascal's law**, which states that a liquid under pressure in a closed container transmits pressure undiminished to all parts of the enclosing wall. If the 10-lb weight is acting on an area of 1 in², then a pressure of 10 lb [68.95 kPa] is transmitted to every square inch of the enclosing wall. This principle is the key to the use of fluids to transmit power or force from one point to another.

The terms **area**, **pressure**, and **force** are mathematically related. This relationship establishes the foundation upon which hydraulic systems are based. It permits the engineer to determine the operating pressures required for certain units in a system, the size of pump required, and the material strength needed in system units.

Consider the relationship of force, pressure, and area. If any two of these factors are known, it is possible to calcu-

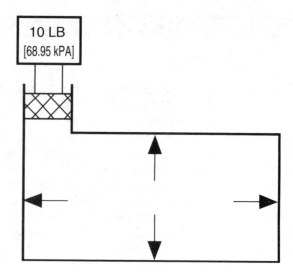
FIGURE 2–27 Pressure transmitted by a liquid.

late the third. Force equals pressure times area (F = P × A), pressure equals force divided by area (P = F/A), and area equals force divided by pressure (A = F/P). A simple aid for the solution of problems involving these factors is the diagram shown in Figure 2–28. For example, suppose a force of 25 lb is exerted on a piston whose area is 5 in². That pressure is the amount of force per unit of area expressed in pounds per square inch; therefore, on each square inch of the piston there is 5 lb of force, or 5 psi.

Great force can be applied by means of a hydraulic system merely by selecting the size of piston to produce the de-

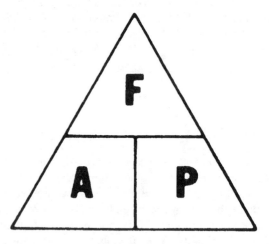
FIGURE 2–28 Device for determining the arrangement of the force, pressure, and area formulas.

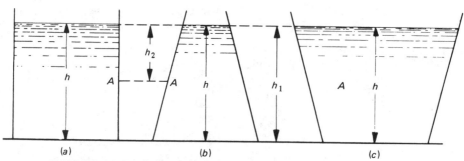
FIGURE 2–26 Total effect of liquid pressures on differently shaped vessels.

sired force. The multiplication of force through the use of hydraulic pistons is illustrated in Figure 2–29. If a force of 10 lb is applied on the small piston that has an area of 1 in², the pressure of the fluid becomes 10 psi. This pressure is transmitted undiminished to the large piston, which has an area of 50 in². Then 50 × 10 = 500 lb [2224 N], which is the force exerted by the large piston.

In the landing-gear actuating system for a large airplane, the operating pressure of the hydraulic system is generally 3000 psi [20 685 kPa]. If this pressure is applied to a piston with an area of 15 in², a force of 45 000 lb [200 160 N] is developed to raise the landing gear. In practice, this amount of force is not actually required, although the system is capable of developing it.

THE NATURE AND LAWS OF GASES

There are three states of matter: **solid**, **liquid**, and **gas**. All three are collections of **molecules** bound together by forces of varying strengths. There is a difference in the freedom of movement which molecules have that depends on whether they are part of a solid, a liquid, or a gas; the three states of matter can be defined in terms of the freedom of motion of their molecules. If the molecules are restrained so that the distance between them is constant as is their relative position, then the material is called a solid. If the molecules remain more or less at a constant distance apart but are free to change their relative position, then the material is called a liquid. If neither the relative position nor the distance between the molecules is maintained constant, the material is called a gas.

The molecules in a gas are constantly in motion, and the velocity of the motion is dependent upon the temperature of the gas. Each molecule of a gas travels in a straight line until it strikes another molecule, and at this time both of the two colliding molecules continue their travels in different directions. The movement of molecules in a gas causes the gas to dissipate quickly when it is not confined. Because of the movement of the molecules in a gas, the gas will always completely fill any container in which it is placed; that is, the molecules will distribute themselves evenly throughout the space in the container.

A gas can be easily compressed, and as it is compressed its pressure increases as its volume decreases, with the tem-

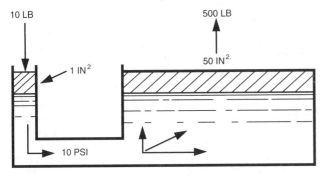

FIGURE 2–29 Multiplication of force by means of hydraulic pistons.

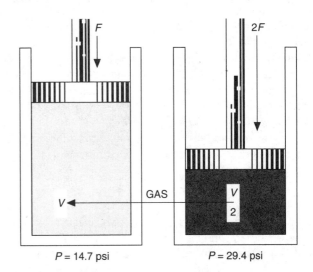

FIGURE 2–30 Gas compressed in a cylinder.

perature remaining constant. This is in accordance with **Boyle's law**, which states that the volume of a confined body of gas varies inversely as its absolute pressure varies, the temperature remaining constant. This can be expressed by the following equation:

$$\frac{V_1}{V_2} = \frac{P_2}{P_1} \text{ (temperature constant)}$$

In this equation V is the symbol for volume and P is the symbol for pressure. The subscript figures identify the first volume and pressure and the second volume and pressure.

Absolute pressure means the pressure above zero pressure, keeping in mind that the pressure of the atmosphere at sea level is approximately 14.7 psi [101.325 kPa], or 29.92 in [76 cm] of mercury (Hg). Therefore, if a confined gas is in a cylinder with the gas at atmospheric pressure and then the gas is compressed to one-half its former volume, as is shown in Figure 2–30, the pressure exerted by the gas will then be approximately 29.4 psi [202.65 kPa]. In this example it is assumed that the temperature remains constant, although under normal conditions, when a gas is compressed, its temperature increases.

Just as changes in gas volume are related to pressure changes, so are they also related to temperature changes. This characteristic of a gas is expressed by the law attributed to the French physicist Jacques A. C. Charles (1746–1823). **Charles' law** states that the volume of a gas varies in direct proportion to the absolute temperature. Absolute temperature is temperature related to absolute zero, or the condition where there is a complete absence of heat. Absolute zero is −460°F [−273°C]. The application of Charles' law to a gas requires that the pressure remain constant. If a gas is confined so that the volume remains constant, it will be found that the pressure varies with absolute temperature. Charles' law can be expressed by the following equation:

$$\frac{V_1}{V_2} = \frac{T_1}{T_2} \text{ (pressure constant)}$$

The equation relating to the pressure of a gas where the volume is constant is

$$\frac{P_1}{P_2} = \frac{T_1}{T_2} \text{ (volume constant)}$$

In each of the preceding equations remember that the temperature must be expressed in absolute units, that is, in degrees above absolute zero.

In order to determine the amount of expansion in a gas as the result of an increase in temperature, the coefficient of expansion must be known. This value is approximately the same for all gases and is found to be $\frac{1}{273}$, or 0.00366, for each degree Celsius with the gas at 0°C.

The **general gas law** is derived by combining Boyle's law and Charles' law. It is expressed by the equation

$$\frac{P_1 V_1}{T_1} = \frac{P_2 V_2}{T_2}$$

This equation can be used to determine a change in volume, pressure, or temperature of a gas when the other conditions are changed. Remember that the pressures and temperatures expressed in this equation must be stated in absolute values.

SOUND

The Nature of Sound

We can define **sound** in a number of ways, but the simplest definition is that which can be heard. Since sound is actually a vibration of a substance (solid, liquid, or gas), a sound can exist even though there may be no human ear in the vicinity to hear it. However, sound cannot travel in a vacuum.

Vibration

Vibration is a rhythmic motion back and forth across a position of equilibrium. The particles of a fluid or of an elastic solid vibrate when its equilibrium has been disturbed. Such a state is clearly demonstrated in the plucking of a string on a musical instrument. The effects of vibration in mechanical devices create many of the problems that plague engineers in the design of such devices. It is therefore necessary in many instances to conduct vibration studies before the design of a particular aircraft or engine can be approved.

To obtain a clear picture of a simple vibratory or harmonic motion, a device such as that shown in Figure 2–31 can be used. A T-shaped bar with a slotted head is mounted so that a pin on the rim of a wheel will fit into the slot in the bar. The bar is mounted in guides so that its motion is limited to two directions, up and down. When the wheel is rotated at a constant speed, the movement of the bar up and down will be a harmonic motion. If a marking pen is attached to the end of the bar so that it will mark on a strip of paper moved at a uniform speed under the pen, then the pen will describe a **sine curve** on the paper. It can be seen that the point P will move with a constantly changing speed, with the velocity being zero at points Y and Y^1 and maximum at the midpoint M.

Vibratory motion is **periodic** in nature and has characteristics by which it can be described. The time required for the motion to complete one cycle (one rotation of the wheel in Figure 2–31) is called the **period**. The number of complete cycles occurring per second is the **frequency** of the vibration. The unit for frequency is the hertz (Hz). One hertz is

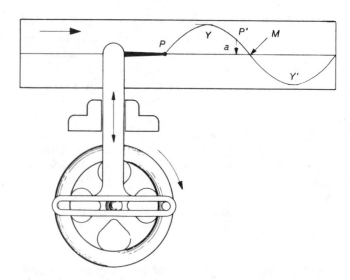

FIGURE 2–31 Demonstration of harmonic motion.

equal to one cycle per second. The **amplitude** of the vibration is the distance from the midpoint of the swing to the point of maximum displacement. **Displacement** is the distance of the vibrating point from the midpoint of vibration at any particular time. In Figure 2–31 the displacement of the point P^1 is a.

The wavelength is the distance measured along the direction of propagation between two corresponding points of equal intensity that are in phase on adjacent waves. This length can be represented by the distance between the adjacent maximum rarefaction (expansion) points in the traveling sound wave (see Figure 2–32).

Wave Motion

To understand sound it is first necessary to examine wave motion because sound travels in waves. Sound is produced by initiating a series of compression waves in a medium capable of transmitting the vibrational disturbance. The particles of the medium acquire energy from the vibrating source and enter the vibrational mode themselves. As they do, they pass on the energy to adjacent particles. If the energy source continues to vibrate, a train of periodic waves travels through the medium, and a transfer of energy takes place.

Almost everyone has seen waves in water resulting from a disturbance in or on the water. The effects of sound in the atmosphere are similar to disturbances in water, but the difference in the compressibility of the two media also makes a difference in the nature of the waves. Figure 2–33 shows

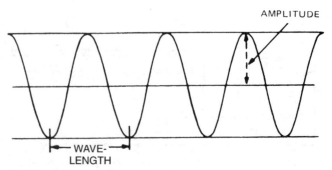

FIGURE 2–32 Sound wavelength.

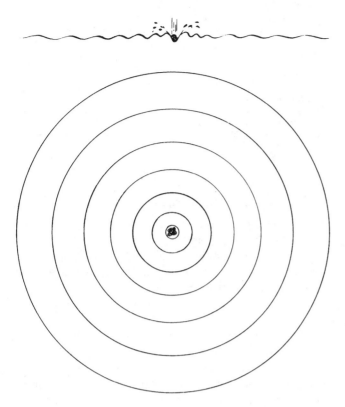

FIGURE 2–33 Wave motion in a liquid.

how waves emanate from a point in water where an object, such as a small stone, has been dropped. The illustration above the series of circles shows how a cross section of the water surface looks when an object is dropped into the water.

Waves occur as two different kinds of motion. Wave movements are therefore described as transverse and compressional (longitudinal). If a rope is tied to a stationary point and then, after being stretched to its full length, the free end is moved up and down rapidly with uniform motion, **transverse waves** will be produced in the rope (see Figure 2–34). From this, it will be seen that a transverse wave is one in which the material moves back and forth, sideways, or up and down from a zero reference line. The surface waves on water are of the transverse type.

One of the simplest methods for illustrating a **compressional wave** is to use a long coil spring stretched between two stationary points. After the spring is stretched, if one end is compressed and then released, a compression wave

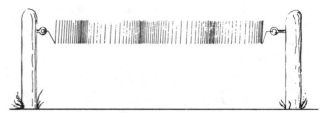

FIGURE 2–35 Wave motion in a coiled spring.

will travel the length of the spring. The compression wave is formed by a series of alternately compressed and expanded coils of the spring, as illustrated in Figure 2–35. Sound travels through matter in the form of compressional wave motions. With compressional waves, the greater the amplitude (energy), the greater the amount of compression for each wave.

If a piece of spring steel is secured solidly at one end as shown in Figure 2–36, harmonic motion can be observed when the open end of the strip of steel is pulled to point A and released. On the return swing, the end of the strip will spring back to point B, which is almost as far from the center point M as was point A. Without added energy to maintain the vibration, the amplitude of the swing will decrease rapidly, but the frequency will remain constant. This requires that the speed of the motion also decrease. The frequency of the vibration of the steel spring in Figure 2–36 depends upon the length of the strip, l, and the mass. If a weight was mounted on the end of the strip, the frequency of vibration decreases.

As the spring steel moves from A to B, it does work on the gas molecules to the right by compression; the spring steel thus transfers energy to the molecules in the direction in which the compression occurs. At the same time, the gas molecules to the left expand into the space behind the spring steel as it moves, and they then become rarefied. This motion also represents energy which is transferred to other molecules in the medium to the left of the spring steel. The combined effect of the simultaneous compression and rarefaction transfers energy to the molecules in both directions of the motion of the spring steel.

At the same time, of course, a corresponding series of

FIGURE 2–34 Wave motion demonstrated with a rope.

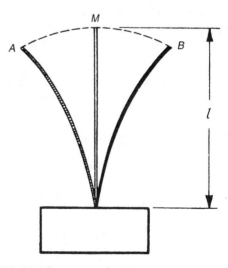

FIGURE 2–36 Vibratory motion.

rarefaction and compressions is produced to the right. The vibration of the spring steel thus generates longitudinal trains of waves in which vibrating gas molecules move back and forth along the path of the traveling waves, receiving energy from adjacent molecules nearer the source and passing it on to adjacent molecules farther from the source.

Sound Transmission

Sound must travel through some form of matter; it cannot travel through empty space. Most sounds come to us through the air, which acts as the transmitting medium. Sound is a series of expansions and compressions in the molecules of the air. It is of a similar nature when passing through a liquid or solid; however, the velocity is different. The velocity of sound through a substance depends upon both the density and the elasticity of the substance that is conducting the sound.

Sound is transmitted better at low altitudes than at high altitudes because the air is less dense at higher altitudes. In a vacuum, sound will not be transmitted at all. Liquids are better transmitters of sound than gases because they have higher elastic moduli and transmit the sound energy more readily. In general, because of their still higher elastic moduli, solids are better transmitters of sound than are liquids or gases.

The speed of sound in air is about 331.5 m/s at 0°C. As temperature increases, the speed of sound increases about 60.96 cm/s for each degree Celsius rise in temperature. At very high altitudes where the temperature is many degrees below zero, the speed of sound is much lower than it is at sea level. The speed of sound in water is about 4 times that in air; in water at 25°C sound travels about 1500 m/s. In some solids, the speed of sound is even greater. In a steel rod, for example, sound travels approximately 5000 m/s—about 15 times the speed in air. In general, the speed of sound varies with the temperature of the transmitting medium. Table 2–9 gives the speed of sound through several common substances at the indicated temperatures.

Measurement of Sound Intensity

The **intensity** of sound is defined in terms of the energy being carried by the sound wave. It was pointed out previously that the vibrating source transmits motion, or kinetic

TABLE 2–9   Speed of Sound in Various Substances

Medium	Temperature, °C	Speed	
		ft/s	m/s
Air	0	1 087	331.4
Hydrogen	0	4 220	1 286.7
Oxygen	0	1 040	317.1
Aluminum	0	16 700	5 091.8
Copper	0	13 000	3 963.7
Glass	0	17 000	5 183.3
Iron, cast	0	14 200	4 329.6
Lead	0	4 040	1 231.8
Steel	0	16 000	4 878.4
Water	15	4 760	1 451.3

TABLE 2–10   Intensity Levels of Sounds

Type of sound	Intensity, dB
Threshold of hearing	0
Whisper	10–20
Very soft music	30
Average residence	40–50
Conversation	60–70
Heavy street traffic	70–80
Thunder	110
Threshold of pain	120
Jet engine	170

energy, to the particles near it and that the wave is the flow of this energy away from the source. The vibrating source does work on the surrounding air, and this work appears as the energy of the sound wave. Therefore, the energy carried away from the source by the sound wave in each second, the power, is actually a measure of the energy carried by the wave.

The loudness, or **volume**, of a sound is determined by the amplitude of the sound waves. A larger amplitude means that matter is compressed more as the sound wave passes through it. As the amplitude of the sound waves increases, the loudness of the sound increases. When the sound waves strike the eardrum, the amount of compression is perceived as loudness. Sound intensity levels are measured by a unit called the **decibel (dB)**. The decibel is defined according to a logarithmic scale. Intensity is related to the pressure of sound on the eardrum. Sounds that are intense enough to cause pain in the ear have an intensity of about 120 decibels. The intensity levels of a number of familiar sounds are listed in Table 2–10.

Resonance

Another interesting wave phenomenon is that of **resonance**. Resonance can be observed if two objects have the same natural vibrational frequency. Resonance can occur in an airplane by matching the vibration of the aircraft structure with the engine vibration. In the case of resonance, the air molecules will transmit their vibrations from one to another and eventually to the second object as well. This may allow vibration levels to build to dangerous and even destructive levels. For this reason aircraft undergo extensive vibration testing.

Doppler Effect

Today, the term **Doppler effect** is often used in aviation in discussions of electronic navigation and control systems as well as in discussions of sound. This is possible because both electromagnetic energy and sound travel in waves. The Doppler effect is observed in sound when the source of a sound wave changes its direction with respect to the hearer so that the number of sound waves per second reaching the ear is changed. Assume that an aircraft is emitting a sound with a frequency of 1100 hertz (Hz) and that the aircraft is approaching the listener at a speed of 100 ft/s [30.5 m/s]. Assume also that the temperature is such that the speed of sound is 1100 ft/s [33.5 m/s]. Then, with the frequency at 1100 Hz and the speed of sound 1100 ft/s, there will be one sound wave (cycle) for each foot distance from the sound source. Since the sound source is approaching the listener at 100 ft/s, the listener will hear 1100 + 100, or 1200, Hz; that is, the sound will have a higher pitch than that at which it is emitted. When the sound source reaches and then goes away from the listener, the pitch will suddenly change, so that the listener will hear a pitch of 1000 Hz. This apparent change in pitch is illustrated in Figure 2–37 and is called the Doppler effect. The formula for determining the change in frequency as a result of the Doppler effect is

$$p = f \ \frac{V}{(V - S)} \text{ (source moving toward listener)}$$

$$p = f \ \frac{V}{(V + S)} \text{ (source moving away from listener)}$$

where

    p = apparent frequency of sound heard by listener
    f = frequency of sound at source
    V = speed of sound
    S = speed of source

The Doppler effect principle is used in electronic indicating systems because electronic signals are transmitted by means of waves, and the apparent frequency of a signal from an approaching signal source will be higher than the frequency of a signal from a signal source that is moving away from the receiver. One of the principal applications of the Doppler effect in electronics is in navigation radar equipment.

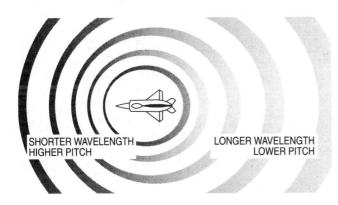

SHORTER WAVELENGTH
HIGHER PITCH

LONGER WAVELENGTH
LOWER PITCH

FIGURE 2–37 The Doppler effect.

REVIEW QUESTIONS

1. What is the basis for the nautical mile?
2. What is the advantage of the metric system of measurements over the English system?
3. State the universal law of gravitation.
4. Compare the terms *mass* and *weight*.
5. Define *density*.
6. Define *specific gravity*.
7. What is the difference between the terms *velocity* and *speed*?
8. State Newton's first law of motion.
9. Define *inertia*.
10. What is the acceleration of gravity?
11. What is the definition of *friction*?
12. State Newton's third law of motion.
13. Define *thrust*.
14. What is the difference between linear and angular momentum?
15. Explain the difference between centripetal and centrifugal forces.
16. Describe a vector.
17. Define *work*.
18. Define *energy*.
19. Differentiate between potential energy and kinetic energy.
20. Define *power*.
21. What is meant by the term *mechanical advantage*?
22. Define a British thermal unit.
23. Define the phrase *coefficient of linear expansion*.
24. List three methods of heat transfer.
25. What is meant by the term *viscosity*?
26. State Archimedes' principle.
27. What three laws are used to understand the behavior of gases?
28. What is meant by the term *frequency of vibration*?
29. What is the unit of measure for sound intensity?

3 Basic Aerodynamics

INTRODUCTION

An understanding of the basic principles of aerodynamics is as important to the aviation maintenance technician as it is to the pilot and the aerospace engineer. The technician is concerned with the strength of an aircraft because of the stresses applied through the forces of aerodynamics when the aircraft is in flight. Often responsible for the repair or restoration of aircraft structures, the technician must know that the repair work will restore the required strength to the parts that are being repaired. There are certain physical laws which describe the behavior of airflow and define the various aerodynamic forces acting on a surface. These principles of aerodynamics provide the foundations for a good understanding of what may be termed the "theory of flight."

The study of moving air and the force that it produces is referred to as **aerodynamics**. As studied by the engineer or scientist, aerodynamics involves the use of advanced mathematics and physics; however, this chapter presents only the basic principles of the subject and their application to the flight of aircraft, without the necessity of advanced mathematical analysis. The subject can therefore be more easily understood by you, the student whose primary concern lies with the maintenance, operation, and repair of the aircraft.

PHYSICAL PROPERTIES OF THE AIR

Atmosphere

The aerodynamic forces acting on a surface are due in great part to the properties of the air mass in which the surface is operating.

Air is a mixture of several gases. For practical purposes, it is sufficient to say that air is a mixture of one-fifth oxygen and four-fifths nitrogen. Pure, dry air contains about 78% (by volume) nitrogen, 21% oxygen, and 0.9% argon. In addition, air contains about 0.03% carbon dioxide and traces of several other gases, such as hydrogen, helium, and neon. The distribution of gases in the air is shown in Figure 3–1.

Static Pressure

The atmosphere is the whole mass of air extending upward hundreds of miles. It may be compared to a pile of blankets. The air in the higher altitudes, like the top blanket of the pile, is under much less pressure than the air at the lower altitudes. The air at the earth's surface may be compared to the bottom blanket because it supports the weight of all the layers above it. The **static pressure** of the air at any altitude results from the mass of air supported above that level.

The term **pressure** may be defined as force acting upon a unit area. For example, if a force of 5 lb is acting against an area of 1 in, there is a pressure of 5 psi (pounds per square inch); if a force of 20 lb is acting against an area of 2 in, the pressure is 10 psi. Air is always pressed down by the weight of the air above it. The atmospheric pressure at any place is equal to the weight of the column of air above it and may be represented by a column of water or mercury of equal weight. If the cube-shaped box shown in Figure 3–2 has dimensions of 1 in^2 on all sides and is filled with mercury, the weight of the mercury will be 0.491 lb [222.72 g], and a force of 0.491 lb will be acting on the square inch at the bottom of the box. This means that there will be a pressure of

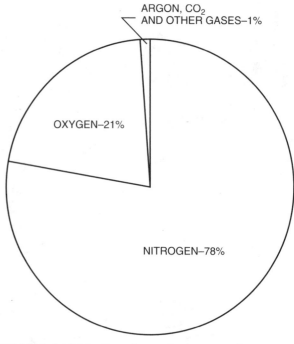

FIGURE 3–1 Distribution of gases in the atmosphere.

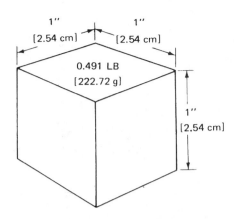

FIGURE 3–2 Weight of a 1-in cube of mercury.

0.491 psi on the bottom of the box. If the height of the box were extended to 4 in with the cross-sectional area remaining at 1 in², the pressure at the bottom would be 4 × 0.491 psi, or 1.964 psi. The pressure, as measured, per square inch exerted by a column of mercury does not change with the area of the cross section. If a 1- in column of mercury has a cross-sectional area of 10 in², the pressure will be 0.491 psi even though the total volume of mercury weighs 4.91 lb. Likewise, if the 1 in column of mercury has a cross-sectional area of $\frac{1}{4}$ in², the pressure will still be 0.491 psi.

A "standard" atmosphere was adopted by the National Advisory Committee for Aeronautics (now the National Aeronautics and Space Administration, or NASA). This standard atmosphere is entirely arbitrary, but it provides a reference and standard of comparison and should be known by all persons engaged in work involving atmospheric conditions.

Atmospheric pressure at sea level under standard conditions is 29.92 inches of mercury (inHg), or 14.69 psi. Remember that 1 in of mercury produces a pressure of 0.491 psi; therefore 29.92 inHg will produce a pressure of 14.69 psi (0.491 × 29.92 = 14.69).

Atmospheric pressure may be designated by a number of different units. Those more likely to be encountered are inches of mercury, millibars (mbar), pounds per square inch, kilopascals, and millimeters of mercury (mmHg). Standard atmospheric pressure at 59°F [15°C] is approximately as follows in the units just described:

29.92 inHg
1013 mbar (0°C)
14.69 psi
101.04 kPa (60°F) [15.56°C]
760 mmHg

The effect of atmospheric pressure was demonstrated early in the seventeenth century by the Italian mathematician and scientist Evangelista Torricelli (1608–1647). Torricelli had worked with Galileo and had noted his theories regarding the "law" that nature abhors a vacuum. To explore the idea, Torricelli filled a long glass tube, having one end closed, with mercury. He then placed his thumb over the open end of the tube. Holding the tube in a vertical position with the closed end up, he placed the open end of the tube in a container of mercury and removed his finger from the end of the tube. Some of the mercury immediately flowed out of the tube into the container, leaving a vacuum in the upper end of the tube, as indicated in Figure 3–3. The height of the column of mercury remaining in the tube was measured and found to be approximately 30 in [762 mm]. At sea level under standard conditions, the height of such a column of mercury is 29.92 in [760 mm]. Therefore we say that standard atmospheric pressure at sea level is 29.92 in high. Barometers and sensitive altimeters are scaled to provide pressure information in inches of mercury.

As just mentioned, in Figure 3–3 the space above the mercury in the tube is a vacuum; this means that the pressure at this point is 0 psia. **Psia** indicates "pounds per square inch absolute." Any gauge marked for psia measures pressure from absolute zero rather than from ambient pressure zero.

Atmospheric pressure pressing down on the surface of any liquid will cause the liquid to rise in an evacuated tube in the same manner as mercury; however, the height to which a liquid will rise depends upon the density or specific gravity of the liquid. For example, water will rise to approximately 33.9 ft [10.34 m] in a completely evacuated tube. Sometimes pressure gauges are scaled for inches of water (inH$_2$O) rather than for inches of mercury because such a gauge is more sensitive and will measure lower pressure differences.

A mercury barometer is essentially a mercury-filled glass tube scaled to show the height of a mercury column. The upper end of the tube is sealed, and the lower end is exposed to the pressure being measured. The barometer can be scaled for pounds per square inch, inches of mercury, or other unit of pressure. On weather maps, the unit of pressure is the **millibar** (mbar), which is approximately one-thousandth of a bar. For standard purposes, the sea-level pressure is set at 1013 mbar (standard conditions). The bar is therefore the approximate atmospheric pressure at sea level. 1 inHg equals 33.86 mbar.

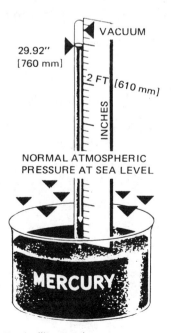

FIGURE 3–3 Torricelli's experiment.

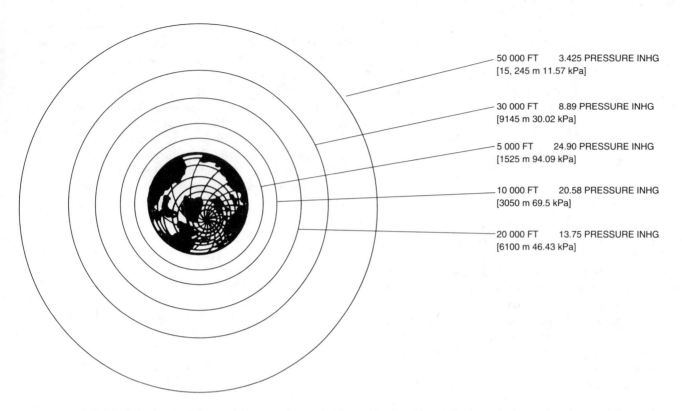

FIGURE 3–4 Pressure of the earth's atmosphere at various altitudes. (*Note:* Metric equivalents given in round figures.)

Since air has weight, it is easy to recognize that the pressure of the atmosphere will vary with altitude. This is illustrated in Figure 3–4. Notice that at 20 000 ft [6097.56 m] the pressure is less than half the sea-level pressure. This means that more than half the atmosphere lies below the altitude of 20 000 ft even though the "outer" half extends hundreds of miles above the earth. Table 3–1 shows the pressures and temperatures at various altitudes above the earth. This table is based upon standard conditions established by the International Civil Aviation Organization (ICAO). The table also shows the density of the air in slugs and the speed of sound at each altitude.

Air Temperature

Under standard conditions, temperature decreases at approximately 1.98°C for each increase of 1000 ft [304.88 m]

TABLE 3-1 ICAO Standard Atmosphere

Altitude		t		P			$\rho \times 10^3$, slugs/ft^3	c_s, ft/s
ft	m	F	C	inHg	kPa	lb/ft^2		
−2 000	−609.76	66.1	18.9	32.100	108.40	2273.70	2.520	1124.54
0	0.0	59.0	15.0	29.920	101.04	2116.20	2.380	1116.89
1 000	304.88	55.4	13.0	28.860	97.46	2040.80	2.310	1113.05
2 000	609.76	51.9	11.0	27.820	93.95	1967.70	2.240	1109.19
3 000	914.63	48.3	9.1	26.820	90.57	1896.60	2.180	1105.31
4 000	1 219.51	44.7	7.1	25.840	87.26	1827.70	2.110	1101.43
5 000	1 524.39	41.2	5.1	24.900	84.09	1760.80	2.050	1097.53
10 000	3 048.78	23.3	−4.8	20.580	69.50	1455.30	1.760	1077.81
15 000	4 573.17	5.5	−14.7	16.890	57.04	1194.30	1.500	1057.73
20 000	6 097.56	−12.3	−24.6	13.750	46.43	972.50	1.270	1037.26
25 000	7 621.95	−30.2	−34.5	11.100	37.46	785.30	1.070	1016.38
30 000	9 146.34	−48.0	−44.4	8.890	30.02	628.40	0.890	995.06
36 089	11 002.74	−69.7	−56.5	6.680	22.56	472.70	0.710	968.46
40 000	12 195.12	−69.7	−56.5	5.540	18.71	391.70	0.5850	968.46
50 000	15 243.90	−69.7	56.5	3.425	11.57	242.20	0.3620	968.46
60 000	18 292.68	−69.7	−56.5	2.118	7.15	149.80	0.2240	968.46
70 000	21 341.46	−69.7	−56.5	1.322	4.46	93.52	0.1388	968.46

t Standard temperature
P Pressure, lb/ft^2 or inHg
ρ Density
c_s Standard speed of sound

of altitude until an altitude of 38 000 ft [11 585.44 m] is reached. Above this altitude the temperature remains at approximately −56.5°C.

Textbooks on meteorology often state that the temperature normally decreases with altitude at a rate of approximately 0.5°C per 100 m, or about 1°F per 300 ft. This amounts to a decrease of about 1.52°C for each increase of 1000 ft, which is different from the decrease under standard conditions. Remember that the textbooks using the foregoing values are discussing **average** rather than standard conditions.

Adiabatic Lapse Rate

As shown in Table 3–1, the temperature of the air decreases as pressure decreases with an increase in altitude. This decrease of temperature with altitude is defined as the **lapse rate.** An **adiabatic** temperature change means that the temperature of the air has changed, but the air has neither gained nor lost heat energy. The temperature change in such a case is due to a change in pressure.

Atmospheric pressure differences cause the air to flow from an area of higher pressure to an area of lower pressure. Thus air may flow up and over mountains or from higher elevations down into valleys. As air flows to higher altitudes, it becomes cooler, and as it flows to lower altitudes, it becomes warmer. This is in accordance with Charles' law (explained in Chapter 2). The **adiabatic lapse rate** is the increase or decrease in the temperature of the air for a given change in altitude. The adiabatic lapse rate varies from 3°F [1.67°C] per 1000 ft [304.88 m] for moist air to more than 5°F [2.78°C] per 1000 ft for very dry air. The standard rate shown in the ICAO chart is approximately 3.5°F per 1000 ft.

Keep in mind that the temperature of the air often does not conform to standards. For example, sometimes the air temperature 1000 ft or more above the surface of the earth is higher than it is at the surface. This condition is called an **inversion**. Mountains, clouds, surface winds, bodies of water, and sunshine all affect the temperature of the air.

Density

The **density** of the air is a property of great importance in the study of aerodynamics. *Density* has been defined previously; however, additional discussion is given here to relate density to the study of aerodynamics. Air is compressible, as illustrated in Figure 3–5. As the air is compressed, it becomes more dense because the same quantity of air occupies less space. Density varies directly with pressure, with the temperature remaining constant. In Figure 3–5, the air in cylinder *B* has twice the density of the air in cylinder *A*.

For the purposes of aerodynamic computations, air density is represented by the Greek letter ρ (rho), indicating mass density in slugs per cubic foot. The **slug** is a unit of mass with a value of approximately 32.175 lb [14.59 kg] under standard conditions of gravity. The word **mass** designates the pull in standard gravitational units exerted by the earth upon a piece of matter. The difference between mass

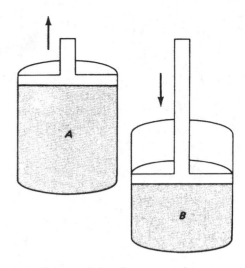

FIGURE 3–5 Air expanded and compressed.

and weight is explained in Chapter 2. Since the slug is used to indicate the density of air, it is sufficient for the technician to know that the value of ρ can be found in standard atmospheric tables.

Air at standard sea-level conditions weighs 0.0765 lb/ft³ and has a density of 0.002378 slug/ft³. At an altitude of 40 000 ft [12 192 m], the air density is approximately 25% of the sea-level value.

The **general gas law** defines the relationship of pressure, temperature, and density when there is no change of state or heat transfer. Simply stated, this law says that density varies directly with pressure and inversely with temperature. On a hot day, air expands, becoming "thinner," or less dense; conversely, on a cold day, the air contracts, becoming more dense.

Changes in air density affect the flight of an airplane. With the same thrust, an airplane can fly faster at a high altitude, where the density is low, than at a low altitude, where the density is greater. This is because the air offers less resistance to the airplane when it contains a smaller number of particles of air per unit volume. This concept is illustrated in Figure 3–6. However, an often-encountered problem is an inability to hold the thrust constant as altitude increases. Generally, engine performance will decrease with altitude.

Humidity

The condition of moisture or dampness in the air is called **humidity**. The maximum amount of water vapor that the air can hold depends on the temperature of the air; the higher the temperature of the air, the more water vapor it can absorb. By itself, water vapor weighs approximately five-eighths as much as an equal volume of perfectly dry air. Therefore, when air contains 5 parts of water vapor and 95 parts of perfectly dry air, it is not as heavy as air containing no moisture. This is because water is composed of hydrogen (an extremely light gas) and oxygen. Air is composed principally of nitrogen, which is almost as heavy as oxygen.

Assuming that the temperature and pressure remain the same, the density of the air varies with the humidity. On damp days the density of air is less than it is on dry days.

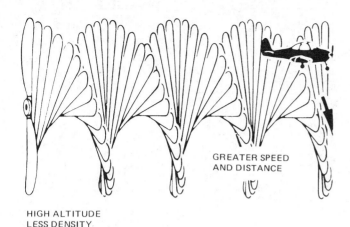

GREATER SPEED
AND DISTANCE

HIGH ALTITUDE
LESS DENSITY

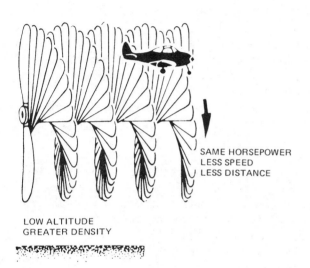

SAME HORSEPOWER
LESS SPEED
LESS DISTANCE

LOW ALTITUDE
GREATER DENSITY

FIGURE 3–6 Effect of air density on an aircraft in flight.

LIFT

The unique feature of an aircraft as compared with all other types of transportation vehicles is its ability to lift into the air. The force of **gravity** acts on all bodies on or near the surface of the earth and results in the weight of an object. In order for an aircraft to fly, a force must be created that will overcome the force of gravity. This force is called **lift**.

The Physics of Lift

Some basic principles of physics can be used to understand how lift is created. The physical forces that support an aircraft in flight may be explained principally by two basic laws of physics: Newton's third law of motion and Bernoulli's principle.

As a wing moves through the air, the airflow will be divided to flow over and under the wing, as is shown in Figure 3–7. When there is a positive angle between the wing and the direction of the airstream, the air is forced to change direction. If the wing is tilted upward against the airstream, the air flowing under the wing is forced downward. The wing therefore applies a downward force to the air, and the

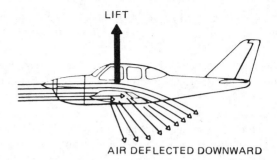

LIFT

AIR DEFLECTED DOWNWARD

FIGURE 3–7 Wing deflecting the air downward.

air applies an equal and opposite upward force to the wing. This is the creation of lift and can be explained by **Newton's third law of motion**, which states: For every action there is an equal and opposite reaction. The angle through which an airstream is deflected by any lifting surface is called the **downwash angle**. It is especially important when control surfaces are studied, because they are normally placed to the rear of the wings where they are influenced by the downward-deflected airstream known as the **downwash**.

The lift or support the airfoil receives from deflecting the air downward can vary from 0 to 100% of the total lift required. On the other hand, if the wing is nosed downward, it may scoop the air upward, causing a down load to be placed on the wing.

Lift can also be created by the way that the air flows around an airfoil. In order to understand how this lift is created, it is first necessary to understand **Bernoulli's principle**. Daniel Bernoulli, a Swiss scientist of the eighteenth century, discovered that as fluid (air) velocity increases, the pressure decreases, and as the velocity decreases, the pressure increases.

This principle can be written mathematically as

$$p + \tfrac{1}{2}\rho V^2 = \text{constant}$$

where

$$p = \text{pressure}$$
$$\rho = \text{air density in slugs}$$
$$V = \text{velocity}$$

The equation can be simplified as the following:

$$\text{pressure} + \tfrac{1}{2} \times \text{density} \times \text{velocity squared} = \text{constant}$$

Actually, in technical language, Bernoulli's principle states that the total energy of a particle in motion is constant at all points on its path in a steady flow. In the preceding equation, the term *pressure* refers to the static pressure of the air. The term $\frac{1}{2} \times density \times velocity \ squared$ is referred to as the dynamic pressure. The term **static** means "still," and the term **static pressure** refers to the pressure exerted by a mass of stationary air equally on all of the walls of a container. **Dynamic pressure** is the pressure associated with moving air (velocity). The dynamic pressure is the pressure that would be exerted if the moving air were brought to a stop. The most appropriate means of visualizing the effect of airflow and the resulting aerodynamic pressures is to study the fluid flow within a closed tube.

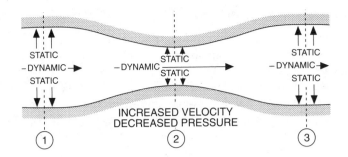

FIGURE 3-8 Fluid flow within a closed tube.

Suppose a stream of air is flowing through the venturi tube shown in Figure 3-8. The airflow at station 1 in the tube has a certain velocity and static pressure. As the airstream approaches the constriction at station 2, certain changes must take place. According to the law of conservation of matter, the mass flow at any point along the tube must be the same, and the velocity or pressure must change to accommodate this continuity of flow. As the flow approaches the constriction of station 2, the velocity increases to maintain the same mass flow. As the velocity increases,

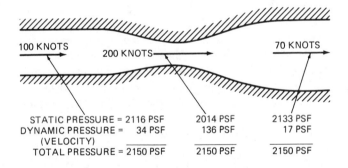

FIGURE 3-9 Bernoulli's principle.

the static pressure will decrease. The total energy of the airstream in the tube is unchanged. However, the airstream energy may be in two forms. The airstream may have a po-

tential energy, which is related by the static pressure, and a kinetic energy, represented by its dynamic pressure (velocity). Because the total energy is unchanged, an increase in velocity (dynamic pressure) will be accompanied by a decrease in static pressure (velocity). Therefore, it can be said that the sum of static and dynamic pressure in the flow tube remains constant.

static pressure + dynamic pressure = constant

Figure 3-9 illustrates the variation of static, dynamic, and total pressure of air flowing through a closed tube. Note that the total pressure is constant throughout the length and that any change in dynamic pressure produces the same magnitude change in static pressure.

In Figure 3-10a the air streams moving through a venturi tube are indicated by arrows. Notice that close to the venturi wall the air flow conforms to the shape of the walls. In the center of the tube the airflow is straight. In Figure 3-10b the walls of the venturi have been moved farther apart. Notice that the airflow streams close to the walls still follow the wall contours and that those farther away gradually straighten out. Before the airflow enters the venturi tube, the streamlines are equally spaced, indicating a uniform flow velocity. Adjacent to the wall of the tube the streamlines come closer together, indicating that the velocity of the flow is greater at that point. In Figure 3-10c the upper wall has been removed. The flow lines immediately adjacent to the bottom wall still follow the contour and are spaced closely together. It is therefore indicated that the velocity immediately adjacent to a curved surface will increase. According to Bernoulli's principle, this increase will bring about a similar decrease in pressure. Notice the similarity between the shape of the remaining wall of the venturi in Figure 3-10c and the shape of a wing.

The effect produced by a wing moving through the air is illustrated in Figure 3-11. When the air strikes the leading edge of the wing, the passage of the air is obstructed and its velocity is reduced. Some of the particles of air flow over the upper surface and some flow under the lower surface, but all separating particles of air must reach the trailing edge of the wing at the same time. Those particles that pass over the upper surface have farther to go and therefore must move faster than those passing under the lower surface. In accordance with Bernoulli's principle, the increased velocity

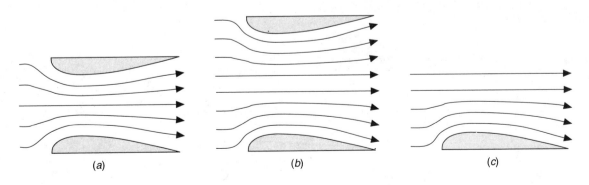

FIGURE 3-10 Airflow in a venturi.

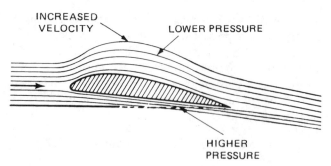

FIGURE 3–11 Pressure differential created by a wing in flight.

above the wing results in a lower static pressure than that existing below the wing.

If an area of low pressure can be produced on the top of a wing, the higher pressure below the wing will create an upward force, which is lift. This pressure differential is quite small, perhaps as small as 1%. Even this small difference, however, can provide adequate lift, as demonstrated by the following example: Assume that we have an atmospheric pressure of 14.70 pounds per square inch on the under surface of a wing and a wing curvature and airspeed combination producing a pressure of 14.49 pounds per square inch on the upper surface of the wing. This situation provides a pressure differential between the upper and lower surfaces of 0.21 psi, acting upwards. This pressure is equal to 30.34 pounds per square foot (psf) (0.21 × 144), a wing loading that is fairly common for many types of aircraft. The total lift produced by the wing can be calculated by multiplying 30.34 psf by the square footage of the total wing area.

AIRFOILS

The structure which makes flight possible is the **airfoil**. An airfoil is technically defined as any surface, such as an airplane aileron, elevator, rudder, or wing, designed to obtain a useful reaction from the air through which it moves. An airfoil section is a cross section of an airfoil, which can be drawn as a silhouette. If the wing of an airplane were sawed through from the leading edge to the trailing edge, the side view of the section through the wing at that point would be its airfoil section. An **airfoil profile** is merely the outline or shape of an airfoil section. The word *airfoil* is often used when "airfoil section" or perhaps "airfoil profile" is meant.

Airfoil Terminology

Since the shape of an airfoil and its angle to the airstream are so important in determining its performance, it is necessary to understand airfoil terminology. Figure 3–12 shows a typical airfoil and illustrates various airfoil-related terms.

The **chord line** is a straight line connecting the leading edge (the forward-most tip) and the trailing edge of the airfoil. The distance between the leading edge and the trailing edge is referred to as the **chord**.

Notice that in Figure 3–12 there is more area above the chord line than below it. This is typical of most airfoils. The **mean camber line** is a line drawn halfway between the upper and lower surfaces. This line is also referred to as the **mean line** or **mid line**. Any point on this mean line should be the same distance from the upper and lower surfaces. The mean camber is the curvature of the mean line of an airfoil profile from the chord.

Camber is defined as the curvature of an airfoil surface or an airfoil section from the leading edge to the trailing edge. The perpendicular distance between the chord line and the mean camber line is camber. The degree or amount of camber is expressed as the ratio of the maximum departure of the mean camber line from the chord to the chord length. Figure 3–12 shows an airfoil that has a double convex curvature, which means that it has camber above and below the chord line. Upper camber refers to the curve of the upper surface of an airfoil, and lower camber refers to the curve of the lower surface. Camber is positive when the departure from the straight line is upward, and negative when it is downward. When the upper and lower camber of an airfoil are the same, the airfoil is said to be symmetrical.

The shape of the mean camber line is very important in determining the aerodynamic characteristics of an airfoil section. The **maximum camber** (the maximum displacement of the mean line from the chord line) and the location of the maximum camber help to define the shape of the mean camber line. These quantities are expressed as fractions or percentages of the basic chord dimension. A typical low-speed airfoil may have a maximum camber of 4% located 40% aft of the leading edge. The maximum camber is sometimes referred to simply as the camber.

The **thickness** and **thickness distribution** of the profile are important properties of an airfoil. The distance between the upper and lower surfaces is called, simply, the thickness.

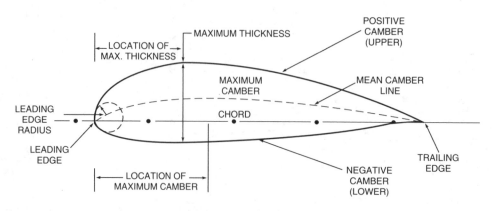

FIGURE 3–12 Airfoil terminology.

The maximum thickness and the location of the maximum thickness are expressed as fractions of the percentage of the chord. A typical low-speed airfoil may have a maximum thickness of 12% located 30% aft of the leading edge.

The **leading-edge radius** of the airfoil is the radius of curvature given the leading-edge shape. It is the radius of the circle centered on a line tangent to the leading-edge camber connecting tangency points of upper and lower surfaces with the leading edge. Typical leading-edge radii are 0 (knife edge) to 4 or 5%.

Figure 3–13 illustrates five airfoil profiles of different shapes together with their chords. In the figure, profile A has a double convex shape. The chord is simply the straight line from the leading edge to the trailing edge. Profile B, which is designed to produce high lift, has a convex upper curvature and a concave lower curvature. The chord is the straight line connecting the imaginary perpendiculars erected at the leading and trailing edges. Profile C has a flat lower surface; therefore the chord is the straight line connecting the leading and trailing edges. Profile D resembles profile B, and again the chord is the straight line connecting imaginary perpendiculars erected at the leading and trailing edges. Profile E is designed for supersonic flight and is almost symmetrical.

Airfoils in Motion

When an airfoil is moved through the air, a stream of air flows around the airfoil. If the airfoil is set at the proper angle and has sufficient velocity, enough lift will be produced to sustain the heavier-than-air craft in flight. In examining the flow of air around the airfoil, assume that this is a no-wind day. Any airflow, or wind, will be a **relative wind**, that is, a wind created by the movement of an object through still air. A relative wind flows opposite the direction of the object in motion. The velocity of its flow around or over the object in motion is the object's airspeed.

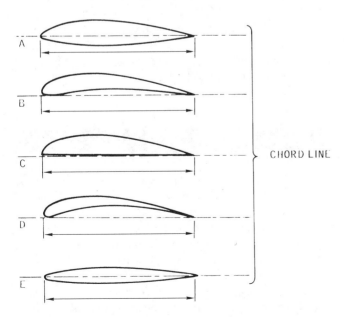

FIGURE 3–13 Airfoil profiles of different shapes.

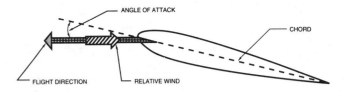

FIGURE 3–14 Airfoil in motion.

In Figure 3–14 a symmetrical airfoil is shown with its chord extended through the leading edge. As the airfoil moves through the air, a relative wind is created which flows opposite the flight direction of the airfoil. The velocity of the relative wind is in direct proportion to the velocity at which the airfoil is being moved through the air. Figure 3–14 also shows that an acute angle is formed between the chord of the airfoil and the relative wind. This angle is called the **angle of attack** (AOA). The Greek letter α (alpha) is used to denote this angle.

A cambered airfoil can produce lift at 0° AOA because there is more cross-sectional area above the chord line than below it, resulting in a greater velocity and lower static pressure of the airflow above the airfoil. At 0° AOA, the airflow will divide at the leading edge. With a positive angle of attack, such as is illustrated in Figure 3–15, the airflow no longer divides right at the tip of the leading edge but at a point farther down on the nose. The point where the airflow divides is called the **stagnation point**. The stagnation point is shown in Figure 3–15 with a small x. Immediately in front of the stagnation point the air separates, with the upper part of it forced to flow up over the top of the leading edge, and the lower part flowing under the wing. In Figure 3–15 the shaded arrow shows the direction of the relative wind, and airflow lines are shown steaming over and under the airfoil. The distance A that the air must flow from the point of impact to the trailing edge over the top of the airfoil is much greater than the distance B that the air must flow from the point of impact to the trailing edge. According to Bernoulli's principle, this greater distance on the top surface will result in the creation of lift. As the angle of attack increases, the stagnation point moves farther down on the airfoil.

Critical Angle of Attack

Beginning with small angles of attack, the lift increases as the angle of attack increases, until an angle of attack is reached where the lift has a maximum value. This angle is the angle of attack at which the streamline flow of air begins to break down over the upper surface of the airfoil and **burbling** begins at the trailing edge of the airfoil. This breakdown and separation of the airflow is attributed to the fact that as the stagnation point moves further down on the leading edge, the airflow over the top has an increasingly longer

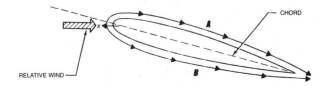

FIGURE 3–15 Airflow around an airfoil.

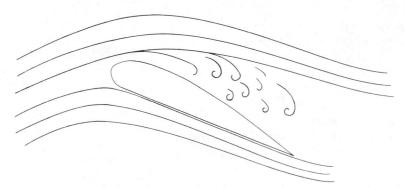

FIGURE 3–16 Airfoil in a stall.

path to travel. As air flows over a surface, a certain amount of friction is developed. As the path gets longer, the frictional force continues to build until the energy available in the airstream is no longer sufficient to overcome it. At this point, the airflow will detach itself, as is shown in Figure 3–16. With the loss of a smooth airflow over the top surface, pressure is no longer being reduced to create lift. This angle of attack is called the **stalling angle**. At angles greater than the angle of maximum lift, the lift decreases rapidly, as is shown in Figure 3–17.

For each critical angle of attack there is a corresponding airspeed, assuming that other conditions, such as wing area and air density, remain constant. As the critical angle of attack increases, the corresponding airspeed decreases; therefore, the lowest possible airspeed exists at the angle of maximum lift (the stalling angle). Thus another name for the angle of maximum lift is the **angle of minimum speed**. The **stalling speed** of an airplane is the minimum speed at which the wing will maintain lift.

Velocity and Lift

A positive angle of attack causes increased velocity and decreased pressure on the upper surface of a wing and decreased velocity and increased pressure on the lower surface. If the air flows slowly around the airfoil, a certain amount of lift is generated. If the velocity of the airstream increases, the pressure differential increases and the lift increases. Lift does not vary in direct proportion to speed; lift varies as the square of the speed. An aircraft traveling at 100 knots has four times the lift it would have at 50 knots.

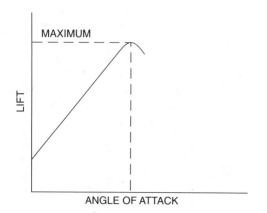

FIGURE 3–17 Critical angle of attack.

Air Density and Lift

You have seen that the amount of lift generated depends upon the shape of the airfoil, the angle of attack, and the airspeed. Another factor that affects lift is the air density. Lift varies directly with air density. At 18 000 feet, where the density of air is just half as much as at sea level, an aircraft would have to travel 1.414 (the square root of 2) times as fast as it would at sea level to maintain altitude.

On hot days, the density of the air is less than on cold days; on wet days, the density is less than on dry days. Also, density decreases with altitude. When the density is low, the lift will also be comparatively lower.

If an airplane flies at a certain angle of attack at sea level and then flies at the same angle of attack at a higher altitude, where air density is less, the airplane must be flown faster. On hot days, when the density is less, the airplane must be flown faster for the same angle of attack than on cold days, when the density is greater. Therefore, the airspeed must increase as the density decreases in order to maintain the airplane at the same angle of attack in level flight.

Area and Lift

One of the factors that determines the total lift of an airfoil is the area of the surface exposed to the airstream. Lift varies directly with the area, other factors being equal. A wing with an area of 200 square feet will lift twice as much as a wing of only 100 square feet, providing other factors remain the same. Later in this text, we will explain why more lift can be obtained from a long, narrow wing than from one in which the width more closely approaches the length.

DRAG

As explained previously, air has mass. When an airplane flies through air, the air is moved. When any mass is moved or accelerated, force is required, and the application of force produces an equal and opposite force. This is in keeping with Newton's law of motion. The impact of the air against the surfaces of the airplane applies force, which tends to hold the airplane back. This is **drag**. Specifically, drag is a retarding force acting upon a body in motion. There are several different types of drag, which are classified according to their origin.

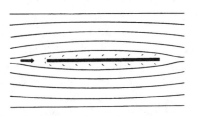

FIGURE 3–18 Skin friction.

Skin Friction and Air Viscosity

Figure 3–18 illustrates **skin friction**. In the figure, a thin, flat plate is held edgewise to an airstream. The particles of air separate at the leading edge and flow over the upper surface and under the lower surface, reuniting behind the trailing edge. The resistance of skin friction is caused by a tendency of the particles of air to cling to the surface of the plate. There are two reasons for this tendency. First, the plate has a certain amount of roughness, relatively speaking. It is impossible to make it perfectly smooth.

The second reason why air tends to cling to the surface is the **viscosity** of the air. Technically, viscosity is the resistance offered by a fluid to the relative motion of its particles, but the term is most commonly used to describe the adhesive or sticky characteristics of a fluid. Even though it is not always apparent, air does have "thickness," as does oil. Viscosity may best be visualized by thinking of the difference between syrup and water: the syrup is considerably more viscous than water. The viscosity of gases is unusual in that the viscosity is generally a function of temperature alone; a decrease in temperature increases the viscosity (thickness). Therefore, viscosity will generally increase with altitude.

Boundary Layer

The term **laminar flow** describes the situation when air is flowing in thin sheets, or layers, close to the surface of a wing with no disturbance between the layers of air; that is, there is no cross-flow of air particles from one layer of air to another. Also, there is no sideways movement of air particles with respect to the direction of airflow.

Laminar flow is most likely to occur where the surface is extremely smooth and especially near the leading edge of an airfoil. Under these conditions the boundary layer will be very thin. The **boundary layer** is that layer of air adjacent to the airfoil surface. The air velocity in the boundary layer varies from zero on the surface of the airfoil to the velocity of the free stream at the outer edge of the boundary layer. This is illustrated in Figure 3–19. The boundary layer is caused by the viscosity of the air sticking to the surface of the wing and the succeeding layers of air.

Ordinarily, the airflow at the leading edge of a wing will be laminar, but as the air moves toward the trailing edge of the wing, the boundary layer becomes thicker and laminar flow diminishes. The area where the airflow changes from laminar to turbulent is called the transition region. This is illustrated in Figure 3–20. It is desirable to keep a laminar flow over the airfoil as much as possible.

Reynolds Number

Osborne Reynolds studied the flow of liquids in pipes and found that at a low speed the flow is smooth but at a high speed the flow is turbulent. By experimenting with pipes of various sizes and with different liquids, he found a value which he called the **critical Reynolds number** (R_e). The flow was laminar (smooth) for values below the critical R_e and turbulent for the values above the critical R_e.

Whether a laminar or turbulent boundary layer exists around an airfoil depends on the combined effects of velocity, viscosity, density, and the size of the chord. It is the combined effects of these important parameters which produce the Reynolds number.

While the actual magnitude of the Reynolds number has no physical significance, the quantity is used as an index to predict various types of airflow. At low Reynolds numbers, the flow is laminar, and at high Reynolds numbers the flow becomes turbulent. High Reynolds numbers are obtained with large chord surfaces, high velocities, and low altitudes; low Reynolds numbers result from small chord surfaces, low velocities, and high altitudes. Since a different Reynolds number would be obtained for each location downstream from the leading edge, it is common to specify a characteristic length and airspeed to define the Reynolds number for a particular airfoil. The length commonly used

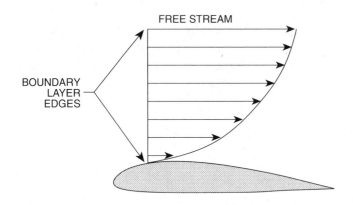

FIGURE 3–19 Increasing airspeed velocities in a boundary layer.

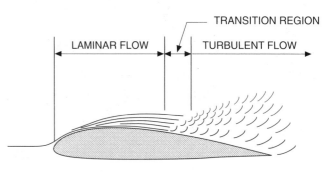

FIGURE 3–20 Boundary layer airflow.

is the average wing chord. The formula used for calculating the Reynolds number is

$$R_e = \frac{V \times d}{v}$$

where

V = fluid velocity
d = the distance downstream from the leading edge
v = kinematic viscosity of the fluid

Using this formula, an airplane with a 60-in average wing chord, an airspeed of 100 mph, operating at sea level on a standard day would have a Reynolds number of 4 700 000. For transport-category aircraft, a Reynolds number in the millions is very common.

A term frequently used in discussing airplane design is **scale effect**. This is the change in any force coefficient, such as a drag coefficient, due to a change in the value of a Reynolds number.

Parasite Drag

The term **parasite drag** describes the resistance of the air produced by any part of the airplane that does not produce lift. Parasite drag can be further classified into pressure drag, skin friction drag, and interference drag.

One type of drag, **pressure drag**, is caused by the frontal area of the airplane components being exposed to the airstream. A similar reaction is illustrated in Figure 3–21, where the side of the airfoil is exposed to the airstream. The pressure against the front side is much greater than that formed on the back side, in the wake. This drag is caused by the form (shape) of the airfoil and is the reason streamlining is necessary to increase airplane efficiency and speed.

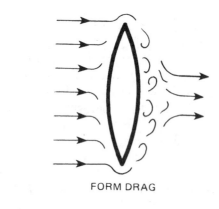

FORM DRAG

SKIN FRICTION DRAG

FIGURE 3–21 Form drag and skin friction drag.

Figure 3–21 also illustrates that when the leading edge of the airfoil is parallel to the airstream, the pressure drag is reduced. However, this does not eliminate all of the drag. The air flowing along the surface of the airfoil creates a frictional force on the body. This force is called **skin friction drag**. Skin friction drag is caused by air passing over the airplane's surfaces, and it increases considerably if the airplane surfaces are rough and dirty.

Most parts of an airplane, such as the fuselage, cowlings, landing-gear struts, and other components, will have both thickness and surface area, resulting in both pressure and friction drag.

Pressure drag and friction drag are both components of parasite drag. However, in calculating the total parasitic drag force of an aircraft, another type of drag must also be considered. This type of drag is called **interference drag** and is caused by the interference of the airflow between adjacent parts of the airplane, such as the intersection of wings and tail sections with the fuselage. Fairings are used to streamline these intersections and decrease interference drag.

Several factors affect parasite drag. When each factor is considered independently, it must be assumed that the other factors remain constant. These factors are (1) the more streamlined an object is, the less the parasite drag; (2) the more dense the air moving past the airplane, the greater the parasite drag; (3) the larger the size of the object in the airstream, the greater the parasite drag; and (4) as speed increases, the amount of parasite drag increases.

Induced Drag

The term **induced drag** describes the undesirable but unavoidable by-product of lift. The pressure differential between the upper and lower surfaces of the wing result in a vortex being formed at each wing tip, causing a downward push on the air leaving the trailing edge. This downward component is termed **downwash** and produces a rearward component to lift. The lift component is most effective acting vertically. When the lift vector is tilted rearward due to the downwash, lift is lost. This loss of lift is induced drag.

Since induced drag is the direct result of wing-tip vortices, the aspect ratio (the wingspan to chord ratio), which is discussed in Chapter 4, has a great effect on the amount of induced drag produced. Induced drag increases in direct proportion to increases in the angle of attack. The greater the angle of attack, up to the critical angle, the greater the amount of lift developed and the greater the induced drag. The amount of air deflected downward increases greatly at higher angles of attack; therefore, the higher the angle of attack, the greater the induced drag. However, an important point to remember is that induced drag decreases with velocity.

Total Drag

The sums of both the induced drag and the parasite drag is the **total drag**.

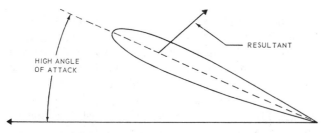

FIGURE 3–22 The resultant force.

total drag = induced drag + parasite drag

In designing or repairing an airplane, the drag forces exerted on the structure of the airplane should be considered. As the airplane flies through the air, the effect is as if millions of tiny particles were striking against the forward parts of the airplane. If the total force of these particles becomes too great for the strength of the structure, damage will result. It is therefore necessary to make sure that the leading edges of the wings, the leading edges of the stabilizers, and the forward parts of the engine cowling are constructed of material with sufficient strength to withstand the maximum impact forces that will ever be imposed upon them. Furthermore, the devices by which the parts are attached to the main structure must have sufficient strength to withstand maximum drag forces.

Lift and Drag Components

Lift and drag are components of the total aerodynamic force acting upon the wing. This total force is called the **resultant**. Each tiny portion of the wing in flight has a small force acting upon it. The force acting on one small portion of the wing is different in magnitude and direction from all the other small forces acting upon all the other portions of the wing. By considering the magnitude, direction, and location of each of these small forces, it is possible to add them all together into one resultant force. This resultant force has magnitude, direction, and location with respect to the wing.

The resultant force on an airfoil flying at a specified speed and angle of attack can be shown as a single entity possessing both magnitude and direction, such as in Figure 3–22. It is also possible to break the resultant down into two

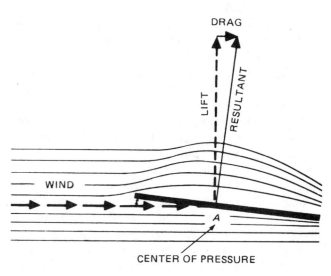

FIGURE 3–24 Center of pressure.

major components (lift and drag), with magnitudes in two directions. In aerodynamics these forces are discussed as having directions perpendicular and parallel to the relative wind. The component of the resultant force which acts perpendicular to the relative wind is lift. The component of the resultant force which acts parallel to the relative wind is called drag. Figure 3–23 illustrates the resultant force broken down into the separate component forces. The lengths of the component arrows are in proportion to the magnitudes of the forces they represent. For example, if lift is two times as great as drag, the arrow representing lift must be drawn twice as long as the arrow representing drag.

Center of Pressure

The **center of pressure** (CP) is the point at which the resultant force intersects the chord of an airfoil. The center of pressure is shown at A in Figure 3–24. Lift acts from the center of pressure, or, stated another way, the center of pressure is the center of lift.

The location and direction in which the resultant will point depends upon the shape of the airfoil section and the angle at which it is set to the airstream. Throughout most of the flight range, that is, at the usual angles of attack, the CP moves forward as the angle of attack increases and backward as the angle of attack decreases. As is illustrated in Figure 3–25 the resultant intersects the chord line or center of pressure at progressively forward locations as the angle of attack is increased. The center of pressure is generally located at approximately the 25% chord position for most air-

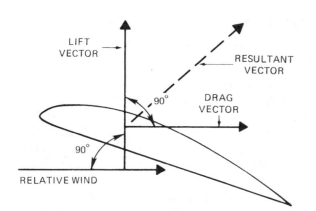

FIGURE 3–23 Relationship between relative wind, lift, and drag.

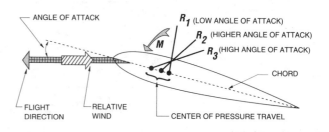

FIGURE 3–25 Center of pressure travel.

foils. On an airfoil with a 60-inch chord, this would locate the center of pressure at 15 inches aft from the leading edge. While the CP travel will generally be restricted to the 25% (±10%) chord area, the CP can travel forward or backward from these usual positions. For example, at a low angle of attack, the CP may run off the trailing edge and disappear because there is no more lift.

The pressure distribution is different on the top and bottom of an airfoil. This results in the center of pressure being different on the two surfaces, with the center of pressure on the lower surface generally being forward of that on the upper surface. Because the resultant forces on the two surfaces are not acting at the same chordwise location, a rotation of the airfoil will result. This rotation situation is illustrated in Figure 3–25 by the letter *M* and the corresponding arrow. This tendency for rotation is called the pitching moment. Due to the pitching moment, most airfoils are unstable by themselves and require some additional method to stabilize them, such as the installation of a horizontal stabilizer.

One of the reasons for studying CP travel is that the CP is the point at which the aerodynamic forces can be considered to be concentrated; therefore the airplane designer must make provisions for the CP travel by preparing a wing structure that will meet any stress imposed upon it. Technicians and inspectors cannot change the design, but they can perform their duties better if they know the characteristics and limitations of an airplane.

HIGH-SPEED FLIGHT

So far we have discussed airflow and aerodynamic principles with respect to subsonic airspeeds only. Developments in aircraft and power plants, however, have produced high-

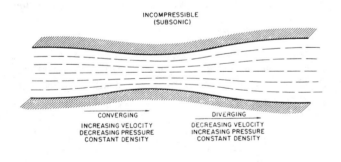

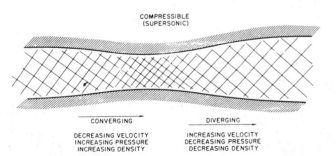

FIGURE 3–26 Comparison of compressible and incompressible airflows.

performance airplanes with capabilities for very high-speed flight. The study of aerodynamics at these very high flight speeds has many significant differences from the study of low-speed aerodynamics. The behavior of an airfoil under subsonic conditions is easily predictable. However, when operating at transonic and supersonic speeds, the reaction of an airfoil is altogether different from that which is found at subsonic speeds. The reason for this difference is the reaction of the air itself.

Compressibility

At low flight speeds, the study of aerodynamics is greatly simplified by the fact that air may experience relatively small changes in pressure with only negligible changes in density. This airflow is termed **incompressible** since the air may undergo changes in pressure without apparent changes in density, as is shown in Figure 3–26. Such a condition of airflow is similar to the flow of water, hydraulic fluid, or any other incompressible fluid. However, at high flight speeds, the pressure changes that take place are quite large, and significant changes in air density occur, as is also illustrated in Figure 3–26. The study of airflow at high speeds must account for these changes in air density and consider the fact that the air is **compressible**.

Speed of Sound

A factor of great importance in the study of high-speed airflow is the speed of sound. The **speed of sound** is the rate at which small pressure disturbances will be spread through the air. The aerodynamic effects of pressure are carried through the air at the same rate as that of sound disturbances.

The speed at which sound travels in air under standard sea-level conditions is 1116 ft/s [340.24 m/s], or 761 mph, or 661 kn. The speed of sound is not affected by a change in atmospheric pressure because the density also changes. However, a change in the temperature of the atmosphere changes the density without appreciably affecting the pressure; therefore, the speed of sound changes with a change in temperature. The speed of sound can be calculated with the equation

$$a = 49.022 \ \sqrt{T}$$

where

a = speed of sound, ft/s
T = absolute temperature, °R

The temperature of the air decreases with an increase in altitude up to an altitude of about 37 000 ft [11 280 m], and it is then constant to an altitude of more than 100 000 ft [30 488 m]. Therefore, under standard atmospheric conditions, the speed of sound decreases with altitude to about 37 000 ft and then remains constant to more than 100 000 ft. For example, at 30 000 ft [9146.34 m] the temperature of standard air is −48°F [−44.44°C], and the speed of sound is 995 ft/s [303.35 m/s], or 589 kn. Table 3–2 illustrates the variation of the speed of sound in the standard atmosphere.

TABLE 3–2 Variation of Speed of Sound with Temperature

Altitude, ft	Temperature °F	Temperature °C	Speed of sound, kn
Sea level	59.0	15.0	661.7
5 000	41.2	5.1	650.3
10 000	23.3	−4.8	638.6
15 000	5.5	−14.7	626.7
20 000	−12.3	−24.6	614.6
25 000	−30.2	−34.5	602.2
30 000	−48.0	−44.4	589.6
35 000	−65.8	−54.3	576.6
40 000	−69.7	−56.5	573.8
50 000	−69.7	−56.5	573.8
60 000	−69.7	−56.6	573.8

Mach Number

Because of the relationship between the effect of high-speed air forces and the speed of sound and because the speed of sound varies with temperature, it is the ratio of the speed of the aircraft to the speed of sound that is important, rather than the speed of the aircraft with respect to the air. This ratio, called the **Mach number** (M), is the true airspeed of the aircraft divided by the speed of sound in the air through which the aircraft is flying at the time.

$$M = \frac{\text{true airspeed}}{\text{speed of sound}}$$

Thus, Mach 0.5 at sea level under standard conditions (558 ft/s [170 m/s]) is faster than Mach 0.5 at 30 000 ft (497 ft/s [151.52 m/s]).

Types of High-Speed Flight

Airflow speeds under Mach 1 are termed **subsonic**, airflow at Mach 1 is considered **sonic**, and airflow speeds in excess of Mach 1 are called **supersonic**. It is important to note that compressibility effects are not limited to flight speeds at and above the speed of sound.

The speed of the air flowing over a particular part of the aircraft is called the **local speed**. The local speed of the airflow may be higher than the speed of the aircraft. For example, the speed of the air across the upper portion of the wing is accelerated to produce lift. Thus, an aircraft flying in the vicinity of Mach 1 will have local airflows over various parts of the aircraft at speeds both above and below Mach 1. Since there is the possibility of having both subsonic and supersonic flows existing on the aircraft, it is convenient to define certain regimes of flight. These regimes are defined as follows:

Subsonic. Mach numbers below 0.75.

Transonic. Mach numbers from 0.75 to 1.20.

Supersonic. Mach numbers from 1.20 to 5.00.

Hypersonic. Mach numbers above 5.00.

The flight Mach numbers used to define these areas of flight are approximate. A more accurate way to define these ranges for a particular aircraft is

Subsonic. The aircraft maximum Mach number that all local speeds will be less than Mach 1.

Transonic. The regime where local speeds are greater and less than Mach 1.

Supersonic. The aircraft's minimum Mach number when all local speeds are greater than 1.

The airspeed when local flows start to reach Mach 1 is very critical to aircraft performance. For this reason, the free-stream Mach number that produces the first evidence of local sonic flow is called the **critical Mach number**. This is shown in Figure 3–27. The critical Mach number is the boundary between subsonic and transonic flight and is an important point of reference for all compressibility effects encountered in transonic flight.

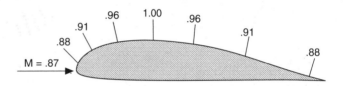

FIGURE 3–27 Critical Mach number.

Shock Wave Formation

At speeds less than 300 mph [483 km], the airflow around an aircraft behaves as though the air were incompressible. Pressure disturbances, or pressure pulses, are formed ahead of the parts of the aircraft, such as the leading edge of the wing. These pressure pulses travel through the air at the speed of sound and, in effect, serve as a warning to the air of the approach of the wing. As a result of this warning, the air begins to move out of the way. Evidence of this "pressure warning" is similar to waves spreading across water. If pebbles were dropped into a smooth pond at the rate of one per second, at the same point, waves would spread out as is shown in Figure 3–28a. In Figure 3–28b, instead of dropping the pebbles in the water at the very same location, the point of dropping is slowly moved to the left each time. Each pebble still produces a circular wave, but the spacing between the rings is no longer uniform and is more closely spaced in the direction of movement. As already described, these wave disturbances provide a warning of the approaching aircraft. Notice that the warning time (the distance between the waves), however, will decrease as speed increases. If the speed of movement continues to increase between dropping the pebbles, a wave pattern such as is shown in Figure 3–28c will develop. Notice that no advance

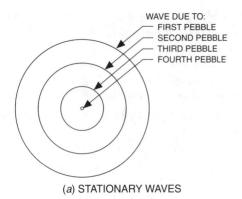

(a) STATIONARY WAVES

WAVE DUE TO:
FIRST PEBBLE
SECOND PEBBLE
THIRD PEBBLE
FOURTH PEBBLE

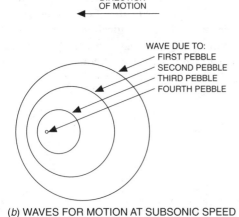

(b) WAVES FOR MOTION AT SUBSONIC SPEED

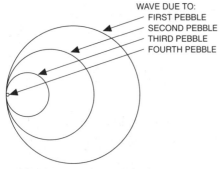

(c) WAVES SIMULATING MACH 1.0 MOTION

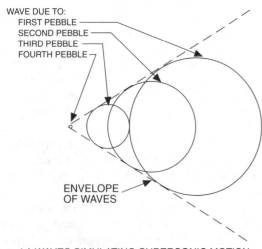

(d) WAVES SIMULATING SUPERSONIC MOTION

FIGURE 3–28 Wave formation.

warning is being sent out by the wave pattern. This is similar to an aircraft flying at the speed of sound. If the speed of movement continues to increase between dropping the pebbles, the smaller wave circles are no longer completely inside the next larger ones. The circles are now within a wedge-shaped pattern, such as is shown in Figure 3–28d. This is similar to an aircraft flying at speeds above Mach 1 and illustrates how a shock wave is formed.

Transonic Flight

As has been previously discussed, an aircraft in the transonic flight area can be expected to have local velocities which are greater than the free-stream airspeed. Therefore the effects of compressibility can be expected to occur at flight speeds less than the speed of sound.

Using a conventional airfoil shape, such as is shown in Figure 3–29, and assuming the aircraft is flying at Mach .5, the maximum local velocity will be greater than the free-stream speed but most likely will be less than Mach 1. Notice that there is upwash and flow direction change well ahead of the leading edge.

Assume that a speed increase to Mach .72 will produce the first evidence of sonic flow. This means that M = .72 is the critical Mach number for this aircraft. As the critical Mach number is exceeded, an area of supersonic airflow is created and a normal shock wave forms as the boundary between the supersonic airflow and the subsonic airflow on the aft portion of the airfoil surface. The acceleration of the airflow from subsonic to supersonic is smooth and unaccompanied by shock waves if the surface is smooth and the transition gradual. However, transition of airflow from supersonic to subsonic is always accompanied by a shock wave and, when there is no change in direction of the airflow, the wave formed will be a normal shock wave.

One of the principal effects of the normal shock wave is to produce a large increase in the static pressure of the airstream behind the wave. If the shock wave is strong, the boundary layer may not have sufficient energy to withstand the adverse pressure gradient of the wave, and airflow sepa-

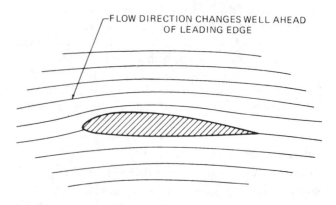

FIGURE 3–29 Typical subsonic flow pattern.

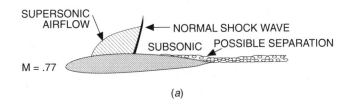

(a)

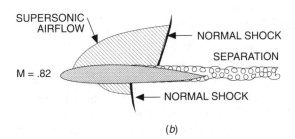

(b)

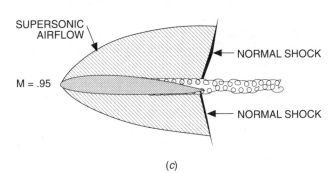

(c)

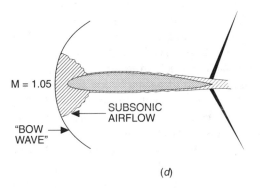

(d)

FIGURE 3–30 Transonic flow patterns.

ration will occur. At speeds that are only slightly beyond the critical Mach number, the shock wave formed is not strong enough to cause separation or any noticeable change in the aerodynamic handling characteristics of the aircraft. Figure 3–30a shows the normal shock wave being formed at Mach .77. As the Mach number continues to increase to Mach .82, the supersonic flow area gets larger on the upper surface and an additional area of supersonic flow and normal shock wave forms on the lower surface.

As the flight speed approaches the speed of sound (Mach = .95), the areas of supersonic flow enlarge and the shock waves move nearer the trailing edge. The boundary layer may remain separated or may reattach depending primarily upon the airfoil shape.

The magnitude and location of these shock waves are constantly changing. Airflow separation will occur with the formation of shock waves, resulting in the loss of lift. Other phenomena that may be associated with transonic flight are aircraft buffeting, trim and stability changes, and a decrease in control-surface effectiveness. These forces and the turbulence that accompanies transonic flight may cause the pilot to lose control, especially if the airplane is not designed to operate under transonic conditions.

When the flight speed exceeds the speed of sound (M = 1.05), the **bow wave** forms at the leading edge. The typical flow pattern is shown in Figure 3–30d. If the speed is increased to some higher Mach number, the oblique portions of the waves incline more greatly and the detached normal shock wave moves closer to the leading edge. At supersonic speeds the aerodynamic control conditions become predictable and orderly again. A typical supersonic flow pattern is shown in Figure 3–31. The airflow ahead of the object is not influenced until the air particles are suddenly forced out of the way by the concentrated pressure wave set up by the object. Therefore the airflow does not change direction ahead of the airfoil as occurred in the subsonic flow shown in Figure 3–29.

When supersonic flow is clearly established, all changes in velocity, pressure, density, and flow direction take place quite suddenly and in relatively confined areas. The areas of

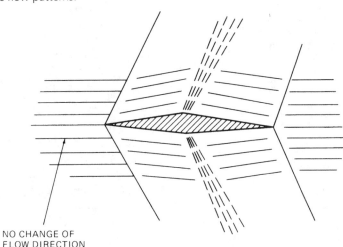

FIGURE 3–31 Typical supersonic flow pattern.

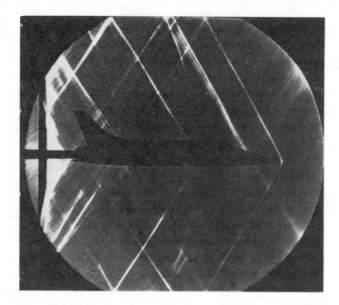

FIGURE 3–32 Photograph of shock waves.

flow change are very distinct and these areas are referred to as **wave formations**.

The wave formations formed at supersonic speeds are not imaginary or theoretical, and in a suitably arranged high-speed wind tunnel they can be photographed. Figure 3–32 is a photograph showing shock waves formed on a wind-tunnel model at low supersonic speed.

Various types of waves can occur in supersonic flow, and the nature of the wave formed depends upon the airstream and the shape of the object causing the flow change. Essentially, there are three fundamental types of waves formed in supersonic flow: (1) the oblique shock wave, (2) the normal shock wave, and (3) the expansion wave.

Oblique Shock Wave

A typical case of **oblique shock wave** formation is that of a wedge pointed into a supersonic airstream. An oblique shock wave will form on each surface of the wedge, as shown in Figure 3–33. A supersonic airstream passing through the oblique shock wave will experience these changes:

1. The airstream is slowed down; the velocity and Mach number behind the wave are reduced, but the flow is still supersonic.

2. The flow direction is changed to flow along the surface of the airfoil.

3. The static pressure of the airstream behind the wave is increased.

4. The density of the airstream behind the wave is increased.

5. Some of the available energy of the airstream (indicated by the sum of dynamic and static pressure) is dissipated and turned into unavailable heat energy. Therefore, the oblique shock wave is wasteful of energy.

Normal Shock Wave

If a blunt-nosed object is placed in a supersonic air stream, the shock wave which is formed will be detached from the leading edge. Whenever the shock wave forms perpendicular to the upstream flow, the wave is termed a **normal shock wave**, and the flow immediately behind the wave is subsonic. Any relatively blunt object, such as is shown in Figure 3–34, placed in a supersonic airstream will form a normal shock wave immediately ahead of the leading edge, slowing the airstream to subsonic so that the airstream may feel the presence of the blunt nose and flow around it. Once past the blunt nose, the airstream may remain subsonic or accelerate back to supersonic, depending on the shape of the nose and the Mach number of the free stream. This type of normal shock wave which is formed ahead of an object in a supersonic airstream is called a bow wave.

In addition to the formation of normal shock waves just described, this same type of wave may be formed in an entirely different manner. Figure 3–30 illustrates the way in which an airfoil at high subsonic speeds has local flow velocities which are supersonic. As the local supersonic flow moves aft, a normal shock wave forms, slowing the flow to subsonic. The transition of flow from subsonic to supersonic is smooth and is not accompanied by shock waves if the transition is made gradually with a smooth surface. The transition of flow from supersonic to subsonic without direction change always forms a normal shock wave.

A supersonic airstream passing through a normal shock wave will experience these changes:

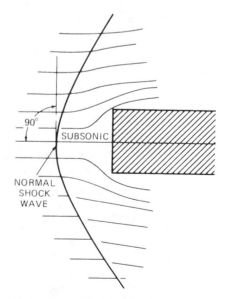

FIGURE 3–34 Normal shock wave formation.

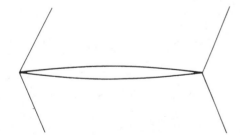

FIGURE 3–33 Shock waves at the edges of an airfoil.

1. The airstream is slowed to subsonic.
2. The airflow direction immediately behind the wave is unchanged.
3. The static pressure of the airstream behind the wave is increased greatly.
4. The density of the airstream behind the wave is increased greatly.
5. The energy of the airstream (indicated by total pressure, dynamic plus static) is greatly reduced. The normal shock wave is very wasteful of energy.

Expansion Wave

If a supersonic airstream were to flow "around a corner" as shown in Figure 3–35, an **expansion wave** would form. An expansion wave does not cause sharp, sudden changes in the airflow except at the corner itself and thus is not actually a shock wave. A supersonic airstream passing through an expansion wave will experience these changes:

1. The airstream is accelerated; the velocity and Mach number behind the wave are greater.
2. The flow direction is changed to flow along the surface, provided separation does not occur.
3. The static pressure of the airstream behind the wave is decreased.
4. The density of the airstream behind the wave is decreased.
5. Since the flow changes in a rather gradual manner, there is no shock and no loss of energy in the airstream. The expansion wave does not dissipate airstream energy.

Drag Divergence

As previously described, the operation of aircraft can be both unpredictable and unstable in the transonic flight area. The formation of shock waves can create significant problems with airflow separations and the resulting rapid rise in drag. At speeds only slightly above the critical Mach number, the shock wave formation is usually not strong enough

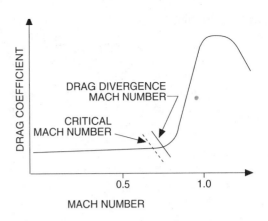

FIGURE 3–36 Drag divergence Mach number.

to cause flow separation, and therefore drag changes only slightly. However, as the speed continues to increase and the shock wave grows in strength, airflow separation will occur. This separation will result in a rapid increase in the airfoil drag coefficient. This point is called the **drag divergence Mach number**. It is also sometimes referred to as the **critical drag Mach number**. In referring to Figure 3–36, note that from the drag divergence Mach number to Mach 1 the drag rises sharply and operation within this range is not desirable. Also notice that at speeds higher than Mach 1 the drag drops off sharply.

Sonic Booms

When an airplane is in level supersonic flight, a pattern of shock waves is developed. Although there are many shock waves coming from an aircraft flying supersonically, these waves tend to combine into two main shocks, one originating from the nose of the aircraft and one from the tail.

If these waves (pressure disturbances) extend to the ground or water surface, as shown in Figure 3–37, they will be reflected, causing a **sonic boom**. An observer would actually hear two booms. The time between the two booms and their intensity is primarily a function of the distance the airplane is from the ground. The lower the aircraft is, the closer together and louder the two booms will be.

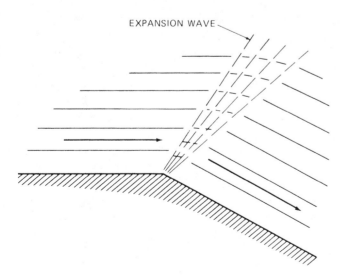

FIGURE 3–35 Expansion wave formation.

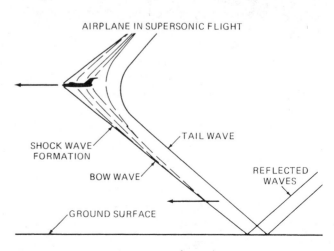

FIGURE 3–37 Sonic boom wave formation.

Hypersonic Flight

When the speed of an aircraft or spacecraft is five times the speed of sound or greater, the speed is said to be **hypersonic**. Practical experience with such speeds has been gained by engineers working in the space program under the direction of the National Aeronautics and Space Administration (NASA), with the result that certain hypersonic vehicles are now practical, with others to be developed. The principal hindrance to hypersonic flight is the extreme temperature generated by air friction at hypersonic speeds. New materials and cooling methods are being developed to overcome this problem.

The high-velocity test of a space shuttle model is shown in Figure 3–38. The shuttle enters the earth's atmosphere at hypersonic speed. The model shown in Figure 3–38 was tested in a Mach 20 helium tunnel at NASA's Langley Research Center. The configuration developed by the NASA Manned Spacecraft Center is being tested at a 20° angle of attack at simulated Mach 20 reentry speed. Row shock patterns from the nose and fixed straight wing are made visible by exciting the helium flow with a high-energy electron beam.

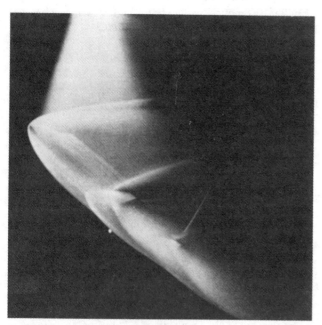

FIGURE 3–38 Hypersonic wind-tunnel test. *(NASA)*

REVIEW QUESTIONS

1. What are the approximate percentages of the principal gases in our atmosphere?
2. What is the atmospheric pressure at sea level under standard conditions in pounds per square inch and inches of mercury?
3. What effect does temperature have on the density of the air?
4. In simple terms, explain Bernoulli's principle.
5. Define *airfoil*.
6. Define *chord line* and *chord*.
7. Define *camber*.
8. Explain the term *relative wind*.
9. Explain *angle of attack*.
10. What is the stagnation point of an airfoil?
11. Discuss the effect of air density with respect to lift.
12. How does wing area affect lift?
13. What is the cause of skin friction on a surface moving through the air?
14. What does the term *viscosity* mean?
15. Describe laminar flow.
16. What is the boundary layer?
17. Define *parasite drag*.
18. Define *induced drag*.
19. Define *resultant force*.
20. Explain what is meant by the term *center of pressure*.
21. How does the angle of attack affect the CP on a wing?
22. Define the term *speed of sound*.
23. What is a Mach number?
24. Explain the term *critical mach number*.
25. What are the three types of shock waves?

Airfoils and Their Applications 4

INTRODUCTION

A general knowledge of the nature of airfoils and the factors affecting their performance is of value to many technicians, particularly those involved in the structural repair of airfoils, and to others who may be interested in building their own airplanes. The previous chapter discussed basic aerodynamic principles and their application to airfoils in producing lift.

Since the early days of aircraft research when the Wright brothers tested airfoil shapes in a small wind tunnel, literally thousands of different airfoil shapes have been developed and tested. These range from the types that operate at low subsonic speeds to those designed for supersonic and hypersonic speeds. In this chapter the basic elements of airfoil design and the characteristics that determine airfoil selection for different aircraft applications will be discussed.

AIRFOIL PROFILES

An **airfoil** is any surface, such as an airplane wing, aileron, or rudder, designed to obtain reaction from the air through which it moves.

An **airfoil profile** is the outline of an airfoil section. An **airfoil section** is a cross section of an airfoil parallel to the plane of symmetry or to a specified reference plane. If the airfoil shown in Figure 4–1 is a cross section of an actual airplane wing, it is correct to call it an airfoil section; if it is merely the outline, it should be called an airfoil profile. The terms airfoil, airfoil profile, and airfoil section are used interchangeably in conversation, even by people who know the technical distinctions. Airfoil profiles can be made up of certain forms of thickness distributed about certain mean lines. The major shape variables then become two: the thickness form and the mean-line form. The thickness form is of particular importance from a structural standpoint. On the other hand, the form of the mean line determines some of the most important aerodynamic properties of the airfoil section. Slight variations in either one of these factors will alter the airfoil characteristics in some manner.

Over the years, several thousand different airfoils have been developed and tested. The Wright brothers tested several hundred airfoils themselves in a homemade wind tunnel. For many years airplanes relied upon considerable external bracing to provide the necessary support to a wing. This bracing took different forms, including struts and wires. As stronger materials were developed, an attempt was made to eliminate the external bracing in order to reduce drag. To accomplish this, it was necessary to make the airfoils thicker, and they evolved into the shape shown in Figure 4–1. Early textbooks on the theory of flight use this **Clark Y airfoil** to illustrate basic principles of airfoils, and this practice has continued. However, the airfoil profiles developed by **NACA** (National Advisory Committee for Aeronautics) were described in detail in NACA Report 460, published November 1933 and entitled *The Characteristics of the Seventy-Eight Related Airfoil Sections from Tests in the Variable Density Wind Tunnel*. NACA was a government agency which performed thousands of tests on airfoil shapes to develop information regarding which were most efficient for various flight conditions. NACA was the forerunner of **NASA**, the National Aeronautics and Space Administration, which has continued airfoil development and testing. The NACA and NASA airfoil reports do not contain information that a technician would normally use, but they are of interest to those who wish to widen their knowledge of aerodynamics.

NACA developed and tested hundreds of airfoils. NACA airfoil profiles are designated by a standardized numbering system consisting of four, five, or six digits. The first series of airfoils developed was the four-digit-numbered airfoils. The first digit indicates the camber of the mean line in percentage of the chord, the second digit shows the position of the maximum camber of the mean line in tenths of the chord from the leading edge, and the last two digits indicate the maximum thickness in percentage of the chord. Thus the NACA2315 profile has a maximum mean camber of 2% of the chord at a position three-tenths of the chord from the leading edge and a maximum thickness of 15% of the chord. Likewise, the NACA0012 airfoil is a symmetrical airfoil having a maximum thickness of 12% located at 30% of the chord, as is shown in Figure 4–2.

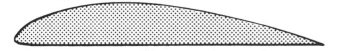

FIGURE 4–1 Clark Y airfoil profile.

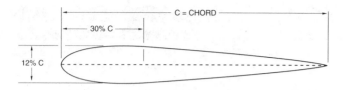

FIGURE 4–2 NACA0012 symmetrical airfoil.

Figure 4–3 shows how to draw the NACA2421 profile. First, draw a base line and call it the chord. Second, divide the chord into 20 equal divisions; each division will represent 5% of the chord. Draw vertical lines of indefinite length at each of these divisions and number them, starting with 0 at the leading edge and ending with 100 at the trailing edge. These lines are called **stations**. Then draw vertical lines at the other stations shown on the table. For this particular airfoil the additional stations are 1.25, 2.5, and 7.5% of the chord behind the leading edge. Third, lay out the points for the upper and lower contour lines from the ordinates given in the second and third columns of the table shown in the figure. Notice that those in the second column are positive ordinates and those in the third column are negative or-

STA	UP' R	L'W'R.
0	—	0
1.25	3.87	− 2.82
2.5	5.21	− 4.02
5.0	7.00	− 5.51
7.5	8.29	− 6.48
10	9.28	− 7.18
15	10.70	− 8.05
20	11.59	− 8.52
25	12.15	− 8.67
30	12.38	− 8.62
40	12.16	− 8.16
50	11.22	− 7.31
60	9.79	− 6.17
70	7.94	− 4.87
80	5.74	− 3.44
90	3.18	− 1.88
95	1.76	− 1.06
100	(.22)	(−.22)
100	—	0

L. E. RAD.: 4.85
SLOPE OF RADIUS
THROUGH END OF
CHORD: 2/20

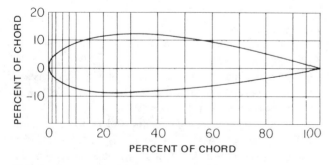

FIGURE 4–3 Method for drawing an airfoil profile.

dinates. Positive ordinates are measured upward from the chord at their stations, and negative ordinates are measured downward from the chord at their stations in percentage of chord. Fourth, using a spline or a French curve, connect all these points with a smooth line. This provides the airfoil section, except that the nose requires more work.

Notice that the bottom of the table in Figure 4–3 says, "L.E. rad.: 4.85. Slope of radius through end of chord: 2/20." Refer now to Figure 4–4, which shows how to draw the nose. Through the 0% station of the chord, that is, the leading edge, construct the line OC at an angle that will have a slope of 2:20. This is the same as a slope of 1:10, which means that for every 10 units of horizontal distance there will be a rise of 1 unit of vertical distance. The slope can be established by measuring 10 in along the chord, erecting a perpendicular 1 in above the chord, and connecting the top of that perpendicular with the leading edge.

On the line OC, locate the center of the leading-edge radius by measuring 4.85% of the chord length from zero, the leading edge. Set a compass for 4.85 percent of the chord length, place the point of the compass at the point found to be the center of the leading-edge radius, and inscribe an arc that passes through the leading edge. This provides the nose. Now, with a spline or a French curve, connect the leading-edge curve so that it blends smoothly into the curve previously constructed. (In practice, the nose is usually drawn before the ends of the ordinates are connected, but this part of the instruction was postponed so that the principal work could be explained before discussing the details.)

It is important to remember that all the airfoil's characteristics are based on the chord length. The NACA number gives information only on the shape of the airfoil, not the size. To determine the dimensions, the chord length must be given.

Airfoil Development

Further airfoil development required that additional information be included in the numbering system. While this numbering system is still similar to the four-digit system, it is more complicated and will not be explained here. Airfoil numbers with five and six digits can presently be found in the reports of NACA and NASA, along with an explanation of the significance of the digits.

Although general-aviation aircraft have been operating satisfactorily for many years with a variety of proven airfoil designs, research and development have been conducted to improve the efficiency of airfoils used for light aircraft. Much research has been conducted at the NASA Langley Research Center using information and technology derived

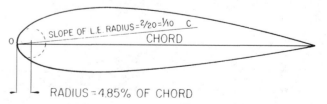

FIGURE 4–4 How to draw the nose of an airfoil profile.

FIGURE 4–5 Beechcraft Skipper using the GAW-1 airfoil profile. *(Beech Aircraft Co.)*

from the supercritical airfoils (discussed later in this chapter) developed by Richard Whitcomb and designed for aircraft which operate at speeds just below the speed of sound.

The first of the new derivatives taken from the supercritical concept but designed for lower speed, light aircraft is called the **GAW airfoil** (GAW stands for General Aviation-Whitcomb). The GAW-1 airfoil is now being used on some aircraft, such as the Beechcraft Skipper shown in Figure 4–5. Its profile is approximately as shown in Figure 4–6 and is identified by the concave shape of the aft lower surface. This airfoil has a thickness ratio of 17%, and the design has improved the lift/drag ratio during climb by 50% when compared to data from the wing which the GAW-1 replaced.

A newer design, the GAW-2 airfoil, has a thickness ratio of 13% and has a lower drag and a higher C_L (coefficient of lift) than the GAW-1. These features provide an increase in the rate of climb and a decrease in stall speed. The two airfoils can be used to advantage in a tapered wing with the GAW-1 at the root, where the extra thickness provides for greater structural strength, and the GAW-2 at the outer portion of the wing to give increased aerodynamic efficiency. The approximate profile of the GAW-2 airfoil is shown in Figure 4–7.

Airfoil Selection

The selection of the best airfoil for an airplane requires a careful consideration of the many factors that may conflict with one another, with the result that the final decision is usually a compromise. Some important factors to be considered in airfoil selection are (1) airfoil characteristics, (2) airfoil dimensions, (3) airflow about the airfoil, (4) the speed at which the aircraft is designed to operate, and (5) flight operating limitations.

It is not unusual for an airplane to use more than one airfoil profile in a wing. Often a manufacturer will use one air-

FIGURE 4–6 GAW-1 airfoil profile.

FIGURE 4–7 GAW-2 airfoil profile.

foil profile for the inboard wing area, a second profile for the center section, and a third for the wing tip. While using more than one profile may increase aerodynamic performance, it also adds greatly to the construction costs.

PERFORMANCE OF AIRFOILS

Airfoil Characteristics

A particular airfoil, that is, one having certain definite dimensions, has specific lift, drag, and center-of-pressure (CP) position characteristics during flight. These features are collectively known as **airfoil characteristics** and they are classified as follows:

1. Lift coefficient
2. Drag coefficient
3. Lift/drag ratio
4. Center-of-pressure position

In place of the CP position, highly technical publications may use some equivalent characteristics, such as *the moment of aerodynamic force about the leading edge* or the *aerodynamic center*; it is sufficient for the technician, however, to consider only the CP position.

Symbols Relating to Airfoils

Aerodynamicists employ certain letter symbols to represent various quantitative factors in performing mathematical calculations on airfoil performance. The following are those applicable to the material in this chapter:

AR	Aspect ratio
C_D	Coefficient of drag
C_L	Coefficient of lift
D	Drag
L	Lift
M	Pitching moment
R_e	Reynolds number
S	Wing area
V	Speed or velocity
a	Wing lift-curve slope
b	Wing span
c	Wing chord
c_r	Root chord
c_t	Tip chord
x	Longitudinal coordinate (fore and aft)
y	Lateral coordinate
z	Vertical coordinate
α	(alpha) Angle of attack
ρ	(rho) Air density

Fundamental Equation for Lift

The term **lift** has been defined as the net force developed perpendicular to the relative wind. The aerodynamic force of lift on an airplane results from the generation of a pressure distribution on the wing.

The lift of an airfoil can be determined mathematically by using given known values in the fundamental equation for lift. This formula is as follows:

$$L = C_L \frac{\rho}{2} V^2 S$$

where

L = lift, lb
C_L = lift coefficient
ρ = mass density, slugs/ft^3
V = velocity of wind relative to the body, ft/s
S = airfoil area, ft^2

Coefficient of Lift

A coefficient is a number or symbol which acts as a multiplier to a variable or unknown quantity. For example, 3 is a coefficient in 3ab; x is a coefficient in x(y + z). Thus, a coefficient is a number used as a multiplier.

The fundamental equation for lift can be used to determine the coefficient of lift (C_L) by rewriting it as follows:

$$C_L = \frac{L}{\frac{1}{2} \rho V^2 S}$$

This equation says that lift is equal to the lift coefficient times the dynamic pressure times the wing area. The lift coefficient is usually determined in wind-tunnel testing. The **coefficient of lift** is a measure of how efficiently the wing is changing velocity into lift. High coefficient-of-lift numbers indicate a more efficient airfoil design. The coefficient of lift is a function of the airfoil shape and the angle of attack. For a given shape, the coefficient of lift varies with the angle of attack; therefore, when the fundamental equation for lift is used, the angle of attack must be specified to make the computation meaningful. A certain airfoil may have a C_L of 0.4 at a 4° angle of attack and a C_L of 1.2 at a 16° angle of attack. The angle of attack must be known before the answer to the lift equation has a usable value.

To calculate the amount of lift that can be obtained from an airplane wing having an area of 180 ft^2 [16.7 m^2], a velocity of 120 mph [53.6 m/s], an altitude of 1000 ft [304.8 m], and a C_L of 0.4 at a 4° angle of attack, the following steps can be used: First the velocity is converted into feet per second. A velocity of 120 mph is equal to 176 ft/s. Next, by reference to the NASA standard atmospheric tables, it is found that the value of ρ (density) is 0.002 309. Substitute values in the fundamental equation for lift where

$$L = C_L \frac{\rho}{2} V^2 S$$

$$L = 0.4 \times \frac{0.002\ 309}{2} \times (176)^2 \times 180$$

or

$$L = 2574.85 \text{ lbs}$$

Fundamental Equation for Drag

The fundamental equation for drag is found to be almost identical with the equation for lift except that drag and the coefficient of drag are used in place of the values for lift. The equation for drag is then

$$D = C_D \frac{\rho}{2} V^2 S$$

The symbol meanings are the same as those stated previously in the equation for lift except for D (drag) and C_D (coefficient of drag).

The equation for drag can be rewritten to obtain the coefficient of drag, as follows:

$$C_D = \frac{D}{\frac{1}{2} \rho V^2 S}$$

The **drag coefficient**, like the coefficient of lift, is a measure of how efficient the wing is. However, while a higher number is desirable for lift coefficients, a low drag coefficient number indicates a more efficient airfoil. Similarly to the coefficient of lift, the coefficient of drag is a function of the airfoil shape and the angle of attack. For a given shape, the coefficient of drag varies with the angle of attack. For example, a certain airfoil has a C_D of 0.11 at a 4° angle of attack, but at a 16° angle of attack it has a C_D of 0.24.

To calculate the amount of drag that is produced from an airplane wing having an area of 180 ft^2 [16.7 m^2], a velocity of 120 mph [53.6 m/s], an altitude of 1000 ft [304.8 m], and a C_D of 0.4 at a 4° angle of attack, the following steps can be used: First the velocity is converted into feet per second. As was calculated in the previous lift problem, a velocity of 120 mph is equal to 176 ft/s. The value of ρ (density) is 0.002 309. Substitute values in the fundamental equation for drag where

$$D = C_D \frac{\rho}{2} V^2 S$$

$$D = 0.11 \times \frac{0.002\ 309}{2} \times (176)^2 \times 180$$

or

$$D = 707.55 \text{ lbs}$$

Lift/Drag Ratio

The **lift/drag ratio** is the ratio of the lift to the drag of any body in flight. This ratio is a measure of the effectiveness of an airfoil because the lift is the force required to support the weight while the drag is a necessary nuisance that must be accepted to obtain lift. For any angle of attack, the C_L divided by the C_D will give the L/D ratio for an airfoil. For example, using a C_L of 0.4 and a C_D of 0.11, the L/D ratio would be calculated as follows:

$$\frac{L}{D} = \frac{0.4}{0.11} = 3.63$$

The maximum value of the L/D ratio for the wing is always more than the maximum value of the L/D ratio for the complete airplane, because the drag for the complete airplane includes not only the drag of the wing but also the drag contributed by the rest of the airplane. This is based upon the assumption that all the lift of a conventional airplane is obtained from the wing. The complete airplane L/D can be represented in the formula

$$\text{airplane } \frac{L}{D} = \frac{\text{lift from wing}}{\text{wing drag}} + \text{other drag}$$

Center-of-Pressure Coefficient

It has been explained that the CP of an airfoil is a point on the chord of the airfoil that is at the intersection of the chord and the resultant force. The CP coefficient is the ratio of the distance of the CP from the leading edge to the chord length. In other words, the CP is given by stating that it is a certain percentage of the chord length behind the leading edge. For example, if the chord is 6 ft [1.82 m] long and the CP is 30% of the chord, then it is 30% of 6 ft, or 1.8 ft [0.546 m] behind the leading edge at that given angle of attack. Since the pressure distribution varies along the chord with changes in the angle of attack, the CP, which is the point of application of the resultant, moves accordingly. It has been previously explained that the CP generally moves

forward as the angle of attack increases and backward as it decreases, although there are exceptions to this rule.

Characteristic Curves

At the beginning of this section, airfoil characteristics such as the lift coefficient, drag coefficient, lift/drag ratio, and CP position were explained. In order to better visualize the characteristics of a particular airfoil, these characteristics may be displayed graphically. **Characteristic curves** are graphical representations of airfoil characteristics for various angles of attack. It is important to understand that the values of all the airfoil characteristics vary with the angle of attack and that they are different for different airfoil sections.

Figure 4–8 is a coefficient-of-lift diagram for the NACA2421 airfoil. In this diagram the angle of attack is plotted horizontally, and the value of the lift coefficient is plotted vertically. Notice that a horizontal line is drawn and along this line are located points which correspond to the various angles of attack from −8 to 32°. These points are marked with the angles represented. At each of these points, the person preparing the illustration measures vertically upward a distance that represents, to a suitable scale, the coefficient of lift for that particular angle of attack and makes a

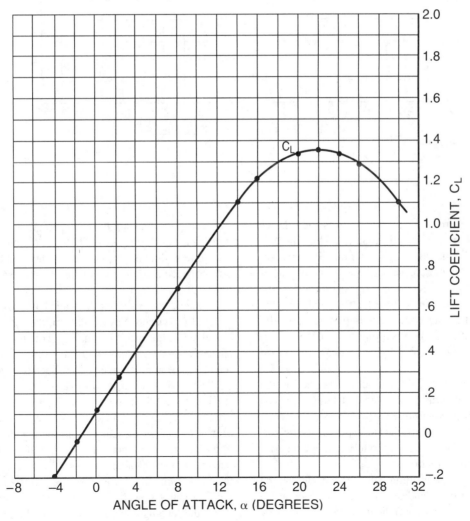

FIGURE 4–8 Coefficient-of-lift curve.

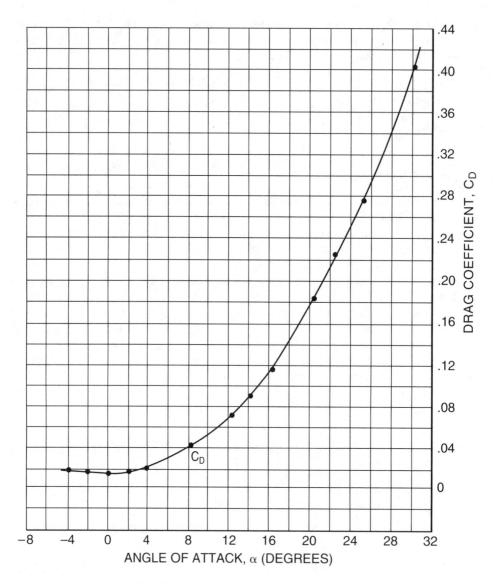

FIGURE 4–9 Coefficient-of-drag curve.

dot or draws a tiny circle. A smooth curve is then drawn through the points located in this manner. This curve is called the **lift-coefficient curve**, or simply the **lift curve**, for the particular airfoil being represented.

Figure 4–9 is a coefficient-of-drag diagram for the NACA2421 airfoil. The angle of attack is again plotted horizontally, and the drag coefficient is plotted vertically for each angle of attack. The resulting curve is called the **coefficient-of-drag curve**, or simply the **drag curve**, for the airfoil.

Figure 4–10 shows the lift and drag curves for the NACA2421 airfoil drawn on the same diagram. This is common practice because it presents two types of information on one drawing. The coefficient-of-lift and coefficient-of-drag curves were shown here separately to make it easier to understand their construction.

Figure 4–11 shows the lift/drag ratio curve drawn on the same diagram with the lift and drag curves. For each angle of attack the corresponding coefficient of lift is divided by the corresponding coefficient of drag to obtain the L/D ratio

for that particular angle of attack. A distance is measured vertically upward to represent the L/D ratio value, and a dot or circle is drawn. The dots or circles found in this manner are connected by a smooth line to represent the L/D ratio curve on the diagram.

To understand how the L/D ratio curve is obtained from the C_L and C_D curves on the diagram, the following example is provided: If the airfoil is at an angle of attack of 12°, the coefficient of lift, C_L, is approximately 0.95, and the drag coefficient, C_D, is approximately 0.07. Then the L/D ratio is C_L/C_D = 0.95/0.07 = 13. The lift/drag or L/D ratio is equal to 13. Examine the L/D ratio curve on Figure 4–11 and you will see that at a 12° angle of attack this curve has a value of 13. Figure 4–12 shows the curve representing the position of CP drawn on the same diagram with the lift, drag, and L/D ratio curves. For each of the angles of attack represented on the horizontal line, the location of the corresponding CP is plotted vertically upward in terms of the percentage of chord from the leading edge, and the points located are connected by a smooth line to provide the curve.

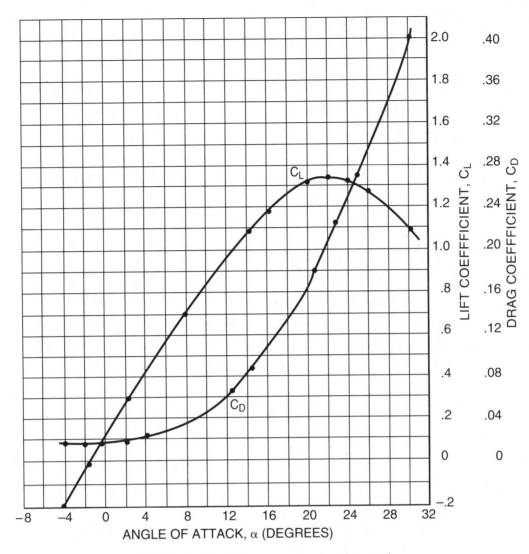

FIGURE 4–10 Lift and drag curves on the same graph.

SHAPES AND DIMENSIONS OF AIRFOILS

So far, our discussions of aerodynamic forces have been limited to how they affect airfoil profiles. Since an airfoil profile has no span, airflow has been examined in two dimensions only. The wing span (length) and the planform of the wing also affect the aerodynamic characteristics. **Planform** is the shape of the wing as viewed from the top or bottom. Some typical planforms are shown in Figure 4–13 on page 84. When the effects of wing planform are introduced, attention must be directed to the existence of airflow in the spanwise direction. In other words, airfoil section properties deal with flow in two dimensions, while actual wings have flow in three dimensions.

The wing area, aspect ratio, taper ratio, and sweepback of a planform are the principal factors which determine the aerodynamic characteristics of a wing. These same quantities also have a definite influence on the structural weight and stiffness of a wing.

Wing Area

As has been previously explained, the lift of an airfoil varies directly with the area. The **wing area**, S, is simply a measure of the total surface of the wing. Although a portion of this area may be covered by the fuselage or the nacelles, pressure is still acting on it; therefore, it is included in the calculation of the total wing area. The wingspan, b, is measured tip to tip. The average wing chord, c, is simply a geometric average of the wing chords. As an example, a pointed-tip delta wing would have an average chord equal to one-half of the root chord. As shown in Figure 4–14 on page 84, the product of the span and the average chord is the wing area ($b \times c = S$).

Effects of Wing Planform

Air flows over and under the wing of an airplane, as shown in Figure 4–15 on page 84. This produces a region of relatively low pressure above the wing and a region of relatively high pressure under the wing. If the airfoil were of infinite span, that is, if it had no ends, the airflow would be direct from the

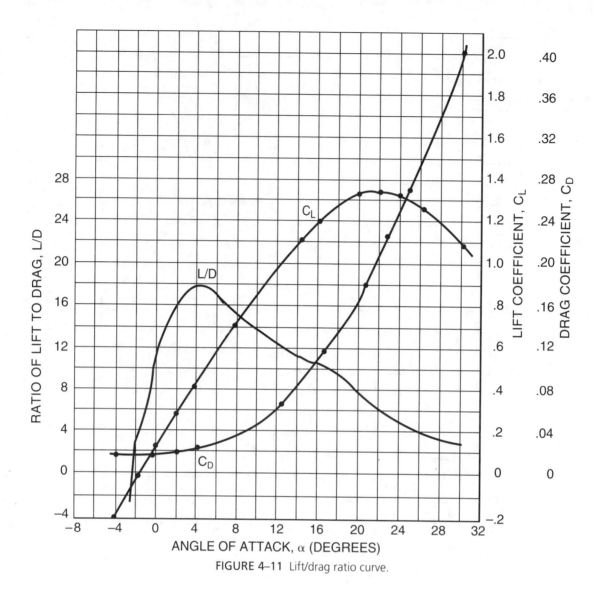

FIGURE 4–11 Lift/drag ratio curve.

leading edge to the trailing edge, as shown in Figure 4–15, because there would be no way for the air in the region of high pressure below the wing to flow into the region of low pressure above the wing. However, the airfoil is of finite area and span; that is, it has ends. Therefore the air under the wing will seek the region of low pressure above the wing. This tendency will result in the high-pressure air on the underside of the wing moving outboard and "spilling over" the tips, as shown in Figure 4–16. Eddies, or regions of turbulence, are formed in this manner at the tips, causing the streamlines to form a vortex on each wingtip. These vortices are like tiny tornados, such as those illustrated in Figure 4–17. The turbulence produced absorbs energy and increases the induced drag. Induced drag is that part of the drag caused by lift, that is, drag caused by the change in the direction of airflow. The vortices tend to exert a downward motion to the airflow leaving the trailing edge. If the airstream is tilted downward by the vortices, the lift vector is also tilted somewhat aft. The amount of movement of the lift vector from the vertical is referred to as the induced angle of attack, and is the amount of lift lost due to the wingtip vortices and resulting downwash.

All aircraft have wingtip vortices; however, as lift in-

creases, the strength of the vortices also increases. Therefore, since heavy aircraft require more lift, stronger wingtip vortices will be associated with them. These vortices are often referred to by pilots as *wake turbulence*.

Aspect Ratio

If wingtip vortices could be eliminated or reduced, the associated induced drag would be proportionately reduced. To totally eliminate the wingtip vortices, a wing of infinite span would have to be used. However, it is possible to reduce the induced drag by changing the dimensions of a wing. The **aspect ratio** (*AR*) of an airfoil of rectangular shape is the ratio of the span to the chord. Thus, airfoil *A* in Figure 4–18 has a span of 24 ft and a chord of 6 ft, and the aspect ratio is 4. Likewise, airfoil *B* of the same figure has a span of 36 ft and a chord of 4 ft; therefore, the aspect ratio is 9. The formula for aspect ratio can be written as follows:

$$AR = \frac{\text{span}}{\text{chord}}$$

However, for aerodynamic reasons, airfoils are hardly ever designed with a rectangular planform. The aspect ratio

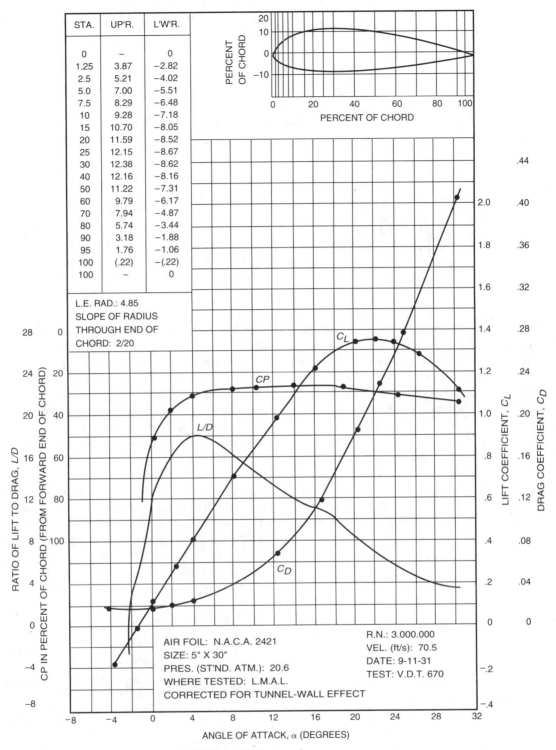

STA.	UP'R.	L'W'R.
0	–	0
1.25	3.87	−2.82
2.5	5.21	−4.02
5.0	7.00	−5.51
7.5	8.29	−6.48
10	9.28	−7.18
15	10.70	−8.05
20	11.59	−8.52
25	12.15	−8.67
30	12.38	−8.62
40	12.16	−8.16
50	11.22	−7.31
60	9.79	−6.17
70	7.94	−4.87
80	5.74	−3.44
90	3.18	−1.88
95	1.76	−1.06
100	(.22)	−(.22)
100	–	0

L.E. RAD.: 4.85
SLOPE OF RADIUS THROUGH END OF CHORD: 2/20

AIR FOIL: N.A.C.A. 2421
SIZE: 5" X 30"
PRES. (ST'ND. ATM.): 20.6
WHERE TESTED: L.M.A.L.
CORRECTED FOR TUNNEL-WALL EFFECT

R.N.: 3.000.000
VEL. (ft/s): 70.5
DATE: 9-11-31
TEST: V.D.T. 670

FIGURE 4–12 Center-of-pressure curve.

for nonrectangular airfoils is defined as the span squared divided by area. If the area is represented by the letter S and the span by the letter b, the formula for nonrectangular airfoils can be expressed thus:

$$AR = \frac{b^2}{S}$$

The formula for the aspect ratio of a nonrectangular airfoil could be applied to a rectangular airfoil, if any such object were in existence, since S, the area, is a product of the width c and the length b. Typical aspect ratios vary from 35 for a high-performance sailplane to 3.5 for a jet fighter.

Effects of Aspect Ratio

The effect of aspect ratio on the lift and drag characteristics of a wing is shown in Figure 4–19. The effect of increasing aspect ratio is principally to reduce induced drag for any given coefficient of lift. This improves the L/D ratio.

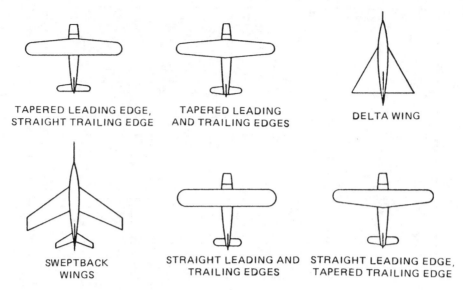

TAPERED LEADING EDGE, STRAIGHT TRAILING EDGE

TAPERED LEADING AND TRAILING EDGES

DELTA WING

SWEPTBACK WINGS

STRAIGHT LEADING AND TRAILING EDGES

STRAIGHT LEADING EDGE, TAPERED TRAILING EDGE

FIGURE 4–13 Wing planforms.

Since wingtip vortices exert their influence for a distance inboard from the tips in any given airfoil, the percentage of area so affected is less for a long, narrow airfoil than it is for a short, wide airfoil. This is shown in Figure 4–20, which illustrates the area affected by wingtip vortices. The shape in the upper-left-hand corner of Figure 4–20 is short and wide, the center shape is long and relatively narrow, and the lower-right shape is still longer and narrower. Although the area affected by wingtip vortices remains the same for all these shapes, a smaller proportion of the total area is affected when the airfoil is long and narrow.

The relationship of wingtip vortices to induced drag emphasizes the need of a high aspect ratio for an airplane which is continually operated at high lift coefficients. In other words, airplane configurations designed to operate at high lift coefficients during the major portion of their flight, such as sailplanes, demand a high-aspect-ratio wing to minimize the induced drag. While the high-aspect-ratio wing will minimize induced drag, long, thin wings increase structural weight and have relatively poor stiffness characteristics. This fact will temper the preference for a very high aspect ratio.

Since a limit will be reached where it is no longer practical to increase the aspect ratio of a wing, alternative methods have been researched that can reduce wingtip vortices. These methods are used to create an **effective aspect ratio**, which is greater than the physical aspect ratio. Since the induced drag is created by the cross-flow at the wingtips, anything that can be done to reduce this flow will help in reducing the drag. One method employed on many aircraft

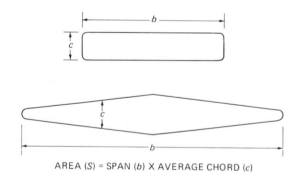

AREA (S) = SPAN (b) X AVERAGE CHORD (c)

FIGURE 4–14 Area of a wing.

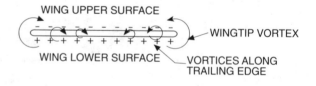

FIGURE 4–15 Airflow over and under the wing.

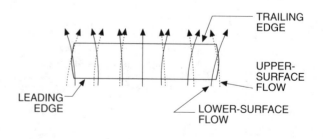

FIGURE 4–16 High-pressure air on wing bottom "spilling over" to low-pressure area on top.

FIGURE 4–17 Wingtip vortices.

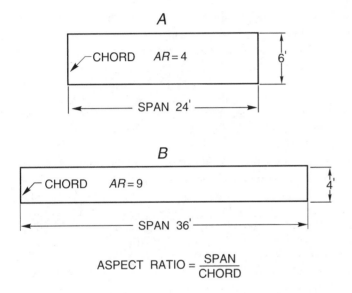

$$ASPECT\ RATIO = \frac{SPAN}{CHORD}$$

FIGURE 4–18 Aspect ratio.

is modifying the shape of the wing tips. One of the most effective wingtip designs is a drooped wingtip such as the one illustrated in Figure 4–21. Drooped wingtips give the outward flow of air a downward direction as it attempts to flow around the tip. This downward movement of the air results in the vortex being formed farther outboard than it would otherwise be, in effect increasing the aspect ratio of the wing.

Another method that has been used on several aircraft is the installation of an **endplate** such as the ones shown in Figure 4–22. The endplate has had only moderate success, with the result in many installations being that the reduction in induced drag is offset by a corresponding increase in parasitic drag. A variation of the endplate concept is the use of fuel tanks on the wingtips such as the ones illustrated in Figure 4–23.

One of the most successful attempts at increasing the effective aspect ratio has resulted in the development of a device called a **winglet**, shown in Figure 4–24, which looks like a small jib sail mounted at and above the wingtips. The inward flow of air from the wingtip vortex acts on the winglet, much like wind acting on a jib sail, and produces a forward component of lift. A winglet also reduces the cross-flow on the wing, which in turn reduces the trailing vortex in strength. This reduction of wingtip vortices causes a corresponding reduction in drag, thereby increasing the wings' lift/drag ratio. A winglet, however, also has some disadvantages. It will add weight to the aircraft and produce parasitic drag. Winglets are most effective at high lift coefficients.

Taper

An airfoil is tapered when one or more of its dimensions gradually decreases from the root to the tip. When the airfoil decreases from the root to the tip in both thickness and chord, the airfoil is said to have **taper in plan and thickness**. This is shown in the bottom drawing of Figure 4–25. If the thickness and the chord remain the same from the root to the tip, there is no taper.

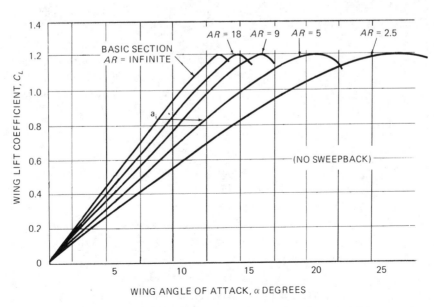

FIGURE 4–19 Effect of aspect ratio on wing characteristics.

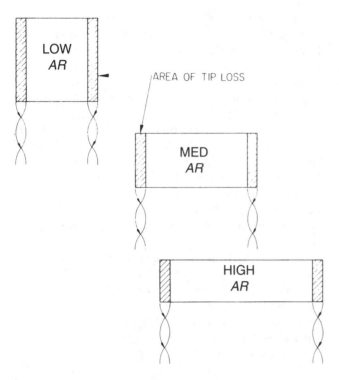

FIGURE 4–20 How increased aspect ratio reduces the effect of wingtip vortices.

FIGURE 4–21 Drooped wingtip.   (Cessna Aircraft Co.)

FIGURE 4–23 Fuel-tip tanks.   (Cessna Aircraft Co.)

When there is a gradual change (usually a decrease) in the chord length along the wing span from the root to the tip, with the wing sections remaining geometrically similar, the airfoil is said to have **taper in plan only**, as shown in the upper-right-hand drawing of Figure 4–25. When there is a gradual change in the thickness ratio along the wing span, with the chord remaining constant, the airfoil is said to have **taper in thickness ratio only**, as shown in the middle drawing of Figure 4–25.

The **taper ratio**, λ (lambda), is the ratio of the tip chord to the root chord. Thus the airfoil in Figure 4–26 has a root chord (C_r) of 6 ft [1.8 m], a tip chord (C_t) of 3 ft [0.9 m], and a taper ratio of 0.5. A rectangular wing has a taper ratio of 1.0, and the pointed tip delta wing has a taper ratio of 0.0.

The taper ratio affects the lift distribution and the structural weight of the wing. When a wing is tapered in thickness in such a manner that the thickness near the tip is 60% of the thickness at the root and it is compared with an airfoil of constant section (not tapered) equal to the mean (average) section of the tapered wing, the following characteristics are observed on certain airfoils: (1) the CP moves less for changes in angle of attack; (2) the maximum C_L is greater and the peak of the characteristic curve is flatter because all of the wing does not attain the maximum C_L at the same time (that is, each section reaches its maximum C_L at a different angle of attack from any other section); (3) the C_D values are lower, the most noticeable decrease being at the low angle of attack from any other section; and (4) the max-

FIGURE 4–22 End plates installed on wingtips.

FIGURE 4–24 Winglets on a Learjet Model 55. *(Gates Learjet Corporation)*

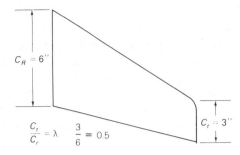

FIGURE 4–26 Taper ratio.

imum $\frac{L}{D}$ ratio is larger at small angles of attack. It is interesting to note that a tapered wing may also have a constantly changing airfoil section from the root of the wing to the tip.

When a wing is tapered in planform and is compared with a rectangular airfoil that has an equivalent aspect ratio, the following characteristics are observed on certain airfoils: (1) the CP moves more for changes in angle of attack; (2) there is a greater maximum C_L; (3) there are lower values of C_D, especially at low angles of attack; and (4) the L/D ratio is greater throughout the flight range, especially at the higher angles of attack. When a wing or any airfoil is tapered in both thickness and planform, it is possible to take advantage of the best aerodynamic features of an airfoil tapered in thickness only and an airfoil tapered in planform only.

When the distribution of the area of a tapered wing places the resultant force near the center line, it may be possible to build a wing of relatively light weight, having the thicker, heavier, and stronger portions near the root, where the greatest stresses normally occur. On the other hand, in a tapered airfoil, the spars must be tapered and different jigs must be used for building the ribs. For this reason the construction of the wing tapered in both planform and thickness becomes considerably more costly than the construction of other types of wings.

Sweep Angle

A wing is *swept* when the leading edge of the wing angles backward or forward from the fuselage. When wing sweep angle is being discussed, generally it is assumed that the wing is swept backwards, although a few new-technology aircraft employ forward-swept wings. The **sweep angle**, Λ, is usually measured as the angle between the line of 25% chord and a perpendicular to the root chord, as shown in Figure 4–27. The sweep of a wing causes definite changes in compressibility, maximum lift, and stall characteristics. The principal reason for sweeping a wing is to increase the critical Mach number of the aircraft. If a wing is swept, the airflow will no longer be in a direct chordwise flow across the wing. A component of the airflow will also travel spanwise. It is this spanwise direction of the airflow that raises the critical Mach number. As an example, assume that the critical Mach number for a wing is .85, and the airfoil is swept 25°. The component of the airflow that will travel chordwise across the wing is equal to the cosine of 25°, which is 0.9063, multiplied by the free-stream Mach number. If the aircraft speed is increased to Mach .90, as is shown in Figure 4–27, then the airflow component flowing

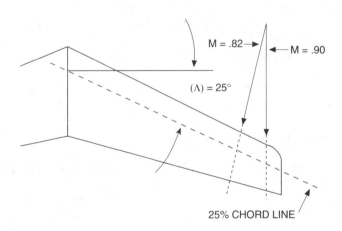

FIGURE 4–27 Aircraft sweep angle.

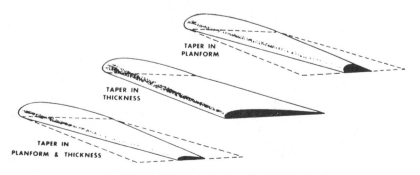

FIGURE 4–25 Taper in aircraft wings.

chordwise across the wing is Mach .82 (.90 × 0.9063), which is still below the critical free-stream Mach number of .85.

Sweeping a wing back, however, also has some undesirable characteristics. The component of the airflow that flows in the spanwise direction does not contribute to the production of lift. This results in higher stall speeds for swept-wing aircraft. Low-speed aircraft may have wings swept slightly aft in order to shift the center-of-lift point on the wing aft and move it closer to the center of gravity in order to improve the stability and loading properties of the aircraft.

Mean Aerodynamic Chord

Certain aerodynamic and weight-and-balance characteristics are referenced as a percent of the wing chord. However, when a wing is tapered, the chord is not uniform across the entire wing span. For this reason these characteristics are referenced as a percent of the **mean aerodynamic chord** (MAC). The mean aerodynamic chord is the chord drawn through the center of the area of the airfoil; that is, equal amounts of wing area will lie on both sides of the MAC. Often, the MAC is confused with the average chord. As an example, the pointed-tip delta wing shown in Figure 4–28, would have an average chord equal to one-half the root chord but a MAC equal to two-thirds of the root chord.

Airflow Control Devices

You have seen that the way in which the air flows across a wing has a direct result on the lift that is produced. Items such as camber, aspect ratio, and laminar flow are all important in the generation of lift. The ability to vary these characteristics results in an aircraft that has more desirable aerodynamic characteristics over a wider operating range. There are many different types of devices that can either increase or decrease lift, such as flaps, slots, slats, and spoilers. There are also devices that affect the airflow as it passes over the wing, such as wing fences and vortex generators.

Wing Flaps

A **wing flap** is defined by NASA as a hinged, pivoted, or sliding airfoil, usually near the trailing edge of the wing. It is designed to increase the lift, drag, or both when deflected and is used principally for landing, although large airplanes

use partial flap deflection for takeoff. Most flaps are usually 15 to 25% of the airfoil's chord. The deflection of a flap produces the effect of adding a large amount of camber well aft on the chord. The more camber that the airfoil has results in a greater pressure differential and the creation of more lift. This makes it possible for the airplane to have a steeper angle of descent for the landing without increasing the airspeed. Flaps are normally installed on the inboard section of the wing trailing edge.

Some of the basic types of flap design are illustrated in Figure 4–29. An airfoil without a flap is shown at the top.

The second drawing shows a **plain flap**. The plain flap, in effect, acts as if the trailing edge of the wing were deflected downward to change the camber of the wing, thus causing a significant increase in both the coefficients of lift and drag. If the flap is moved down sufficiently it becomes an effective air brake.

The **split-edge flap** is usually housed flush with the lower surface of the wing immediately forward of the trailing edge. This flap is illustrated in the middle of Figure 4–29. The split-edge flap is usually nothing more than a flat metal plate hinged along its forward edge. The split-edge flap produces a slightly greater change in lift than the plain flap. However, a much larger change in drag results from the great turbulent wake produced by this type of flap.

Figure 4–29 also illustrates the **Fowler flap**, which is constructed so that the lower part of the trailing edge of the wing rolls back on a track, thus increasing the effective area of the wing and at the same time lowering the trailing edge. The flap itself is a small airfoil that fits neatly into the trail-

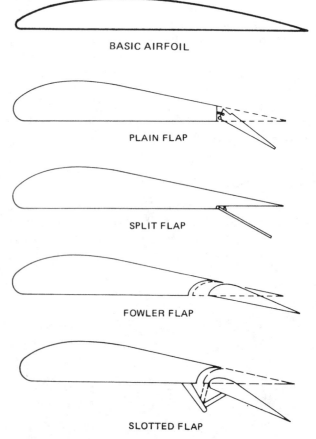

FIGURE 4–29 Basic types of flaps.

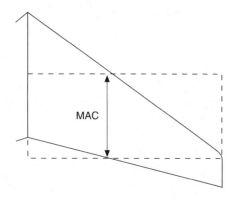

FIGURE 4–28 Mean aerodynamic chord.

FIGURE 4–30 Operation of the Fowler flap.

ing edge of the main wing when closed. As shown in Figure 4–30, when the flap opens, the small airfoil slides downward and backward on tracks until it reaches the position desired, thus providing a wing with a variable coefficient of lift and a variable area. With the Fowler flap, the wing area can be increased, causing large increases in lift with minimum increases in drag, the exact amount of increase of each depending upon the angle to which the flap is lowered. The Fowler flap is one of the designs which are particularly well adapted for use at takeoff as well as landing.

The **slotted flap**, shown in the bottom drawing of Figure 4–29, is similar to a plain flap except that as the flap is extended, a gap develops between the wing and the flap. The slots allow air from the bottom of the wing to flow to the upper portion of the flap and downward at the trailing edge of the wing. This aids in delaying airflow separation and creates a downward flow of air, which produces an upthrust to the wing.

A variation, and improvement, to the basic Fowler and slotted flaps is the **slotted Fowler flap**. When such flaps are initially extended, they move aft on their track. Once past a certain point on the track, further aft movement is accompanied by a downward deflection, which opens up one or more slots. A triple-slotted Fowler flap is shown in Figure 4–31.

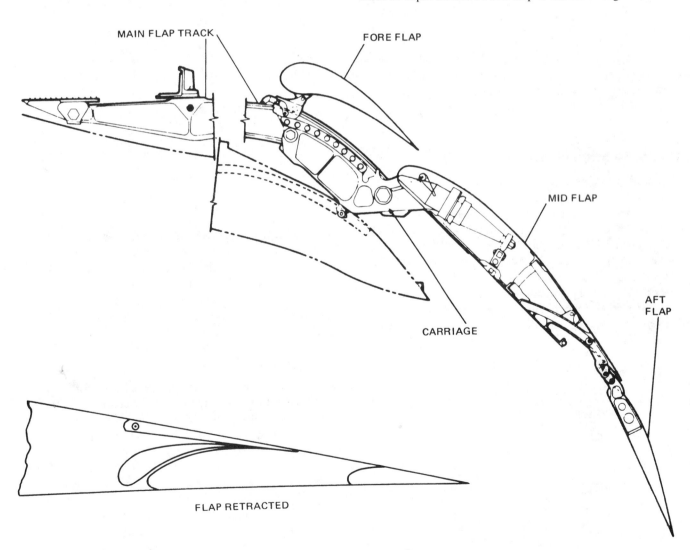

FIGURE 4–31 Slotted Fowler flap arrangement. *(Boeing Commercial Airplane Co.)* .

The slotted flap can provide much greater increases in lift than the plain or split flap, and corresponding drag changes are much lower. This type of flap requires the installation of a rather complicated structure. The slotted Fowler flap is usually used on the trailing edge of most turbine transport-category aircraft.

Effects of Flaps

At normal flying speeds, when flaps are fully retracted, that is, when they are all the way up, they have no effect on the lift characteristics of the wing. On the other hand, when they are lowered for landing, there is increased lift for similar angles of attack of the basic airfoil, and the maximum lift coefficient is greatly increased, often as much as 70%, with the exact amount of increase depending upon the type of flap installed.

The effectiveness of flaps on a wing configuration depends on many different factors. One important factor is the amount of the wing area affected by the flaps. Since a certain amount of the span is reserved for ailerons, the actual wing maximum lift properties will be less than that of the flapped two-dimensional section. If the basic wing has a low thickness, any type of flap will be less effective than on a wing of greater thickness. The curves of Figure 4–32 illustrate the lift characteristics of a wing with and without flaps. With the increase of lift comes a decrease in landing speed; there is also an increase of drag when the flap is down, however, and this requires a steeper glide to maintain the approach speed. The increase of drag also acts as a brake when the airplane is rolling to a stop on the landing strip.

Leading-Edge Flaps

While flaps are generally located on the trailing edge of a wing, they can also be placed on the leading edge. Leading-edge flaps are normally used only on large transport-category aircraft that need large amounts of additional lift for landing. A leading-edge flap is a high-lift device which reduces the severity of the pressure peak above the wing at high angles of attack. This enables the wing to operate at higher angles of attack than would be possible without the flap.

One method for providing a wing flap is to design the wing with a leading edge that can be drooped, as shown in

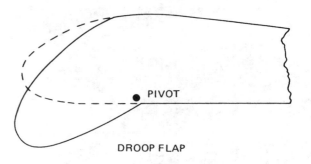

DROOP FLAP

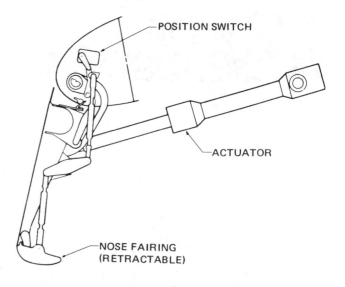

KREUGER FLAP

FIGURE 4–33 Types of leading-edge flaps.

the top drawing of Figure 4–33. Another method for providing a leading-edge flap is to design an extendable surface known as the Krueger flap that ordinarily fits smoothly into the lower part of the leading edge. When the flap is required, the surface extends forward and downward, as shown in the second drawing of Figure 4–33.

Slots and Slats

Another device that is used on the leading edge of a wing is a **slot**. A slot is also a high-lift device because it improves lift. It is a nozzle-shaped passage through a wing designed to improve the airflow conditions at high angles of attack and slow speeds. As the angle of attack of the wing increases, air from the high-pressure region below the wing flows to the low-pressure area above the wing, as shown in the bottom drawing of Figure 4–34. This flow of air postpones the breakdown of streamline flow that accompanies an increase in the angle of attack. A slot is normally placed very near the leading edge. Slots are illustrated in Figures 4–34 and 4–35.

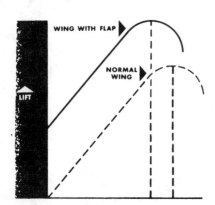

FIGURE 4–32 Effect of flaps on lift.

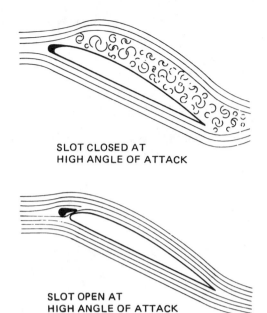

SLOT CLOSED AT
HIGH ANGLE OF ATTACK

SLOT OPEN AT
HIGH ANGLE OF ATTACK

FIGURE 4–34 Effect of a slot on wing performance.

There are two general types of slots: the fixed and the automatic. When the fixed type is used, the airflow depends on the angle of attack. The disadvantage of a fixed slot is that it adds excessive drag at low angles of attack. The automatic slot is formed by having a leading-edge airfoil that will separate from the main leading edge to form a slot. This auxiliary airfoil is commonly referred to as a slat. A **slat** is a movable auxiliary airfoil attached to the leading edge of the wing which, when closed, falls within the original contour of the wing and which, when opened, forms a slot.

The automatic slot is nested into the leading edge of the wing while the wing is at low angles of attack but is free to move forward a definite distance from the leading edge at high angles of attack. This forms a slot through which a portion of the airstream flows and is deflected along the upper surface of the wing, thus maintaining a streamline flow around the wing. Figure 4–34 shows the effect of the airstream diverted by a slot and the advantage gained by its use. The top picture shows the airfoil with its slot closed at a high angle of attack. The airfoil is shown in a stalling position because the burbling of the air reaches almost the leading edge of the wing.

The automatic slot has disadvantages as well as advantages. The number of moving parts and the weight of the wing are increased. The slots must be installed properly and operate equally well on both wings or they are useless. If a slot on one wing opens before the slot on the opposite wing does so, disastrous results could occur. The usual location of slots is such that they are subjected to ice formation, and in spite of any anti-icing or deicing equipment, they may fail to function. If any of these factors causes a lack of balance, lateral control may be impaired. For these reasons, a device is usually provided for locking slots in a closed position if they do not function properly.

Figure 4–35 illustrates the effect of a slot on the lift coefficient. Notice that at angles where the slot is opened, the lift is greater and the maximum C_L occurs at a much higher angle of attack. This indicates that an airplane with a slotted wing has a lower stalling speed than one without slots, other things being equal.

Figure 4–36 illustrates the effect of having a combination of slots and flaps. With this arrangement, it is possible to have a much lower landing speed, better control of the flight path, and at least a partial elimination of the nose heaviness

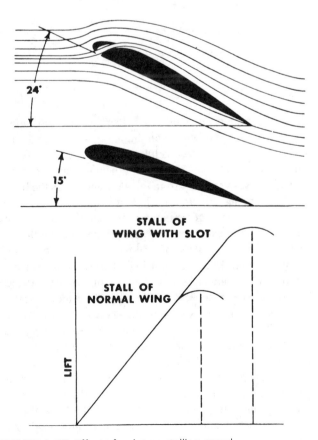

FIGURE 4–35 Effect of a slot on stalling speed.

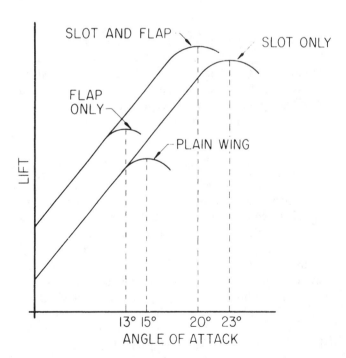

FIGURE 4–36 Effect of the combination of slots and flaps.

Shapes and Dimensions of Airfoils **91**

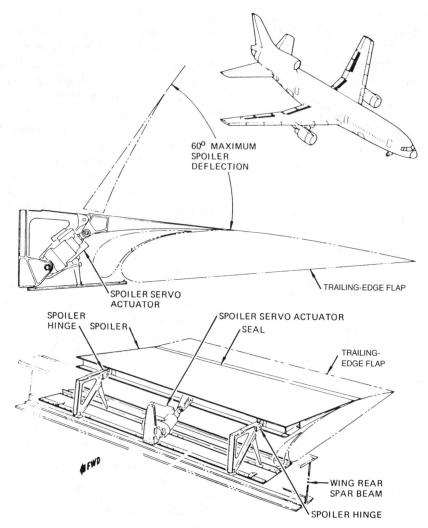

60° MAXIMUM
SPOILER
DEFLECTION

TRAILING-EDGE FLAP

SPOILER SERVO
ACTUATOR

SPOILER
HINGE SPOILER

SPOILER SERVO ACTUATOR
SEAL

TRAILING-
EDGE FLAP

FWD

WING REAR
SPAR BEAM

SPOILER HINGE

FIGURE 4–37 Spoilers are commonly hinged at their leading edge.

that may result from the use of flaps alone. It should be understood that Figure 4–36 is based upon a particular set of conditions and does not illustrate the effect produced by various airfoils and combinations of different flaps and slots. Other types of flaps and combinations with slots will produce values differing from those shown in this figure.

Spoilers

While flaps, slats, and slots are devices that are designed to greatly increase the lift that an airfoil creates, it is sometimes desirable to quickly and effectively decrease the lift on an airfoil. A device designed to reduce the lift on a wing is called a **spoiler**. The spoiler is the opposite of a high-lift device and derives its name from the fact that its purpose is to "spoil" the lift of the wing. Spoilers are located on the upper surface of wings and are one of two basic configurations. The more common configuration on jet transports, shown in Figure 4–37, has a flat-panel spoiler laying flush with the surface of the wing and hinged at the forward edge.

When the spoilers are deployed, the surface rises up and reduces the lift. The other configuration, shown in Figure 4–38, is common among sailplanes and has the spoiler located inside the wing structure. When the spoiler is deployed, it rises vertically from the wing and spoils the lift.

Flight spoilers are used in flight to reduce the amount of lift that the wing is generating to allow controlled descents without gaining excessive air speed. Depending on the aircraft design, the spoilers may be used as the aircraft's primary roll control. Instead of using ailerons, as explained in Chapter 5, an outboard spoiler on the wing can be deflected into the airstream to destroy lift and induce the aircraft to roll. The principal reason for using spoilers for roll control is that it frees the entire trailing edge of the wing for flap use. Longer flap spans mean more of the wing's camber can be changed and higher lift coefficients can be obtained.

Ground spoilers are only used when the aircraft is on the ground and are employed along with the flight spoilers to greatly reduce the wing's lift upon landing. They also increase the aerodynamic drag of the aircraft after landing to aid in slowing the aircraft.

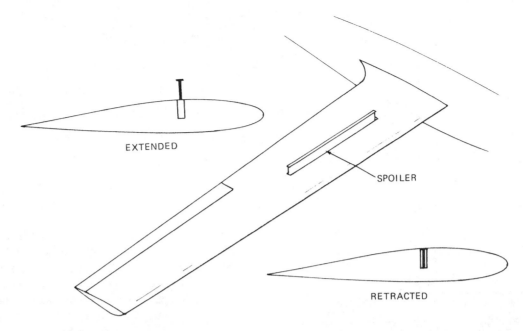

FIGURE 4–38 Some aircraft, such as sailplanes, have spoilers that rise vertically out of the wing.

Angle of Incidence

The term *angle of attack* has been discussed in a previous chapter of this text. A related but different term is *angle of incidence*. The **angle of incidence** of a wing is the angle formed by the intersection of the wing chord line and the horizontal plane passing through the longitudinal axis of the aircraft. Angle of incidence is illustrated in Figure 4–39. Airplanes are usually designed with a positive angle of incidence in which the leading edge of the wing is slightly higher than the trailing edge. The correct angle of incidence is essential for low drag and longitudinal stability.

Many airplanes are designed with a greater angle of incidence at the root of the wing than at the tip; this wing characteristic is called **washout**.

The term **washin** refers to an increase in the angle of incidence of the wing from the root to the tip. If a wing has an angle of incidence of 2° at the root and an angle of 3° at the outer end, it has a washin of 1°.

A difference in the washout and washin of the right and left wings of an aircraft is used to compensate for **propeller torque**. Propeller torque causes the aircraft to roll in a direction opposite that of the propeller rotation. To compensate for this, the right wing is rigged or designed with a smaller angle of incidence at the tip than that of the left wing. Thus the right wing is washed out more than the left. Information

on adjusting the angle of incidence of a wing can be found in an accompanying text in this series, *Aircraft Maintenance and Repair*.

STALLS AND THEIR EFFECTS

Conditions Leading to a Stall

In Chapter 3, it was pointed out that a stall occurs when the angle of attack becomes so great that the laminar airflow separates from the surface of an airfoil, leaving an area of burbling that destroys the low-pressure area normally existing at the upper surface of a wing in flight. Figure 4–40 illustrates this condition, which also represents the maximum coefficient of lift.

When an airplane is in flight, there are a number of flight conditions that may lead to a stall. First, if an airplane is pulled up sharply until its forward speed diminishes to a point where lift is less than gravity, the airplane will begin to lose altitude. The angle of attack increases, and when it reaches the stalling value (about 20°), the wing stalls and the airplane stops flying. If the stall is balanced on both sides of the airplane, it will pitch forward and may soon regain flying speed.

Stalls may also occur at high speeds. Stalls occurring under these conditions are called **high-speed stalls**, and they occur when an airplane is pulled up so abruptly that the angle of attack exceeds the stall angle. This type of stall is not often encountered because under ordinary conditions it is not necessary to pull an airplane up sharply enough to cause a stall.

Stalls are more likely to occur during turns than in level flight. This is because greater lift is required to maintain level flight in a turn.

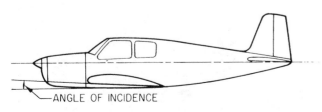

FIGURE 4–39 Angle of incidence.

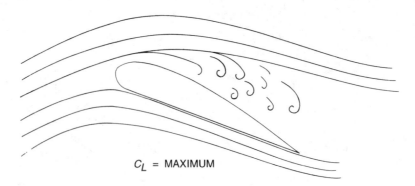

C_L = MAXIMUM

FIGURE 4–40 Airfoil in a stall.

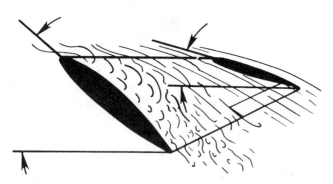

FIGURE 4–41 View of wingtip twist. Ailerons are still effective even though wing root is in the stalled condition.

Stall Warning

The experienced pilot can usually sense when a stall is about to happen because of the "feel" of the airplane controls and the reactions of the airplane. Often the airplane will start to shake or buffet because of the flow separation on the wing and the turbulent air buffeting the tail surfaces. The controls become "sloppy" and do not have the solid feel of normal flight.

Most airplanes are equipped with stall-warning devices. Typical of such devices is a small vane mounted near the leading edge of the wing and arranged so that it will actuate a switch when it rises as a result of an excessive angle of attack. The switch causes a warning horn to sound when the angle of attack approaches maximum, usually about 5 to 10 kn [2.57 to 5.14 m/s] above stalling speed.

Effect of Wing Design on Stall

A desirable stall pattern can be accomplished by (1) designing the wing with a twist so that the tip has a lower angle of

incidence (washout) and, therefore, a lower angle of attack when the root of the wing approaches the critical angle of attack (see Figure 4–41); (2) designing slots near the leading edge of the wing tip to allow air to flow smoothly over that part of the wing at higher angles of attack, therefore stalling the root of the wing first (see Figure 4–34); and (3) attaching stall strips on the leading edge near the wing root (see Figure 4–42).

The **stall strip** is a triangular strip mounted on the leading edge of the wing at the inboard end. At high angles of attack where stalling would be likely to occur, the strip causes the inboard portion of the wing to stall before the outer portion. This enables the pilot to maintain control of the aircraft with the ailerons, and the airplane does not "fall off" on one wing.

Laminar Flow Control

Today's civil transports typically cruise at about 500 mph (800 km/h). One problem of traveling at such a speed occurs in the **boundary layer**, a thin sheet of flowing air that moves along the surfaces of the wing, fuselage, and tail of an airplane.

As discussed in Chapter 3, at low speeds this layer follows the aircraft contours and is smooth, a condition referred to as **laminar**. At high speeds, the boundary layer changes from laminar to turbulent, creating friction and drag that waste fuel. This is illustrated in Figure 4–43. Many experiments have been carried out in an effort to control the boundary layer and increase laminar flow.

The **laminar flow control system** calls for removing the turbulent boundary layer by suction, thus maintaining laminar flow, as is shown in Figure 4–44. Basically, this system includes a suction surface through which a portion of the boundary-layer air is taken into the airplane, a system for metering the level and distribution of the ingested flow, a ducting system for collecting the flow, and pumping units

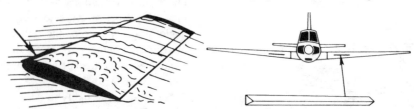

FIGURE 4–42 The stall strip ensures that the root section stalls first.

FIGURE 4-43 Development of the turbulent boundary layer as a result of skin friction.

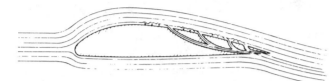

FIGURE 4-44 One method of boundary-layer control.

which provide sufficient compression to discharge the suction flow at a velocity at least as high as the airplane velocity. The effect of this system is to keep the boundary layer thin and permit laminar flow to continue.

Vortex Generators

Even though most turbine transport-category aircraft do not fly at the speed of sound (Mach 1), there are certain areas on the airplane where the airflow velocity will be greater than Mach 1. This is particularly true at the upper surface of parts of the wing where, because of the curvature of the wing, the air velocity must increase substantially above the airspeed of the airplane. This is illustrated in Figure 4-45, which shows an airfoil profile moving through the air at high subsonic speed. A short distance back from the leading edge of the wing and above the top surface, the air reaches supersonic speed. At the rear part of the supersonic area where the airflow returns to subsonic speed, a shock

wave is formed, as was discussed in Chapter 3. To the rear of this shock wave the air is very turbulent, and this area of the wing is, in effect, partially stalled. This, of course, causes a substantial increase in drag, which increases as airspeed increases.

In order to reduce the drag caused by supersonic flow over portions of the wing, small airfoils called **vortex generators** are installed vertically into the airstream. Although commonly used on the upper inboard surfaces of a cambered wing, vortex generators may be installed anywhere that airflow separation creates a problem, including on tail surfaces and in engine ducts. Because of the low aspect ratio of the vortex generators, they develop strong tip vortices. The tip vortices cause air to flow upward and inward in circular paths around the ends of the airfoil, as is shown in Figure 4-46. The vortices generated have the effect of drawing high-energy air from outside the boundary layer into the slower-moving air close to the skin. The strength of the vortices is proportional to the lift developed by the generators. To operate effectively, the generators are mounted forward of the point where separation begins.

Drag reduction achieved by the addition of vortex generators can be seen in the drag-rise curve. Since the generators effectively reduce the shock-induced drag associated with the sharp rise in the curve at speeds approaching Mach 1.0, the curve is pushed to the right, as shown in Figure 4-47.

The addition of the vortex generators actually increases overall drag very slightly at lower speeds. However, the gains at cruise speeds more than balance out the losses at lower speeds. Since the airplane spends most of its flight time at cruise speeds, the net gain is significant.

Wing Fences

Ideally, air would always flow chordwise over a wing; however, as has been discussed, air will tend to flow spanwise toward the tip. Spanwise flow is particularly a problem on swept wings. This spanwise flow of air may be partially

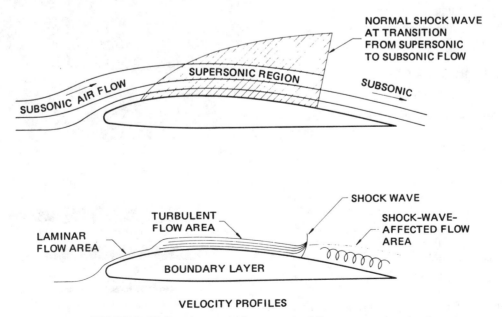

FIGURE 4-45 Development of supersonic airflow over a subsonic wing.

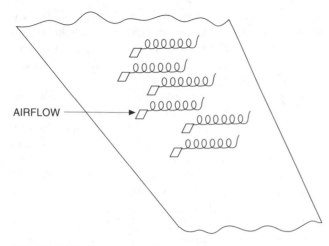

FIGURE 4–46 Vortex generators.

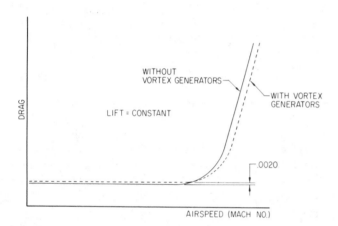

FIGURE 4–47 Drag reduction achieved by vortex generators.

controlled by the use of a wing (flow) fence, such as is illustrated in Figure 4–48. A **wing fence** is a stationary vane, projecting from the upper surface of an airfoil, which is used to prevent the spanwise flow of air. Flow fences are often located in alignment with trailing-edge control surfaces, such as ailerons, to improve the chordwise flow and thereby the effectiveness of the control surfaces.

Planforms for High-Speed Flight

Because of the complex nature of supersonic phenomena, aircraft and spacecraft designers have resorted to designs which at first seem rather unconventional. Many of these designs are decidedly inferior to the more familiar aircraft forms at subsonic flight speeds. For example, when the airflow across a wing is considered, as a wing becomes thicker the increase in local speed across the curved areas will become greater. But if an increase in the critical Mach number needs to be obtained, in order for aircraft to operate at speeds approaching Mach 1, then wings that are as thin as possible are desirable. This presents difficulties, however, because a very thin wing does not have great strength and also has poor lift at slow speeds.

The ideal airfoil shape for a supersonic airfoil is a biconvex design with a sharp leading edge. However, such a de-

FIGURE 4–48 Wing fences. *(Gates Learjet Corporation)*

sign has very poor low-speed characteristics and is not practical on most aircraft designs.

Another method for increasing the critical Mach number is to sweep back the wing. This will improve the critical Mach number, but it will also present a problem in stability. If an aircraft with a pronounced sweepback changes heading as a result of rudder movement or a disturbance in the air, the wing that moves forward will have much greater lift than the other wing and the airplane will have a tendency to roll. At supersonic speed and higher, the advantage of the sweptback wing begins to decrease, and at Mach 2, the straight wing is superior.

Another method for increasing the critical Mach number is to reduce the aspect ratio. This is not a completely satisfactory solution, because a low aspect ratio adversely affects flight at low speeds. The Concorde, pictured in Figure 4–49, cruises at speeds of over Mach 2, yet its design, which incorporates a delta planform wing with an aspect ratio of 1.7, is still capable of adequate performance at subsonic speeds.

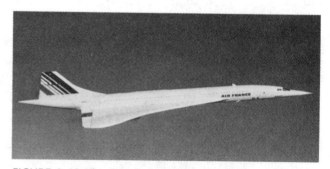

FIGURE 4–49 The Concorde, designed for flight speeds of over Mach 2. *(Air France)*

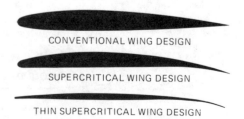

FIGURE 4–50 Supercritical wings.

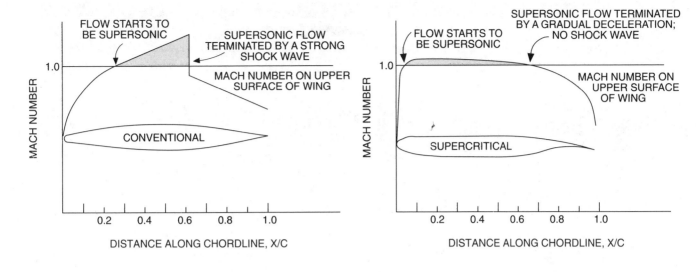

FIGURE 4–51 Comparison of supersonic airflow pattern across conventional and supercritical airfoils.

Supercritical Wing

Recently an airfoil has been developed that has a critical Mach number close to 1. This airfoil is referred to as the **supercritical** design and was developed by Dr. Richard Whitcomb of NASA.

The supercritical airfoil, illustrated in Figure 4–50, has a very slight curvature on the upper surface and the maximum thickness is much farther back than normal. The airfoil curves downward at the trailing edge. This design prevents the rapid pressure rise normally associated with a more cambered airfoil. It also delays and softens the onset of shock waves on the upper surface of a wing. The shock wave is far less severe than on a conventional wing, as is shown in Figure 4–51, and fuel efficiency is substantially improved. This design is being adopted on many transport-category and business-jet aircraft.

REVIEW QUESTIONS

1. Explain the meaning of the digits in the NACA four-digit airfoil number.
2. What is the principal difference in design between a GAW-1 and GAW-2 airfoil?
3. Name four airfoil characteristics.
4. What do the following symbols indicate: AR, C_D, C_L, D, L, S, V.
5. What is the fundamental equation for lift?
6. What is the coefficient of lift?
7. What is meant by *lift/drag ratio*?
8. Define the term *center-of-pressure coefficient*.
9. What are characteristic curves for an airfoil?
10. What is meant by the *planform* of an airfoil?
11. What are wingtip vortices?
12. Define *aspect ratio* and explain how it may be determined.
13. What is the effect of increasing the aspect ratio of a wing?
14. Give some methods used for increasing the effective aspect ratio of a wing.
15. Describe what is meant when a wing is *tapered*.
16. How is the sweep angle of a wing usually measured?
17. Define the term *mean aerodynamic chord*.
18. What is the result of flap deflection?
19. Describe a wing slot.
20. Describe a slat.
21. Describe a spoiler.
22. For what purposes can flight spoilers be used?
23. Why are vortex generators employed on some airfoils?
24. What is the purpose of a wing fence?
25. Describe a supercritical airfoil.

5 Aircraft in Flight

INTRODUCTION

An aircraft must have satisfactory handling properties in addition to adequate performance. The aircraft must have sufficient stability to maintain a uniform flight condition and recover from various disturbing influences.

Any person associated with aviation should have some understanding of the actual processes of flight. An individual who is responsible for the design, construction, and operation or maintenance of aircraft must be thoroughly familiar with the forces acting on an aircraft in flight, the components of the aircraft that control the flight forces, and the reactions of the aircraft to this control. Understanding why the aircraft is designed with a particular type of primary and secondary control system is essential in maintaining today's complex aircraft.

This chapter includes information concerning the stability and control of aircraft that today's technician needs to possess in order to make intelligent decisions affecting the flight safety of both airplanes and helicopters.

FORCES ON THE AIRPLANE IN FLIGHT

Lift and Weight

The four forces acting on an airplane in flight are **lift**, **weight**, **drag**, and **thrust**. The weight acts vertically downward from the **center of gravity** (CG) of the airplane. The lift acts in a direction perpendicular to the direction of the relative wind from the center of pressure (CP). In straight and level flight the lift and weight must be equal. When the airplane is flying at a constant speed, the thrust must equal the drag. Thrust, of course, is provided by the engine and propeller or by the high-velocity gases ejected from the tail pipe of a turbine engine.

Figure 5–1 illustrates the forces acting on an airplane in straight and level flight at a constant speed. In this case all the forces are in balance.

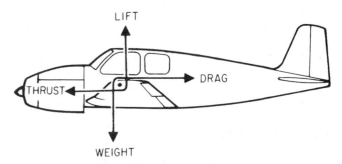

FIGURE 5–1 Forces on an airplane in flight.

Drag and Thrust

Thrust is defined as the forward-directed pushing or pulling force developed by an aircraft engine. This includes reciprocating engines, turbojet engines, turboprop engines, or rocket engines.

The **thrust line** is an imaginary line passing through the center of the propeller hub perpendicular to the plane of the propeller rotation, as illustrated in Figure 5–1. In the case of a jet-propelled airplane, the thrust line is parallel to the path of the ejected gases from the jet engine.

In general terms, **drag** is the force which opposes the forward motion of the airplane. Previously, when drag was discussed, it was divided into two types, induced drag and parasite drag. The total drag of the airplane includes the wing drag and a number of other forms of drag, including that caused by the fuselage, the landing gear, and the other parts of the airplane. The total drag of the airplane is in opposition to thrust, as illustrated in Figure 5–1.

Assuming that the airplane is flying straight and level in calm air, the drag acts parallel to the direction of the relative wind. As long as thrust and drag are equal, the airplane flies at a constant speed. If the engine power is reduced, the thrust is decreased and the speed of the airplane is reduced. If the thrust is lower than the drag, the speed of the airplane becomes less and less until it finally is lower than the speed required to maintain level flight, and the airplane will descend. On the other hand, if the power of the engine is increased, the thrust is increased and the airplane will accelerate because the thrust is greater than the drag. While the airplane accelerates, the drag increases until it eventually equals the thrust. Then the airplane flies at a constant speed.

These facts can be summarized by saying that in straight

and level flight at a constant speed, the sum of the components of the forces acting on an airplane equals zero.

Loads and Load Factors

During level flight the forces exerted on an airplane are at a minimum; however, they still exist. The drawings of Figure 5–2 show the forces acting on an airplane in level flight. It will be noted that impact pressures exist at all points where the air strikes the surfaces from a forward direction. Negative pressures exist above and to the rear of the cabin, above the wings, and below the horizontal stabilizer. The total upward force (lift) exerted on the airplane is equal to the weight of the airplane; therefore, the airplane flies in a level flight path.

The internal structure of the airplane wing shown in Figure 5–2 must be such that it can withstand the severe bending moments imposed by the combination of weight and lift. During flight, the wings of an airplane will support its maximum allowable gross weight. As long as the airplane is moving at a steady rate of speed and in a straight line, the load imposed upon the wings will remain constant.

A change in speed during straight flight will not produce any appreciable change in load, but when a change is made in the airplane's flight path, an additional load is imposed upon the airplane structure. This is particularly true if a change in direction is made at high speeds with rapid, forceful control movements.

According to certain laws of physics, a mass (an airplane, in this case) will continue to move in a straight line unless some force intervenes, causing the mass (the airplane) to assume a curved path. During the time the airplane is in a curved flight path, it still attempts, because of inertia, to force itself to follow straight flight. This tendency to follow straight flight rather than curved flight generates a force

known as centrifugal force, which acts toward the outside of the curve.

Any time the airplane is flying in a curved flight path with a positive load, the load the wings must support will be equal to the weight of the airplane plus the load imposed by centrifugal force. A **positive load** occurs when back pressure is applied to the elevator, causing centrifugal force to act in the same direction as the force of weight. A **negative load** occurs when forward pressure is applied to the elevator control, causing centrifugal force to act in a direction opposite to that of the force of weight.

Curved flight producing a positive load is a result of increasing the angle of attack and, consequently, the lift. Increased lift always increases the positive load imposed upon the wings. However, the load is increased only at the time the angle of attack is being increased. Once the angle of attack is established, the load remains constant. The loads imposed on the wings in flight are stated in terms of **load factor**.

Load factor is the ratio of the total load supported by the airplane's wing to the actual weight of the airplane and its contents; that is, the actual load supported by the wings divided by the total weight of the airplane. For example, if an airplane has a gross weight of 2000 lb [907 kg] and, during flight, is subjected to aerodynamic forces which increase the total load the wing must support to 4000 lb [1814 kg], the load factor would be 2.0 (4000/2000 = 2). In this example the airplane wing is producing lift that is equal to twice the gross weight of the airplane. An airplane in straight and level flight would have a load factor of 1.0.

Another way of expressing load factor is the ratio of a given load to the pull of gravity; that is, referring to a load factor of 3 as "3 g's," where g refers to the pull of gravity. In this case the weight of the airplane is equal to 1 g, and if a load of 3 times the actual weight of the airplane were

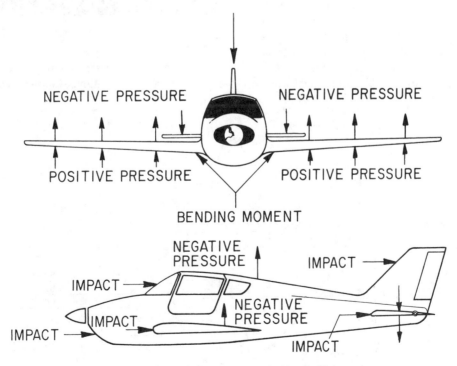

FIGURE 5–2 Forces on an airplane in flight.

imposed upon the wing due to curved flight, the load factor would be equal to 3 g's.

Load Factors and Airplane Design

To be certified by the Federal Aviation Administration, the structural strength (load factor) of an airplane must conform with prescribed standards set forth by Federal Aviation Regulations.

All airplanes are designed to meet certain strength requirements depending upon the intended use of the airplane. Classification of airplanes as to strength and operational use is known as the category system. Aircraft may be type-certificated into normal, utility, or acrobatic categories.

The **normal category** is limited to airplanes intended for nonacrobatic operation and has a positive load-factor limit of 3.8. The **utility category** applies to airplanes intended for limited acrobatic operations and has a positive load factor limit of 4.4. Aircraft in the **acrobatic category** may have a positive load factor limit of 6.0 and are free to operate without many of the restrictions that apply to normal- and utility-category aircraft. Aircraft are constructed to withstand higher positive loads than negative loads. Negative load-factor limits for normal- and utility-category aircraft are not less than 0.4 times the positive load-factor limit and for acrobatic-category aircraft not less than 0.5 times the positive load-factor limit. Small airplanes may be certificated in more than one category if the requirements of each category are met.

The category in which each airplane is certificated may be readily found in the aircraft's Type Certificate Data Sheet or by checking the Airworthiness Certificate found in the cockpit.

An airplane is designed and certificated for a certain maximum weight during flight. This weight is referred to as the **maximum certificated gross weight**. It is important that the airplane be loaded within the specified weight limits because certain flight maneuvers will impose an extra load on the airplane structure which may, particularly if the airplane is overloaded, impose stresses which will exceed the design capabilities of the airplane. If, during flight, severe turbulence or any other condition causes excessive loads to be imposed on the airplane, a very thorough inspection must be given to all critical structural parts before the airplane is flown again. Damage to the structure is often recognized by bulges or bends in the skin, "popped" rivets, or deformed structural members.

Effects of Turns on Load Factor

In turns, the load factors on an airplane increase similarly to those experienced in pulling out of a dive. The reason for increased wing loading in a turn is illustrated in Figure 5–3. In this diagram the airplane is in a turn with a 45° bank. Under these conditions, the gravity will still be pulling the airplane down with a force of 1 g and the centrifugal force will be pulling the airplane horizontally with a force of 1 g. When these forces are resolved, a resultant of 1.41 g is found. This means, of course, that the wing is required to carry 1.41 times the weight of the airplane. In a 60° bank, the load factor is 2.0; in a 70° bank, the load factor is nearly 3.0; and in an 80° bank, it is almost 6.0.

An airplane should not be turned with a bank so steep that the safe load factor of the airplane is exceeded.

Wing Loading

Load factors should not be confused with wing loading. **Wing loading** is the ratio of the total gross weight of the aircraft divided by the total wing area. Wing loading is expressed in pounds per square foot (lb/ft^2). Thus a 2000 lb airplane with a 200-square-foot wing would have a wing loading of 10 lb/ft^2. Wing loading will range from approximately 10 lb/ft^2 for a small single-engine aircraft to around 75 lb/ft^2 for a business jet.

AIRCRAFT STABILITY

Definition

All airplanes must possess stability in varying degrees for safety and ease of operation. **Stability** is the inherent ability

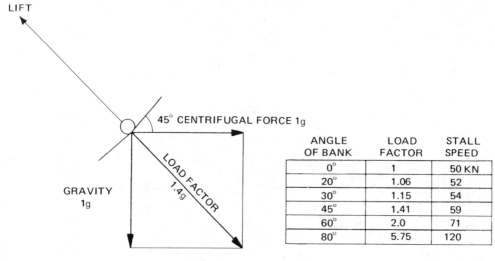

ANGLE OF BANK	LOAD FACTOR	STALL SPEED
0°	1	50 KN
20°	1.06	52
30°	1.15	54
45°	1.41	59
60°	2.0	71
80°	5.75	120

FIGURE 5–3 Load factors in a turn.

of a body, after its equilibrium is disturbed, to develop forces or moments that tend to return the body to its original position. In other words, a stable airplane will tend to return to the original condition of flight if disturbed by a force, such as turbulent air. Airplanes have varying degrees of stability. Aircraft that are inherently unstable require some type of computerized artificial stability system to be flown.

Static Stability

An aircraft is in a **state of equilibrium** when the sum of all forces and all moments is equal to zero. When an aircraft is in equilibrium, there are no accelerations and the aircraft continues in a steady condition of flight. If the equilibrium is disturbed by a gust or a deflection of the controls, the aircraft will experience acceleration because of an unbalance of moment or force.

The **static stability** of an aircraft is defined by the initial tendency to return to equilibrium conditions following some disturbance from equilibrium. If an object is disturbed from equilibrium and has the tendency to return to equilibrium, **positive static stability** exists. If the object has a tendency to continue in the direction of disturbance, **negative static stability**, or **static instability**, exists. If the object subject to a disturbance has neither the tendency to return nor the tendency to continue in the displacement direction, **neutral static stability** exists. This is an intermediate condition

which could occur when an object displaced from equilibrium remains in equilibrium in the displaced position.

These three categories of static stability are illustrated in Figure 5–4. The ball in a trough illustrates the condition of positive static stability. If the ball is displaced from equilibrium at the bottom of the trough, the initial tendency of the ball is to return to the equilibrium condition. The ball may roll back and forth through the point of equilibrium, but displacement to either side creates the initial tendency to return. The ball on a hill illustrates the condition of static instability. Displacement from equilibrium at the hilltop brings about the tendency for greater displacement. The ball on a flat, level surface illustrates the condition of neutral static stability. The ball encounters a new equilibrium at any point of displacement and has neither stable nor unstable tendencies.

The term *static* is applied to this form of stability since the resulting motion is not considered. Only the tendency to return to equilibrium conditions is considered in static stability.

Dynamic Stability

While static stability is concerned with the tendency of a displaced body to return to equilibrium, **dynamic stability** is defined by the resulting motion with time. If an object is disturbed from equilibrium, the time history of the resulting motion indicates the dynamic stability of the system.

Positive dynamic stability is the property which dampens the oscillations set up by a statically stable airplane, enabling the oscillations to become smaller and smaller in magnitude until the airplane eventually settles down to its original condition of flight.

The existence of static stability does not necessarily guarantee the existence of dynamic stability. However, the existence of positive dynamic stability implies the existence of static stability. An aircraft must demonstrate the required degrees of static and dynamic stability if it is to be operated safely. Figure 5–5 illustrates the relationship of dynamic and static stability.

AXES OF THE AIRPLANE

While being supported in flight by lift and propelled through the air by thrust, an airplane is free to revolve or move around **three axes**, namely the longitudinal axis, the lateral axis, and the vertical axis. These are illustrated in Figure 5–6.

The axis which extends lengthwise through the fuselage from the nose to the tail is the **longitudinal axis**. The axis extending through the fuselage from wing tip to wing tip is the **lateral axis**. The axis which passes vertically through the fuselage at the center of gravity is the **vertical axis**.

During flight, an airplane is rotated about the three axes by means of the three primary flight controls. The ailerons control roll about the longitudinal axis, the elevators control pitch about the lateral axis, and the rudder controls yaw about the vertical axis.

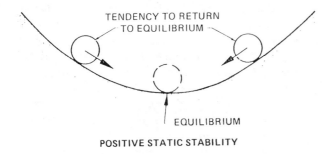

TENDENCY TO RETURN
TO EQUILIBRIUM

EQUILIBRIUM

POSITIVE STATIC STABILITY

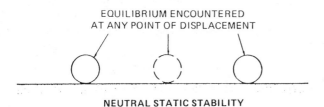

EQUILIBRIUM ENCOUNTERED
AT ANY POINT OF DISPLACEMENT

NEUTRAL STATIC STABILITY

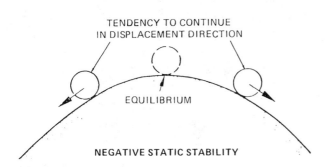

TENDENCY TO CONTINUE
IN DISPLACEMENT DIRECTION

EQUILIBRIUM

NEGATIVE STATIC STABILITY

FIGURE 5–4 Static stability.

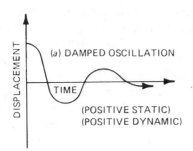

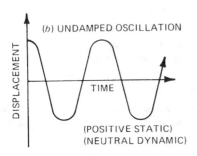

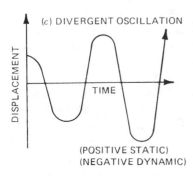

FIGURE 5–5 Relationship of oscillation and stability.

Discussing stability and relating it to the axis of the aircraft is somewhat confusing because stability is referenced to movement along an axis instead of rotation about an axis. For example, stability about the lateral axis (pitch) is referred to as longitudinal stability since it involves rotation of the airplane along the longitudinal axis. Stability around the longitudinal axis (roll) is called lateral stability, and stability around the vertical axis (yaw) is called directional stability.

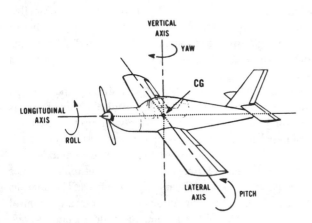

FIGURE 5–6 Axes of an airplane.

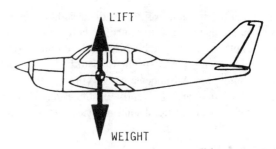

FIGURE 5–7 Neutral longitudinal stability.

Longitudinal Stability

The stability of an airplane about the lateral axis is **longitudinal stability**. If the airplane is put into a dive or climb and then the control is released, the airplane should return to level flight automatically. If the airplane does not have longitudinal stability, it may increase the angle of dive after being placed in a dive, or it may "porpoise," that is, oscillate through a series of dives and climbs (pitch up and down), unless controlled by the pilot.

The location of the center of gravity with respect to the center of lift determines to a great extent the longitudinal stability of the airplane. For example, Figure 5–7 illustrates neutral longitudinal stability. Note that the center of lift is directly over the center of gravity or weight. An airplane with neutral stability will produce no inherent pitch moments around the center of gravity.

Figure 5–8 illustrates the center of lift in front of the center of gravity. This airplane would display negative stability and an undesirable pitch-up moment during flight. If disturbed, the up-and-down pitching moment will tend to increase in magnitude. This condition can occur especially if the airplane is loaded so that the center of gravity is rearward of the airplane's aft loading limits.

Figure 5–9 shows an airplane with the center of lift behind the center of gravity. Again, this produces negative stability. Some force must balance the down force of the weight. This is accomplished by designing the airplane in such a manner that the air flowing downward behind the trailing edge of the wing strikes the upper surface of the horizontal stabilizer. This creates a downward tail force to counteract the tendency to pitch down and provides positive stability.

Lateral Stability

The stability of an airplane about the longitudinal, or roll, axis is the **lateral stability**. An airplane that tends to return

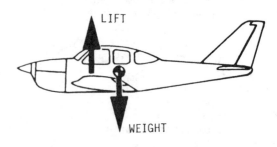

FIGURE 5–8 Negative longitudinal stability.

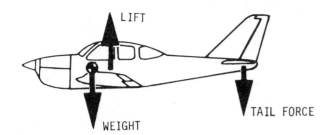

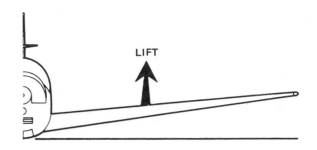

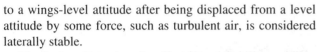

FIGURE 5–9 Positive longitudinal stability.

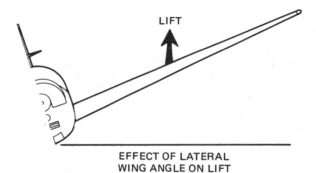

EFFECT OF LATERAL
WING ANGLE ON LIFT

to a wings-level attitude after being displaced from a level attitude by some force, such as turbulent air, is considered laterally stable.

The factors that primarily affect lateral stability are dihedral and sweepback. In aircraft terminology, **dihedral** means the lateral angle of the wing with respect to a horizontal plane; this is illustrated in Figure 5–10. **Positive dihedral** exists when the tip of a wing is above the horizontal plane passing through the root of the wing. **Negative dihedral** exists when the tip of the wing is below the horizontal plane passing through the root of the wing.

The purpose of positive dihedral is to provide lateral stability for the aircraft. The stabilizing effect of dihedral occurs when the airplane sideslips slightly as one wing is forced down in turbulent air. Usually a wing will provide the greatest amount of lift if it is in a perfectly horizontal position laterally. When an airplane is designed with dihedral, both wings will form angles with the horizontal plane. If the aircraft rolls slightly to the right, the tip of the right wing moves downward and the lift of the wing increases. At the same time, the left-wing lift is decreased because the angle with the horizontal plane is increasing. The airplane, therefore, is subjected to more lift from the right wing and less lift from the left wing. This causes the airplane to roll back to the left to resume a level position. These effects are illustrated in Figures 5–11 and 5–12.

The term **sweepback** refers to the angle at which the wings are slanted rearward from the root to the tip. The effect of sweepback in producing lateral stability is similar to that of dihedral but is not as pronounced. If one wing lowers in a slip, the angle of attack on the low wing increases, producing greater lift. This results in a tendency for the lower wing to rise and return the airplane to level flight. Sweepback augments dihedral to achieve lateral stability.

Some lateral stability is also provided by the vertical fin. A sideslip of the aircraft results in an air force being applied

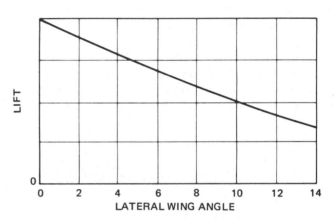

FIGURE 5–11 Effect of lateral wing angle on lift.

against the vertical fin and tends to rotate the aircraft back against the rolling direction.

Directional Stability

The stability of an airplane about the vertical axis is called **directional stability**. This means that the airplane will return to a straight flight path after having been turned (yawed) one way or the other.

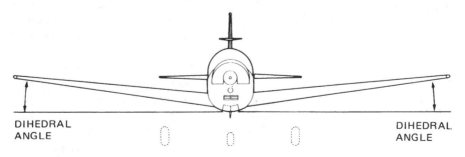

FIGURE 5–10 Dihedral of aircraft wings.

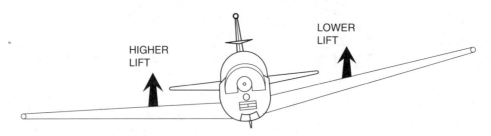

FIGURE 5–12 Effect of dihedral on aircraft in flight.

Directional stability is accomplished by placing a vertical stabilizer, or fin, to the rear of the center of gravity on the upper portion of the tail section. The surface of this fin acts similarly to a weathervane and causes the airplane to pivot into the relative wind. If the airplane is yawed out of its flight path, either by pilot action or by turbulence, the relative wind will exert a force on one side of the vertical stabilizer and return the airplane to its original direction of flight, as is shown in Figure 5–13. The amount of directional stability provided by the vertical fin is proportional to both the size of the fin and the distance it is located aft of the CG. The larger the surface area and the farther aft it is located, the greater the stability. Since the vertical fin is the primary directional stabilizing force, it must be located aft of the CG for a stable aircraft configuration.

Sweptback wings aid in directional stability. If the aircraft yaws from its direction of flight, the wing which is farther ahead offers more drag than the wing which is aft. The effect of this drag is to hold back the wing which is farther

ahead and to let the other wing catch up. This is illustrated in Figure 5–14.

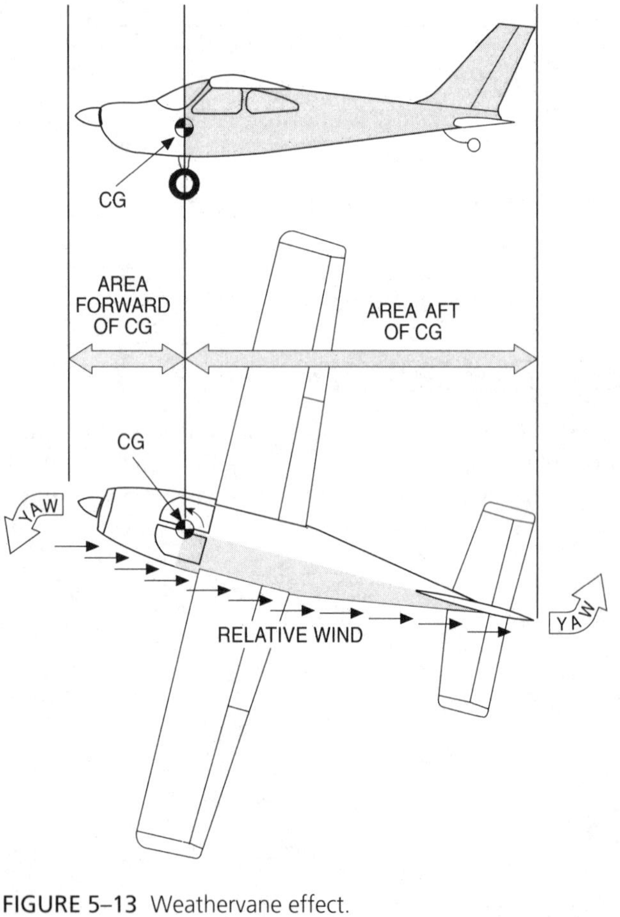

Excessive Stability

It is possible to build into an airplane a degree of stability that reduces or makes difficult control of the aircraft. Stability helps keep the aircraft in a desired flight path; however, sometimes it is desirable to deviate from that path. **Maneuverability** is also an important characteristic of an airplane. This is the ability of an airplane to be directed along a selected flight path. The smooth and easy response of the airplane to its controls is important. If the airplane does not have the proper degree of this quality, it will be difficult and tiring to fly, particularly through maneuvers. A good balance between stability and maneuverability is important in any aircraft design.

AIRCRAFT CONTROL

An airplane is equipped with certain fixed and movable surfaces, or airfoils, which provide for stability and control during flight. These are illustrated in Figures 5–15 and 5–16. Each of the named airfoils is designed to perform a specific function in the flight of the airplane.

Fixed Airfoils

The fixed airfoils are the wings, the horizontal stabilizer, and the vertical stabilizer (fin). The function of the wings has been previously discussed. The tail section of the airplane, including the stabilizers, elevators, and rudder, is commonly called the **empennage**.

Horizontal Stabilizers

As has been discussed, the horizontal stabilizer is used to provide longitudinal pitch stability and is usually attached to the aft portion of the fuselage. It may be located either above or below the vertical stabilizer or at some midpoint of the stabilizer. Conventional tails (horizontal stabilizers) are placed aft of the wing and set at a slight negative angle with respect to the wing chordline. This configuration gives a downward lift force on the tail, as shown in Figure 5–17. The down-lift force is dependent on the size of stabilizer and the distance aft that it is placed from the CG. The horizontal sta-

FIGURE 5–13 Weathervane effect.

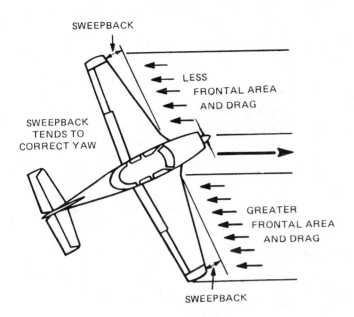

FIGURE 5–14 Effect of sweepback.

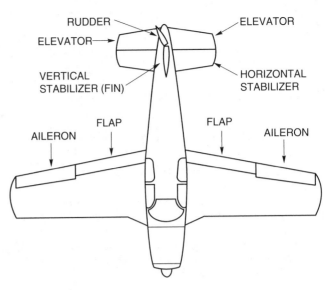

FIGURE 5–15 Control surfaces of an airplane.

bilizer may be designed as a fixed surface attached to the tail or as a movable surface used to provide pitch control.

Vertical Fins

The vertical stabilizer for an airplane is the airfoil section forward of the rudder and is used to provide directional stability (yaw) for the aircraft, as has been previously discussed. This unit is commonly called the *fin*.

A problem encountered on single-engine airplanes is that as the propeller turns clockwise, a rotating flow of air is moved rearward (see Figure 5–18), striking the left side of the fin and rudder, which results in a left-yawing moment.

To counteract this effect, many airplanes have the leading edge of the vertical fin offset slightly to the left, thereby allowing the slipstream to pass evenly around it.

Movable Control Surfaces

The **primary control surfaces** of an airplane include the ailerons, elevators, and rudder. The **secondary control surfaces** include flaps, trim tabs, spoilers, and slats. The principles of the operation of flaps and spoilers are discussed in Chapter 3.

The primary control surfaces are used to "steer" the airplane in flight to make it go where the pilot wishes it to go and to cause it to execute certain maneuvers. The secondary control surfaces are used to change the lift and drag charac-

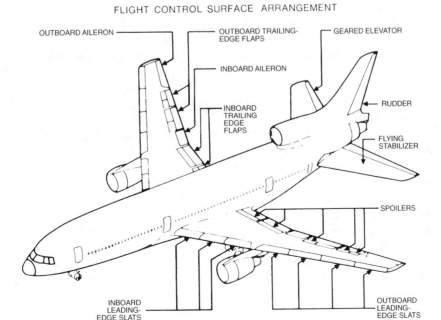

FLIGHT CONTROL SURFACE ARRANGEMENT

FIGURE 5–16 L-1011 flight controls. *(Lockheed Corp.)*

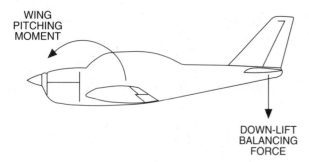

FIGURE 5–17 Horizontal stabilizer down-lift force.

teristics of the aircraft or to assist the primary controls. Large turbine aircraft, gliders, and some other types of aircraft are equipped with lift-control devices called spoilers (see Figure 5–16), which may be used as the primary roll control.

Ailerons

The **ailerons** are the primary flight control surfaces used to provide lateral (roll) control of the aircraft; that is, they control aircraft movement about the longitudinal axis. They are usually mounted on the trailing edge of the wing near the tip, as is shown in Figure 5–15. They are rigged so that when one is moving upward on one wing, the other is moving downward on the opposite wing. This movement changes the camber of the respective wings, with the downward-moving aileron increasing the camber and lift, and the upward-moving aileron decreasing the camber and lift. Since the ailerons are located outboard of the roll axis, this change in camber will result in a rolling motion. This action is illustrated in Figure 5–19.

Large turbine aircraft often employ two sets of ailerons, one set being approximately midwing or immediately outboard of the inboard flaps, and the other set being in the conventional location near the wingtips, as is shown in Figure 5–16. The outboard ailerons become active whenever the flaps are extended beyond a fixed setting. As the flaps are retracted, the outboard aileron control system is "locked out" and fairs with the basic wing shape. Thus, during cruise flight at comparatively high speeds, only the inboard ailerons are used for control. The outboard ailerons are active during landings and other slow-flight operations.

The ailerons are moved by means of a control wheel in the cockpit. If a pilot wants to roll the airplane to the right, he or she turns the wheel to the right. After the desired degree of bank is obtained, the wheel is returned to neutral to stop the roll. During normal turns of an airplane, the movement of the ailerons is coordinated with movements of the rudder and elevators to provide a banked horizontal turn without "slip" or "skid." A **slip**, or **sideslip**, is a movement of an airplane partially sideways. In a turn, the slip is downward and inward toward the turn. A **skid** in a turn is a movement of the airplane sideways and outward from the turn.

Aileron control in an airplane is complicated somewhat by an effect called **adverse yaw**. An aileron that moves down at the trailing edge of a wing creates considerably more drag than the aileron on the opposite wing that moves upward the same amount. Therefore, if the ailerons were

rigged to move the same distance in response to the movement of the cockpit control, the drag of the downward-moving aileron would cause the airplane to turn toward the side on which the downward-moving aileron is located. Thus, a pilot wishing to make a left turn would move the control to the left, causing the right aileron to move downward, but the drag caused by the aileron would cause the airplane to turn to the right, except for strong rudder control. To overcome adverse yaw, the ailerons of an airplane are rigged for differential movement. The differential control causes the up-moving aileron to move a greater distance than the down-moving aileron. The amount of differential is sufficient to balance the drag between the ailerons, thus eliminating the yaw effect. The design for differential control is explained in the associated text *Aircraft Maintenance and Repair*.

The correct rigging of the ailerons is of primary importance. After an airplane has been overhauled and during preflight inspections, the direction of aileron movement with respect to control-stick movement must be carefully noted. If the wheel is moved to the right, the right aileron must move up and the left aileron must move down. Reverse movement of the control should then cause a reverse of the position of the ailerons.

Rudders

The **rudder** is a vertical control surface that is usually hinged to the tail post aft of the vertical stabilizer and designed to apply **yawing moments** to the airplane, that is, to make it turn to the right or the left about the vertical axis. The movement of the rudder is controlled by pedals operated by the feet of the pilot. When the right pedal is pressed, the rudder swings to the right, thus bringing an increase of dynamic air pressure on its right side. This increased pressure causes the tail of the airplane to swing to the left and the nose to turn to the right. The operation of a rudder is shown in Figure 5–20.

Although it appears that the rudder causes the airplane to turn, it must be pointed out that the rudder itself cannot cause the airplane to make a good turn. Newton's first law of motion states that a moving body tends to continue moving in a straight line unless some outside force changes its direction. When rudder is applied to an airplane in flight, the airplane will turn, but it will continue to travel in the same direction as before unless a correcting force is applied. Thus, with rudder only, the airplane will turn sideways and

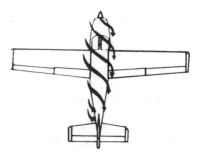

FIGURE 5–18 Slipstream.

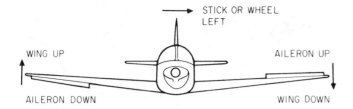

FIGURE 5–19 Effect of ailerons in flight.

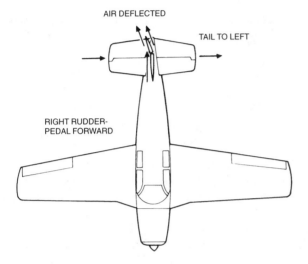

FIGURE 5–20 Action of the rudder.

Elevators

The **elevators** are the control surfaces which govern the movement of the aircraft around the lateral axis (pitch). They are normally attached to hinges on the rear spar of the horizontal stabilizer. When the control wheel in the airplane is pulled back, the elevators are raised. The force of the relative wind on the elevator surfaces tends to press the tail down, thus causing the nose to pitch up and the angle of attack of the wings to increase. The reverse action takes place when the control wheel is pushed forward. The action of the elevators is illustrated in Figure 5–21.

During flight of an airplane the operation of the elevators is quite critical, especially at low speeds. When power is off and the airplane is gliding, the position of the elevators will determine whether the airplane dives, glides at the correct angle, or stalls. Remember that an airplane will not necessarily climb when the control is pulled back. It is the power developed by the engine that determines the rate of climb of an airplane rather than the position of the elevators. If the elevators are held in a fixed position, the throttle alone can be used to make the airplane climb, dive, or maintain level flight. The position of the elevator is important, however, to establish the most efficient rate of climb and a good gliding angle when power is off. It is also most essential for proper control when "breaking the glide" and holding the airplane in landing position.

A special type of elevator that combines the functions of the elevator and the horizontal stabilizer is called a **stabilator**. A stabilator is an all-moving tail that works by changing the angle of attack of the control surface and thereby changing the amount of downward lift that is generated by the tail. When this type of control airfoil is installed on an airplane, there is no fixed horizontal stabilizer. The stabilator is an airfoil that responds to the normal elevator control and serves as an elevator as well as a stabilizer. A stabilator is illustrated in Figure 5–22.

T-Tails

The **T-tail arrangement** positions the stabilizer and elevator at the top of the vertical fin. A T-tail is illustrated in Figure 5–23. The use of a T-tail configuration not only makes the fin and rudder more effective because of the end-plate action of the stabilizer location which act similar to the addition of an end plate on a wingtip, as was discussed in Chapter 4 (see Figure 4–22), but it also positions the hori-

skid. In order to prevent this skid in a turn, the ailerons are used to **bank** the airplane. In a car, a banked turn is much easier to negotiate at comparatively high speeds than a flat turn. It is the same with an airplane. To prevent skidding in a turn, the airplane must be banked.

Too much of a bank without sufficient rudder in a turn will cause slipping; that is, the airplane will slide down toward the inside of the turn. It is therefore necessary that the proper amount of rudder and aileron be applied when entering a turn in order to produce what is termed a **coordinated turn**. Usually, after the airplane is placed in a turn, the rudder pressure is almost neutralized to hold the turn. Likewise, it is necessary to reduce the amount of aileron used to place the airplane in the turn.

Another factor to note concerning turns is that the steeper the turn, the more the elevator will have to be used. Thus a properly executed turn requires the use of all three of the primary controls.

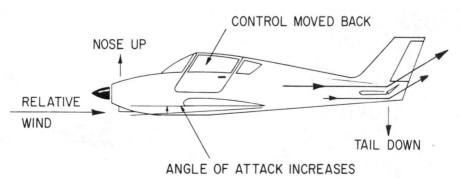

FIGURE 5–21 Action of elevators.

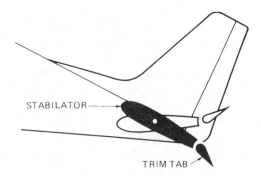

FIGURE 5–22 Stabilator.

zontal tail above wing turbulence. A T-tail structure will be somewhat heavier than a conventional tail arrangement due to combined horizontal tail-and-fin bending loads which must be carried by the fin and the fuselage.

Unusual Controls

Some airplanes have been designed with special types of control surfaces that do not fit into the descriptions of the conventional controls. One such control is called a ruddervator. The **ruddervator** is used on airplanes with a V-tail, and the surfaces serve both as rudders and as elevators. A V-tail has a slight drag reduction due to the reduction of interference drag, since there is one less intersection than on a conventional tail. However, since the total surface area must be the same as on a conventional tail, there is no reduction in skin-friction drag. A disadvantage of the V-tail is that the heavier tail structure necessary to support combined horizontal and vertical surface loading along with a somewhat heavier control system make the V-tail generally as heavy as the conventional design it would replace. The other disadvantage to a V-tail is that it is susceptible to roll tendencies, and the stability characteristics are somewhat less desirable, particularly in rough air.

With a ruddervator, when a pilot wants to increase the angle of attack, he or she pulls back the control wheel and both ruddervators move upward and inward, as shown in Figure 5–24. When the wheel is pushed forward, the ruddervators move downward and outward, as illustrated.

If a pilot wants to turn an airplane with ruddervators and the right rudder is applied, the right ruddervator moves downward and outward while the left ruddervator moves upward

FIGURE 5–23 Beechcraft Duchess. *(Beech Aircraft Co.)*

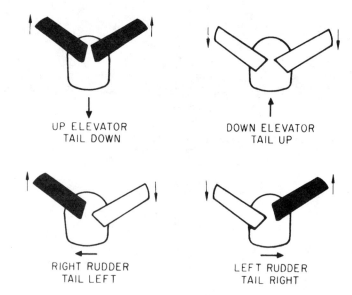

UP ELEVATOR
TAIL DOWN

DOWN ELEVATOR
TAIL UP

RIGHT RUDDER
TAIL LEFT

LEFT RUDDER
TAIL RIGHT

FIGURE 5–24 Operation of ruddervators.

and inward. These movements are in response to the movement of the rudder pedals and provide the forces necessary to rotate the airplane about the vertical axis. The turning action of the ruddervators is also illustrated in Figure 5–24.

Another somewhat unconventional control is the elevon. **Elevons** are combination elevators and ailerons used on the outer tips of some delta wings. When used as elevators, they both move in the same direction; when used as ailerons, they move in opposite directions. Elevons are especially needed for all-wing airplanes, or "flying wings."

Ailerons that are rigged to serve as ailerons or flaps are called **flaperons**. When employed as flaps, flaperons on opposite wings move either upward or downward together. When employed as ailerons, the flaperons move in opposite directions. The use of flaperons allows the wings to vary in camber or curvature. By varying the wings' camber, the pilot gives the aircraft better performance capabilities over a wider operating range. The use of ailerons that can be drooped to change a wing's camber and, in effect, function as flaperons is becoming popular on transport-category aircraft. This allows the entire trailing edge to be equipped with flaps to vary the camber.

Trim Tabs

The term **trim tabs** describes small secondary flight-control surfaces set into the trailing edges of the primary control surfaces. Tabs are used to reduce the work load required to hold the aircraft in some constant attitude by "loading" the control surface to a neutral or trimmed-center position. Figure 5–25 demonstrates the tab action. Tabs can be fixed or

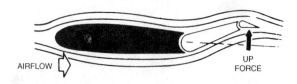

AIRFLOW

UP
FORCE

FIGURE 5–25 Effect of trim tabs.

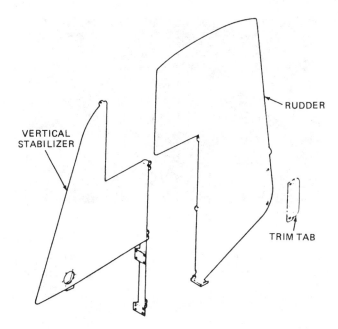

FIGURE 5–26 A fixed trim tab is adjusted on the ground for the average flight condition. *(Ayres Corp.)*

variable, and the variable tabs can be designed to operate in several different manners.

Fixed Trim Tabs

A **fixed trim tab**, such as is shown in Figure 5–26, is normally a piece of sheet metal attached to the trailing edge of a control surface. This fixed tab is adjusted on the ground by bending it in the appropriate direction to eliminate flight-control forces for a specific flight condition. The fixed tab is normally adjusted for zero-control forces in cruise flight. Adjustment of the tab is a trial-and-error process where the aircraft must be flown and the trim tab adjusted based on the pilot's report. The aircraft must then be flown again to see if further adjustment is necessary. Fixed tabs, normally found on light aircraft, are used to adjust rudders and ailerons.

Controllable Trim Tabs

A **controllable trim tab** is illustrated in Figure 5–27. Controllable tabs are adjusted by means of control wheels, knobs, or cranks in the cockpit, and an indicator is supplied to denote the position of the tab.

Controllable trim tabs are found on most aircraft with at least the elevator tab being controlled. These tabs may be operated mechanically, electrically, or hydraulically. When the trim-control system is activated, the trim tab is

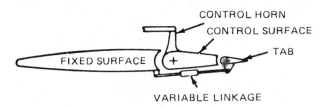

FIGURE 5–27 Controllable trim tab.

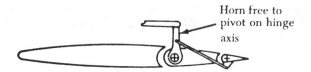

FIGURE 5–28 Servo tab.

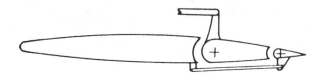

FIGURE 5–29 Balance tab.

deflected in the direction opposite to the desired movement of the control surface. When the trim tab is deflected into the airstream, the air tries to push the tab back flush with the control surface. Since the control mechanism prevents the tab from being pushed back flush, the control surface will be moved.

Servo Tabs

The **servo tabs**, sometimes referred to as the **flight tabs**, are used primarily on the large main control surfaces. A servo tab is one that is directly operated by the primary controls of the airplane. In response to movement of the cockpit control, only the servo tab moves. The force of the airflow on the servo tab then moves the primary control surface. The servo tab, illustrated in Figure 5–28, is used to reduce the effort required to move the controls on a large airplane.

Balance Tabs

A **balance tab** is linked to the airplane in such a manner that a movement of the main control surface will give an opposite movement to the tab. Thus, the balance tab will assist in moving the main control surface. Balance tabs are particularly useful in reducing the effort required to move the control surfaces of a large airplane. A balance tab is illustrated in Figure 5–29.

Spring Tabs

The **spring tabs**, like some servo tabs, are usually found on large aircraft that require considerable force to move a control surface. The purpose of the spring tab is to provide a boost, thereby aiding in the movement of a control surface. On the spring tab, illustrated in Figure 5–30, the control horn is connected to the control surface by springs.

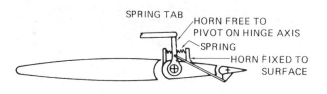

FIGURE 5–30 Spring tab.

AIRCRAFT DESIGN VARIATIONS

There are many shapes and sizes of airplanes, most of which are similar in appearance. In its traditional form, the airplane is marked by an arrangement of clearly distinguishable parts. The traditional design of the fuselage supported by wing lift, stabilized by the tail surfaces, and propelled by the engine in the nose has worked well over the years. However, many variations of the standard design appear to work equally well.

Canard Aircraft

The earliest powered aircraft, such as the *Wright Flyer* (see Figure 5–31), had horizontal surfaces located ahead of the wings. This configuration, also on the Beech Starship (Figure 5–32), which has two lifting surfaces, with the forward airfoil being called a **canard**, is an appealing way to assist in carrying some of the airplane weight to reduce drag and increase cruising speed.

Conventional airplane designs that have tail surfaces located behind the wing use the horizontal tail to balance the wing pitching moment. This means a down load on the tail, as previously discussed, and requires an increase in the lift coefficient to support the added wing load. Since wing drag increases with wing lift, a climb-and-cruise penalty is paid for the stability offered by an aft-located horizontal tail.

With the horizontal stabilizer being mounted forward, a nose-up balancing moment is provided by an upward-lifting force on the canard. The canard airplane has no stabilizing down loads because the canard, being mounted forward, shares the lifting loads with the wing. This lift adds to the wings' lift and results in a higher *L/D* ratio for the airplane. The canard design, with both surfaces providing lift, makes the aircraft somewhat unstable. This instability is referred to as **relaxed static stability**. The forward wing (canard) lifts

FIGURE 5–32 Beech Starship. *(Beech Aircraft Co.)*

a greater share of the total weight per square foot of wing area (i.e., it has a heavier wing loading) than the aft wing. This is achieved by having the center of gravity well ahead of the aft wing. The aft wing pitching moment also adds to the foreplane load.

In a well-designed canard, the forward wing must always stall at a lower angle of attack than the aft wing. If the aft wing were to stall first, the aircraft would pitch up, deepening the stall. With the canard stalling at a lower angle, the aircraft could be flown with the canard alternately stalling and unstalling, the nose bobbing up and down gently in a porpoising mode. The CG location in a canard-equipped aircraft is very critical, with the requirement being that the CG always be located between the canard and the main wing. Center of gravity travel is also very limited.

Forward-Swept Wing

Another concept in aircraft design is that of the **forward-swept wing**, as illustrated in Figure 5–33. Forward-swept wings achieve the same result as aft-swept wings in achieving higher critical Mach numbers; however, forward-swept

FIGURE 5–31 *Wright Flyer.*

FIGURE 5–33 X-29 Forward-Swept Wing. *(Grumman Aircraft Corp.)*

wings do not suffer the problems with spanwise flow and the resulting wingtip stall characteristics. The 30° forward-swept wing of the X-29 provides drag reductions of up to 20% in the transonic maneuvering range, giving it performance equivalent to an aircraft with a more powerful engine. As illustrated in Figure 5–34, air moving over the forward-swept wing tends to flow inward rather than outward, allowing the wing tips to remain unstalled at high angles of attack and therefore easier to control in extreme maneuvers. Forward-swept wings provide less drag, more lift, better maneuverability, and more efficient cruise speed. These improvements in performance are gained at the expense of reduced lateral and longitudinal stability.

To control an aircraft designed with relaxed stability, the flight-control system must provide an artificial stability. This is accomplished with a digital fly-by-wire flight-control system. A fly-by-wire system enables the control surfaces of an airplane to be operated electronically through a computer system. The pilot moves the aircraft's stick, sending a command to the flight-control computer. The computer calculates the control surface movements necessary and sends a command to the actuator to move the control surfaces.

Forward-swept wings present a serious structural problem with the wingtips tending to flex upward as lift is increased. Only with the recent use of composite materials that are capable of absorbing this flexing tendency is forward-swept wing design possible.

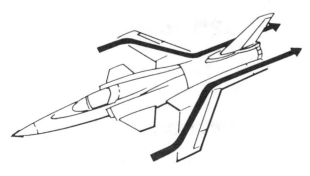

FIGURE 5–34 Airflow over forward-swept wing.

Flying Wing

A concept that has been tried several times but with only limited success until recently is the flying-wing design, which is shown in Figure 5–35. On a conventional aircraft, the fuselage is used to carry passengers and cargo. For the most part, fuselages create no lift, but add greatly to the production of drag. Therefore, if the need for fuselages could be eliminated, the *L/D* ratio of the aircraft could be greatly increased. This is the concept of the flying-wing design, which eliminates the fuselage. Significant performance improvement can be achieved with this design. There are, however, serious stability and control problems that must be overcome. Pitch stability is one of the most serious problems since there is no horizontal stabilizer to overcome the positive pitching moment of the wing. Stabilizing the wing-pitching moment can be achieved by sweeping the wing aft and twisting the wingtips to a negative angle so that they will apply a negative lifting force. Directional stability is also a problem since the effectiveness of a vertical fin is directly proportional to the distance that it is located aft of the CG. Since there is no fuselage located aft on which to locate the vertical fin or rudder, the surface area of these controls must be dramatically increased. Although stability on the flying-wing design is relaxed, the recent advent of computerized artificial stability systems makes the flying wing a viable concept.

AIRFOILS ON BIPLANES

Biplane Pressure Interference

Even though it is recognized that the biplane type of airplane is not used as extensively today as it has been in the past, still there are many such airplanes in operation, and it is important for the technician to understand some of the details of construction and the aerodynamic characteristics of such aircraft.

When air flows over and under the wing of a monoplane, there is decreased pressure above the wing and increased pressure below. The region of high pressure seeks the region of low pressure, but these two regions can merge only at the tips and at the trailing edge. If the upper and lower wings of a biplane were far enough apart, the airflow would be the same for each of the two wings of the biplane as it would be for the wings of the monoplane, disregarding the effect of the struts connecting the wings of the biplane.

In actual practice, however, the wings of the biplane are so close together that the interference of the streamlines reduces the comparatively low pressure on the upper surface of the lower wing, because air always attempts to flow from a region of high pressure to a region of low pressure. The lift of both wings is therefore reduced, but the lower wing loses more lift than the upper wing, and the loss is so great that a biplane is usually less efficient than a monoplane that has an equivalent wing area.

In general, the greater the distance between the wings of a biplane, the smaller the loss of lift due to **interplane**

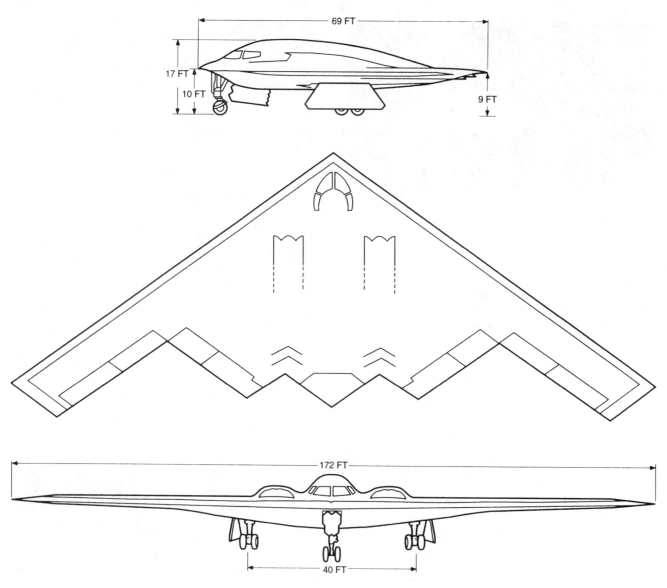

FIGURE 5–35 Northrop B-2 flying-wing design. *(Northrop Corp.)*

interference. The inverse is also true; that is, the closer the wings are together, the greater will be the interplane interference. The gap/chord ratio, stagger, and decalage are all involved in interplane interference.

Gap/Chord Ratio

The **gap** is the distance between the leading edges of the upper and lower wings of a biplane, as shown in Figure 5–36, and is measured perpendicular to the longitudinal axis of the airplane. The gap is sometimes defined as the distance separating two adjacent wings of a multiplane. These definitions mean the same thing.

The **chord** has been defined and explained before in this text. In addition to other definitions, it may be defined as the straight line tangent to the lower surface of the airfoil at two points or as the straight line between the trailing edge and the imaginary perpendicular line at the leading edge. The chord is indicated in Figure 5–36. Instead of the gap of a biplane, it is customary to give the **gap/chord ratio**. If the gap/chord ratio is 1, it means that the gap and the chord have the same length. In practice, the gap/chord ratio is usually close to 1. The principal determining factor for the gap/chord ratio is the interplane interference. The upper and lower wing should be as near to each other as possible for structural reasons and yet still be far enough apart so that interplane interference is kept at a minimum.

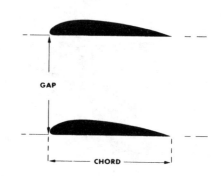

FIGURE 5–36 Gap and chord.

Gap/Span Ratio

Some textbooks on aerodynamics mention the **gap/span ratio**. This is the ratio of the gap to the span, but the gap is always much less than the span; therefore, the ratio is always less than 1.

Stagger

The term **stagger** is defined as the difference in the longitudinal position of the axes of two wings of an airplane. In simple words, stagger is the amount which the leading edge of one wing of a biplane is ahead of the leading edge of the other wing.

The upper biplane of Figure 5–37 has **positive stagger**, because the leading edge of the upper wing is ahead of the leading edge of the lower wing. The lower biplane of Figure 5–37 has a **negative stagger**, because the leading edge of the upper wing is behind the leading edge of the lower wing. Stagger is expressed in inches, percentage of chord length, or degrees.

When stagger is expressed in degrees, a line is drawn between the leading edges, and the angle that this line makes with a line drawn perpendicular to the chord of the upper wing is the **angle of stagger**. For example, in Figure 5–38 the angle formed by a line connecting the leading edges of the wings and a line drawn perpendicular to the chord of the upper wing is 23°. Since the leading edge of the upper wing is ahead of the leading edge of the lower wing, there is 23° positive stagger.

The aerodynamic advantages of stagger are small. A biplane may have stagger to improve the vision of the pilot and provide better access to the cockpit.

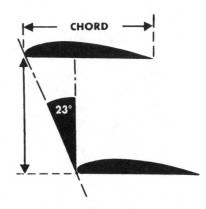

FIGURE 5–38 Stagger expressed in degrees.

Decalage

The term **decalage** describes the angular difference between the mean aerodynamic chords of the wings of a biplane. The **angle of wing setting**, or **chord angle**, is the same thing as the angle of incidence. Therefore, decalage is the difference between the angles of incidence of the wings of a biplane.

The decalage is measured by the angle (less than a right angle) between the chords in a plane parallel to the plane of symmetry. The decalage is considered positive if the upper wing of a biplane is set at the larger angle of incidence. In Figure 5–39, there is an angle of incidence for the upper wing but none for the lower wing; therefore, there is a positive angle of decalage, which, in this particular case, happens to be the same as the angle of incidence of the upper wing.

In Figure 5–40, the lower wing has an angle of incidence but the upper wing has none; therefore, there is a negative angle of decalage. In this particular case, the angle of incidence of the lower wing is the angle of decalage. If the upper wing were set at an angle of incidence of 3° and the lower wing were set at an angle of incidence of 2°, there would be a positive angle of decalage of 1° (3° − 2°).

If the chords of the upper and lower wings of a biplane are parallel, the downwash of the upper wing has the effect of decreasing the angle of attack of the lower wing. Setting the lower wing at a greater angle of incidence will more

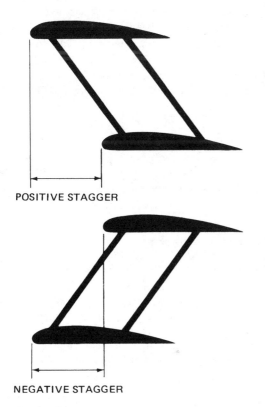

POSITIVE STAGGER

NEGATIVE STAGGER

FIGURE 5–37 Stagger.

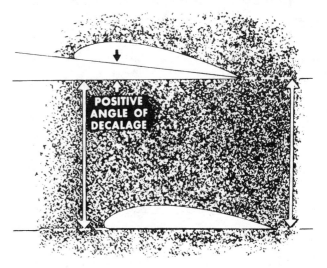

POSITIVE ANGLE OF DECALAGE

FIGURE 5–39 Positive decalage.

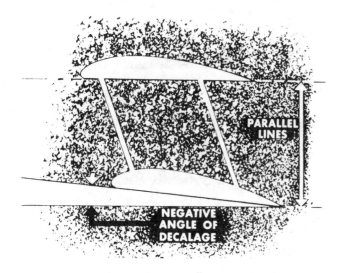

FIGURE 5–40 Negative decalage.

properly distribute the lift between the two wings. Since the upper and lower wings then have different angles of incidence, they have different angles of attack in flight, and there will be a difference of pressure distribution between them. Positive decalage gives the upper wing an increase in load percentage, especially at high speed. Negative decalage gives the lower wing an increase in load percentage. Each of the wings will reach its burble point at a different angle of attack, with a result that a stall will be less abrupt. On lift curves, this condition is shown by a flatter peak.

Biplane Loads

A biplane is constructed with external bracing between the wings to support a large part of the loads that occur during flight and landing. These external supporting members are shown in Figure 5–41. During flight, the **flying wires** are under a high-tension stress and the **interplane struts** are subjected to compression stress. Upon landing, the **landing wires** are under tension and the wing struts are still under compression. The **cabane struts** are always under compression, and the **cross-brace wires** are under tension.

Monoplanes Compared With Biplanes

The early airplane designers and builders reasoned that a biplane would provide more lifting surface than a monoplane of the same span, but they soon found that drag was caused by interplane interference and also by the struts and wires.

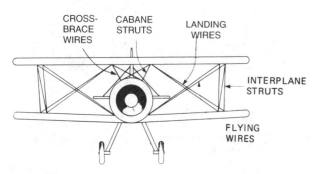

FIGURE 5–41 Load-bearing members in a biplane.

Many designers were aware that a cantilever (unbraced) monoplane with tapered wings was superior on a basis of aerodynamic performance, but they had not yet developed a deep airfoil section that would eliminate the need for external bracing. Furthermore, they lacked materials that would provide the necessary structural strength and rigidity for an unbraced wing.

The early biplane designers also argued that even if they had a deep airfoil section and the proper materials for building cantilever wings, the biplane was structurally more efficient and was more maneuverable. There were other arguments that were extremely important, although they were seldom mentioned. One was that those making biplanes had a vast amount of money invested in their equipment. Another was that the engineers and mechanics were reluctant to change from familiar procedures to untried methods. For all these reasons, the conversion from biplanes to monoplanes was a slow process. Nevertheless, today the monoplane is the accepted type of aircraft, although a few new biplanes are still being produced for specialty uses, such as acrobatic or agricultural use.

HELICOPTERS

One of the most versatile and useful aircraft for a wide variety of applications is the helicopter. The main difference between a helicopter and an airplane is the main source of lift. The airplane derives its lift from a fixed airfoil surface, while the helicopter derives lift from a rotating airfoil called the rotor.

The word **helicopter** is derived from the Greek words meaning "helical wing" or "rotating wing." The **rotating wing** (main rotor) of a helicopter has two or more blades, depending upon the design and size of the helicopter. Each blade is an airfoil having a profile similar to that of an airplane wing. The same laws of aerodynamics that apply to other airfoils apply to the helicopter rotor blades.

The great usefulness of the helicopter is its ability to fly straight up, sideways, forward, or backward, or to remain still in a hovering position. Because of its ability to land in almost any small, clear area, the helicopter is used for air-taxi service, police work, intercity mail and passenger service, power-line patrolling, construction work, fire fighting, agricultural work, air-sea rescue, and a variety of other services.

Helicopter Flight

The helicopter flies in accordance with the same laws of aerodynamics that govern the flight of a conventional airplane. In the helicopter, however, these laws are applied differently. The helicopter is subject to the same four forces that affect other aircraft: lift, weight, thrust, and drag. **Lift** supports the **weight** of the helicopter, and **thrust** overcomes the **drag** and moves the helicopter in the direction desired.

If the lift vector is broken down into components perpendicular and parallel to the ground, as shown in Figure 5–42, there will be a vertical lift component to support the heli-

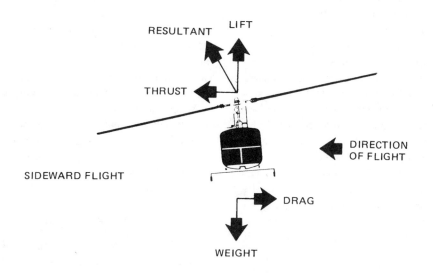

SIDEWARD FLIGHT

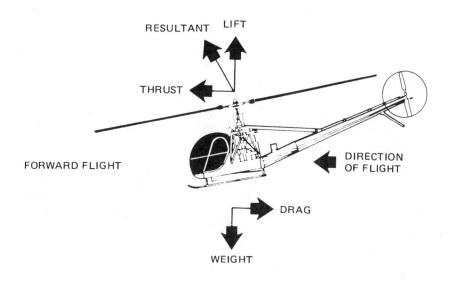

FORWARD FLIGHT

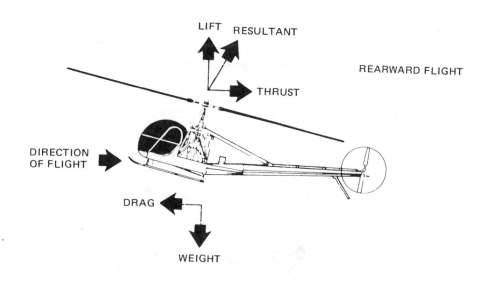

REARWARD FLIGHT

FIGURE 5–42 Conditions for horizontal flight in a helicopter.

copter and a horizontal thrust component producing directional flight. The helicopter will move in the direction of rotor tilt. A helicopter can fly in any direction with reference to the heading. For example, the helicopter can be headed north and flying south, east, west, or any other direction. Furthermore, it can move straight up or straight down, or it can remain stationary. When the helicopter is in stationary flight, it is said to be **hovering**. If the helicopter is hovering in a no-wind condition, the plane of rotation of the rotor is horizontal, or parallel, with the level ground. During hovering, the sum of the lift and thrust of the helicopter is equal to the sum of the weight and drag. This is illustrated in Figure 5–43. During vertical flight, when the sum of the lift and thrust is greater than the sum of the weight and drag, the helicopter will rise. If the sum of the weight and drag is greater than the sum of the lift and thrust, the helicopter will descend.

For forward, rearward, or sideways flight, the rotor must be tilted in the direction in which the pilot wants to fly. The

conditions for various types of horizontal flight are illustrated in Figure 5–42. When flying forward, the rotor is tilted forward, as shown in the illustration. The forces of flight are resolved into vertical and horizontal components. If the helicopter is flying forward in a straight and level path, the thrust is equal to or greater than the drag in the horizontal direction, and the lift is equal to the weight in the vertical direction.

When the helicopter is flying rearward, sideways, or in any other horizontal direction, the thrust is acting in the direction in which the vehicle is moving, and the drag is acting in the opposite direction. Lift and weight always act in the vertical direction.

Conditions Affecting Rotor Operation

The main rotor, often called a **rotary wing**, is subject to a number of special conditions that must be given consideration in the design of a helicopter. Among these are centrifugal force, dissymmetry of lift, gyroscopic precession, torque, and Coriolis effect.

Before a helicopter takes off and when the rotor is turning, **centrifugal force** is the main force acting on the rotor blades. At this time the rotor blades are in a horizontal position. As the blade pitch is increased and power is applied to the rotor, the lift of the blades increases and the helicopter rises. The lift force of the blades causes the blade tips to rise above the horizontal plane and rotate through a conical path. This effect is called **coning** of the rotor blades and is illustrated in Figure 5–44. The balancing force is the centrifugal force generated by rotor rotation. As rpm is increased, the centrifugal force increases. The centrifugal force seeks to hold the blade in a horizontal plane, while lift exerts an upward force, as is shown in Figure 5–45. Centrifugal force is predominant and the blade will assume a position closer to horizontal than vertical. The lift created by one blade results in a force equal to approximately 10% of the centrifugal force. As a result, the blades will attain a coning angle at a balance point of the two forces. The coning angle will vary with the blade loading. The coning angle of a lightly loaded

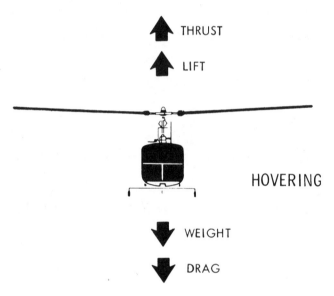

FIGURE 5–43 Forces acting on a helicopter while hovering.

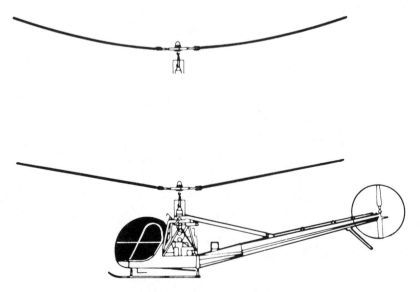

FIGURE 5–44 Coning of rotor blades.

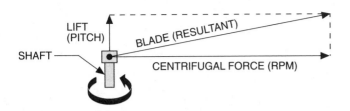

FIGURE 5–45 Resultant of lift and centrifugal forces. (*Sikorsky Aircraft*)

helicopter will not be as great as the angle for a heavily loaded helicopter, where greater lift is required.

When a helicopter is equipped with an **articulated rotor** having flapping hinges, the blades remain straight as they assume the coning position. If the rotor does not have flapping hinges, as in a rigid or semirigid rotor, the blades bend a limited amount when the helicopter is in flight. The plane of rotation of the rotor is often referred to as the **tip-path plane**.

Dissymmetry of Lift

If a helicopter is hovering and no surface wind is blowing, the tip speed of the rotor is uniform at all points, as is shown in Figure 5–46. However, as a helicopter is moved in directional flight, this is not the case. Assuming that the tip speed of the rotor blades is 500 mph and the helicopter is flying in a forward direction at 100 mph, a relative wind of 100 mph is flowing into the rotor, as is shown in Figure 5–47. Notice that over the right side of the helicopter the rotor blade has a 500-mph airspeed, which is caused by its rotational velocity at that point. In addition, the aircraft's forward-velocity relative wind is flowing over the blade at that point. The relative wind has a value of 100 mph. The air flowing over the blade at that point is thus traveling at 600 mph. Over the nose and tail of the aircraft the airspeed has returned to approximately 500 mph. On the left side of the helicopter the rotor blade has a 500 mph rotational velocity. The 500 mph airspeed that this rotational velocity would normally pro-

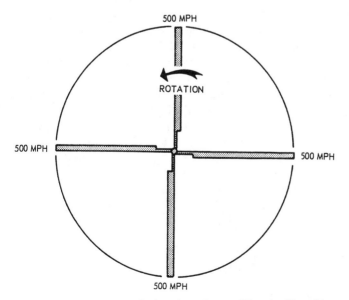

FIGURE 5–46 Tip speed when hovering. (*Sikorsky Aircraft*)

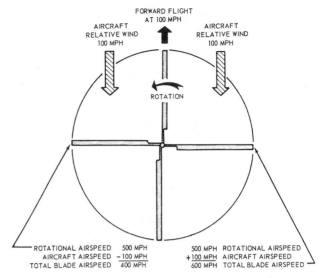

FIGURE 5–47 Tip velocity in forward flight. (*Sikorsky Aircraft*)

duce has been modified, however, by the fact that the aircraft relative wind, flowing at 100 mph, is flowing in the same direction the blade is moving. The 100 mph aircraft airspeed must therefore be subtracted from the normal 500 mph rotational-velocity airspeed. This results in a total blade airspeed of only 400 mph over the left side of the helicopter. It should be noted that a blade traveling from the tail to the nose around the right side of the helicopter is termed the **advancing blade** because it is advancing into the relative wind. The blade traveling from the nose to the tail around the left side of the helicopter is called the **retreating blade**.

The difference in airspeed of the advancing and retreating rotor blades is referred to as **dissymmetry of lift**. Dissymmetry of lift gives the helicopter a tendency to pitch up and roll to the left. This effect is compensated for by the design of the rotor, which permits **blade flapping**, and by the design of the cyclic-pitch-control system. An **articulated rotor** is designed with hinges at the root of each blade, which permits the blades to move up and down. When the helicopter is flying horizontally, the increased lift on an advancing blade causes it to flap up. The upward movement of the blade reduces the angle of attack because the relative-wind direction with respect to the blade is more downward than before. This decreases the lift of the blade. At the same time, the retreating blade of the rotor flaps downward, thus producing a greater angle of attack and increased lift. The combined effect of decreased lift on the advancing blade and increased lift on the retreating blade tends to equalize the lift on the two sides of the rotor disk. A **semirigid rotor** is not hinged at the hub, but the hub itself can tilt as required to produce the flapping action. Thus, as the advancing blade flaps up, the retreating blade flaps down. The design of the **cyclic-pitch-control system** also is such that it decreases the angle of attack of the advancing blade and increases the angle of attack of the retreating blade. The total effect is to make the lift of one side of the rotor equal to the lift of the other side. Additional information on rotor-head design can be found in an accompanying text in this series, *Aircraft Maintenance and Repair*.

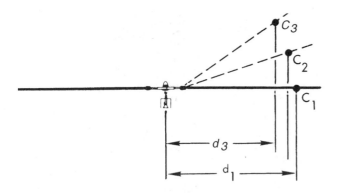

FIGURE 5–48 Movement of center of mass as blades rise.

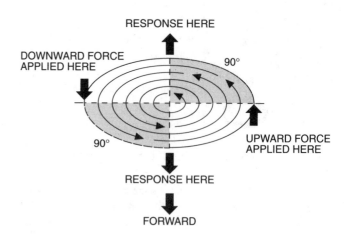

FIGURE 5–49 Gyroscopic precession.   (Sikorsky Aircraft)

The flapping of rotor blades produces what is known as the **Coriolis effect**, which occurs as the center of mass of the rotor moves closer to the center of rotation when the blades rise. An examination of Figure 5–48 shows why the center of mass (C_1, C_2, C_3) moves toward the center of the rotor disk as the blades rise. The Coriolis effect is the tendency of a rotor to increase the speed of rotation as the distance of the center of mass of each blade from the center of rotation decreases.

The result of the Coriolis effect on a rotor is that the advancing blade moves forward (leads) and the retreating blade moves aft (lags) in respect to the blade attach point. The direction of this **leading and lagging** (hunting) is parallel to the plane of the rotor disk. The movement is absorbed by the vertical drag hinges or by dampers and the structure of the blades. Two-blade rotors that are underslung and semirigid are not greatly influenced by the Coriolis effect.

Gyroscopic Precession

The term **gyroscopic precession** describes an inherent quality of rotating bodies in which an applied force is manifested 90° in the direction of rotation from the point where the force is applied. Since the rotor of a helicopter has a relatively large diameter and turns at several hundred revolutions per minute, precession is a prime factor in controlling the rotor operation.

The **cyclic-pitch control** causes a variation in the pitch of the rotor blades as they rotate about the circle of the tip-path plane. The purpose of this pitch change is, in part, to cause the rotor disk to tilt in the direction in which it is desired to make the helicopter move. When only the aerodynamic effects of the blades are considered, it would seem that when the pitch of the blades is high, the lift would be high, and the blade would rise. Thus, if the blades had high pitch as they passed through one side of the rotor disk and low pitch as they passed through the other side of the disk, the side of the disk having the high pitch should rise, and the side having the low pitch should fall. This would be true except for gyroscopic precession.

Gyroscopic precession is caused by a combination of a spinning force and an applied acceleration force perpendicular to the spinning force. Figure 5–49 is a drawing of a spinning disk that represents the main rotor of a helicopter. If the disk is spinning in the direction indicated by the arrow and a force is applied upward at 3:00, the disk will precess

(move) in the direction shown at 12:00. Thus, if a force is applied perpendicular to the plane of rotation, the precession will cause the force to take effect 90° from the applied force, in the direction of rotation.

As a result of the foregoing principle, if the pilot wants the main rotor of a helicopter to tilt in a particular direction, the applied force must be at an angular displacement of 90° ahead of the desired direction of tilt. The required force is applied aerodynamically by changing the pitch of the blades through the cyclic-pitch control. When the cyclic control is pushed forward, the blade at the left increases in pitch as the blade on the right decreases in pitch. This applies an "up" force to the left-hand side of the rotor disk, but the up movement takes place 90° in the direction of rotation. The up movement is therefore at the rear of the rotor plane, and the rotor tilts forward. This applies a forward thrust and causes the helicopter to move forward. The action is illustrated in Figure 5–50.

Blade Stall

In studying the effect of forward aircraft speed on the helicopter rotor, there is a limit of speed beyond which the helicopter cannot fly. As the forward speed of the helicopter increases, the relative-wind velocity on the retreating rotor blade decreases, and in order to maintain its share of the lift, it must continue to increase its angle of attack. In addition, the load is shifting toward the tips of the blades. At a certain relative-wind velocity, the retreating blade will stall, owing to the high angle of attack the blade requires to maintain lift equal to that of the advancing blade. The stall begins at the tip of the blade and works inward as forward speed increases. A typical blade-tip stall pattern is shown in the shaded area of Figure 5–51. A helicopter stall is typically evidenced first by a vibration. If the stall is severe enough, the helicopter will pitch, nose up, because even though the helicopter has lost lift on the left-side retreating blade (due to gyroscopic precession), the effect will take place approximately over the tail rotor.

Ground Effect

When a helicopter is hovering near the ground, the downward stream of air strikes the ground and does not escape

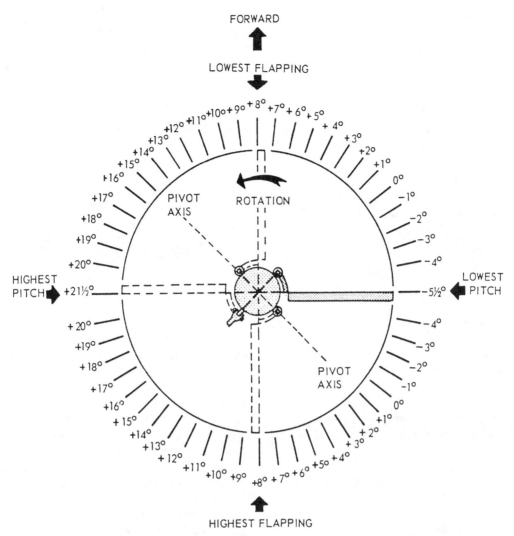

FIGURE 5–50 Gyroscopic effect on cyclic-pitch input. *(Sikorsky Aircraft)*

from beneath the helicopter as rapidly as it is being driven toward the ground. This causes a buildup of air pressure below the helicopter, which acts as a cushion to help support the machine in the hovering position. The ground cushion is usually effective to a height of approximately one-half the diameter of the main rotor while the helicopter is hovering. When the helicopter moves horizontally at 3 to 5 kn [1.54 to 2.57 m/s], the ground cushion is left behind.

Translational Lift

A helicopter requires a good portion of available power to maintain a hover. The power is used to give momentum or acceleration to the mass of air moving through the rotor system, which, in turn, produces an upward thrust. When a helicopter is moving horizontally in flight at more than 15 kn [7.7 m/s], the performance of the main rotor improves, owing to the increased volume of air passing through it. This effect is called **translational lift** because the lift of the rotor increases. Less engine power is required to maintain flight when the helicopter is flying horizontally than when it is hovering.

The Tail Rotor

According to Newton's third law of motion, for every force there is an equal and opposite force or reaction. The torque force applied to the rotor shaft of a helicopter to turn the rotor causes an equal and opposite **torque force**, which would turn the fuselage of the helicopter in the opposite direction unless measures were taken to prevent it. This is the function of the **antitorque rotor**, or **tail rotor**. Since the main rotor turns to the left (counterclockwise), as viewed from the top on American-made helicopters, the torque force causes the helicopter to turn to the right, as is shown in Figure 5–52. Tail-rotor force must therefore be applied to the right to keep the heading steady.

The combined effect of the tail-rotor thrust and the main-rotor torque is to apply a net force that causes the helicopter to drift to the right, as is illustrated in Figure 5–53. This force is called the **lateral drift tendency** or translating tendency. This drift is corrected by rigging the main-rotor mast so that the main rotor will apply a force slightly to the left, or by designing the cyclic-pitch control to provide a slight tilt of the tip-path plane to the left.

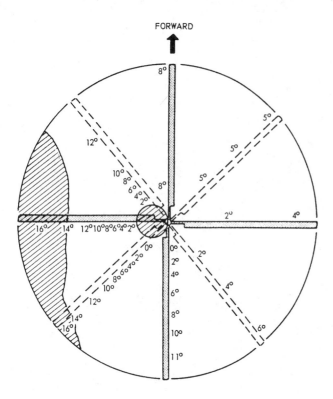

FIGURE 5–51 Blade-stall pattern.   *(Sikorsky Aircraft)*

HELICOPTER CONTROLS

The **collective-pitch control** increases or decreases the pitch of all the main-rotor blades simultaneously or collec-

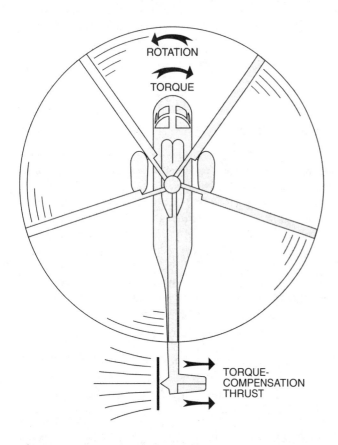

FIGURE 5–52 Torque compensation.   *(Sikorsky Aircraft)*

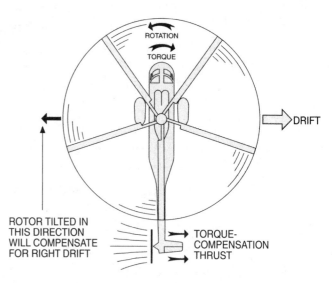

FIGURE 5–53 Lateral-drift compensation.   *(Sikorsky Aircraft)*

tively. Therefore, when a pilot wants to raise the helicopter from the ground, he or she increases the collective pitch, thereby increasing lift to all blades evenly. The collective-pitch control is a lever (stick) that is usually situated at the pilot's left. The control is coordinated with the throttle control of the engine so that engine power will be increased as the lever is raised to increase the collective pitch. In addition, the lever has a motorcycle-type grip that can be rotated to make additional adjustments to engine power. The grip is rotated counterclockwise to decrease power and clockwise to increase power.

The **cyclic-pitch control** causes a variation of the blade pitch as each blade rotates through the tip-path plane. This is accomplished through the tilting of the stationary star (swashplate) shown in Figure 5–54. The purpose of the cyclic-pitch control, as mentioned previously, is to cause the tip-path plane of the main rotor to tilt as required to provide for movement of the helicopter in a desired direction.

The collective and cyclic controls operate through a mixing box where the two control inputs are integrated (mixed) together. These inputs are then transmitted to a star, or swashplate, assembly. The bottom stationary star is linked to the levers of the control mixing box. Through the linkage, the star assembly is raised, lowered, and tilted forward and rearward or to either side. The top star assembly turns with the main rotor. When the star assembly is parallel to the plane of rotation, the pitch of the rotor blades is equal and uniform throughout rotation. When the star assembly is raised or lowered, the pitch of all the blades is changed uniformly, as is shown in Figures 5–55 and 5–56. In Figure 5–55 the collective stick would be at full low pitch, and the cyclic stick would be centered. The star assembly is at the low position and level. In Figure 5–56 the collective stick has been raised to its full high position, and the cyclic stick is still centered. The star assembly is at its highest point and the blades are at high pitch.

When the star assembly is tilted, the pitch of the blades changes throughout the circle of rotation. On one side of the rotor disk the pitch will be decreasing, and on the other side

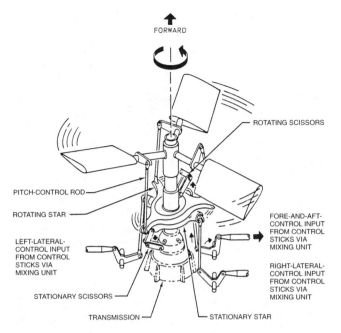

FIGURE 5–54 Cyclic-pitch control. *(Sikorsky Aircraft)*

of the disk the pitch will be increasing. If the cyclic control stick is moved aft, the star assembly is tilted. The lowest point of the tilted star assembly is over the stationary-scissors assembly, while the highest is over a position 180° opposite, or at the fore-and-aft rod. This means that the rotating star is now rotating in an inclined circle and, as the blade-pitch control rods rotate, they are constantly undergoing an up-and-down motion. Because the rods are attached to the rotor blades, the blades will constantly be changing pitch, changing from a high pitch value to a low pitch value each revolution. Because of the use of the mixing unit, both the collective and cyclic controls may be used in any combination or mixture desired.

The helicopter can be moved in any direction desired merely by moving the cyclic-pitch control in that direction.

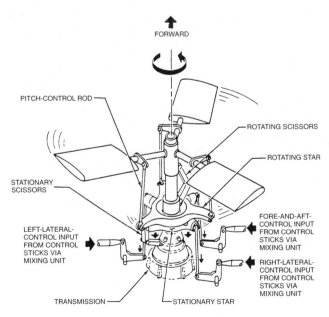

FIGURE 5–55 Low pitch. *(Sikorsky Aircraft)*

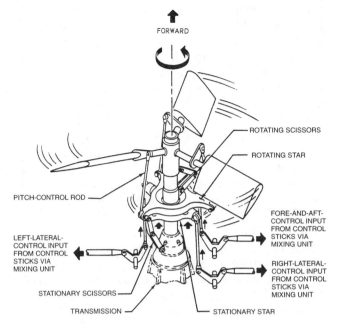

FIGURE 5–56 High pitch. *(Sikorsky Aircraft)*

The main rotor tilts in the direction called for by the control, and the helicopter moves as directed. If the control is in neutral, the helicopter will hover, that is, remain stationary in the air. In a wind, the helicopter will drift in the direction in which the wind is blowing unless sufficient rotor thrust is applied to cancel the effect of the wind.

The **heading control** for a helicopter is similar to the rudder control for a conventional aircraft. The "rudder" pedals in a helicopter are not really rudder pedals because they control the pitch of the tail rotor rather than the deflection of a rudder. The tail rotor is often called an **antitorque rotor**, and the pedals are called **antitorque pedals**.

During straight and level flight, the pitch of the tail rotor is such that the rotor provides a thrust that exactly counterbalances the torque of the main rotor. If the pilot wants to turn the helicopter to the right, she or he decreases the pitch of the tail rotor by pressing the right pedal. The torque of the main rotor then causes the helicopter to turn to the right. If the pilot wants to turn to the left, she or he presses the left pedal and the tail-rotor pitch is increased. This causes the additional thrust needed to push the tail to the right, and the helicopter then turns left.

It must be emphasized that the tail-rotor pedals in a helicopter do not control the direction in which the helicopter is flying. They control only the direction in which the fuselage is headed. Direction of flight is controlled through the cyclic-pitch system, as previously described.

Autorotation

The helicopter must incorporate a safety feature to provide for the condition that exists in the event of power failure. This feature is called **autorotation** and is required before a helicopter can be certificated by the Federal Aviation Administration. If power failure occurs, the engine is automatically disengaged from the rotor system through a free-wheeling device associated with the transmission. This disconnect device, the first step in the autorotation safety feature, will eliminate the engine drag from the rotor system as well as preventing further damage to the engine.

The second step required for autorotation is to provide for adequate windmilling of the rotor during descent and to create enough inertia in the rotor system so the pilot can apply sufficient collective pitch to cushion the landing. Autorotation is accomplished by aerodynamic forces resulting from an upward rotor inflow created by the descent of the helicopter. The turning of the rotor generates lift, which makes it possible to continue controlled flight while descending to a safe landing. Remember that during autorotation, aerodynamic force, not engine force, is driving the rotor.

If engine failure occurs, the pilot immediately lowers the collective-pitch control, thus reducing the pitch of all rotor blades simultaneously. The cyclic-pitch control is moved forward to establish the best forward speed for autorotation. Each helicopter has a characteristic forward speed, which produces maximum lift and lowest rate of descent.

Once the collective pitch is at the low-pitch limit, the rotor revolutions per minute can be increased only by a sacrifice in altitude or airspeed. If insufficient altitude is available to exchange for rotor speed, a hard landing is inevitable. Sufficient rotor rotational energy must be available to permit adding collective pitch to reduce the helicopter's rate of descent before final ground contact.

At low altitudes and low forward velocities, power failure in a helicopter is hazardous because of the difficulty in establishing sufficient autorotational lift to make a safe landing. Manufacturers provide **airspeed-versus-altitude limitations** charts to inform the pilot regarding the combinations of safe altitudes and speeds. A typical chart is shown in Figure 5–57. Note that it is comparatively safe to hover and fly at low speeds at very low altitudes. After attaining an indicated airspeed of 50 mph, or 44 kn [22 m/s], it is comparatively safe to fly at any altitude above 50 ft [16 m] because there is sufficient time to make the transition to the autorotation mode.

During autorotation, the outer 25% of the blades produces the lift, the section between 25 and 70% of the distance from the tip of the blades produces the driving force that keeps the rotor turning, and the inner 25 to 30% produces neither lift nor drive in any measurable degree.

Ground Resonance

There are two types of vibrations found in helicopters: ordinary and self-excited. Self-excited vibrations need no periodic external forces to start or maintain the vibration. Self-excited vibrations can cause the helicopter to rock fore and aft, or sideways with increasing magnitude, with the rotor blades weaving back and forth in the plane of rotation. This phenomenon is known as **ground resonance** because the vibration occurs when the aircraft is on the ground. Two conditions must exist to cause ground resonance. First, there must be some abnormal lead/lag condition which dynamically unbalances the rotor. Secondly, there must be a reac-

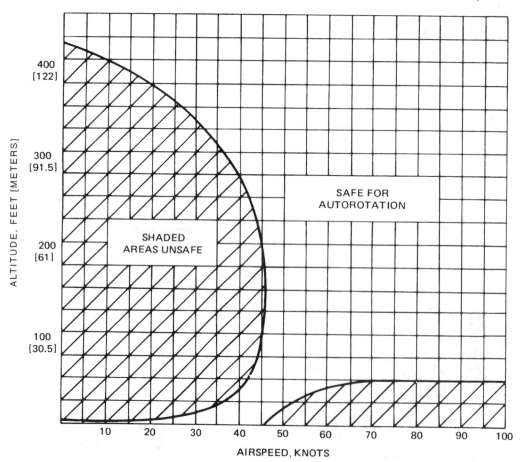

FIGURE 5–57 Airspeed-versus-altitude limitations chart.

FIGURE 5–58 Sikorsky's 76 Mark II Helicopter. *(Sikorsky Aircraft)*

tion between the helicopter and the ground through the tires or struts which would aggravate and further unbalance the rotor.

When resonance is suspected and rotor rpm is within operating range, the immediate application of power and a rapid takeoff will stop the condition.

HELICOPTER CONFIGURATIONS

As in the case of fixed-wing aircraft, helicopters have many different configurations. The most popular helicopter arrangement is that of the **single rotor** using a tail rotor, such as is illustrated in Figure 5–58. The single-rotor helicopter is relatively lightweight when compared to other configurations. This is due to its fairly simple design, with one rotor, one main transmission, and one set of controls.

The disadvantages of the single-rotor machine are its limited lifting and speed capabilities and a severe safety hazard during ground operation, with the tail rotor positioned several feet behind the pilot and out of the line of vision. In addition, a portion of the available engine power must be diverted from producing lift to provide antitorque correction.

Tail-Rotor Designs

The conventional tail rotor consists of a rotor of two or more blades located at the end of the tailboom. As already discussed, the tail rotor is used with single-rotor helicopters to counteract yawing movement resulting from the torque effect of the engine-transmission-main-rotor system. It does this by generating a side thrust at the end of the tailboom which tries to rotate the helicopter about its vertical axis. This yawing thrust is varied to counter the yawing movement imposed by the main-rotor torque. During cruising flight and hovering changes in altitude, the tail rotor corrects for torque. The tail rotor can also be used to change the heading of the helicopter when in a hover.

The Fenestron tail-rotor system uses what might be called a ducted-fan antitorque system. This system uses a multibladed fan mounted in the vertical fin on the end of the tailboom. This arrangement reduces aerodynamic losses due to blade-tip vortices, reduces the blanking of the antitorque control system by the vertical stabilizer, and reduces the chance of people walking into the tail rotor.

The **ring guard** is an adaptation of the conventional tail-rotor system which has a ring built around the tail rotor. This ring acts as a duct for the tail rotor, increases the safety of the tail rotor for ground personnel, and eliminates the vertical-stabilizer-blanking effect. These three types of antitorque controls are shown in Figure 5–59.

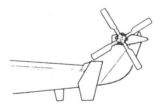

CONVENTIONAL TAIL ROTOR

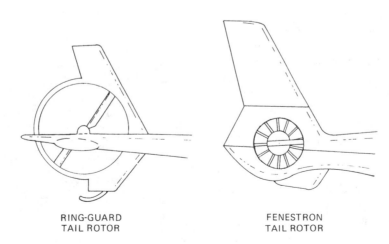

RING-GUARD
TAIL ROTOR

FENESTRON
TAIL ROTOR

FIGURE 5–59 Various types of tail rotors presently in use.

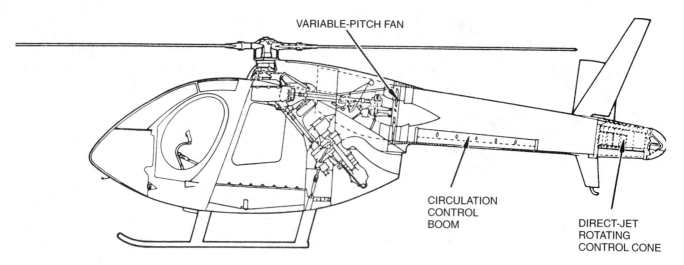

VARIABLE-PITCH FAN

CIRCULATION CONTROL BOOM

DIRECT-JET ROTATING CONTROL CONE

FIGURE 5–60 A NOTAR helicopter uses a flow of air instead of a tail rotor to control yawing motion.

The NOTAR system (NO Tail Rotor), illustrated in Figure 5–60, was developed by McDonnell-Douglas Helicopter Company to eliminate the hazards, maintenance, and noise of the conventional tail rotor. This system uses a ducted airflow in a tail cone to generate a predetermined amount of lift on one side of the tail cone and a controllable rotating cold-air exhaust duct to counteract variations in the main-rotor-system torque.

Tandem-Rotor Helicopters

A **tandem-rotor helicopter** manufactured by Boeing Vertol, a Division of Boeing Company, is shown in Figure 5–61. This helicopter uses two synchronized rotors turning in opposite directions. The opposite rotation of the rotors causes one rotor to cancel the torque of the other, thus eliminating the need for an antitorque rotor. Each rotor is fully articulated and has three blades. Climb or descent is accomplished by means of the collective-pitch control. When the collective-pitch lever is raised, the pitch of all six rotor blades is increased simultaneously, causing the helicopter to ascend. This control, therefore, operates similarly to the collective-pitch control for a single-rotor helicopter. Descent is accomplished by lowering the collective pitch control.

Directional control is achieved by tilting the plane of rotation of the rotors. Control motions to accomplish a turn are imparted by the control stick (cyclic-pitch control), the directional pedals, or both. These controls tilt the swashplates in the rotor controls which, in turn, raise or lower the pitch links. The pitch links vary the pitch of the rotor blades during the rotation cycle. Since the lift of the rotor blades is increased through part of the cycle and decreased through another part of the cycle, the plane of rotation is tilted. When the pilot applies directional-pedal-control movement in one direction, the plane of rotation of the forward rotor is tilted downward in that direction and the plane of rotation of the aft rotor is tilted downward in the opposite direction. This causes the helicopter to make a hovering turn around the vertical axis, as illustrated in Figure 5–62. The tandem helicopter is capable of lifting large loads, since these loads may be distributed be-

tween the two rotors. A disadvantage of the tandem-rotor helicopter is that it is not efficient in forward flight because one rotor is working in the wake of the other. This loss of lift may be minimized by placing the rear rotor above the main rotor.

Side-by-Side Rotor Helicopters

The **side-by-side helicopter** has two main rotors mounted on pylons or wings positioned out from the sides of the fuselage, as illustrated in Figure 5–63. The side-by-side configuration has the rotors turning in opposite directions, which eliminates the need for a tail rotor.

The advantages of the side-by-side configuration are that it has excellent stability and that the rotors are more efficient in forward flight than in the tandem arrangement, due to the fact that one rotor is not in the wake of the other. The side-by-side helicopter has the disadvantages of having high parasitic drag and high structural weight, both resulting from the structure necessary to support the main rotors.

Coaxial Rotor Helicopters

In the **coaxial helicopter**, illustrated in Figure 5–64, fuselage torque is eliminated by using two counter-rotating rigid

FIGURE 5–61 Tandem-rotor helicopter. *(Boeing Vertol, Division of Boeing Commercial Aircraft Co.)*

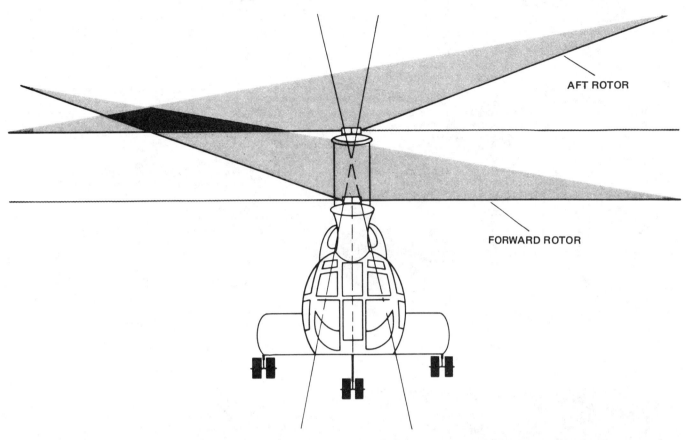

FIGURE 5–62 Rotor-plane movement during a turn.   (*Boeing Vertol, Division of Boeing Commercial Aircraft Co.*)

main rotors mounted one above the other on a common shaft. This type of configuration is also referred to as the **advancing-blade concept** because the lift load at high forward speed is carried primarily by the advancing blades. It combines the advantages of a low-speed helicopter with those of a high-speed aircraft, without the need for a wing and without the need for a tail rotor.

By avoiding retreating-blade stall, which is a key limiting factor in the speed and maneuverability of a pure helicopter, the coaxial helicopter has been made faster and more maneuverable. This design has successfully demonstrated forward speeds of more than 250 kn [129 m/s].

Tilt-Rotor Aircraft

After many years of research, the technology in the areas of composites and electronic flight-control systems has finally been developed to make a functional **tilt-rotor aircraft** a reality. The tilt rotor has the ability to combine the vertical takeoff low-speed capabilities of the helicopter with the high-speed performance of a turboprop airplane.

The V-22 Osprey, illustrated in Figure 5–65, resembles a twin turboprop with large-diameter propellers, or rotors, mounted on wingtip nacelles. The engines rotate from a vertical position for helicopter flight to a horizontal position for cruise flight.

Lift in the vertical, or hover, position is provided totally from the 38-ft- [11.6-cm-] diameter **prop rotors**. When in the horizontal, or cruise, position, lift is provided by the wings. The tilt rotor operates effectively with the engines between the vertical and horizontal positions, which provides for a wide range of lift and speed combinations. When operated in this conversion configuration, lift is provided from both the wings and the prop rotors. Full conversion from the hover to the fixed-wing mode can be accomplished in approximately 12 s.

X-Wing Aircraft

The **X-wing aircraft** illustrated in Figure 5–66 is a concept model for a vertical takeoff and landing aircraft which uses a four-bladed, helicopter-like rotor system that rotates for hover and low-speed flight and stops at approximately 200 kn [103 m/s] to become a fixed-wing aircraft for high-speed flight. The vehicle then flies as a fixed-wing aircraft while accelerating to speeds of approximately 450 to 500 kn [232 to 257 m/s] using an auxiliary propulsion system.

In addition to employing computerized flight controls and the latest in composite technology, the X-wing aircraft relies on an air-circulation system, which is essential to the X-wing concept. This system is comprised of a compressor, control valves, and ducts to the leading and trailing edges of the

Helicopter Configurations   **125**

FIGURE 5–63 Russian Mi-12 helicopter. *(General Electric)*

FIGURE 5–64 Sikorsky S-69 Advancing-Blade-Concept Demonstrator. *(Sikorsky Aircraft)*

FIGURE 5–66 Sikorsky X-Wing Aircraft. *(Sikorsky Aircraft)*

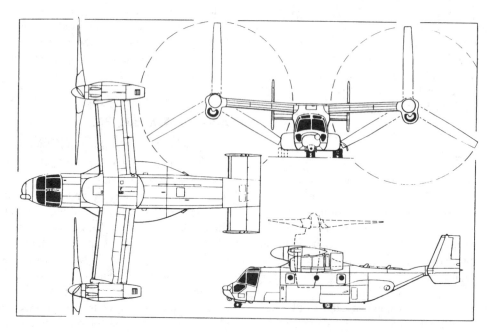

FIGURE 5–65 V-22 Osprey tilt-rotor aircraft.

rotor/wing. The compressor feeds pressurized air through hollow leading and trailing edges to slots in the rotor/wing. The air is blown out the slots over the rounded leading and trailing edges of the symmetrically shaped rotor/wing, providing circulation-controlled lift. Lift is controlled by altering the airflow through the slots, thus allowing the air-circulation system to substitute for collective control and cyclic control as well as for flaps and ailerons. During transition from a rotor/wing to a fixed-wing aircraft, all lift must be generated by the air-circulation system.

REVIEW QUESTIONS

1. What are the four forces that are applied to an airplane in flight?
2. How is an airplane's load factor determined?
3. In what categories may airplanes be type-certificated?
4. What is wing loading?
5. Define *static stability*.
6. Define *dynamic stability*.
7. Describe the three axes of an airplane.
8. Define *dihedral*.
9. How is directional stability accomplished on an aircraft?
10. What are the three primary control surfaces of an airplane?
11. Which primary control affects aircraft movement around each of the three axes?
12. What is a stabilator?
13. Explain the use of trim tabs.
14. How do the loads that are imposed on a canard vary from those that are imposed on a conventional tail?
15. Explain the term *stagger* as applied to a biplane.
16. Define the term *decalage* as applied to a biplane.
17. What is meant by the term *blade coning*?
18. What is the cause of dissymmetry of lift in a helicopter?
19. Describe the Coriolis effect and explain how it affects the rotor blades.
20. What is gyroscopic precession?
21. What factor limits the forward speed in a helicopter?
22. How is torque counteracted in a single-rotor helicopter?
23. What does the collective-pitch control do in a helicopter?
24. What does the cyclic-pitch control do in a helicopter?
25. Why is a tail rotor not required on a tandem-rotor helicopter?

6 Aircraft Drawings

INTRODUCTION

Engineering drawings and prints (copies of drawings) are essential tools in the design and manufacture of aircraft. An engineering drawing is used to describe an object by means of lines and symbols. With the use of a print or a drawing, an engineer can convey to those who build, inspect, operate, and maintain aircraft and spacecraft the necessary instructions for ordering the materials, making the parts, assembling the units, and finishing the surfaces. Drawings constitute the abbreviated written language of the aerospace industry, a shorthand method for presenting information that would take many pages of manuscript to transmit. The aviation maintenance technician must be able to correctly interpret the information on many types of drawings in technical reference manuals. In addition to engineering drawings, the technician will encounter schematic diagrams, installation and location drawings, and wiring charts. Reproduction of drawings was originally done by a process that produced a print of white lines on a blue background. The term *blueprint* was used for these prints. Today, the term **blueprint** is commonly used to refer to many types of drawings and prints, without regard to the production or copying process.

TYPES OF DRAWINGS

Production Drawings

A major use of engineering drawings is for the fabrication or assembly of components. Drawings used for this purpose are also called **production drawings** or **working drawings**. Production drawings can be categorized as detail drawings, assembly drawings, or installation drawings.

A **detail drawing** can consist of one part or several parts of an entire assembly. The detail drawing will provide, by the use of lines, notes, and symbols, all the specifications (size, shape, and material) needed to make the part. An example of a detail drawing is shown in Figure 6–1.

An **assembly drawing** is used to show how parts produced from detail drawings fit together to form a compo-

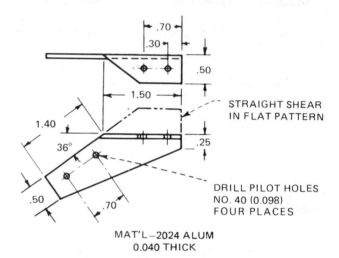

FIGURE 6–1 Detail drawing for an angle bracket.

nent. The assembly drawing does not show the dimensions of the detail parts except as necessary for location purposes. Figure 6–2 shows how the parts of an electrical connector are assembled. Figure 6–3 is an assembly drawing of a structural-panel fastener. The upper part of the drawing is a **pictorial drawing** of the three detail parts. A pictorial drawing is similar to a photograph and shows the parts as they appear to the eye. The pictorial portion of the drawing is often referred to as an **exploded view**. The detail parts are shown individually but arranged so as to indicate how they are assembled. The lower part of the drawing shows the panel fastener correctly installed into the structure. Illustrated parts catalogs use exploded-view assembly drawings for identification of parts and part numbers.

An **installation drawing** shows how a part or a component is installed in the aircraft. Figure 6–4 shows a bracket

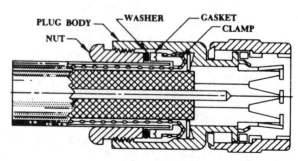

FIGURE 6–2 Assembly drawing of a cable connector.

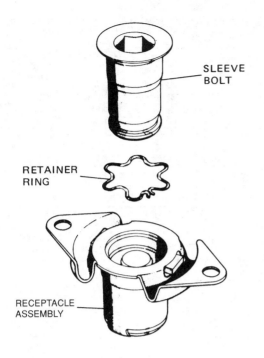

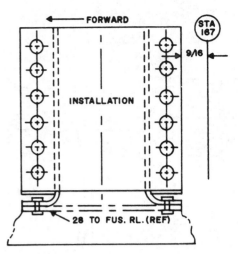

FIGURE 6–4 Fuselage bracket installation.

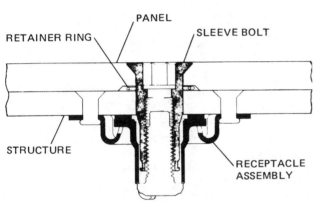

FIGURE 6–3 Assembly drawing of a structural-panel fastener. *(VOI-SHAN Division of VSI Corp.)*

installed at a specific point in an aircraft fuselage. Dimensions and directional information are provided in this drawing. Figure 6–5 shows the location and directional orientation of a door latch. Dimensional information is not provided, nor is it necessary, on this drawing. The individual parts of the latch are to be attached to existing structural assembly. The dimensional information for the installation and assembly of the structure is provided on other drawings.

Block Diagrams

A **block diagram** is a special drawing used to simplify the explanation of complex circuits. Block diagrams are widely used for electronic circuits but can be used for any type of aircraft system. A block diagram allows nonspecialized personnel to understand the function and relationship of various systems within a component. Various shapes may be used within the diagram to help explain function. Figure

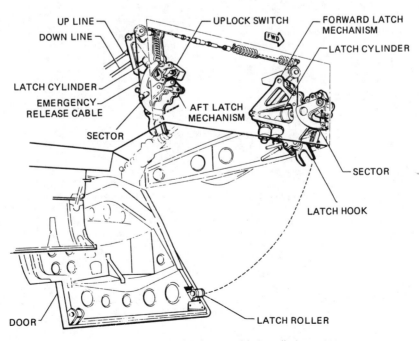

FIGURE 6–5 Door latch assembly installation.

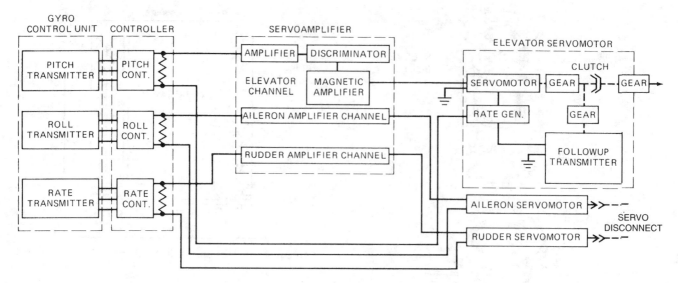

FIGURE 6–6 Block diagram of an autopilot system.

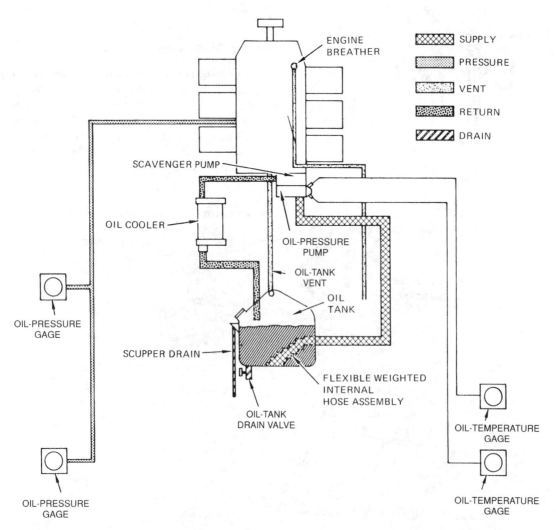

FIGURE 6–7 Lubrication system schematic.

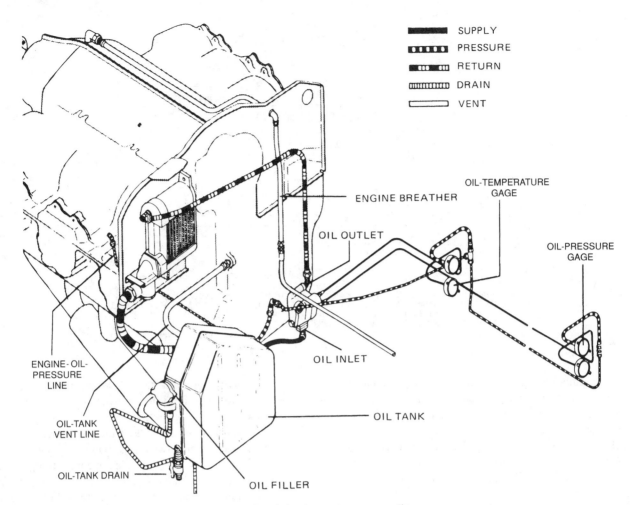

SUPPLY
PRESSURE
RETURN
DRAIN
VENT

ENGINE BREATHER

OIL-TEMPERATURE GAGE

OIL OUTLET

OIL-PRESSURE GAGE

ENGINE-OIL-PRESSURE LINE

OIL INLET

OIL-TANK VENT LINE

OIL TANK

OIL-TANK DRAIN

OIL FILLER

FIGURE 6–8 Lubrication system perspective.

6–6 is a block diagram of an autopilot system. Block diagrams are very useful for troubleshooting. By knowing the input and output for each component, a technician can identify and isolate those contributing to a malfunction.

Schematic Diagrams

A **schematic diagram** is also used to explain a system. A simple schematic shows the functional location of components within a system. This is done without regard to the physical location of the components in the aircraft. The flow of fluid in a lubrication system is indicated by the use of various types of shading, as shown in Figure 6–7. While some schematics are colored, black-and-white diagrams provide better copies and are usable for microfilm and microfiche.

Figure 6–8 is an illustration of the same lubrication system as in Figure 6–7. In this drawing, called a perspective drawing, the components and lines are shown in relation to their physical location in the aircraft installation. While the same items are shown in each drawing, it is apparent that tracing the system's flow is easier with the schematic. The perspective view is useful for locating the component in the aircraft once it has been identified from the schematic.

Shop Sketches

A **shop sketch** may be anything from a simple line drawing, such as that shown in Figure 6–9, to a rather complex and detailed drawing, such as a standard engineering drawing. The purpose of a shop sketch is to convey information concerning the repair of a part or structure, to illustrate a proposed modification, to provide information for engineering drafters from which they can make standard engineering drawings, and for various other uses where an illustration is necessary to convey technical information. Due care should be exercised in the preparation of a shop sketch so that it presents a good appearance, contains complete and accurate information, and conveys the information that it is intended to convey.

The aviation maintenance technician should develop as much skill as possible in preparing shop sketches. Such sketches are often necessary in preparing a repair proposal for approval by the FAA. The sketches become a part of the maintenance and repair records for repairs that are not documented by manufacturers or FAA publications.

Drawings for Electrical and Electronic Systems

Today's aircraft are equipped with extensive electrical circuitry and electronic units. The proper documentation of

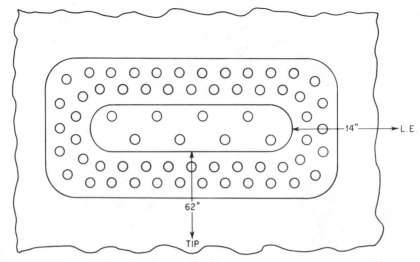

REPAIR OF CUT — BOTTOM, LEFT WING — 14" REAR OF L.E.–62" INBOARD OF WING TIP –
CUT 3 7/8" – DAMAGED MATERIAL CUT OUT–SKIN MATERIAL – 2024-T3 –0.032" ALCLAD –
PATCH SAME MATERIAL–RIVETS–AN 426-AD-4-3 –PATTERN EXCEEDS ORIGINAL IN STRENGTH
PATCH INSIDE WING–SKIN MADE FLUSH WITH PLUG

FIGURE 6–9 Shop sketch.

these circuits requires numerous and complex engineering drawings, wiring drawings, schematics, and so on. This section shows typical electrical- and electronic-system and circuit drawings that will be encountered by the maintenance technician. The use of wiring diagrams and schematic circuits in troubleshooting is covered in the associated text called *Aircraft Electricity and Electronics*.

Wiring Diagrams. The purpose of a wiring diagram is to show all the wires, wire segments, and connections in an electrical system or circuit. Figure 6–10 is a sample wiring diagram produced in accordance with Air Transport Association (ATA) specifications. It should be noted that every wire segment is identified by an alphanumerical code. The letters and numbers shown in the diagram are stamped on the wires in the aircraft at intervals of 15 in [38.1 cm] or less. The connections, to connector plugs or electrical units, are identified by letters or numbers.

Electrical Schematic Diagrams. An electrical schematic diagram from a manufacturer's service manual is shown in Figure 6–11. The wires and units are identified so relation of all the units of the electrical system may be easily observed.

The wire identification in this schematic uses letters and numbers, as shown in the chart of Figure 6–12. This system makes it possible to identify every segment of every wire in a circuit. If a particular unit malfunctions, the technician can quickly determine which wires are involved and make appropriate tests.

Logic Circuitry for Electronic Systems. Because of the complexity of electronic systems and the use of logic

systems involving solid-state electronic units, it has been necessary to simplify system drawings using symbols that have meaning to the electronic technician. Logic systems involve the use of binary mathematics. The binary system of mathematics uses only two digits, 1 and 0. If a circuit is conducting, the signal is 1, and if it is not conducting, the signal is 0. Thus a switch, transistor, or some other unit or combination of units can be used as a "gate" to provide the correct signal for the function involved.

As shown in Figure 6–13 on page 136, if an OR gate is in the circuit, the first input to become 1 produces a 1 in the output. For the symbols shown in the illustration, the input is on the left and the output is on the right. A NOR gate produces a 1 in the output if there are no 1s in the input. With the AND gate, all the inputs must be 1s to produce a 1 in the output. The NAND gate will produce a 1 signal in the output if any of the inputs is 0.

A diagram showing the use of logic symbols for a system is provided in Figure 6–14 on page 137. This illustration is included so you will be able to identify drawings containing logic circuits. Logic systems are covered in the associated text called *Aircraft Electricity and Electronics*. The diagram shown is called an integrated block test (IBT) and should be easily interpreted by a qualified technician.

DRAFTING TECHNIQUES

An aircraft maintenance technician does not have to be a skilled draftsperson. However, in order to correctly read and interpret drawings, a knowledge of the techniques used to graphically communicate technical information is essential. These techniques include the use of different types of views,

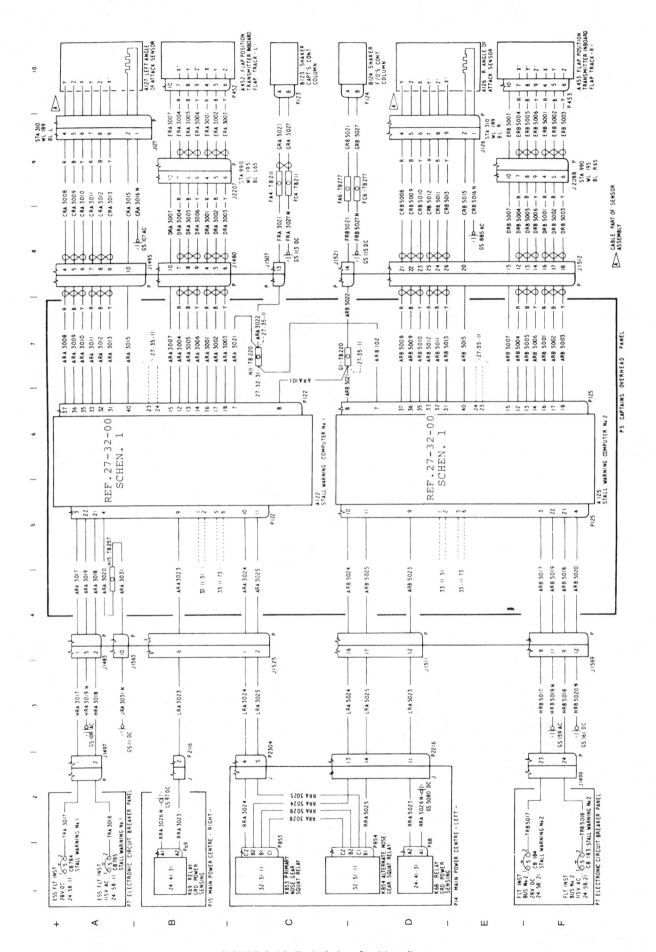

FIGURE 6–10 Typical aircraft wiring diagram.

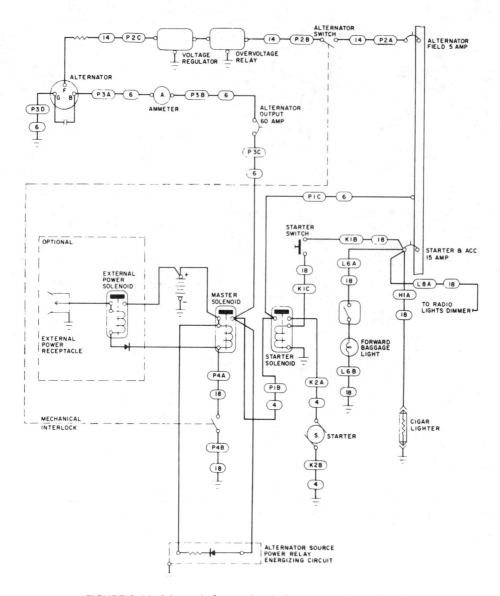

FIGURE 6–11 Schematic for an electrical system. *(Piper Aircraft Co.)*

lines, symbols, standard dimensioning practices, and title-block information.

Projections

Perspective Drawings. Figure 6–15 on page 138 is a photograph of an aircraft. Compare the photograph with the perspective drawing in Figure 6–16 on page 138 of the same aircraft. There is not any difference in the shape or arrangement of parts, but there is a difference in shading. Both illustrations show the aircraft as it would look to the eye, but they often do not present the detailed information needed by the aircraft maintenance technician. In a **perspective drawing**, parallel lines are shown as converging for the same reason that the rails of a railroad seem to meet in the distance. This produces an effect known as *foreshortening*. As a result of foreshortening, the various parts of the aircraft are not drawn in a true scale with each other. The perspective drawing is best used for overall views and for providing location information, such as in the lubrication system shown in Figure 6–8.

Oblique Views and Isometric Projections. Figure 6–17 on page 138 shows two different views of the same object, one perspective and the other oblique. The **oblique view** is similar to the perspective view, but the lines are drawn parallel and the length of each line is true to scale. Foreshortening has been eliminated, but optical illusion is still a problem. The lines are parallel and of the same length, yet the farthest one from the observer appears to be longer.

An **isometric projection** is similar to an oblique drawing but with less optical illusion than in the perspective oblique view. In this type of drawing, the object appears distorted, but does show equal distances on the subject as equal distances on the drawing, thus making dimensions more clear to the reader.

Figure 6–18 on page 138 is an isometric drawing. All vertical lines are drawn as verticals and all horizontal lines are drawn at an angle of 30°. It shows that certain lines are parallel to each other and at right angles to other lines. There is no foreshortening, and there is no convergence of lines as in the true perspective drawing. The isometric drawing provides a bird's-eye view, gives proportions, and is often very

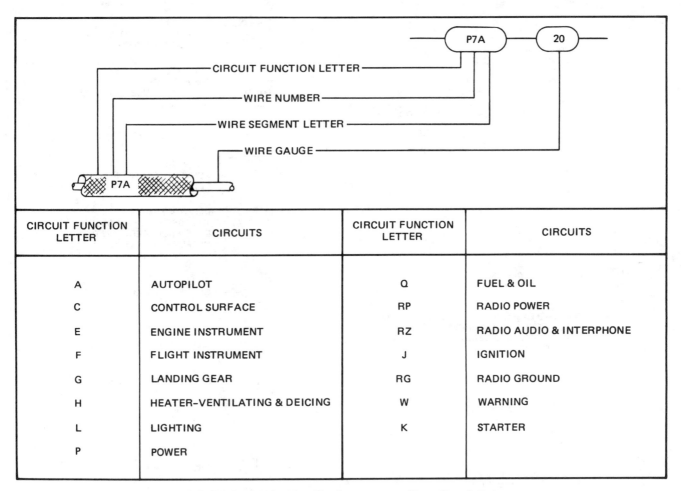

CIRCUIT FUNCTION LETTER	CIRCUITS	CIRCUIT FUNCTION LETTER	CIRCUITS
A	AUTOPILOT	Q	FUEL & OIL
C	CONTROL SURFACE	RP	RADIO POWER
E	ENGINE INSTRUMENT	RZ	RADIO AUDIO & INTERPHONE
F	FLIGHT INSTRUMENT	J	IGNITION
G	LANDING GEAR	RG	RADIO GROUND
H	HEATER–VENTILATING & DEICING	W	WARNING
L	LIGHTING	K	STARTER
P	POWER		

FIGURE 6–12 Wire identification system.   *(Piper Aircraft Co.)*

useful in explaining the design and construction of complicated assemblies. Its faults are that the shape of the object is distorted and the angles do not appear in their true size.

The term *isometric* is derived from two Greek words, *isos* and *metron*, meaning "equal" and "measure." Thus we see that isometric means "of equal measure."

Orthographic Projections. The three types of drawings we have already discussed (perspective, oblique, and isometric) all give a bird's-eye view of an object, but a print often must have more than one view, especially if the object shown is three-dimensional. It is also desirable that each view be presented without distortion in most cases.

Figure 6–19 on page 138 is an **orthographic projection** of a bracket. The word *orthographic* means "the projections of points on a plane by straight lines at right angles to the plane" and is derived from the Greek word *orthos*, meaning "straight." There is a front view, a top view, and a right-side view. The **front view** is what you would see if you were directly in front of the bracket; it would be impossible to see any of the top, the bottom, or the sides. The **top view** is what you would see if you were looking directly down on the bracket; you would not see the front, the bottom, or any of the other surfaces. The **right-side view** is what you would see if you looked directly at that side; you would not see the top, the bottom, or any other surface except the right side.

Figure 6–20 demonstrates three views for an orthographic projection. There are actually six views that can be used.

Not shown are the back, the left-side, and bottom views. For most objects, the three views shown are adequate to describe the object. By studying the three different views it is possible to determine the configuration of the object drawn. In order to interpret the views, one must understand the use of various types of lines.

The Meaning of Lines

Standards have been set for lines to be used in drafting, although minor variations will be found in the use of these standards among various companies or industries. Figure 6–21 on page 139 shows the types and forms of lines that are used for drawings. Figure 6–22 on page 139, an illustration of a flap-control unit, illustrates the use of many of these lines.

Most drawings use three widths, or intensities, of lines: **wide**, **medium**, and **narrow**. These line widths may vary somewhat on different drawings, but on any one drawing there will be a noticeable contrast between a wide line and a narrow line, with the medium line somewhere between.

The **visible outline (object line)** is a medium-to-wide line which should be the outstanding feature of the drawing. The thickness may vary to suit the drawing, but it should be at least 0.015 in [0.038 cm] wide. This line represents edges and surfaces that can be seen when the object is viewed directly.

The **invisible outline**, often called a **hidden line**, is a

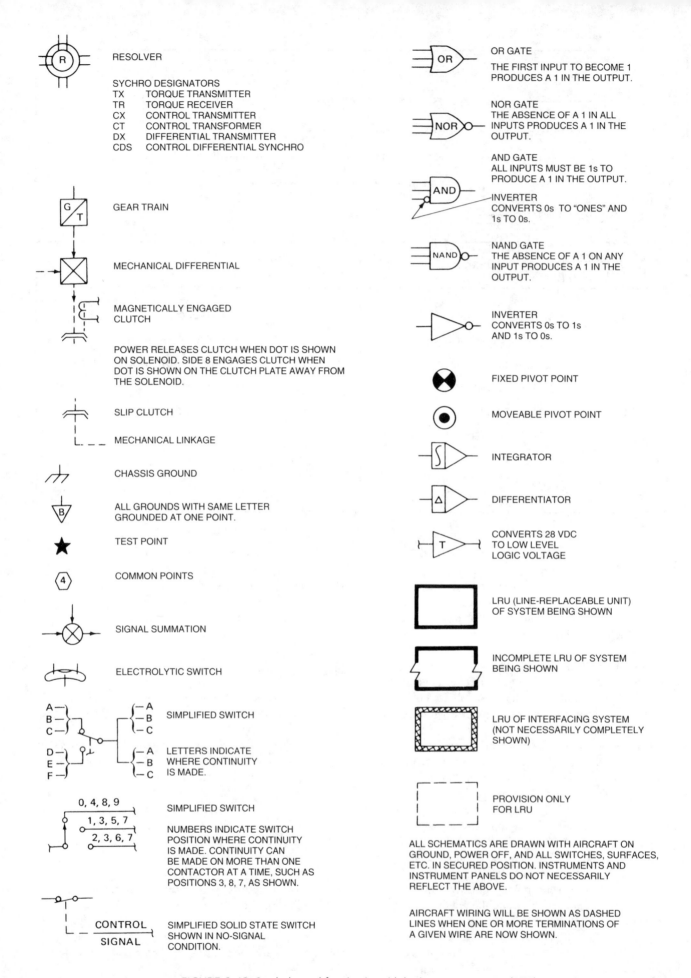

RESOLVER

SYCHRO DESIGNATORS
TX   TORQUE TRANSMITTER
TR   TORQUE RECEIVER
CX   CONTROL TRANSMITTER
CT   CONTROL TRANSFORMER
DX   DIFFERENTIAL TRANSMITTER
CDS   CONTROL DIFFERENTIAL SYNCHRO

GEAR TRAIN

MECHANICAL DIFFERENTIAL

MAGNETICALLY ENGAGED CLUTCH

POWER RELEASES CLUTCH WHEN DOT IS SHOWN ON SOLENOID. SIDE 8 ENGAGES CLUTCH WHEN DOT IS SHOWN ON THE CLUTCH PLATE AWAY FROM THE SOLENOID.

SLIP CLUTCH

MECHANICAL LINKAGE

CHASSIS GROUND

ALL GROUNDS WITH SAME LETTER GROUNDED AT ONE POINT.

TEST POINT

COMMON POINTS

SIGNAL SUMMATION

ELECTROLYTIC SWITCH

SIMPLIFIED SWITCH

LETTERS INDICATE WHERE CONTINUITY IS MADE.

SIMPLIFIED SWITCH

NUMBERS INDICATE SWITCH POSITION WHERE CONTINUITY IS MADE. CONTINUITY CAN BE MADE ON MORE THAN ONE CONTACTOR AT A TIME, SUCH AS POSITIONS 3, 8, 7, AS SHOWN.

SIMPLIFIED SOLID STATE SWITCH SHOWN IN NO-SIGNAL CONDITION.

CONTROL
SIGNAL

OR GATE
THE FIRST INPUT TO BECOME 1 PRODUCES A 1 IN THE OUTPUT.

NOR GATE
THE ABSENCE OF A 1 IN ALL INPUTS PRODUCES A 1 IN THE OUTPUT.

AND GATE
ALL INPUTS MUST BE 1s TO PRODUCE A 1 IN THE OUTPUT.

INVERTER
CONVERTS 0s TO "ONES" AND 1s TO 0s.

NAND GATE
THE ABSENCE OF A 1 ON ANY INPUT PRODUCES A 1 IN THE OUTPUT.

INVERTER
CONVERTS 0s TO 1s AND 1s TO 0s.

FIXED PIVOT POINT

MOVEABLE PIVOT POINT

INTEGRATOR

DIFFERENTIATOR

CONVERTS 28 VDC TO LOW LEVEL LOGIC VOLTAGE

LRU (LINE-REPLACEABLE UNIT) OF SYSTEM BEING SHOWN

INCOMPLETE LRU OF SYSTEM BEING SHOWN

LRU OF INTERFACING SYSTEM (NOT NECESSARILY COMPLETELY SHOWN)

PROVISION ONLY FOR LRU

ALL SCHEMATICS ARE DRAWN WITH AIRCRAFT ON GROUND, POWER OFF, AND ALL SWITCHES, SURFACES, ETC. IN SECURED POSITION. INSTRUMENTS AND INSTRUMENT PANELS DO NOT NECESSARILY REFLECT THE ABOVE.

AIRCRAFT WIRING WILL BE SHOWN AS DASHED LINES WHEN ONE OR MORE TERMINATIONS OF A GIVEN WIRE ARE NOW SHOWN.

FIGURE 6–13 Symbols used for circuits with logic components. *(ATA)*

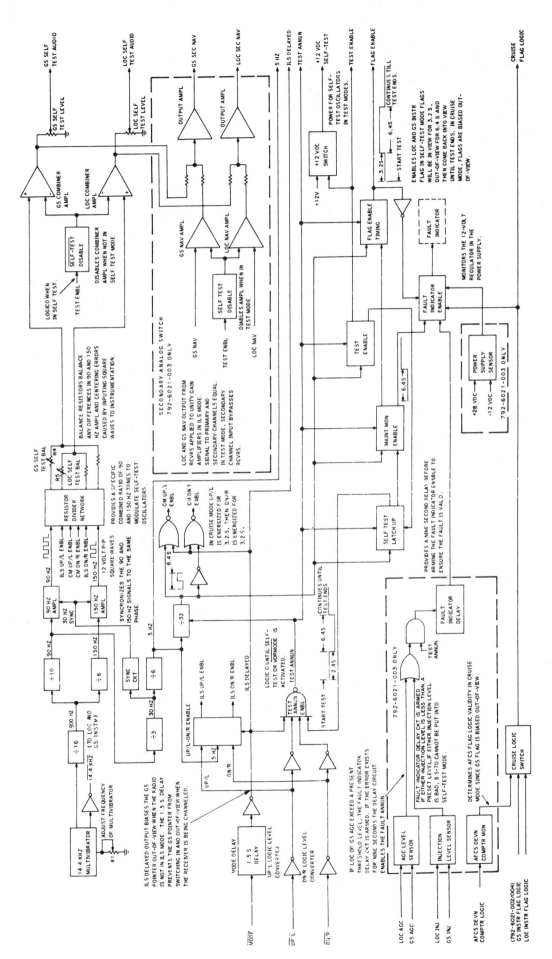

FIGURE 6–14 Integrated-block-test diagram. *(ATA)*

FIGURE 6–15 Photograph of an aircraft. *(Beech Aircraft Co.)*

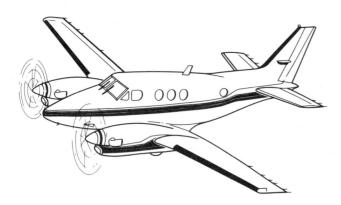

FIGURE 6–16 Perspective drawing of an aircraft.

medium-width line made up of short dashes. It represents edges and surfaces behind the surface being viewed and therefore not visible to the observer.

The **center line** is drawn narrow, or thin, and consists of alternate long and short dashes. It shows the location of the center of a hole, rod, symmetrical part, or symmetrical section of a part. Center lines are usually drawn first, and they provide the basic reference for the rest of the drawing.

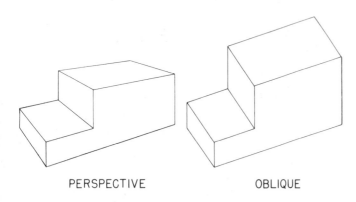

PERSPECTIVE   OBLIQUE

FIGURE 6–17 Perspective and oblique views of the same object.

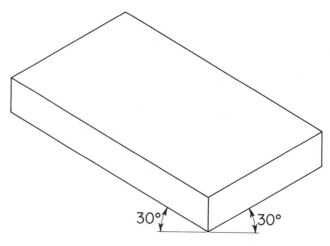

30°   30°

FIGURE 6–18 Isometric drawing.

The **adjacent-part line**, or **alternate-position line**, may be used to indicate the position of an adjacent part or to show alternate positions for an installed part. A **phantom line** is very similar and used on some drawings for the same general purpose. This line is a medium-width line and is usually made up of a series of long dashes with two short dashes between the long dashes. In some cases this line will be represented by a medium, broken line made up of long dashes. Examples of both types are shown in Figures 6–21 and 6–22. In Figure 6–22, two alternate positions are shown for the lever, using dashed lines. Lines made up of long and short dashes are used to illustrate the structure adjacent to the flap-control unit.

The **dimension line** is a narrow line and is unbroken except where a dimension is written in. Having determined the general shape of an object, the person using the blueprint also wants to know the size. The length, width, or height of a dimensioned part is customarily shown by a number placed in a break in the line.

An **extension line**, or **witness line**, is used to extend the line indicating an edge of the object for the purpose of dimensioning. The extension line is drawn very narrow, or thin.

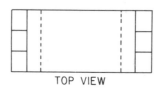

TOP VIEW

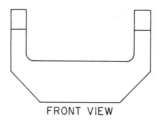

FRONT VIEW

RIGHT SIDE VIEW

FIGURE 6–19 Orthographic projection of a bracket.

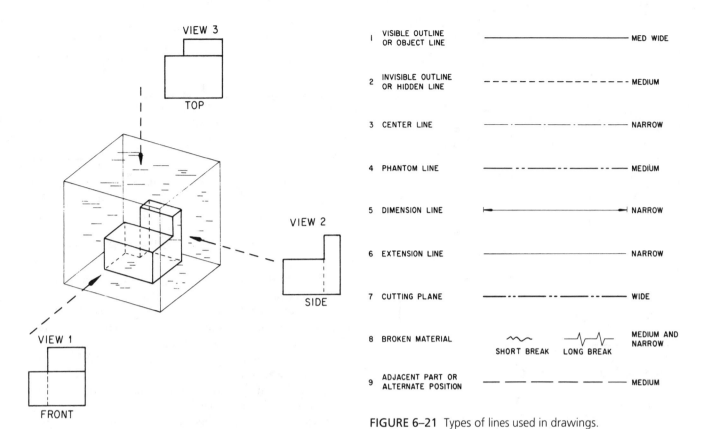

VIEW 3

TOP

VIEW 2

SIDE

VIEW 1

FRONT

FIGURE 6–20 Demonstration of an orthographic projection.

1	VISIBLE OUTLINE OR OBJECT LINE		MED WIDE
2	INVISIBLE OUTLINE OR HIDDEN LINE		MEDIUM
3	CENTER LINE		NARROW
4	PHANTOM LINE		MEDIUM
5	DIMENSION LINE		NARROW
6	EXTENSION LINE		NARROW
7	CUTTING PLANE		WIDE
8	BROKEN MATERIAL	SHORT BREAK LONG BREAK	MEDIUM AND NARROW
9	ADJACENT PART OR ALTERNATE POSITION		MEDIUM

FIGURE 6–21 Types of lines used in drawings.

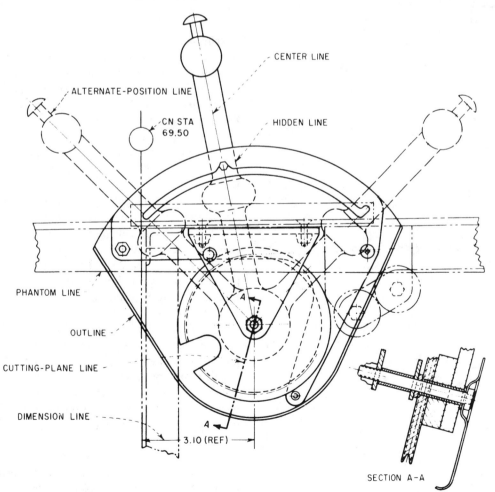

CENTER LINE

ALTERNATE-POSITION LINE

CN STA 69.50

HIDDEN LINE

PHANTOM LINE

OUTLINE

CUTTING-PLANE LINE

DIMENSION LINE

3.10 (REF)

SECTION A–A

FIGURE 6–22 Drawing showing the use of lines.

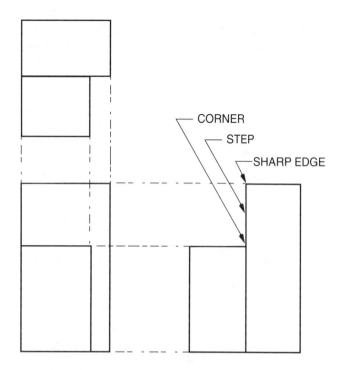

FIGURE 6–23 Projection of edges and corners.

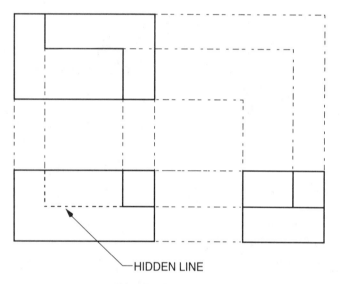

FIGURE 6–24 Use of hidden lines in a drawing.

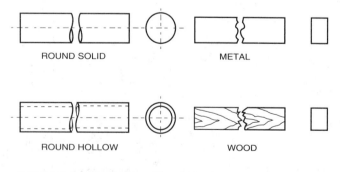

FIGURE 6–25 Conventional breaks in drawings.

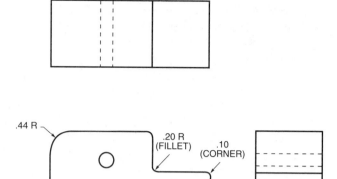

FIGURE 6–26 Curved surfaces in a three-view drawing.

The **cutting-plane line** is a heavy, wide, broken line usually made up of one long and two short dashes, alternately spaced. In some cases a series of heavy dashed lines are used for this purpose. It is used to refer to another view and direct attention to a cross section, or "slice," that reveals the interior. The cutting-plane line is easily recognizable because of the arrowhead at each end accompanied by a letter. The letters at the end of the line are used to reference the correct cross-sectional view.

When a surface has been cut away to reveal a hidden or inner feature of an object, **section lines** are used. These are narrow, solid lines spaced evenly to present a shaded effect. Figure 6–22 illustrates the use of section lines in section A-A. The shading or line pattern used to indicate a cutaway section usually varies according to the material from which the object is constructed. Some of the patterns used for various materials are shown in Figure 6–35 on page 146.

The **broken-material line**, often simply called a **break line**, can be used to save space on the drawing. Where the drawing would run off the paper if full size, a medium line is used to show that a portion of the drawing has been reduced. A narrow, ruled line with zigzags is used for long breaks, with a wider, freehand, wavy line used for shorter breaks. Other variations of break lines are shown in Figure 6–25.

The break line is also used to avoid the necessity of repeating the second half of a symmetrical part. This saves drafting time and has been done in the case of the pulley in Figure 6–22, section A-A. There is enough room to complete the drawing, but it is quicker not to, and sufficient information is given as shown.

Visible and Hidden Lines. Figure 6–23 shows that all edges or sharp corners are projected into other views as outlines or visible edges. Figure 6–24 shows that any line or edge that cannot be seen from one particular view may be shown as a hidden line (invisible edge) in that view by means of the appropriate line symbol. Hidden lines are often omitted for clarity if the drawing is clear without them.

Conventional Breaks. A pipe, tube, or a long bar having a uniform cross section is not always drawn for its entire length. When one or more pieces of the object are broken out and the ends moved together, a larger and more legible scale can be used. The true length will not be shown, but dimensions give the measurements to be used for the work. Different types of breaks may be used for different shapes and materials. Figure 6–25 shows the conventional breaks used in drawing a round, solid object; a round, hollow object; a metal object; and a wood object.

Curved Surfaces. Figure 6–26 illustrates the general rule that a curved surface or a circle appears in one view only in that form. In other views, the curve or the circle is shown as a straight line or by a pair of straight lines. For example, there is a hole through the object in Figure 6–26 which appears as a circle in the front view, as two parallel lines in the top view, and as two parallel lines in the right-side view. In the top and right-side views, the lines representing the hole are dashed lines because the hole cannot be seen from the top or the side of the object.

The fillet in Figure 6–26 is shown as a curved line in the front view and as a straight line in each of the other views. Since it is visible, it is represented by a solid, straight line in the top and right-side views and by a solid, curved line in the front view.

Selection of Views

Two-View Drawings. It is general practice to present three views of an object, but it is also common to illustrate such simple parts as bolts, collars, studs, and simple castings by means of only two views.

Figure 6–27 includes drawings of two different objects, with two views of each object. The one at the top is a simple

casting, illustrated by what may be called a front view and a top view, although the naming of the views may vary. The second drawing shows a tubular object, such as a bushing. In both drawings, a circle is shown in only one view, in accordance with the general rule previously explained.

Figure 6–28 also includes drawings of two different objects. The object at the top has invisible (hidden-line) circles. It is customary to use two-view drawings such as these to illustrate objects having either drilled holes or hidden circles. The object at the bottom shows the drilled holes as small circles in the front view and as dashed lines in the side view.

Single-View Drawings. Some objects are so simple that their shape can be shown by only one view. Figure 6–29 shows two objects of this type. The object at the top of the drawing is a cylindrical part with a groove that must be made according to the directions shown near the root of the arrow. The letter D indicates that the long cylindrical section has a uniform diameter throughout its length.

The object at the bottom of the illustration is a piece of sheet metal of regular shape; therefore, only one view is needed. Since the kind of material to be used and its thickness are important, these are given in coded language in a note at the bottom of the drawing.

Sectional Views. A **sectional view** is obtained by cutting away part of an object to show the shape and/or construction at the cutting plane. The surfaces that have been

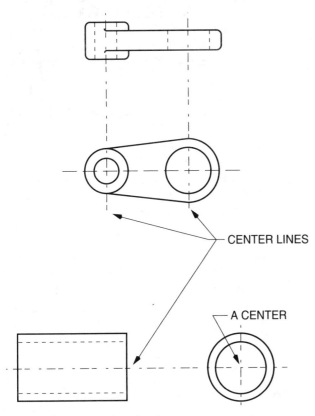

CENTER LINES

A CENTER

FIGURE 6–27 Two-view drawings.

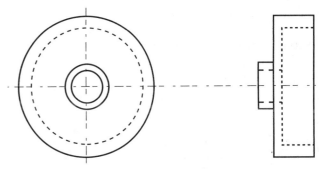

INVISIBLE CIRCLES

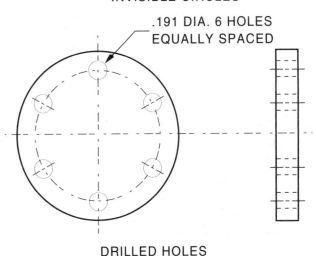

.191 DIA. 6 HOLES
EQUALLY SPACED

DRILLED HOLES

FIGURE 6–28 Two-view drawings showing hidden circles and drilled holes.

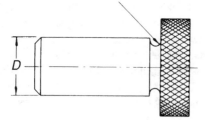

RECESS 3/32"DEEP X 1/8"

D

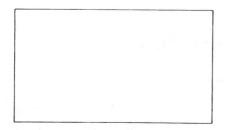

NOTE : MATERIAL .040 2024 - T3

FIGURE 6–29 Single-view drawings.

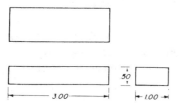

50

3.00

1.00

FIGURE 6–30 Dimension lines for a simple object.

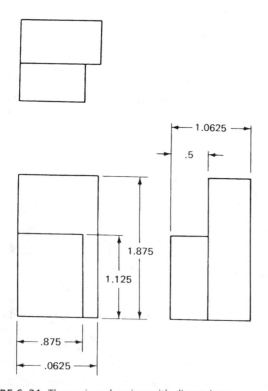

1.0625

.5

1.875

1.125

.875

.0625

FIGURE 6–31 Three-view drawing with dimensions.

cut away are indicated with section lines. Figure 6–2 on page 128 shows a sectioned view of an electrical connector. When the whole view has been sectioned, it is called a **full section**. A **half section** has the cutting plane extended only halfway across the object, leaving the other half in exterior view.

Detail Views. A **detail view** shows only a portion of the object but in greater detail than the principal view. The detail view may consist of a larger scale or a sectional view. The portion involved in the detail view will be indicated on the principal view. This indication may take the form of a letter or a cutting-plane line, in the case of a sectional detail view. The flap-control assembly drawing in Figure 6–22 shows a detailed sectional view of the shaft and pulley area.

Dimensions

Dimensions are required on any drawing used to fabricate or repair parts. Dimensions can be broken down into two principal classes, according to their purpose. When locating holes for drilling or positions for slots, the technician uses **location dimensions**. When cutting a piece of stock to the size and shape for a part, **size dimensions** are used.

Standards have been developed for the placement and use of dimensions on drawings, although minor variations will be found. Detailed knowledge of such standards is the responsibility of the person producing the drawings. The maintenance technician needs to have adequate knowledge of such practices to be able to interpret the drawings correctly.

Placement of Dimensions. Figure 6–30 shows how dimensions are located for a simple drawing. A **dimension line** has an arrowhead at each end, which shows where the dimension begins and ends. The dimension is shown by the number at the break in the line. The short lines at right angles to the arrowheads are extension lines, which help identify the part of the drawing that the dimension applies to.

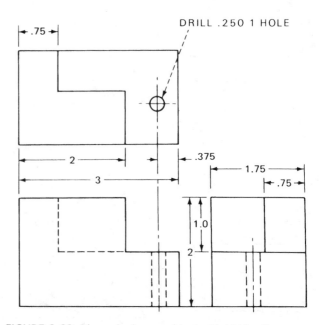

DRILL .250 1 HOLE

.75

2

.375

3

1.75

.75

1.0

2

FIGURE 6–32 Dimensioning an object with hidden lines.

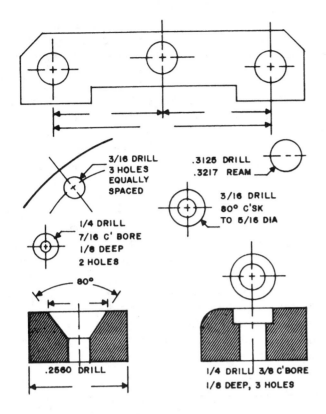

FIGURE 6–33 Dimensioning drilled holes.

Dimensions which are common to two views are usually placed between the views. On drawings of simple parts, the length is given under the front view, while the width and thickness are given in the right-side view.

Figure 6–31 shows three views of an object that is solid but can be regarded for purposes of discussion as two blocks joined together. The front view shows the length of the small block as 0.875 in [2.22 cm] and the length of the large block as .0625 in [2.70 cm]. The front view also shows the height of the small block as 1.125 in [2.857 cm] and the height of the large block as 1.875 in [4.762 cm]. The right-side view gives the width of the small block as 0.5 in [1.27 cm] and the width of the whole object as 1.0625 in [2.70 cm].

A slightly more complex object is shown in Figure 6–32. A general rule of dimensions being used only with visible lines is illustrated. The three views show that there is a cutaway section and a hole in the object. The hole is dimensioned in the top view, where it is indicated that the hole is 0.375 in [0.952 cm] from the right side. The right-side view shows that the top of the hole is 1 in [2.54 cm] below the upper surface of the object.

Dimensions are usually given in inches, but the inch symbol (") is often omitted from aircraft drawings to save space and time. When dimension lines and extension lines cross each other, they are often broken at the intersection.

Holes to be drilled in a part are located by dimensions to the centers, as shown in Figure 6–33. The hole will often be dimensioned by the size of the drill to be used (e.g., "#7 Drill," "Drill F," or "$\frac{1}{4}$-in drill"). Unless it is otherwise indicated, a hole dimension will include the letter D to show that the dimension is a diameter. Where accuracy is impor-

tant, the hole dimensions may be given in decimals. Notes are used to indicate the depth of holes and counterbores. Notes may also be used to indicate if the part has a number of equally sized or equally spaced holes.

Curved surfaces are often dimensioned as a radius from a specified point. The dimension is shown with a view that shows the curve as an arc, and it is followed by the letter R, for radius. Figure 6–26 illustrates this practice.

Limits, Tolerance, and Allowance. The dimensions on a drawing that represent the nominal size are sometimes called the **basic dimensions**. For example, the basic dimension for the length of an object might be 4 in.

All basic dimensions will have **limits**, or applicable maximum and minimum sizes. For example, if the limits for a 4-in dimension are 3.995 and 4.005 in, the part could be made 0.005 in shorter or 0.005 in longer than the basic dimension and still pass inspection. Limits are usually written with one over the other. For example, a drawing may read, "DRILL 0290/0293." For external dimensions the maximum dimension is placed above the line, and for internal dimensions the minimum value is placed above the line.

The **tolerance** is the total permissible variation of a size. It is the acceptable difference between the dimension limits. For example, if the limits are ±0.005, then the plus limit is the dimension plus 0.005, the minus limit is the dimension minus 0.005, and the tolerance is 0.010.

An **allowance** is a prescribed difference between the maximum material limits of **mating parts**. It represents the condition of the tightest permissible fit for two parts that are assembled into a common component. The American National Standards Institute (ANSI) has established standard **classes of fit**, ranging from large positive allowances (loose fit) to negative allowances (tight or shrink fit). For example, when an assembly is being put together, it may be desirable to have a driving, or force, fit, which may be described as a tight fit. On the other hand, if clearance is required between moving parts, a loose fit is needed.

The ANSI classes of fits are used to determine the limits or tolerances to be used for two parts. The class of fit is not ordinarily used as a designation on a drawing. Actual sizes will usually be "called out," or indicated on the drawing.

Title Blocks

The title block is the index to the drawing. It provides all necessary information that is not shown in or near the actual drawing. It is usually located in the lower right-hand corner of the drawing, although it may occupy the whole bottom portion of a drawing made on a small sheet of paper.

Figure 6–34 is a small drawing prepared by the engineering department of an aerospace company. The title block is typical, but there will be variations within the industry. All manufacturers do not prepare their drawings exactly alike, and even within the same organization it is impossible and often undesirable to standardize completely the manner of preparing drawings. Most manufacturers will have standardized specifications that set forth their procedures for producing drawings. The person reading the print must be familiar with the practices used by that manufacturer.

A number of features are included in the title block of Figure 6–34. Each of these will be discussed separately.

Drawing, or Print, Number. The **drawing, or print, number** is customarily printed in large numerals in the lower right-hand corner of the title block and repeated elsewhere on the same drawing. This keeps the number available when the drawing is folded and even if one of the numbers should be torn off or otherwise obliterated. In some companies, the drawing number is repeated in the upper left-hand corner or in some other location where it will not interfere with other information.

When a drawing consists of more than one sheet, each sheet is identified by the basic drawing number and the sheet number. The sheet number is entered in the number block.

In recent years a variety of number systems have been used by different companies. To understand any particular system, it is necessary to consult the specifications for the company concerned.

Part Number. Each part of an aerospace vehicle always has a number of its own. On a print, the part number and the print number may be the same; if they are not, then the part number is a dashed number shown in the title block.

Right-Hand and Left-Hand Parts. A profile or silhouette drawing of a person's right shoe is the same as the outline of the left shoe viewed from the side, even though the shoes have different shapes as viewed from the top. In like manner, many parts of large assemblies are right-hand and left-hand mirror images of each other and not two identical parts.

An aircraft has many such mirror-image parts. To reduce the amount of paper work, it is common practice to show the part for one side in the drawing and indicate the other side as opposite. A common practice is to show the left-hand part and add −1 to the part number. The right-hand part would carry the same number but with −2 as a dashed number. This is not a universal system, as some manufacturers will apply separate part numbers to opposite parts. The system used, if any, can usually be determined from examining the drawing.

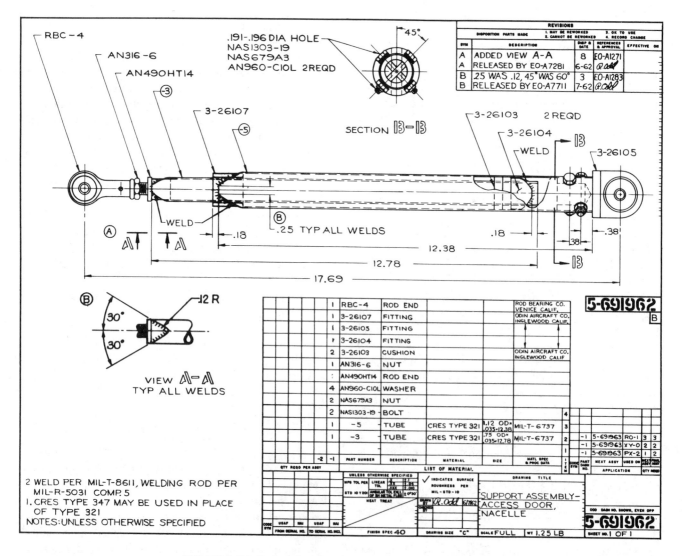

FIGURE 6–34 Typical drawing showing details of an assembly.

Scale. The word *scale* has many meanings. One of them is the proportion in dimensions between a drawing, map, or plan and the actual object that is represented on paper. For example, a map may be drawn on a scale of 1 in to 1 mi, which means that 1 in on the map represents 1 mi on the earth. In a like manner, a drawing from which a blueprint is to be made may be prepared to any desired scale, and the scale is shown in the title block on the drawing. Most aerospace drawings are made full scale, $\frac{1}{2}$ size, $\frac{1}{4}$ size, or $\frac{1}{10}$ size.

Name of Part, Unit, or Assembly. The name of the part, unit, or assembly is given first and then followed by descriptive terms, just as the family name is given in a telephone directory, followed by the given or Christian name, such as "Jones, John J." In like manner, in the aerospace industry, the name of a carburetor air-intake flange would read, "Flange, Carburetor Air Intake." Likewise, the assembly shown in Figure 6–34 is "Support Assembly, Access Door, Nacelle." In other words, the noun is given first and is followed by the descriptive terms.

The location may be a part of the name, such as "Elevator and Tab Assembly—Left," showing that it belongs on the left side of the airplane. Another example is "Installation—C.N. sta. 77 flap control cable pul. brkt. assemb." This translates as an installation, located at crew nacelle station 77, and is a flap-control pulley bracket assembly.

Revisions. On the drawing of Figure 6–34 there is a space in the upper right-hand corner labeled "Revisions." Note that two revisions have been made to the drawing. The disposition and date of the revision, the name or initial of the person making it, an indication of approval, and the serial number on which it is to be effective are recorded. The disposition of previously made parts may also be indicated.

Dimensions and Limits. In Figure 6–34 there are spaces for the dimensions of the standard stock from which the part is made. In this case the standard stock is tubing, MIL-T-6737. Limits and tolerances are shown in a supplementary block adjacent to the basic title block.

Station Numbers. A station numbering system can be used to help find the location of fuselage frames, wing frames, and stabilizer frames. For example, the nose of an airplane or some other point that can be easily identified is designated as the zero station, and other stations are located at measured distances in inches behind the zero station. Thus, when a print reads, "Fuselage Frame-Sta. 182," that means that the frame is 182 inches in back of station zero. In a similar manner, the fore-and-aft center line of an airplane may be a zero station for objects on its right and left. Thus, the wing and stabilizer frames can be located a certain number of inches to the right or left of the airplane center line. On some aircraft the firewall is a zero station, and on others the leading edge of the wing can be used as a zero station for certain purposes.

Model and Next Assembly Information. When an object shown on a print is to be a part of an assembly, the next assembly number is given in the title block to indicate the drawing number of another blueprint that gives the necessary information for completing the assembly. The number of parts required, such as one for the right hand and one for the left hand, is also given. The model designation, if any, will usually be found in this section.

Material Listing. Every drawing carries information regarding the material and specifications to be used. For example, a drawing may have "C.M. Sh't" in the material space, indicating the use of chrome-molybdenum steel sheet. In Figure 6–34, the material list is shown in the upper portion of the title block. The material list is sometimes referred to as the **bill of materials**. The specification space may show that this is SAE4130 or some other particular type of chrome-molybdenum steel sheet. In a similar manner, a drawing may show that 0.040 sheet aluminum (0.040 in thick) is required and then specify that it must be 2024-T3 aluminum alloy.

The material specification is coded when it is for the Army, Navy, or Air Force. For example, it may read, "ANQQ—," or it might have some other combination of letters and numerals that have a definite meaning to those handling military contracts. The finish, or protective coating, may be indicated in clear language, or it may be coded. Materials and specification codes are covered in other chapters of this text.

Weight. In a space in the title block for a drawing, the weight of the object shown in the blueprint is given. This quantity may be the calculated weight, the actual weight, or both. This information is particularly useful for the weight-and-balance engineers and others who are concerned with the balance of the device involved.

Notes. Information that cannot be given completely and yet briefly in the title block is placed on the drawing in the form of notes, which do not duplicate information given elsewhere in the same drawing. Notes may be used to tell the size of a hole, the number of drill to be used in making the hole, the number of holes required, and similar information, but only when it cannot be conveyed in the conventional manner or to avoid crowding the drawing. If the notes apply to specific places on the part, they are placed on the face of the drawing. If they apply to the part in general, they go in the "General Notes" at the bottom and to the left of the title block.

Symbols and Abbreviations

A **symbol** is a visible sign used instead of words to represent ideas, operations, quantities, qualities, relations, or positions. It may be an emblem, such as a picture of a lion to represent courage or an owl to represent wisdom. Likewise, the cross represents Christianity, the Star of David represents Judaism, and the crescent stands for Islam. A symbol may be an abbreviation, a single letter, or a character. In other words, symbols constitute picture writing.

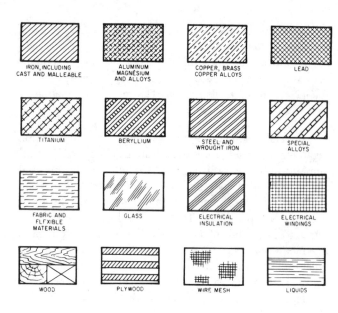

FIGURE 6–35 Symbols used for materials. *(MIL-STD-1A)*

Symbols and abbreviations are used extensively in place of long explanatory notes on drawings and prints. A few of these symbols and abbreviations are common to every trade, whereas special trades have special symbols and abbreviations of their own.

Material Symbols. The ANSI has standardized certain symbols used to represent materials in section views. The military services have adopted some of these symbols and added others of their own, and these are all included in the design handbooks used by aerospace manufacturers. Figure 6–35 shows an important group of such symbols.

When used, these symbols are intended only for general information and not to indicate specific types of materials. For example, in the upper left-hand corner of Figure 6–35, there is a symbol for iron, including cast iron and malleable iron, but it does not tell the specific type of iron to be used. Such information would appear elsewhere on the print. To simplify changes, the symbol for cast iron and malleable iron is often used by many companies to refer to all metals.

Process Code. Manufacturers in the aerospace industry use symbols for process codes. Examples include the following: A, anodize; B, chromodize; C, cadmium plate; D, dichromate (Dow No. 7); H, degrease in vapor degreaser; and S, sand blast. These code letters simplify the instructions required on drawings and consume less space than would otherwise be required.

REVIEW QUESTIONS

1. Why is a drawing often called a blueprint?
2. What is the difference in use between a detail drawing, an assembly drawing, and an installation drawing?
3. What is the purpose of a block diagram?
4. What is the function of a schematic diagram?
5. In what cases may shop sketches be used?
6. How may a wiring diagram be used by a technician?
7. What is a perspective drawing?
8. How does an isometric view vary from a perspective view?
9. Describe what is meant by *orthographic projection*.
10. Explain the importance of the width and type of lines used for aircraft drawings.
11. Under what conditions may two-view and single-view drawings be used?
12. When is a detail view used?
13. Compare location dimensions and size dimensions.
14. Where is a dimension placed when it is common to two views?
15. What unit of measurement is generally used for dimensions of aircraft drawings?
16. What is a limit when used with a dimension?
17. Explain what is meant by *tolerance*.
18. What type of information is contained in the title block of a drawing?
19. What is the significance of the drawing number?
20. What is meant by *right-hand* and *left-hand* parts?
21. With respect to a drawing, what is meant by *scale*?
22. How are modifications to a drawing noted?
23. What is another term used for the material listing of a drawing?
24. Why are notes found on a drawing?
25. What is meant by *process code*

Weight and Balance 7

INTRODUCTION

Aviation has been one of the most dynamic industries since its beginning. New aircraft are continually being developed with improvements over previous models. Improvements in design have, in many cases, tended to increase the importance of the proper loading and balancing of today's airplanes. Weight-and-balance calculations are performed according to exact rules and specifications and must be prepared when aircraft are manufactured and whenever they are altered, whether the airplane is large or small. The constantly changing conditions of modern aircraft operation present more complex combinations of cargo, crew, fuel, passengers, and baggage. The necessity of obtaining maximum efficiency for all flights has increased the need for a precise system of controlling the weight and balance of an aircraft.

FUNDAMENTAL PRINCIPLES

In a previous chapter, the laws of physics were discussed. Included were discussions of specific gravity and balance, together with explanations of levers. These principles form the basis for computing weight-and-balance data for an airplane and will be reviewed briefly here.

Force of Gravity

Every body of matter in the universe attracts every other body with a certain force that is called **gravitation**. The term *gravity* is used to refer to the force that tends to draw all bodies toward the center of the earth. The weight of a body is the result of all gravitational forces acting on the body.

Center of Gravity

Every particle of an object is acted on by the force of gravity. However, in every object there is one point at which a single force, equal in magnitude to the weight of the object and directed upward, can keep the body at rest, that is, can keep it in balance and prevent it from falling. This point is known as the **center of gravity (CG)**.

The CG might be defined as the point at which all the weight of a body can be considered concentrated. Thus, the CG of a perfectly round ball would be the exact center of the ball, provided that the ball was made of the same material throughout and that there were no air or gas pockets inside (see Figure 7–1). The CG of a uniform ring would be at the center of the ring but would not be at any point on the ring itself (see Figure 7–2). The CG of a cube of solid material would be equidistant from the eight corners, as shown in Figure 7–3. In airplanes or helicopters, ease of control and maneuverability require that the location of the CG be within specified limits.

Location of the CG

Since the CG of a body is that point at which its weight can be considered to be concentrated, the CG of a freely suspended body will always be vertically beneath the point of support when the body is supported at a single point. To locate the CG, therefore, it is necessary only to determine the point of intersection of vertical lines drawn downward from two separate points of support employed one at a time. This technique is demonstrated in Figure 7–4, which shows a flat, square sheet of material lettered A, B, C, and D at its four corners, suspended first from point B and then from point C. The lines drawn vertically downward from the point of suspension in each case intersect at the CG.

The CG of an irregular body can be determined in the same way. If an irregular object, such as the one shown in

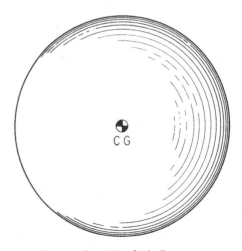

FIGURE 7–1 Center of gravity of a ball.

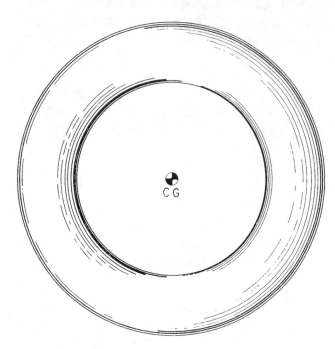

FIGURE 7–2 Center of gravity of a ring.

Figure 7–5, is suspended from a point *P* in such a manner that it can turn freely about the point of suspension, it will come to rest with its CG directly below the point of suspension, *P*. If a plumb line is dropped from the same point of suspension, the CG of the object will coincide with some point along the plumb line; a line drawn along the plumb line passes through this point. If the object is suspended from another point, which will be called *A*, and another line is drawn in the direction indicated by the plumb line, the intersection of the two lines will be at the CG. In order to verify the results, the operation can be repeated, this time with the object suspended from another point, called *B*. No matter how many times the process is repeated, the lines should pass through the CG; therefore, it can be shown that the CG of the object lies at the point of intersection of these lines of suspension. Therefore, any object behaves as if all its weight were concentrated at its CG.

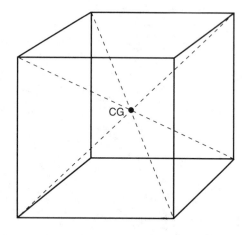

FIGURE 7–3 Center of gravity of a cube.

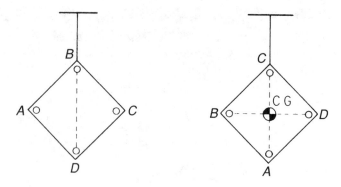

FIGURE 7–4 Location of the CG.

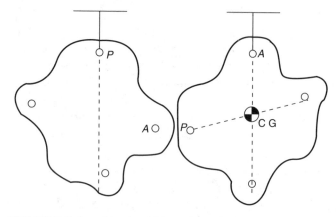

FIGURE 7–5 Locating the CG in an irregular body.

The General Law of the Lever

In Chapter 2 the law of levers was explained; it will be repeated briefly here to show how it relates to the weight and balance of an airplane.

Wrenches, crowbars, and scissors are levers used to gain mechanical advantage, that is, to gain force at the expense of distance or to gain distance at the expense of force. A lever, in general, is essentially a rigid rod free to turn about a point called the **fulcrum**. There are three types of levers, but the study of weight and balance is principally interested in the type known as a **first-class lever**. This type has the fulcrum between the applied effort and the resistance, as shown in Figure 7–6.

In Figure 7–6 the fulcrum is marked *F*, the applied effort is *E*, and the resistance is *R*. If the resistance, *R*, equals 10 lb [4.535 kg], and it is 2 in [5.08 cm] from the fulcrum, *F*, and if the effort, *E*, is applied 10 in [25.4 cm] from the fulcrum, it will be found that an effort of 2 lb [0.907 kg] will balance the resistance, *R*. In other words, when a lever is balanced, the product of the effort and its lever arm (distance from the fulcrum) equals the product of the resistance and its lever arm. The product of a force and its lever arm is called the **moment** of the force.

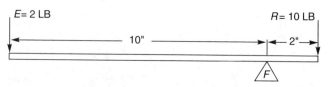

FIGURE 7–6 First-class lever.

The general law of the lever is as follows: *If a lever is in balance, the sum of the moments tending to turn the lever in one direction about an axis equals the sum of the moments tending to turn it in the opposite direction.* Therefore, if the lever is in balance, and if several different efforts are applied to the lever, the sum of the moments of resistance will equal the sum of the moments of effort.

Moment of a Force and Equilibrium

The tendency of a force to produce rotation around a given axis is called the **moment of the force** with respect to that axis.

The amount and direction of the moment of a force depend upon the direction of the force and its distance from the axis. The perpendicular distance from the axis to the line of the force is called the **arm**, and the moment is measured by the product of the force and the arm. Thus, a force of 10 lb [4.536 kg] acting at a distance of 2 ft [0.6096 m] from the axis exerts a turning moment of 20 ft-lb [2.765 kg-m].

In order to avoid confusion between moments tending to produce rotation in opposite directions, those tending to produce a clockwise rotation are called positive and those tending to produce counterclockwise rotation are called negative. If the sum of the positive, or clockwise, moments equals the sum of the negative, or counterclockwise, moments, there will be no rotation. This is usually expressed in the form $\Sigma M = 0$. The symbol Σ is the Greek letter sigma, and ΣM means the sum of all the moments, M, both positive and negative.

In Figure 7–7 a moment diagram is shown, with moments about the point A. M_1 acts in a counterclockwise direction, with a force of 1 lb [0.4536 kg] at a distance of 3 ft [0.9144 m]; therefore, the value of M_1 is -3 ft-lb [-0.4148 kg-m]. M_2 acts in a counterclockwise direction, with a force of 2 lb [0.9072 kg] at a distance of 2 ft [0.6096 m], thus producing a moment of -4 ft-lb [-0.5528 kg-m]. M_3, acting in a counterclockwise direction with a force of 1 lb [0.4537 kg] at a distance of 1 ft [0.3048 m], produces a moment of -1 ft-lb [-0.1383 kg-m]. M_4 acts in a clockwise direction, with a force of 4 lb [1.814 kg] at 2 ft, which makes a moment of $+8$ ft-lb [$+1.105$ kg-m]. Thus, $-3 - 4 - 1 + 8 = 0$. The sum of the negative moments is equal to the positive moment; therefore, there is a condition of **equilibrium**, and there is no rotation about point A.

There is a total force of 8 lb [3.629 kg] acting downward, and unless the axis is supported by an upward force of 8 lb, there will be downward movement but no rotation.

Aircraft CG Range and Limits

The first-class lever is in balance only when the horizontal CG is at the fulcrum. However, an aircraft can be balanced in flight anywhere within certain specified forward and aft limits if the pilot operates the trim tabs or elevators to exert an aerodynamic force sufficient to overcome any static unbalance. CG locations outside the specified limits will cause unsatisfactory or even dangerous flight characteristics.

The allowable variation within the CG range is carefully determined by the engineers who design an airplane. The CG range usually extends forward and rearward from a point about one-fourth the chord of the wing, back from the leading edge, provided that the wing has no sweepback. The exact location is always shown in the **Aircraft Specifications** or the **Type Certificate Data Sheet**. Heavy loads near the wing location are balanced by much lighter loads at or near the nose or tail of the airplane. In Figure 7–8, a load of 5 lb [2.268 kg] at A will be balanced by a load of 1 lb [0.4536 kg] at B because the moments of the two loads are equal.

Since the CG limits constitute the range of movement that the aircraft CG can have without making it unstable or unsafe to fly, the CG of the loaded aircraft must be within these limits at takeoff, in the air, and on landing. In some cases, the takeoff limits and landing limits are not exactly the same, and the differences are given in the specifications for the aircraft.

Figure 7–9 shows typical limits for the CG location in an airplane. As previously stated, these limits establish the **CG range**. The CG of the airplane must fall within this range if the airplane is to fly safely; that is, the CG must be to the rear of the forward limit and forward of the aft limit.

CG and Balance in an Airplane

The CG of an airplane may be defined, for the purpose of balance computations, as an imaginary point about which the nose-heavy ($-$) moments and tail-heavy ($+$) moments are exactly equal in magnitude. Thus, the aircraft, if suspended from that point (CG), would have no tendency to rotate in either direction (nose-up or nose-down). This condition is illustrated in Figure 7–10. As stated previously, the weight of the aircraft can be assumed to be concentrated at its CG.

The CG with the aircraft loaded is allowed to range fore and aft within certain limits that are determined during the flight tests for type certification. These limits are the most forward- and rearward-loaded CG positions at which the aircraft will meet the performance and flight characteristics required by the FAA. These limits may be expressed in

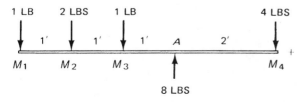

FIGURE 7–7 Moment diagram.

FIGURE 7–8 Balancing of the load.

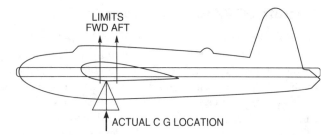

FIGURE 7–9 Center-of-gravity limits.

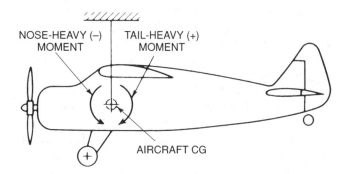

FIGURE 7–10 Airplane suspended from the CG location.

terms of a percentage of the mean aerodynamic chord (MAC) or in inches forward or to the rear of the datum line.

The relative positions of the CG and the center of lift of the wing have critical effects on the flight characteristics of the aircraft. Consequently, relating the CG location of the chord of the wing is convenient from a design-and-operations standpoint. *Normally, an aircraft will have acceptable flight characteristics if the CG is located somewhere near the 25% average chord point.* This means the CG is located one-fourth of the total distance back from the leading edge of the average wing section (see Figure 7–11). Such a location will place the CG forward of the aerodynamic center for most airfoils.

The mean aerodynamic chord (MAC) is established by the manufacturer. If the wing has a constant chord, the straight-line distance from the leading edge to the trailing edge (the chord) would also be the MAC. However, if the wing is tapered, the mean aerodynamic chord is more complicated to define. The MAC is the chord of an imaginary airfoil which has the same aerodynamic characteristics as the actual airfoil. The MAC is established by the manufacturer, who defines its leading edge (LEMAC) and trailing edge (TEMAC) in terms of inches from the datum. The CG location and various limits are then expressed in percentages of the chord.

FIGURE 7–11 Percent of mean aerodynamic chord.

The MAC is usually given in the aircraft's Type Certificate Data Sheet when it is required for weight-and-balance computations; therefore the person working on the airplane is expected to have only a general understanding of its meaning. For simplicity purposes, most light-aircraft manufacturers express the CG range in inches from the datum, while transport-category aircraft are expressed in terms of percentages of the MAC.

WEIGHT-AND-BALANCE TERMINOLOGY

Before proceeding with explanations of the methods for computing weight-and-balance problems, it is important to have a good understanding of the words and terms used.

Arm. The arm is the horizontal distance in inches from the datum to the center of gravity of the item. The algebraic sign is plus (+) if measured aft of the datum and minus (−) if measured forward of the datum (see Figure 7–12).

Center of gravity (CG). The CG is a point about which the nose-heavy and tail-heavy moments are exactly equal in magnitude. If the aircraft were suspended from this point it would be perfectly balanced. Its distance from the reference datum is found by dividing the total moment by the total weight of the airplane.

Center of gravity range. The operating CG range is the distance between the forward and rearward limits within which the airplane must be operated. These limits are indicated on pertinent FAA Aircraft Type Certificate Data Sheets (see Figure 7–13) or in aircraft weight-and-balance records, and they meet the requirements of the Federal Aviation Regulations (FARs).

Datum (reference datum). The datum is an imaginary vertical plane or line from which all horizontal measurements of arm are taken (see Figure 7–12). The datum is established by the manufacturer. Once the datum has been selected, all moment arms must be taken with reference to that point. The location of the datum may be found in the aircraft's Type Certificate Data Sheet (see Figure 7–13).

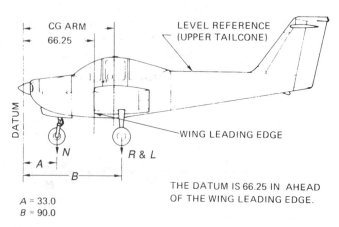

A = 33.0
B = 90.0

THE DATUM IS 66.25 IN AHEAD OF THE WING LEADING EDGE.

FIGURE 7–12 Leveling diagram. *(Piper Aircraft Corp.)*

Empty weight (EW). The empty weight of an aircraft includes the weight of the airframe, power plant, and required equipment that has a fixed location and is normally carried in the airplane. For aircraft certificated under FAR Part 23, the empty weight also includes unusable fuel and full-operating fluids necessary for normal operation of aircraft systems, such as oil and hydraulic fluid. For older aircraft not certificated under FAR Part 23, in place of full oil, only the undrainable oil is included in the empty weight. The current aircraft empty weight must be kept as a part of the permanent weight-and-balance records.

Empty-weight center of gravity (EWCG). The empty-weight CG is the CG of the aircraft in its empty condition and is an essential part of the weight-and-balance record that must be kept with the permanent aircraft records.

Empty-weight CG range. The EWCG range is established so that when the EWCG falls within this range, the aircraft-operating CG limits will not be exceeded under standard loading conditions. The EWCG range shown for many light airplanes is listed in the aircraft specifications or the Type Certificate Data Sheet and may eliminate further calculations by technicians making equipment changes (see Figure 7–13).

Fleet empty weight. The fleet empty weight is used by air carriers as an average basic empty weight which may be used for a fleet or group of aircraft of the same model and configuration. The weight of any fleet member can-

DEPARTMENT OF TRANSPORTATION
FEDERAL AVIATION ADMINISTRATION

TYPE CERTIFICATE DATA SHEET

		Utility Category	Acrobatic Category
Engine:	Lycoming 10360 B2F with "Christen" inverted oil system, or Lycoming AEI0360-B2F fuel injected.		
Fuel:	91/96 minimum aviation grade gasoline.		
Engine Limits:	For all operations 2700 rpm (180 hp).		
Propeller:	Hoffman HO 29-180-170		
Propeller Limits:	(Utility and Acrobatic Categories) Statis rpm at maximum permissible throttle setting—2250 ± 50 Diameter 70.9 in. No cutoff permitted.		
Airspeed Limits	Never exceed	211 mph (183 kts)	211 mph (183 kts)
	Max. structural cruising	186 mps (162 kts)	186 mph (162 kts)
	Maneuvering	124 mph (108 kts)	146 mph (127 kts)
	Flaps extended	99 mph (86 kts)	99 mph (86 kts)
	See NOTE 3 for acrobatic maneuvers.		
Flight Maneuvering Load Factor (g's)	Flaps up	+4.4 −1.8	+6.0 −3.0
	Flaps down	+2.0 −1.8	+2.0 −2.0
CG Range:	Forward limit	+10.6 in (18% MAC)	10.6 in (18% MAC)
	Aft limit	+17.7 in (30% MAC)	15.3 in (26% MAC)
Datum:	Wing leading edge at 51 in from airplane center line. (Length of wing chord at datum 59 in).		
Leveling Means:	Longitudinal: Left canopy rail Lateral: Top of bulkhead #2.		
Empty-Weight CG Range:	+8.6 in – +10.2 in		
Maximum Weight:	Takeoff	1829 lb	1675 lb
	Landing	1763 lb	1675 lb
No. of Seats:		2 at +22.7 in	2 at +22.7 in
Maximum Baggage:		110 lb at +55.1	None
Fuel Capacity:	Front tank (total)	19.8 (at −6.7 in)	19.8 (at −6.7 in)
	(usable)	19.0	19.0
	Rear tank (total)	20.8 (at +55.1 in)	0
	(usable)	20.6	0
	Rear tank must be empty for operations in acrobatic category. Minimum fuel quantity for acrobatics: 2.6 gal.		
Oil Capacity:	Maximum capacity:	2 gal	
	Minimum:	0.5 gal	
	Maximum oil quantity for acrobatics: 1.5 gal		

FIGURE 7–13 Sample Type Certificate Data Sheet.

not vary more than the tolerance established by the applicable government regulations.

LEMAC. LEMAC is the abbreviation for the leading edge of the mean aerodynamic chord.

Leveling means. Leveling means are the reference points used by the aircraft technician to insure that the aircraft is level for weight-and-balance purposes (see Figure 7–12). Leveling is usually accomplished along both the longitudinal and lateral axis. Leveling means are given in the Type Certificate Data Sheet (see Figure 7–13).

Loading envelope. The loading envelope includes those combinations of airplane weight and center of gravity that define the limits beyond which loading is not approved.

Main-wheel center line (MWCL). The MWCL is a vertical line passing through the center of the axle of the main landing-gear wheel.

Maximum gross weight. The maximum gross weight is the maximum authorized weight of the aircraft and its contents as listed in the Type Certificate Data Sheet (Figure 7–13).

Maximum landing weight. The maximum landing weight is the maximum weight at which the aircraft may normally be landed (see Figure 7–13).

Maximum ramp weight. The maximum ramp weight is the maximum weight approved for ground maneuver. (It includes the weight of the start, taxi, and run-up fuel.)

Maximum takeoff weight. The maximum takeoff weight is the maximum allowable weight at the start of the takeoff run (see Figure 7–13).

Mean aerodynamic chord (MAC). The MAC is the length of the mean chord of the wing as established through aerodynamic considerations. For weight-and-balance purposes it is used to locate the CG range of the aircraft. The location and dimension of the MAC, where used, will be found in the aircraft specifications, the Type Certificate Data Sheet (see Figure 7–13), the flight manual, or the aircraft weight-and-balance record.

Minimum fuel. Minimum fuel for weight-and-balance computations is no more than the quantity of fuel required for $\frac{1}{2}$ h of operation at rated maximum continuous power. It is calculated on the maximum except takeoff (METO) horsepower and is the figure used when the fuel load must be reduced to obtain the most critical loading on the CG limit being calculated. The formula usually used in calculating minimum fuel is $\frac{1}{2}$ METO hp = minimum fuel in pounds (e.g., $\frac{1}{2} \times 360$ hp = 180 lb of fuel).

Moment. The moment is the product of the weight of an item multiplied by its arm. Moments are expressed in pound-inches (lb-in). The total moment of an aircraft is the weight of the aircraft multiplied by the distance between the datum and the CG.

Moment index. The moment index is a moment divided by a constant, such as 100, 1000, or 10 000. The purpose of using a moment index is to simplify weight-and-balance computations of large aircraft where heavy items and long arms result in large, unmanageable numbers.

Standard weights. For general weight-and-balance purposes, the following weights are considered standard:

Gasoline	6 lb/gal [2.75 kg/gal]
Turbine fuel	6.7 lb/gal [3.0 kg/gal]
Lubricating oil	7.5 lb/gal [3.4 kg/gal]
Water	8.3 lb/gal [3.75 kg/gal]
General aviation crew and passengers	170 pounds [77 kg] per person
Air-carrier passenger (summer)	160 pounds [72.5 kg]
Air-carrier passenger (winter)	165 pounds [75 kg]

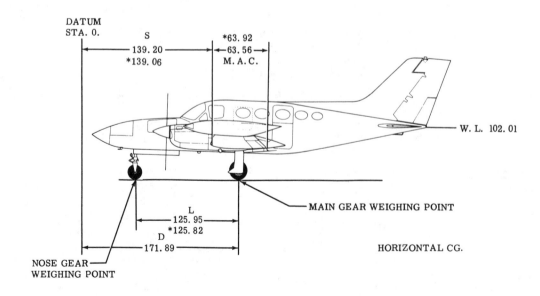

FIGURE 7–14 Weighing and measuring. *(Cessna Aircraft Co.)*

Station. A station is a location along the airplane fuselage given in terms of distance in inches from the reference datum. The datum is, therefore, identified as station zero (see Figure 7–14). The station and arm are usually identical. An item located at station 50 would have an arm of 50 in.

Tare. Tare is the weight of the equipment necessary for weighing the airplane (such as chocks, blocks, slings, jacks, etc.) which is included in the scale reading but is not a part of the actual weight of the airplane. Tare must be subtracted from the scale reading in order to obtain the actual weight of the airplane.

TEMAC. TEMAC is an abbreviation for the trailing edge of the mean aerodynamic chord.

Undrainable oil. That portion of the oil in an aircraft lubricating system that will not drain from the engine with the aircraft in a level attitude is called the undrainable oil. This oil is considered a part of the empty weight of the aircraft.

Unusable fuel. Unusable fuel is the fuel that cannot be consumed by the engine. The amount and location of the unusable fuel may be found in the Type Certificate Data Sheet (see Figure 7–13). Unusable fuel is a part of the aircraft's empty weight.

Usable fuel. Fuel available for flight planning is called usable fuel.

Useful load. The useful load is the weight of the pilot, copilot, passengers, baggage, and usable fuel. It is the empty weight subtracted from the maximum weight.

Weighing point. The weighing points of an airplane are those points by which the airplane is supported at the time it is weighed. Usually the main landing gear and the nose or tail wheel are the weighing points (see Figure 7–14). Sometimes, however, an airplane may have jacking points from which the weight is taken. In any event, it is essential to define the weighing points clearly in the weight-and-balance record.

Weighing the Aircraft

Weighing aircraft with accurately calibrated scales is the only sure method of obtaining an accurate empty weight and CG location. The use of weight-and-balance records in accounting for and correcting the aircraft weight-and-balance location is reliable over limited periods of time. Over extended intervals, however, the accumulation of dirt, miscellaneous hardware, minor repairs, and other factors will render the basic-weight and CG data inaccurate. For this reason, periodic aircraft weighings are desirable; however, they are not required of aircraft operated under FAR Part 91. This is not the case for air-taxi and air-carrier aircraft, which are required by the FARs to be periodically weighed. Aircraft may also be required to be weighed after they are painted; when major modifications or repairs are made; when the pilot reports unsatisfactory flight characteristics, such as nose or tail heaviness; and when recorded weight-and-balance data are suspected to be in error.

Weighing Equipment

The type of equipment which is used to weigh aircraft varies with the aircraft size. Three types of scales are commonly used to weigh aircraft. Each type is equally effective in obtaining accurate results. The three types of scales are platform scales, portable electronic weighing system using load pads, and electronic load cells used in conjunction with jacks.

Light aircraft are often weighed on beam-type **platform scales**, such as those illustrated in Figure 7–15. Platform scales require the use of jacks or ramps to position the aircraft on the scales.

A **portable electronic weighing system** makes it possible to find the weight and balance of large and small aircraft without jacking (see Figure 7–16). The system consists of

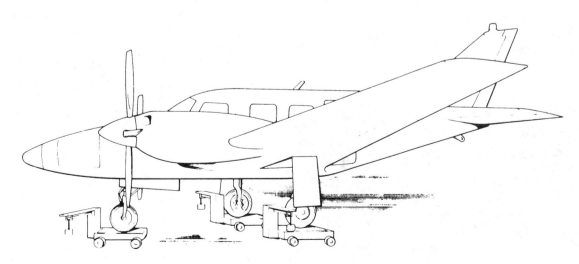

FIGURE 7–15 Weighing an airplane with platform scales. *(Piper Aircraft Corp.)*

FIGURE 7–16 Portable electronic weighing system. (*Evergreen Weigh, Inc.*)

electronic platform scales as necessary to weigh each wheel or pair of wheels on the aircraft, signal amplifiers, a digital CG indicator, a digital gross-weight indicator, and a power panel. Each scale consists of a platform supported by strain-gauge transducers, usually no more than 3 in [7.62 cm] in height. Ramps are supplied with the platforms so that the aircraft can easily be towed to position on the scales. The signals from the scales provide the information that is presented on the digital CG and gross-weight indicators. For larger aircraft the weighing pads may be recessed so that they are level with the floor to facilitate locating the aircraft on the scales.

Another method used to weigh large aircraft is to use **electronic load cells**. These cells are strain gauges whose resistance changes in accordance with the pressure applied to them. A load cell is placed between a jack and a jack point on the aircraft, with particular attention paid to locating the cell so that no side loads will be applied (see Figure 7–17). When weight readings are taken, the entire airplane weight must be supported on the load cells.

The output of the load cells is fed to an electronic instrument that amplifies and interprets the load-cell signals to provide weight readings. The instrument is adjusted to provide a zero reading from each load cell before the aircraft is weighed. After weighing, the cells are checked again and the reading is adjusted to compensate for any change noted.

Whichever type of system is selected, only weighing equipment that is maintained and calibrated to acceptable standards should be used.

FIGURE 7–17 Electronic load cells.

Equipment Preparation

When preparing to weigh an aircraft, the accuracy of the scales must be established. This can be done in accordance with instructions provided by the manufacturer of the scales or by testing the scales with calibrated weights. When there is nothing on the scales, the reading should be zero. *Note:* Most electronic scales require a specified warm-up period.

All the equipment that will be required to perform the weighing procedures should be located prior to beginning the weight check. The following is a list of equipment commonly used when weighing an aircraft:

1. Jacks or ramps
2. Wheel chalks
3. Level
4. Plumb lines
5. Steel measuring tape
6. Hydrometer (for testing the specific gravity of the fuel)
8. Tools and gauges for strut deflation and inflation
9. Nitrogen bottles for strut inflation

Aircraft Preparation

In order to obtain an accurate determination of the aircraft's weight and center of gravity, it is important that the aircraft be properly prepared for weighing.

Specific weighing preparations and procedures will vary with the model of the aircraft being weighed. However, the following information will provide general guidance.

The aircraft should be clean and free from excessive dirt, grease, moisture, or any other extraneous material before weighing. The aircraft should be dry before it is weighed; thus an aircraft should never be weighed immediately after it is washed.

All equipment to be installed in the aircraft and included in the certificated empty weight should be in place for weighing. Each item must be in the location that it will occupy during flight, as shown on the aircraft equipment list. All equipment, such as carpets, seat belts, oxygen masks, and so on, should be placed in their normal location. All tools and other working equipment must be removed before weighing.

Unless otherwise noted in the Type Certificate Data Sheet, the oil system and other operating fluids should be checked to see that they are full. Items that should be filled to operating capacity include lubricating oil, hydraulic fluid, oxygen bottles, and fire extinguishers.

The fuel should be drained from the aircraft unless other instructions are given. Fuel should be drained with the aircraft in the level position to make sure that the tanks are as empty as possible. The amount of fuel remaining in the aircraft tanks, lines, and engine is termed unusable fuel, and its weight is included in the empty weight of the aircraft. In special cases the aircraft may be weighed with full fuel in the tanks, provided that a definite means is available for determining the exact weight of the fuel.

Weighing Area

The aircraft should be weighed inside a closed building to avoid errors that may be caused by wind. Hangar doors and windows should be kept closed during the weighing process. The floor should be level. All fans, air conditioning, and ventilating systems should be turned off.

Positioning the Airplane

The aircraft should be placed in the weighing area. The aircraft's exterior should be checked to see that there is no interference with work stands and other equipment. If the main wheels are used as reaction points, the brakes should not be set because the resultant side loads on the scales or weighing units may cause erroneous readings.

The aircraft should be positioned securely on the scales. If the wheels are used as weighing points, it is advisable to use chocks on the scales both fore and aft so that the aircraft does not roll during the weighing procedure. Remember that items such as chocks and tail stands that are placed on top of the scales during weighing are considered tare weight. Tare weight must be subtracted from the scale readings. Tare weight items are generally weighed on different scales because aircraft scales are likely to be inaccurate in the lower range readings.

An airplane must be level to obtain accurate weighing information. Leveling is usually accomplished along both the longitudinal and the lateral axis. The leveling means are given in the Type Certificate Data Sheet. The leveling means are the reference points used by the aircraft technician to insure that the aircraft is level for weight-and-balance purposes.

One method used on many light aircraft is to set a spirit level on a longitudinal structural member to establish the longitudinal level position and another level across a lateral structural member to establish the lateral level position. This same basic procedure is accomplished in some aircraft by the installation of two nut plates on the side of the fuselage. Screws can be placed in these nut plates and longitudinal level is determined when a spirit level placed on the extended screws is level, as shown in Figure 7–18.

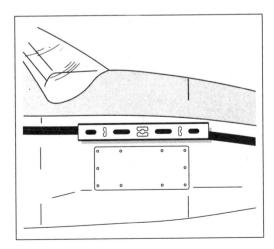

FIGURE 7–18 Leveling longitudinally. *(Piper Aircraft Corp.)*

Some aircraft use a plumb bob and a target to establish the level on both axes. In the DC-10 airplane, an **inclinometer** consisting of a plumb bob and **grid plate** is provided in the right wheel well, and brackets for spirit levels are located in the nose-gear wheel well. In Figure 7–19 locations of the leveling means for the DC-10 are shown.

The inclinometer indicates degrees of roll or pitch. The plumb bob is suspended by a cord and is secured in a stowage clip when not in use. During leveling operations, the plumb bob is released from the clip and is suspended by its cord over the grid plate. The level attitude of the airplane is established by the location of the plumb bob in relation to the grid-plate markings.

When a higher degree of leveling accuracy is required, spirit levels are used. The two sets of brackets provided in the nose-gear wheel well are used to support the levels in both longitudinal and lateral axes.

Weighing Procedure

The scale reading should be given a period of a few minutes to stabilize. The weights of the weighing points should be recorded to provide information needed for the CG determination. Several readings are taken for each reaction point, and the average reading is entered on the aircraft weighing form.

With the aircraft in the level position, it is necessary to measure and record the weigh point locations on the weighing form. On some aircraft the exact location of the weigh points will be provided in the aircraft flight manual or maintenance manual. If the location of weighing points is not provided, the exact location of the weighing points must be accurately measured while the aircraft is in the level position and then recorded for use in the weight-and-balance computation. The location of the datum is provided in the Type Certificate Data Sheet.

For aircraft where the datum passes through the aircraft, a plumb bob is dropped from that point to the floor. For aircraft where the datum is located ahead of the aircraft, a reference point should be located on the aircraft from which a plumb bob can be dropped to locate the datum. Once the datum is located on the floor, the plumb bob is suspended from each of the weighing points. The technician can measure these distances by projecting the required points to the hangar floor. To project these points to the hangar floor, a plumb bob may be suspended so that it is approximately one-half inch above the floor. When the swing of the plumb bob dampens, a cross mark is made on the floor directly under the tip of the plumb bob. The main reaction points are projected to the floor in the same manner. After marking the crosses for the two main gear points, a chalked string is stretched between them. The string is then snapped to the floor, leaving a chalk line between the main reaction points. The nose or tail reaction point is projected to the hangar floor in a similar manner, as is shown in Figure 7–20.

After these points are projected to the floor, it is a simple matter to measure the required dimensions. When measuring these distances, the tape must be parallel to the center line of the aircraft. Measurements made from the main reaction points are taken perpendicular to the chalk line joining these two points. When fuselage and wing jack points are

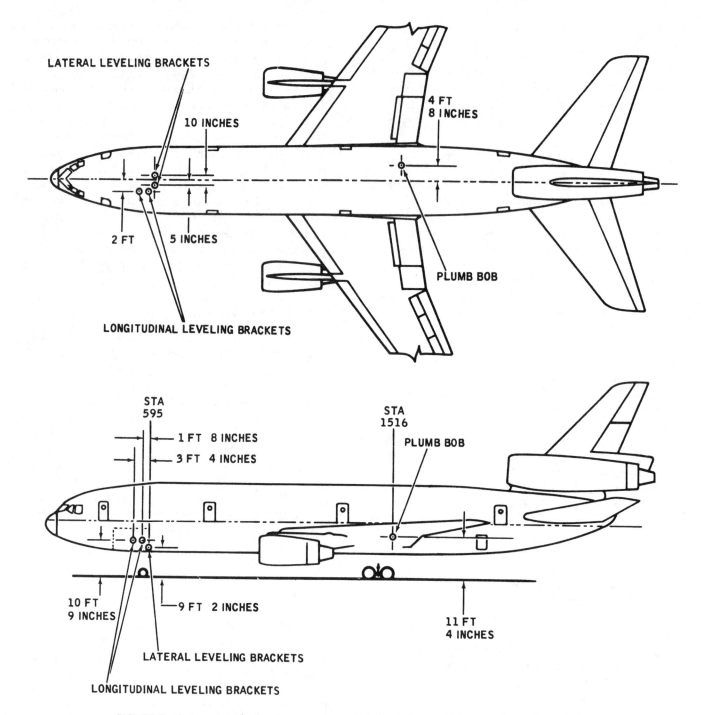

LATERAL LEVELING BRACKETS

10 INCHES

4 FT
8 INCHES

PLUMB BOB

2 FT   5 INCHES

LONGITUDINAL LEVELING BRACKETS

STA
595

1 FT 8 INCHES

3 FT 4 INCHES

STA
1516

PLUMB BOB

10 FT
9 INCHES

9 FT 2 INCHES

11 FT
4 INCHES

LATERAL LEVELING BRACKETS

LONGITUDINAL LEVELING BRACKETS

FIGURE 7–19 Locations for leveling means on a DC-10 airplane.   *(McDonnell Douglas Corp.)*

used as reaction points in weighing the aircraft, it is unnecessary to measure dimensions. These points will remain fixed and their moment arms may be found in the aircraft records. Care must be taken to use the fixed reaction points indicated in the records for the particular aircraft being measured. Because of manufacturing tolerances and minor model changes, the fixed reaction points are not necessarily identical for all aircraft of a particular type.

The weight of the tare should be recorded either before or after weighing the aircraft, and the tare weight should then be subtracted from the total weight obtained from the scales.

When data for comparison are available, an attempt should be made to verify the results obtained from each

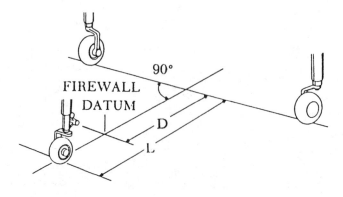

90°

FIREWALL
DATUM

D

L

FIGURE 7–20 Locating weighing points.

weighing. Verification may be made by comparing results with a previous weighing of an aircraft of the same model.

Computing CG Location

After the necessary dimensions and weights have been obtained, the empty weight and the empty weight CG can be calculated. Empty weight is the total of the three scale readings after subtracting the weight of tare items, plus or minus calibration errors. This weight is important for subsequent calculation of maximum weight and also is a necessary factor in the determination of the CG.

Center-of-gravity computations may be figured by several methods. The formulas used in computing the center of gravity are varied. Whenever possible, the manufacturer's weight-and-balance formulas and diagrams should be used, as shown in Figure 7–21. Although most manufacturers use similar formulas, they use different letter designations for different items. If these formulas are not available, a standard formula may be used for the EWCG computation.

Fundamentally, the CG is the point at which all the weights of the aircraft can be considered to be concentrated. The average location of these weights can, therefore, be obtained by dividing the total moment (weight × arm) by the total weight. The process then involves multiplying each measured weight by its arm to obtain a moment and then adding the moments.

Extra care must be taken in these types of empty-weight calculations if one or more of the arms is located ahead of the datum. In this event, the algebraic sign of the arm and moment will be negative. It should be remembered that a positive number (the weight) times a negative number (the arm) results in a negative number (the moment). Following the multiplication step, additional care must be taken when adding wheel moments to obtain the total moment and when dividing the total moment by the total weight to obtain the CG. In all these mathematical operations, the algebraic sign must be observed.

A set of formulas used quite extensively today is contained in the FAA Advisory Circular 43.13.1A. and is

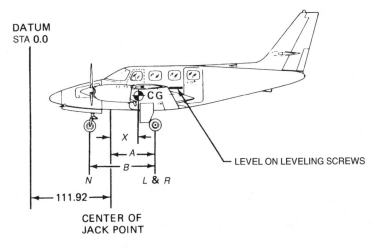

FIGURE 7–21 Sample airplane weighing. *(Cessna Aircraft Co.)*

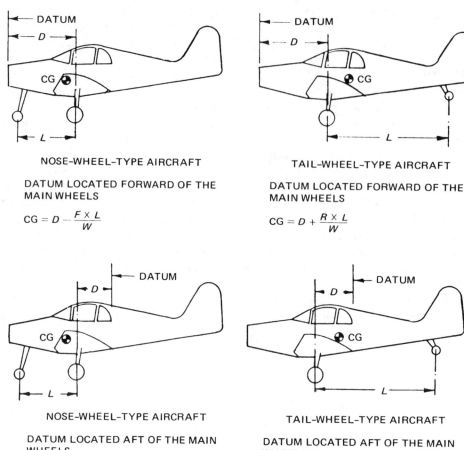

NOSE-WHEEL-TYPE AIRCRAFT

DATUM LOCATED FORWARD OF THE
MAIN WHEELS

$$CG = D - \frac{F \times L}{W}$$

TAIL-WHEEL-TYPE AIRCRAFT

DATUM LOCATED FORWARD OF THE
MAIN WHEELS

$$CG = D + \frac{R \times L}{W}$$

NOSE-WHEEL-TYPE AIRCRAFT

DATUM LOCATED AFT OF THE MAIN
WHEELS

$$CG = -D + \frac{F \times L}{W}$$

TAIL-WHEEL-TYPE AIRCRAFT

DATUM LOCATED AFT OF THE MAIN
WHEELS

$$CG = -D + \frac{R \times L}{W}$$

CG = distance from datum to center of gravity of aircraft
W = weight of aircraft at time of weighing
D = horizontal distance measured from datum to main wheel weighing point
L = horizontal distance measured from main wheel weighing point to nose or tail weighing point
F = weight at nose weighing point
R = weight at tail weighing point

FIGURE 7–22 Different arrangements of the formula for EWCG.

shown in Figure 7–22. This system uses four separate for-mulas. The user selects one of these formulas, depending upon the weighing points and the datum location in refer-ence to the weighing points. These formulas simplify the calculations in several ways. In effect, the datum is mathe-matically moved to the main gear by this process, resulting in relatively small moments, which are easy to handle in weight-and-balance calculations. A major benefit of the use of these formulas is the elimination of multiplication steps that involve negative arms and negative moments. In the first diagram of Figure 7–22, the datum is at the nose of the airplane, and since the airplane is of the tricycle-gear type, the CG must be forward of the MWCL. The part of the for-mula $F \times L/W$ gives the distance of the CG forward of the MWCL. This distance must then be subtracted from the dis-tance D to find the distance of the CG from the datum.

In the second diagram, the airplane is of conventional tail-wheel type, and so the CG must be to the rear of the MWCL. With the datum at the nose of the airplane, it is nec-essary to add the datum-line distance, D, to the $R \times L/W$ distance to find the EWCG from the datum line.

In the third diagram, the CG and the MWCL are both for-ward of the datum line; therefore, both distances are nega-tive. For this reason the CG distance from the MWCL and the datum distance from the MWCL are added together, and the total is given a negative sign.

The fourth diagram shows a condition where the CG is positive from the MWCL but negative from the datum line. The datum to the MWCL is a negative distance, and the CG from the MWCL is a positive distance. Therefore, the EWCG from the datum line is the difference between the two distances and, in this case, carries a negative sign.

Computing EWCG for a Tricycle-Gear Airplane

In Figure 7–23 a tricycle-gear airplane is weighed, and it is found that the nose-wheel weight is 320 lb [145.1 kg], the right-wheel weight is 816 lb [370.1 kg], and the left-wheel weight is 810 lb [367.4 kg]. The datum, which is located at the nose of the airplane, is 40 in [101.6 cm] forward of the nose-wheel center line and 115 in [292.1 cm] forward of the

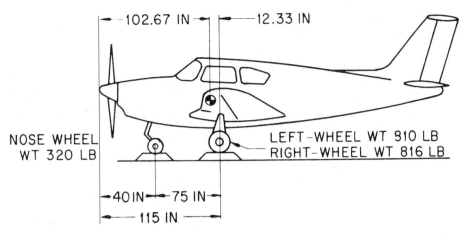

FIGURE 7–23 Quantities required to determine EWCG location for a tricycle-gear airplane.

MWCL. The horizontal distance between the weighing points is 75 in [190.5 cm]. Problem 7–1 shows how EWCG can be computed from these figures.

PROBLEM 7–1

Item	Weight	×	Arm	=	Moment
Right wheel	816		+115		+93 840
Left wheel	810		+115		+93 150
Nose wheel	320		+40		+12 800
	1946				+199 790

Then $\dfrac{+199\,790}{1946} = +102.67$ in [+260.78 cm]

This problem may be reworked using the following formula:

$$\text{CG} = D - \frac{F \times L}{W} = 115 - \frac{320 \times 75}{1946} = 102.67 \text{ in}$$

Care must be taken to ensure that the proper sign is applied to each quantity expressed in a weight-and-balance computation.

Computing EWCG for a Conventional Airplane

Figure 7–24 shows a conventional airplane in position for weighing. The aircraft is weighed with 2 gal [7.57 L] of oil

still in the engine, and the oil arm is given as −20 in the specifications. As shown in the illustration, the weights obtained from the scales are as follows: right wheel, 580 lb [263.1 kg]; left wheel, 585 lb [265.35 kg]; tail wheel, 130 lb [59 kg] (including tare). The computation for the EWCG can be arranged as in Problem 7–2.

PROBLEM 7–2

Item	Weight	Tare	Net Weight	× Arm	= Moment
Right wheel	580	0	580	0	0
Left wheel	585	0	585	0	0
Tail wheel	130	20	110	+160	+17 600
			1275		+17 600

As explained previously, the CG is equal to the total moment divided by the total weight. Therefore,

$$\text{CG} = \frac{17\,600}{1275} = +13.8 \text{ in [35 cm]} \quad \text{(CG aft of MWCL)}$$

The result obtained in the computation is the CG location to the rear of the reference point, which is the center line of the main landing gear. To obtain the CG from the datum line, the distance from the MWCL must be added to the distance of the datum line from the MWCL. Then, 13.8 + 8 = 21.8 in [55.4 cm], which is the distance of the CG from the datum line. If this is an older aircraft not certificated under

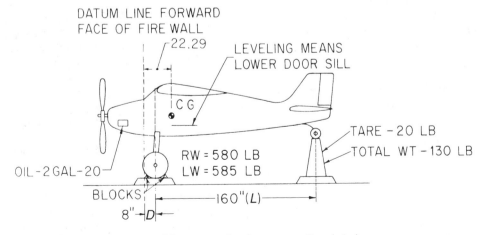

FIGURE 7–24 EWCG computation for a conventional airplane.

FAR Part 23, the empty weight may not include the oil, in which case the calculation for the EWCG has not yet been completed, since the CG obtained includes the weight of the oil that was in the aircraft at the time it was weighed. The oil must be removed by computation to obtain the EWCG (see Problem 7–3).

PROBLEM 7–3

Item	Weight	×	Arm	=	Moment
Airplane	1275		21.8		+27 795
Oil (removed)	−15		−20		+300
	1260				+28 095

$$\text{EWCG} = \frac{+28\ 095}{1260} = +22.29 \text{ in [56.61 cm](EWCG location)}$$

Referring back to Figure 7–24 and disregarding the oil computation, the problem may be solved with the correct formula from Figure 7–22, as is shown here:

$$CG = D + R \times \frac{L}{W} = 8 + 110 \times \frac{160}{1260} = 8 + 13.80$$

$$= 21.80 \text{ in [55.37 cm]}$$

Note that this answer is slightly less than the original computation because the moment of the oil was not considered.

For any computation, it is always a good practice to draw a diagram of the airplane (nose to the left) with the weighing points and the datum. From such a diagram it is easy to determine what formula should be used.

Expressing the CG as a Percentage of the MAC

The center of gravity may be expressed in terms of inches forward or to the rear of the datum line or as percentage of the mean aerodynamic chord (MAC). The MAC is established by the manufacturer, who defines its leading edge (LEMAC) and trailing edge (TEMAC) in terms of inches from the datum. The CG location and various limits are then expressed in percentages of the chord.

The center of gravity is expressed as a percentage and is located aft of LEMAC, as is shown in Figure 7–11.

Assume that the center of gravity for a particular aircraft has been calculated to be located at 130 in aft of the datum. The LEMAC is at station 100, and the TEMAC is at station 250; therefore, the length of the MAC is 250 in − 100 in, or 150 inches in length, as is shown in Figure 7–25.

To calculate the CG as a percentage of the MAC, the following formula can be used:

$$\% \text{ of MAC} = \frac{\text{distance of CG from LEMAC} \times 100}{\text{MAC (length)}}$$

For an example, assume that the MAC is 250 in, the LEMAC is 100 in, and the CG is 130 in.

$$\% \text{ of MAC} = \frac{(130 - 100) \times 100}{150}$$

$$\% \text{ of MAC} = \frac{3000}{150}$$

$$\% \text{ of MAC} = 20\%$$

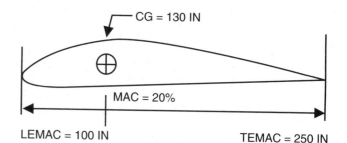

FIGURE 7–25 CG expressed as a percentage of MAC.

Weight-and-Balance Report

After the weight-and-balance calculations are complete, it is important that they be properly recorded and placed in the aircraft weight-and-balance records (a sample form is shown in Figure 7–26). When a new weight-and-balance report is prepared for an aircraft, the previous report should be marked *superseded*, and the date of the new document should be referenced. The series of weight-and-balance documents should start with the manufacturer's data and continue in a chronological order to the latest weight-and-balance report.

AIRCRAFT MODIFICATIONS

During the lifetime of many aircraft, it is often desirable to change the type of equipment that is installed. The owner of an airplane may wish to install new radio equipment, an autopilot, an auxiliary fuel tank, or various other items to make the airplane more serviceable. For each such change, it is necessary to figure the effect on weight and balance. The manufacturer is required to provide documents which show the certified empty weight and the CG for each new aircraft. The continued validity of weight-and-balance records during the life of the aircraft depends upon maintaining a series of similar documents showing the calculations for each successive weight change. It is essential that whenever equipment is added or removed from the aircraft, an entry is made in the airplane's equipment list and permanent weight-and-balance records. Many manufacturers provide a form, such as the one shown in Figure 7–27, that provides for a record of the equipment added or removed as well as a running total of the weight and balance.

The formula used to compute the new EWCG after the addition or subtraction of equipment is

$$CG = \frac{\text{total moment}}{\text{total weight}}$$

In calculating the new EWCG when adding or removing equipment, it is essential that the correct algebraic sign be used. The weight of an airplane is always positive (+). Also, the weight of any item *installed* in the airplane is positive. The weight of any item *removed* from the airplane is negative (−). According to the standard rules of algebra, the product of two positive numbers is positive, the product of two negative numbers is positive, and the product of a positive number and a negative number is negative. This can also be stated: *The product of numbers with like signs is positive; the product of numbers with unlike signs is negative.*

When items of aircraft equipment are added or removed, four combinations are possible. These are as follows:

1. When items are added forward of the datum line, the signs are (+) weight × (−) arm = (−) moment.

2. When items are added to the rear of the datum line, the signs are (+) weight × (+) arm = (+) moment.

3. When items are removed forward of the datum line, the signs are (−) weight × (−) arm = (+) moment.

WEIGHT-AND-BALANCE REPORT

MAKE _____ MODEL _____ S/N _____ N _____

DISTANCE BETWEEN MAIN WHEELS AND TAIL/NOSE WHEEL IS _____ INCHES.

DATUM IS _____ INCHES FORWARD/AFT OF MAIN-WHEEL CENTER-LINE.

1. AIRCRAFT AS WEIGHED

POSITION	SCALE READING	TARE	NET WEIGHT
LEFT WHEEL			
RIGHT WHEEL			
TAIL OR NOSE WHEEL			

2. ITEMS ON BOARD THE AIRPLANE WHEN IT WAS WEIGHED THAT ARE NOT INCLUDED IN THE EMPTY WEIGHT

ITEM	WEIGHT	ARM	MOMENT
FUEL- GAL			
OIL- QT			
TOTAL OF ADDITIONAL ITEMS			

3. ITEMS NOT ON BOARD THE AIRPLANE WHEN IT WAS WEIGHED THAT ARE TO BE INCLUDED IN THE EMPTY WEIGHT

	WEIGHT	ARM	MOMENT
FUEL- GAL			
OIL- QT			
TOTAL OF ADDITIONAL ITEMS			

4. CENTER OF GRAVITY AS CALCULATED

	WEIGHT	ARM	MOMENT
MAIN WHEELS			
TAIL OR NOSE WHEEL			
TOTAL ITEM #2 (REMOVABLE ITEMS)			
TOTAL ITEM #3 (ADDITIONAL ITEMS)			
EMPTY WEIGHT TOTAL			

EMPTY WEIGHT CG _____   SIGNATURE _____

MAXIMUM GROSS WEIGHT _____   CERT. NO. _____

EMPTY WEIGHT _____   DATE _____

USEFUL LOAD _____

FIGURE 7–26 Sample weight-and-balance report.

WEIGHT-AND-BALANCE RECORD

(CONTINUOUS HISTORY OF CHANGES IN STRUCTURE OR EQUIPMENT AFFECTING WEIGHT AND BALANCE)

SERIAL NUMBER		REGISTRATION NUMBER					PAGE NUMBER	
DATE	ITEM	DESCRIPTION OF ARTICLE OR MODIFICATION	WEIGHT CHANGE				RUNNING BASIC EMPTY WEIGHT	
			ADDED (+) OR REMOVED (−)					
	IN	OUT		WT (LB)	ARM (IN)	MOMENT /100	WT (LB)	MOMENT /100
			AS DELIVERED					

FIGURE 7–27 Sample weight-and-balance record.

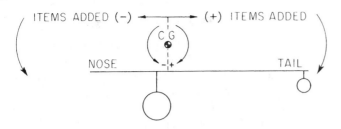

FIGURE 7–28 Effects of weight changes in an airplane.

4. When items are removed to the rear of the datum line, the signs are (−) weight × (+) arm = (−) moment.

A simple diagram will aid in determining the effect of changes in aircraft equipment. In Figure 7–28 a straight line represents the airplane. The nose of the airplane is shown to the left, this being the conventional method for representing aircraft in weight-and-balance diagrams. Using the CG location as a reference, note that any item installed forward of the CG produces a negative moment and causes the CG to move forward. Items added to the rear of the CG produce a positive moment and move the CG rearward. Items removed have an effect opposite to that of items installed.

Observe that the curved arrows shown around the CG location indicate the effects of positive and negative moments. Positive moments are clockwise and cause a tail-heavy force, while negative moments are counterclockwise and cause a nose-heavy force.

Adding Equipment

Let us assume that an owner who has an airplane with an empty weight of 1220 lb [553.4 kg] and an EWCG at +25 wishes to install radio equipment weighing 15 lb [6.8 kg]. In addition to the radio equipment, a larger alternator must be installed in order to provide the additional power required to operate the radio.

First, of course, we must determine where the items of equipment are to be installed and then determine the arm of each item of equipment. Each arm must be measured from the airplane datum line to the CG of the equipment to be installed. It must be pointed out that if the CG of the item of equipment is not given in the accompanying instructions, the CG must be determined by the person making the installation. This is easily done by balancing the item at a single point in the position it will assume in the airplane. The balance point should then be marked or recorded for use in the computation.

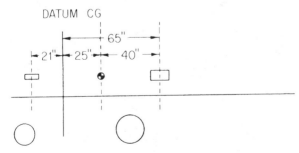

FIGURE 7–29 Installation of equipment.

For the purposes of the problem under consideration, we shall assume that the radio is installed at +65 and the new alternator is installed at −21. These points are shown in Figure 7–29. In order to install a new alternator, the old alternator must be removed. The weight and the arm of the old alternator are given in the aircraft's Approved Flight Manual as 11 lb [5.0 kg] (−21.5). The new alternator weighs 14 lb [6.35 kg], and the arm is found to be −21. Sufficient information is now available to make the computation. Arrange the work as shown in Problem 7–4.

PROBLEM 7–4

Item	Weight	×	Arm	=	Moment
Airplane (empty)	1220		+25		+30 500
Radio	15		+65		+975
Alternator (removed)	−11		−21.5		+236.5
Alternator (installed)	+14		−21		−294
New empty weight	1238				+31 417.5

$$\frac{31\,417.5}{1238} = 25.38 \text{ in (new EWCG)}$$

From the computation it is determined that the EWCG of the airplane has moved rearward 0.38 in [0.96 cm] as a result of the new installation.

Removing Equipment

When removing equipment from an airplane it is just as necessary to make weight-and-balance computations as when installing equipment. Let us assume that the owner of an airplane wishes to remove an extra seat because it is no longer necessary or required for operation. The airplane weighs 1000 lb [589.7 kg] empty (with the seat), and the seat weighs 18 lb [8.16 kg] (see Figure 7–30). The CG of the airplane as equipped is +78, and the arm of the seat is +145. We can now arrange the computation as shown in Problem 7–5.

PROBLEM 7–5

Item	Weight	×	Arm	=	Moment
Airplane (empty)	1300		+78		+101 400
Seat (removed)	−18		+145		−2 610
New empty weight	+1282				+98 790

$$\frac{98\,790}{1282} = +77.06 \text{ in (new EWCG)}$$

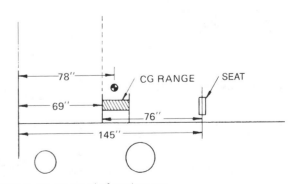

FIGURE 7–30 Removal of equipment.

We observe from the foregoing computation that the removal of the seat causes the EWCG to move forward almost 1 in [2.5 cm]. It should be remembered that any removal of weight aft of the CG will cause the CG to move forward.

If the change in equipment should move the CG outside the CG limits, flight in the airplane would not be safe or legal. It is sometimes necessary to compensate for changes in equipment by changing the baggage-weight allowance, limiting the fuel load, or by adding or removing ballast.

EWCG Range

Some small aircraft are designed so that it is not possible to load them in a condition which will place the CG outside the fore or aft limits if standard load schedules are observed. These aircraft have the seats, fuel, and baggage accommodations located very near the CG limits. They also have **empty-weight CG ranges** listed in their Type Certificates. Loads can be added to or removed from any location within the CG range with complete freedom from concern about CG movement. Such action cannot cause the CG to move beyond the CG limits of these aircraft, but maximum weight limits can still be exceeded.

Most aircraft, however, can be loaded in a manner which will place the CG beyond limits, in which case the manufacturer will be unable to establish a EWCG range. In these instances the EWCG range on the Type Certificate will be listed as "none."

LOADING THE AIRPLANE

The aircraft operator should develop a procedure by which it can be shown that the aircraft is properly loaded and will not exceed authorized weight and balance limitations during operation. Operators of large aircraft must also account for all probable loading conditions which may be experienced in flight and develop a loading schedule which will provide satisfactory weight-and-balance control. Loading schedules may be applied to individual aircraft or to a complete fleet of similar aircraft.

Center-of-Gravity Travel During Flight

On transport-category aircraft the flight manual should provide procedures which fully account for the extreme variations in CG travel during flight caused by any combination of the following variables:

1. The movement of passengers and cabin attendants from their normal seat position in the aircraft to other seats or the lavatory.
2. The loss of weight due to fuel burn.
3. The effect of landing-gear retraction.

Effects of Improper Loading

Improper loading reduces the efficiency of an airplane from the viewpoint of ceiling, maneuverability, rate of climb, and speed. This is the least of the harm that it can cause. The greatest danger is that improper loading may cause the destruction of life and property, even before the flight is well started, because of the stresses imposed upon the aircraft structure or because of altered flying characteristics. Some of the effects of improper loading are illustrated in Figure 7–31.

Overloading. Excessive weight reduces the flying ability of an airplane in almost every respect. The most important performance deficiencies of an **overweight airplane** are

1. Lowered structural safety
2. Reduced maneuverability
3. Increased takeoff run
4. Lowered angle and rate of climb
5. Lowered ceiling
6. Increased fuel consumption
7. Overstressed tires
8. Increased stalling speed

Effects of Adverse Balance. Adverse and abnormal balance conditions affect the flying ability of an airplane with respect to the same flight characteristics as those mentioned for an excess weight condition. In addition, there are two essential airplane attributes which may be seriously re-

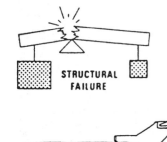
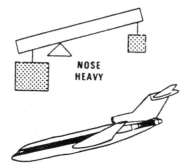
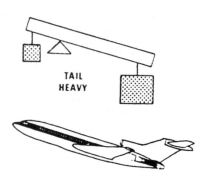

OVERLOADING   CG TOO FAR FORWARD   CG TOO FAR AFT

FIGURE 7–31 Effects of improper loading.

duced by improper balance; these are stability and control. An adversely loaded airplane can become particularly difficult to control during flap operation because of the shift in the center of lift.

CG Too Far Forward.   When too much weight is toward the forward part of the airplane, the center of gravity is shifted forward and any one or a combination of the following conditions may exist:

1. Increased fuel consumption and power settings
2. Decreased stability
3. Dangerous spin characteristics
4. Increased oscillation tendency
5. Increased tendency to dive, especially with power off
6. Increased difficulty in raising the nose of the airplane when landing
7. Increased stresses on the nose wheel

Adverse Rearward Loading.   When too much weight is toward the tail of the airplane, any one or a combination of the following conditions may exist:

1. Increased danger of stall
2. Dangerous spin characteristics
3. Poor stability
4. Decreased flying speed and range
5. Poor landing characteristics

EXTREME WEIGHT-AND-BALANCE CONDITIONS

We have already explained that every aircraft has an approved CG range within which the CG must lie if the aircraft is to be operated safely. In order to determine whether the loaded CG falls within the approved limits, it is necessary to make two computations, one for **most forward loading** and one for **most rearward loading**. These **adverse-loading checks** are a deliberate attempt to load an aircraft in a manner that will create the most critical balance condition while still remaining within the maximum gross weight of the aircraft.

It should be noted that when the EWCG falls within the EWCG range (if one is given), it is unnecessary to perform a forward or rearward weight-and-balance check. In other words, it is impossible to load the aircraft to exceed the CG limits, provided standard loading and seating arrangements are used.

Most Forward Adverse-Loading Checks

To determine the conditions for most forward loading, the Type Certificate Data Sheet is used to determine fuel capacity and arm, oil capacity and arm, passengers and arm, and cargo (baggage) and arm. From this reference it must be determined which items will tend to move the CG forward. Maximum quantities of these items are then included in the

computation. Since some items may have to be included that tend to move the CG rearward, a minimum of these items will be used. *In making a check of the forward CG limit, remember that maximum weights for items forward of the forward CG limit and minimum weights for items to the rear of the forward CG limit are used.*

Information required for a most forward adverse-loading CG check is as follows:

1. Weight, arm, and moment of the empty aircraft
2. Maximum weights, arms, and moments of all items of useful load located ahead of the forward CG limit
3. Minimum weights, arms, and moments of all items of useful load located to the rear of the forward CG limit

In examining the Type Certificate Data Sheet for a certain airplane, the following specifications are given for weight and balance:

CG range	(+85.1) to (+95.9) at 1710 lb
	(+87.0) to (+95.9) at 1900 lb
	(+91.5) to (+95.9) at 2200 lb
Datum	78.4 in forward of wing leading edge
Leveling	Two screws left-side fuselage before window
Max. weight	2200 lb
No. of seats	4 (2 at +85.5, 2 at +118)
Max. cargo	100 lb (+142.8)
Fuel capacity	50 gal (+94)
Oil capacity	2 gal (+31.7)
Engine limits	Maximum except take-off: 2750 rpm, 150 hp
	Take-off: 3000 rpm, 160 hp

For the given CG range a chart can be drawn, such as that shown in Figure 7–32. Note that the values given on the chart agree with the specifications listed. Assuming that the empty weight of the airplane is 1075 lb [487.62 kg] and the EWCG is +84 in [213.36 cm], the airplane can be loaded to determine whether the forward CG is within limits

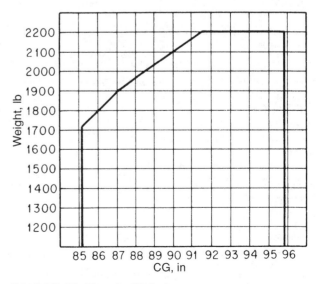

FIGURE 7–32 Chart for CG limits.

with maximum forward loading. *Remember that the purpose is to load the airplane in such a manner that it will result in the most forward CG possible.*

The airplane must have a pilot, so the standard weight of 170 lb [711 kg] is loaded at +85.5 in [217.17 cm]. Since the fuel is at +94 in [238.76 cm], which is substantially to the rear of the forward limit, only the minimum fuel is used. Because the engine of this particular airplane develops 150 hp at METO (maximum except takeoff) power and the formula for minimum fuel is $\frac{1}{2}$ METO hp in lb, 75 lb [34.02 kg] of fuel is included in the computation. This is loaded at +94 in [238.76 cm]. The cargo compartment is at +142.8 in [362.71 cm], and any load at this point will move the CG to the rear; therefore, no cargo will be loaded. Full oil is required for the engine, so 15 lb [6.8 kg] (7.5 lb/gal, or 3.4 kg/L) of oil is loaded at +31.7 in [80.51 cm]. The loading computation will then appear as in Problem 7–6.

PROBLEM 7–6

Item	Weight	×	Arm	=	Moment
Airplane (empty)	1075		+84		+90 300
Pilot	170		+85.5		14 535
Fuel	75		+94		7 050
Oil	15		+31.7		475.5
	1335				112 360.5

Then $\dfrac{112\,360.5}{1335}$ = 84.16 in [213.76 cm](most forward CG)

Checking the result of the foregoing computation against the CG limits for the airplane shows that the CG is located at 0.93 in [2.36 cm] forward of the forward CG limit. To correct this condition, the airplane could carry **ballast** or a certain amount of baggage in the cargo compartment, and a warning placard should be placed on the instrument panel.

Most Rearward Adverse-Loading Checks

To check an airplane for the most rearward CG limit, *maximum weight for items located aft of the rearward CG limit and minimum weight for items forward of the rearward CG limit are used.* Using the specifications for the same airplane as under consideration in the previous problem, it is determined that the cargo compartment and the two rear seats are the only items with locations to the rear of the rearward CG limit. The computation is then arranged as in Problem 7–7.

PROBLEM 7–7

Item	Weight	×	Arm	=	Moment
Airplane (empty)	1075		+84		+90 300
Pilot	170		+85.5		+14 535
Passengers (2)	340		+118		+40 120
Fuel	75		+94		+7 050
Oil	15		+31.7		+ 475.5
Baggage	100		+142.8		+14 280
	1775				+166 760.5

Then $\dfrac{166\,760.5}{1775}$ = 93.9 in [238.51 cm](rearward CG)

Note that the CG could be made to move slightly more toward the rear by including maximum fuel, since the arm of the fuel is +94. This location is still forward of the rear limit, however, and so it could not have moved the CG beyond its rearward limit.

Information required for a most rearward adverse-loading CG check is as follows:

1. Weight, arm, and moment of the empty aircraft
2. Maximum weights, arms, and moments of all items of useful load located to the rear of the rearward CG limit
3. Minimum weights, arms, and moments of all items of useful load located forward of the rearward CG limit

LOADING CONDITIONS

Sample loading conditions are computed as an indication of the permissible distribution of fuel, passengers, and baggage which may be carried in the aircraft at any one time without exceeding either the maximum weight or the CG range. These sample computations should be included in the aircraft's weight-and-balance records or may be posted in the form of a placard. A typical placard may be similar to the one shown in Figure 7–33.

LOADING SCHEDULE		
FUEL	PASSENGERS	BAGGAGE
FULL	2 REAR	100 LB
40 GAL	1 FRONT AND 2 REAR	NONE
FULL	1 FRONT AND 1 REAR	FULL
INCLUDES PILOT AND FULL OIL		

FIGURE 7–33 Loading schedule placard.

Correcting the CG Location

In the example used to compute the forward CG limit of the airplane in the section on "Most Forward Adverse-Loading Checks," it was found that the CG was 0.93 in [2.36 cm] forward of the forward CG limit. It is therefore necessary that a means be found for correcting this condition. To start this computation, the forward CG limit of +85.1 in [216.15 cm] is used as the reference point.

The CG can be corrected by adding **fixed ballast** at a point in the rear of the airplane near the tail, or a fixed ballast could be installed in the cargo compartment. The cargo compartment could also be placarded to the effect that a minimum amount of baggage must be carried under certain conditions. For the purpose of this computation, a fixed ballast will be installed in the cargo compartment.

The cargo compartment has an arm of +142.8. Since the

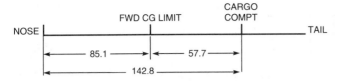

FIGURE 7–34 Correction computation for forward CG limit.

forward CG limit is being used as the reference point, +85.1 must be subtracted from +142.8 to find the arm of the cargo compartment for this computation. Figure 7–34 illustrates that the arm of the cargo compartment from the new reference point is 57.7 in [146.56 cm].

The first step necessary in the computation for the correction of CG location is to determine what moment is necessary to provide the required correction. Since the CG is 0.93 in forward of the forward limit, the moment necessary for correction is 0.93 × 1335 (the weight of the forward-loaded airplane). The product is 1241.6 in-lb [14.31 kg-m].

To determine the weight necessary to provide a moment of 1241.6 in-lb with an arm of +57.7 (the arm of the cargo compartment from the forward CG limit), 1241.6 must be divided by +57.7. The result of this division is 21.5 lb [9.75 kg], which is the weight required in the cargo compartment to correct the CG location. This can be verified by working the original computation with 22 lb [9.98 kg] installed in the cargo compartment (see Problem 7–8).

PROBLEM 7–8

Item	Weight	×	Arm	=	Moment
Aircraft (as loaded)	1335		+84.17		+112 360.5
Ballast	22		+142.8		+3 141.6
	1357				+115 502.1

Then $\dfrac{115\ 502.1}{1357} = 85.1$ in (forward CG limit)

If the ballast in the foregoing problem had been installed near the tail of the airplane, the weight required would have been less. The requirement was to produce a certain moment (1241.6 in-lb), and this could have been done by any combination of weight and arm that would have produced this moment. The longer the arm that is used, the smaller the weight has to be. Note that the moment was computed from the forward CG limit (+85.1). However, after the weight was added to the airplane, the original moment of the airplane and cargo compartment was used.

SIMPLIFIED LOADING METHODS

CG Envelope Charts and Loading Graphs

Because of the many possible loading combinations, especially in airplanes where more than two passengers can be carried, methods have been developed whereby the pilot can quickly determine whether the airplane is loaded within limits without going through a long process of computation. One of these methods involves the use of the CG envelope chart and the loading graph.

The **CG envelope chart** is a graph with airplane weight plotted against index units. The envelope is an area on the graph establishing the combinations of weight and index units where the CG of the airplane will be within limits. A typical CG envelope is shown in Figure 7–35. The index unit used in this example is the moment of the airplane divided by 1000. In this chart it can be seen that there is a satisfactory range of moments for each weight of the airplane. For example, if the airplane is loaded to weight 1650 lb [748.4 kg], the moment can be from 60 000 to about 76 500 in-lb [691 to about 881 kg-m], or from 60 to 76.5 index units. The maximum loaded weight of the airplane is 2200 lb [997.3 kg], and the maximum index number is about 102. If 2200 is divided into 102 × 1000, the rearward CG limit is obtained. The limit is established by the line *AB* in Figure 7–35.

In order to make the CG envelope chart simple to use, a **loading graph** is provided. This graph, illustrated in Figure 7–36, provides for the loading of passengers, fuel, and baggage. Since passengers are loaded at two separate arms, there are two reference lines for passengers. The chart shown is designed for loading the Cessna 170 airplane. The graph plots load weight against index units. The moment of any loaded item may be determined by following the weight line to the right until the position line for the item is intersected and then dropping straight down to the baseline and reading the index unit. The index unit multiplied by 1000 is the moment.

To use the loading graph, proceed as shown in Problem 7–9. Then apply the weight and index number to the CG envelope chart of Figure 7–35, which shows that the point is within the envelope. If the airplane is operated with only a pilot, one passenger, and no baggage, the result will be as shown in Problem 7–10. When these figures are applied to the CG envelope chart, the chart shows that the CG is still within limits, even though it has moved forward.

PROBLEM 7–9

Item	Weight, lb	Index Units
Airplane empty weight	1210	+47.1
Oil	15	−0.3
Pilot and passenger	340	+12.2
Passengers (2)	340	+23.8
Fuel (maximum), 37 gal	222	+10.7
Baggage	70	+6.7
	2197	+100.2

PROBLEM 7–10

Item	Weight, lb	Index Units
Airplane empty weight	1210	+47.1
Oil	15	−0.3
Pilot and passenger	340	+12.2
Fuel	222	+10.7
	1787	+69.7

When computing the CG or loading for a particular airplane, the technician should consult the approved flight manual for that particular aircraft. The method for computing the weight and balance is explained in the manual, and

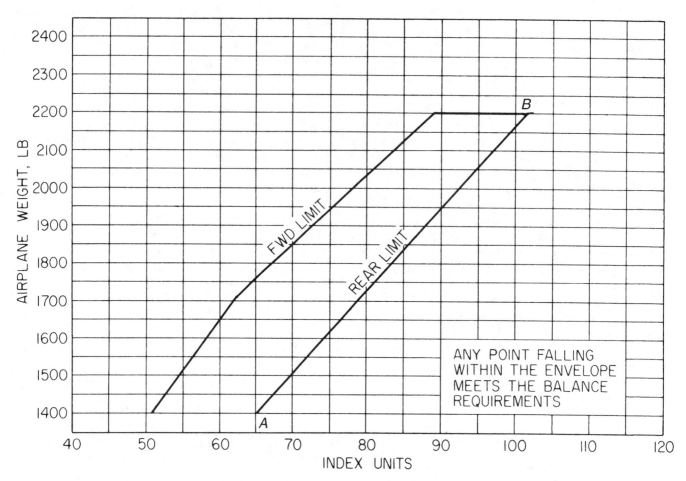

FIGURE 7–35 Center-of-gravity envelope chart.

the charts and graphs used for the airplane CG are also included in the manual. *Remember that the aircraft's weight-and-balance records must include the current empty weight and the empty-weight center of gravity. These two items are necessary to begin all loading computations.*

CALCULATING WEIGHT AND BALANCE FOR LARGE AIRCRAFT

During the flight of a large passenger airplane, the consumption of fuel and the movement of passengers and crew members cause changes in CG location. These changes are compensated for by loading the airplane properly, so that the CG will not move beyond forward or rearward limits regardless of fuel quantity or passenger and crew movements in normal situations. The CG location is calculated before the airplane takes off, and it is adjusted if necessary to assure that changes during flight will not cause the CG to move out of limits. Baggage, cargo, and fuel loading can be used to adjust CG location.

Computer-Calculated Weight and Balance

The principles of weight and balance which have been previously discussed apply to transport-category aircraft as well as smaller, general-aviation aircraft. The general con-

cept of weights, arms, and moments apply regardless of aircraft size. Transport-category aircraft have the same potential for disaster as smaller aircraft when weight and balance limitations are exceeded.

Transport-category aircraft have great flexibility in the configuration that they may be loaded with different combinations of passengers, cargo, and fuel. This flexibility is desirable, but unless adequate consideration is given to weight-and-balance control, such flexibility can easily result in an aircraft being loaded out of balance.

Weight-and-balance control has a direct relationship with the profit or loss made by an air carrier. When extra fuel is required for long trips or to allow for air-traffic-control delays, the payload (passengers, baggage, cargo) must be proportionately reduced to prevent exceeding the maximum weight limits. Since transport-category aircraft generally have a higher maximum certificated ramp weight than takeoff weight, the amount of fuel used for starting and taxiing must be calculated. An example of a form for these calculations is shown in Figure 7–37.

Different methods may be used in providing for proper loading of the aircraft. The system currently in use by most airlines is a computer-programmed weight-and-balance system. This computerized system has the advantage of providing improved load planning, error-free math, and the ability to make adjustments for last-minute changes in the number of passengers and the amount of cargo loaded. The system also has the ability to provide the pilot with a variety of in-

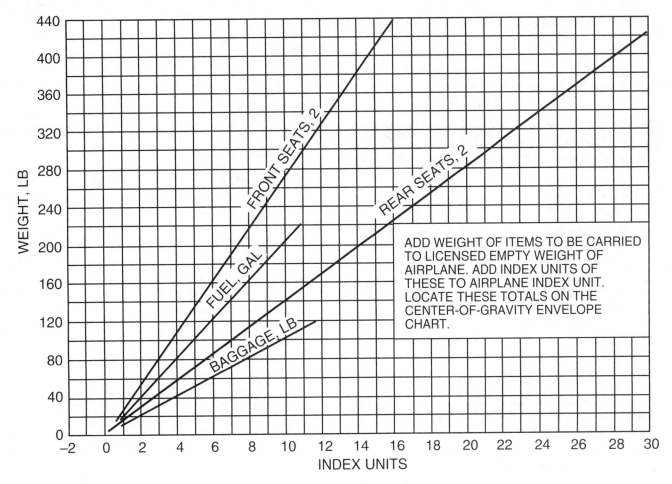

FIGURE 7–36 Loading graph.

formation on the load manifest. This information includes such items as the aircraft's loaded CG, the ramp and takeoff weights, and the payload makeup and its distribution; it also calculates the proper trim setting for takeoff. As with many computer applications, provisions are made for a manual backup system.

For large airliners, rather complex loading charts are prepared. In use, however, these charts greatly simplify the loading process. The charts give the moment arms of the various compartments and fuel tanks and provide an easy method for determining the index of the load in any particular area. The indexes are combined by one of two or three methods, and the CG location is found on the chart. In some cases a special slide rule ("slip stick") is designed to add or subtract moments as the airplane is loaded, thus providing a quick method for computation. This special slide rule is called a **load adjuster** and operates on the principle of index units.

The weight and balance of a large airliner may be determined by means of a **balance computer**. This is a circular chart that is used with an overlay called the **balance planning sheet**. The computer and the balance planning sheet are mounted on a center peg so that the overlay may be rotated over the computer. Each has a vertical index line, and these are superimposed at the start of the computation. A balance computer is shown in Figure 7–38.

The upper portion of the computer has scales to the right and left of the index line representing locations and weights

forward and rearward of the 21% MAC line of the aircraft. As each section of the aircraft is loaded with passengers and cargo, the overlay is rotated and marked. After each section is loaded, the load mark is rotated back to the vertical index line on the computer. When all passengers, cargo, and fuel have been loaded, the index line on the overlay will show whether the loading is within limits. Takeoff weight and CG location in percent of MAC will be indicated.

Automatic Weight-and-Balance Systems

One of the most interesting and useful developments in the weight-and-balance field is the automatic system, or the **on-board aircraft-weighing system (OBAWS)**, designed for such large aircraft as the Boeing 747. This system includes a transducer (strain-gauge unit) installed inside each axle for each wheel of both the main landing-gear wheels and the nose-gear wheels, as shown in Figure 7–39 on page 172. Each transducer generates a signal that can be converted to a weight indication because the transducer senses the shear stress on each axle. The signals from the transducers are sent to a computer, which integrates the weight information from all the axles and sends resulting information to the indicator/control panel and the attitude sensor. A block diagram of this weight-and-balance system is shown in Figure 7–40 on page 172.

The indicator provides a reading of the gross weight of the aircraft and the CG location as a percentage of the

MAC. The flight engineer is therefore always able to determine whether the weight of the aircraft and the location of the CG are within specified limits. The attitude sensor determines whether the aircraft is in the correct attitude (level) for an accurate measurement of CG location.

WEIGHT AND BALANCE FOR A HELICOPTER

The weight-and-balance principles and procedures which have been discussed in connection with airplanes apply generally to helicopters, with one important difference: *Most helicopters have a much more restricted CG range than do airplanes.* In some cases this range is less than 3 in [7.62 cm]. When loading helicopters, it is also often a requirement to calculate the horizontal CG as well as the fore-and-aft CG. The exact location and length of the CG range is specified for each helicopter and usually extends a short distance fore and aft of the main-rotor mast (the center of lift) or in the center of a dual-rotor system. Ideally, the helicopter should have such perfect balance that the fuselage remains horizontal while in a hover and the only cyclic adjustment required should be that made necessary by the wind. The fuselage acts as a pendulum suspended from the rotor. Any change in the CG changes the angle at which it hangs from this point of support. Many recently designed helicopters have loading compartments and fuel tanks located at or near the balance point.

The information in this section applies to a Bell Model 206L Long Ranger helicopter. This information, however, is typical of instructions for leveling, weighing, and computing the CG location for a helicopter.

Leveling

For leveling, a **level plate** is located on the cabin floor approximately 4.0 in [10.16 cm] forward of the aft seat and left of the helicopter center line. This is shown in Figure 7–41. A slotted level plate is located directly above the level plate. The leveling procedure is then as follows:

1. Hang a plumb bob from the small hole in the slotted level plate and suspend it in such a manner that the plumb bob is just above the level plate on the cabin floor.
2. Position the helicopter on a level surface in an enclosed hangar.
3. Position three jacks under the helicopter at the jack and tie-down fittings that are permanently installed. Two forward jack fittings are located at station 55.16, and the aft fitting is located at station 204.92.
4. Adjust the aft jack at the aft jack fitting until the helicopter is approximately level. The forward end of the landing-gear skid tubes should still be in contact with the ground.
5. Adjust all three jacks evenly until the helicopter is level, as indicated when the point of the plumb bob is directly over the intersection of the cross lines of the level plate. This position is shown in Figure 7–41.

REF	ITEM	WEIGHT	MOMENT/ 100
1.	BASIC EMPTY WEIGHT		
2.	PAYLOAD		
3.	ZERO FUEL WEIGHT (SUB-TOTAL) (DO NOT EXCEED MAXIMUM ZERO FUEL WEIGHT)		
4.	FUEL LOADING		
5.	RAMP WEIGHT (SUB-TOTAL) (DO NOT EXCEED MAXIMUM RAMP WEIGHT OF _____ POUNDS)		
6.	LESS FUEL FOR TAXIING		
7.	TAKEOFF WEIGHT (DO NOT EXCEED MAXIMUM TAKEOFF WEIGHT OF _____ POUNDS)		
8.	LESS FUEL TO DESTINATION		
9.	LANDING WEIGHT (DO NOT EXCEED MAXIMUM LANDING WEIGHT OF _____ POUNDS)		

FIGURE 7–37 Typical transport-category aircraft weight-and-balance loading form.

Weighing

A helicopter may be weighed with platform scales or by means of electronic load cells mounted on jacks. The instructions given here are for weighing with scales.

The helicopter should be weighed in a configuration as near empty weight as possible. Empty-weight condition allows for the weight of the basic helicopter together with seats, ballast, special equipment, transmission oil, hydraulic fluid, unusable fuel, and undrainable oil. The baggage compartment should be empty. Weighing is accomplished as follows:

1. Position the scales in an approximately level area and check them for proper adjustment to the zero position. The weighing should be done in an enclosed area to avoid the adverse effects of wind, such as flapping rotors and body sway.

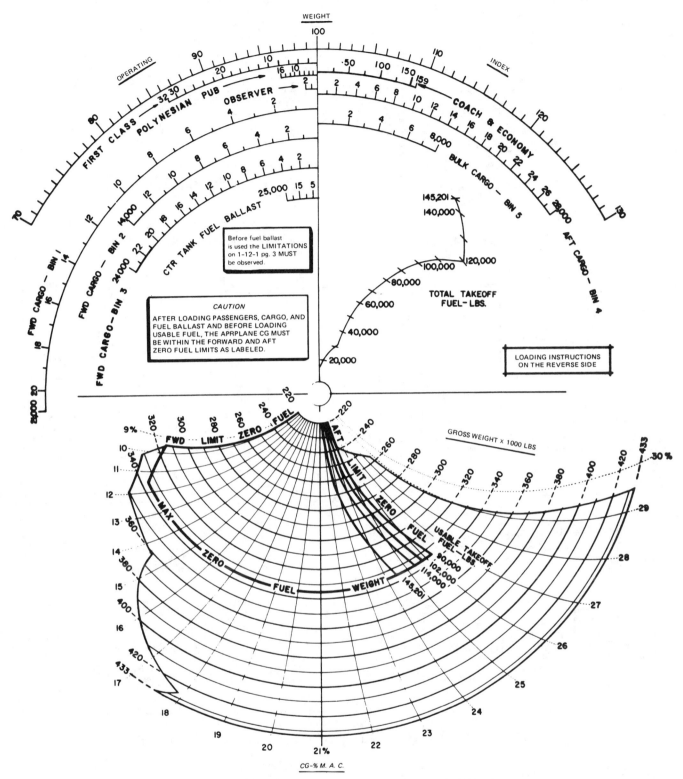

FIGURE 7–38 A balance computer. *(Continental Air Lines)*

2. Position a scale and jack under each jack pad and raise the helicopter clear of the floor.

3. Level the helicopter with the jacks as explained in the section on weighing airplanes.

4. Balance each scale and record its reading.

5. Lower the helicopter to the floor surface and weigh the jacks, blocks, and any other equipment used between the scales and the helicopter. Deduct this tare weight from the scale readings to obtain net scale readings. The total of the net scale readings is the **as-weighed weight** of the helicopter.

A typical example of net weights is 513 lb [232.7 kg] for the forward left scale, 522 lb [236.8 kg] for the forward right scale, and 1063 lb [482.2 kg] for the aft scale. The as-weighed weight is then the sum of the net scale weights, or 2098 lb [951.6 kg].

Weight and Balance for a Helicopter   **171**

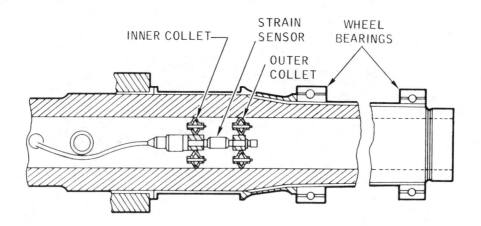

FIGURE 7–39 Transducer installed in the axle of a Boeing 747 airplane.

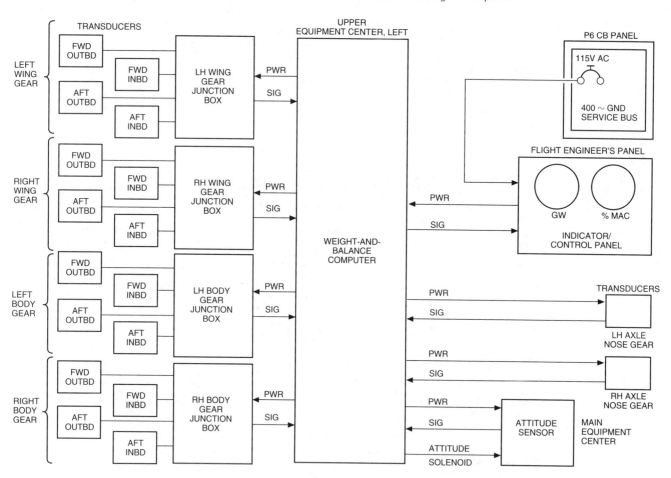

FIGURE 7–40 Block diagram of the weight-and-balance system for a Boeing 747 airplane.

Determining CG Location

The CG location for a helicopter is determined in the same manner as for other aircraft. In the case of the Bell Model 206L, the datum line is at the 0.0 fuselage station, which is just forward of the nose of the helicopter, as shown in Figure 7–42. The CG location aft of the datum line is found as follows:

$$\text{CG location} = \frac{\begin{array}{c} \text{moment of forward weights} \\ + \text{ moment of rear weights} \end{array}}{\text{total net weight}}$$

The location of the forward weighing point is 55.16 in [140.1 cm] aft of the datum line at FS 55.16, and the location of the rear weighing point is at FS 204.92.

The sum of the weights indicated by the forward scales is 1035 lb [469.9 kg], and the moment is 1035 × 55.16 = 57 090.6 in-lb [657.84 kg-m]. The moment of the aft weight is 1063 × 204.92 = 217 829.95 in-lb [2509.9 kg-m]. The total moment is then 274 920.55 in-lb [3167.8 kg-m]. When this is divided by the total net weight of the helicopter, the CG location is found to be 131.04 in aft of the datum line.

If a helicopter when weighed does not include all the

1. SLOTTED LEVEL PLATE
2. AFT JACK FITTING
3. JACKS
4. LEVEL PLATE
5. FORWARD JACK FITTINGS
6. PLUMB BOB

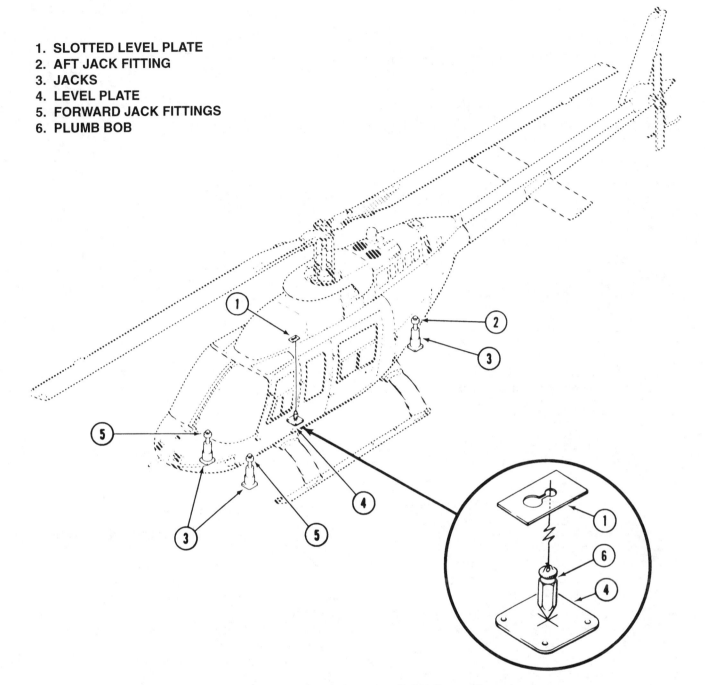

FIGURE 7–41 Leveling a helicopter. *(Bell Helicopter Textron)*

equipment required for the empty-weight condition, these items must be added. The weights must be added to the as-weighed weight, and the moments must be computed and added to the original computed moment. The result is a total weight known as the **derived weight** and a slightly different CG location.

If the final empty-weight CG location does not fall within the limitations set forth in the empty-weight CG location chart, ballast plates are installed either forward or rearward in specified locations. Ballast is never added in both forward and rearward locations. The forward ballast location in the Bell Model 206L helicopter is at +13 (FS 13.0), and the rearward ballast location is at +377.18 (FS 377.18) as

shown in Figure 7–42. The ballast requirement is computed in the way described earlier in the section on correcting the CG location of airplanes.

REVIEW QUESTIONS

1. Define *center of gravity*.
2. What is the general law of the lever?
3. What is the condition that exists when the sum of the positive moments equals the sum of the negative moments?

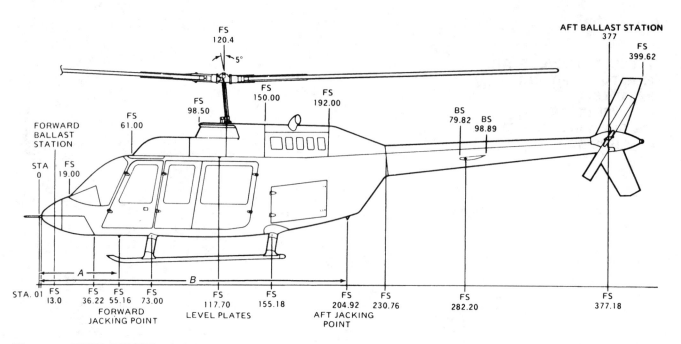

BS = BOOM STATION
CG = CENTER OF GRAVITY
STA = STATION, FUSELAGE
A = FORWARD ARM, 55.16 INCHES
B = AFT ARM, 204.92 INCHES
C = FORWARD SCALE READING (NET)
D = AFT SCALE READING (NET)
AC = FORWARD MOMENT
BD = AFT MOMENT
C+D = TOTAL WEIGHT

$$\frac{AC + BD}{C + D} = \text{CG FROM STA 0}$$

FIGURE 7–42 Datum line and weighing points for a helicopter. *(Bell Helicopter Textron)*

4. Define the term *moment of force*.
5. Define the term *arm*.
6. What is the operating CG range of an aircraft?
7. In what different methods may the location of the CG and CG range be expressed?
8. If the wings of an airplane have no sweepback, at approximately what point should the CG be with respect to the chord of the wing?
9. What is the purpose of the datum?
10. Define *empty weight* as used for weight-and-balance computations.
11. Define *maximum gross weight* for an aircraft.
12. How is the useful load of an aircraft determined?
13. What is meant by the term *unusable fuel*?
14. How much does one gallon of gasoline weigh? One gallon of oil?
15. What types of scales are commonly used for weighing aircraft?
16. List the equipment in addition to the scales that is commonly used to weigh aircraft.

17. How is the leveling means for an aircraft determined?
18. What is tare?
19. What should be done with an old weight-and-balance report when a new weight-and-balance report is computed?
20. List some of the dangers that may exist as a result of overloading an aircraft.
21. What undesirable conditions may exist when an airplane is adversely loaded?
22. What is an adverse-loading check?
23. How is minimum fuel determined for loading checks?
24. In performing a most rearward adverse-loading check, how is it determined whether an item is to be loaded at the maximum or minimum weight?
25. Why is the proper loading of a helicopter particularly critical?

Aircraft Materials 8

INTRODUCTION

Many materials have been used in aircraft. Early aircraft were lightweight assemblies of wood and fabric, kept aloft by engines producing marginal power. Steel-tubing and wooden structural elements with a covering of cotton or linen fabric were used for the first practical aircraft. The development of aluminum-alloy structures, beginning in the late 1930s, resulted in the all-metal designs in use today. By the early 1950s, aircraft development focused on power plants. At that time, aircraft designs were more limited by power considerations than by structural problems.

As more powerful engines were developed, the use of existing materials was pushed to the limit. The development of supersonic aircraft resulted in the need for structural materials that would provide the needed strength at high temperatures, yet be light enough to get off the ground. The use of titanium, corrosion-resistant steels, and metal honeycomb developed in response to this need.

Research and development of structural materials are again the center of attention. Emphasis is being placed upon lighter weight with the development of new metal alloys. In addition, synthetic fibers and resins are being combined to produce composites with very favorable strength-to-weight ratios.

As the twentieth century draws to a close, a revolution is taking place in materials for aerospace structures. A wide variety of material types and designs are being used. The technician of the twenty-first century will no longer be able to perform his or her job by memorizing a few aluminum-alloy designations or the head markings on a few rivets. However, aircraft built with wood, steel tubing, fabric, and conventional aluminum-alloy structures will still be flying and will still require maintenance. The aircraft maintenance technician, unlike maintenance technicians in most industries, cannot ignore the old and move on to the new. The future aviation technician must have a basic knowledge of the properties of a variety of materials.

AIRCRAFT MATERIALS

Early aircraft made extensive use of **wood**. Wood was used for both structural elements and as a cover, or skin. Wood offered a material that was low cost, lightweight, and easily worked. When used within design limitations, wood has a high strength. The Hughes HK-1 Hercules (*Flying Boat*) had both structure and surface built entirely of laminated wood. A large number of all-wood training aircraft were built during World War II. The need to continually protect wood against the elements to prevent decay was a drawback to its continued use for aircraft. The demise of wood as a structural material was also affected by the development of metals for aircraft use.

Materials made of **fabric** are used for covering aircraft structures made from wood or metal. The structural strength and the airfoil shapes are formed by the structural elements. The fabric forms a continuous cover over these parts. The fabric must be treated with a resinous material called aircraft-dope. The dope stiffens the fabric and helps to protect it from the elements. Early fabrics were organic cottons and linens. The strength of cotton deteriorates with age. The need for careful inspection and periodic replacement of the fabric cover is a drawback to its use. Synthetic fabrics with longer service life are replacing cotton as an aircraft covering.

Structures made of **metal** became popular with the development of aluminum alloys usable for aircraft. Their light weight and good formability made aluminum products a natural replacement for wood as a structural material. The weight of iron-based metals has limited their use to areas where high strength is required, such as engine mounts or the steel-tube structure covered with lightweight fabric. Most of the steel, other than engine parts, that the technician will encounter will be in the form of steel tubing. Titanium, relatively new to aircraft use, has weight, strength, and temperature characteristics favorable for high-performance aircraft. Metals exhibit a number of properties, such as formability, that enhance their use in aircraft design. Other properties, like susceptibility to corrosion, limit their use. All metals are subject to and must be protected from corrosion in varying degrees. Magnesium is a strong, lightweight metal, but very susceptible to corrosion. The need for added protection from and constant inspection for corrosion has limited the use of magnesium as an aircraft material. The future use of metals for aircraft structures is heavily dependent upon the development of synthetic and composite materials.

Materials made of **plastic** are synthetic, or man-made. Plastics are manufactured by taking apart the basic elements of a material. These elements are then recombined in a manner which produces a new material with its own properties. Plastics are widely used in aircraft. Familiar plastic products include transparent window materials, fiberglass (polyester-resin) wingtips, ABS-type fairings, and Dacron fabric. Not so apparent are the synthetic materials used for the resins in paints, for Teflon hoses, for the dope used on fabric, for various adhesives and sealants, and for the pulleys used for control cables.

A **composite material** is one which is made of a combination of two or more materials or of a material in two different forms. A major advantage of a composite material is its light weight. A second advantage is the ease with which complex parts can be formed. The use of composite materials is not new. Laminated composite materials have been used for a number of years. For example, control-cable pulleys were once made from a laminated material of phenol-formaldehyde resin reinforced with linen cloth. In recent years the use of laminated and sandwich composites have begun a possible revolution in the materials used for aircraft.

PROPERTIES OF MATERIALS

Materials exhibit a number of characteristics, or **properties**. The degree to which various properties exist in a material will determine its suitability for a specific use. The properties of materials can be categorized as mechanical, physical, or chemical.

Mechanical Properties

The ability of a material to be deformed without rupture or failure is called **plasticity**. Plasticity can take two different forms, ductility and malleability. The **ductility** of a material refers to its plasticity under a tension or a pulling load. Ductility allows materials like aluminum and copper to be drawn into very small wires. **Malleability** is the plasticity exhibited by a material under a pounding or a compression load. The rolling of metal into a thin sheet is possible due to its malleability. A material can have high ductility or high malleability, or both. **Brittleness** is the opposite of plasticity. A brittle material is one that cannot be visibly deformed and will shatter or break under load.

The term **elasticity** describes the ability of a material to deform under load and return to its original shape when the load is removed. The **elastic limit** is the maximum amount of deformation that a material can undergo and still return to its original shape. When the elastic limit is exceeded, permanent, or "plastic," deformation occurs. The elastic limit is also called the **proportional limit**. It has been proven that the amount of deformation in a material is proportional to the stress that causes it, as long as the elastic limit is not exceeded. The proportional principle of stress and strain is referred to as *Hooke's law*.

A material's **hardness** is not a fundamental property but is related to its elastic and plastic properties. An operational definition of hardness is *the resistance to penetration*. In some materials, like steel, the hardness is directly related to the tensile strength. A hardness test can be used to estimate the strength of some materials.

There is no direct and accurate method to measure the **toughness** of a material. Toughness is a desirable characteristic of the material to resist tearing or breaking when it is bent or stretched. Closely related to plasticity, toughness also involves the magnitude of the force causing the deformation.

One of the most important characteristics of a material is **strength**. A simple definition of strength would be the ability of a material to resist deformation. The type of load, or stress, on a material affects the strength it exhibits.

The term **stress** may be defined as an internal force that resists the deformation of a material resulting from an external load. The different types of stress are called tension, compression, bending, torsion, and shear. The five types are illustrated in Figure 8–1. **Tension stress** is the result of a load which tends to pull apart or stretch the material. **Compression stress** occurs when the load presses together or tends to crush an object.

Two layers of a material being pulled apart results in **shear stress**. Shear stress can develop when two pieces of material are bolted or riveted together. If a force is applied such that the two plates tend to slide over one another, shear stress develops in the bolt. If the stress becomes greater than the shear strength of the bolt, it will be cut as with a pair of shears.

The term **bending stress** describes a combination of three types of stress in a bending object. Material on the outside of the bend will tend to stretch and thus have tension stress. Material on the inside of the bend will be compressed. Tension stress and compression stress, acting opposite of each other, create a shear stress where they meet.

The last type of stress, **torsion stress**, is the result of a twisting force. In Figure 8–1 a shaft is clamped solidly on one end and a pulley is mounted on the other end. A cable around the pulley is attached to a weight so that the weight tends to turn the pulley. This action applies a twisting, or torsion, force to the shaft. Stress caused by a torsion force is, like bending, a combination of tension, compression, and shear stress.

The amount of stress developed is a function of the force, or load, and the area upon which the force acts. Stress is expressed in units of pounds per square inch (psi). The amount of stress can be determined by the formula

$$f = \frac{P}{A}$$

where

f = stress, psi
P = force, lb
A = area, in^2

Tension stress tends to pull a member apart. In a member under tension, the force acts upon a plane that is at right angles to the line of the force. A member under compression

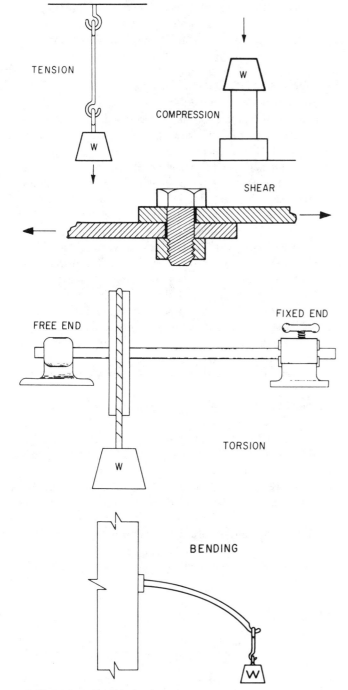

FIGURE 8–1 Five kinds of stress.

elongated or stretched. Strain is measured in units of inches per inch (in/in) by the formula

$$e = \Delta \frac{L}{L}$$

where

e = strain, in/in
ΔL = change in length, in
L = the original length

If a piece of material measures 6.000 in at rest and 6.006 in under load, the strain would be 0.001 in/in. Strain may also be expressed in terms of percentage of elongation.

Stress and strain can be measured by various types of testing equipment. A form of tensile testing is shown in Figure 8–2. Material of a known cross-sectional area is gripped in the machine and a known force, tending to pull it apart, is applied. The strain is measured by marking off a fixed distance before the metal is put under load. The distance between the marks is then measured while under load. The amount of elongation, and thus strain, can be calculated. The load can be increased until failure of the metal occurs. This provides a measure of strength under tension load, or tensile strength, for the material.

A material can be tested under compression with similar equipment. Material under a compression load will usually fail by **buckling**, or bending (Figure 8–3), before it reaches its compressive strength. The load at which a member will buckle under compression will be dependent upon the length of the member, its cross-sectional area, cross-sectional shape, and so on. Because of buckling, compression-test data is not widely used as a measure of the strength of a material. Most materials exhibit their greatest strength when loaded in tension. The amount of shear stress a material can take is always less than the tensile stress.

Stress-Strain Diagrams. Plotting the results of a tensile test produces a graph similar to Figure 8–4. The vertical axis of the graph shows the amount of stress (psi), and the

will also have the affected plane, or area, at right angles to the line of force. Stress components caused by a load acting *normal*, or perpendicular, to the plane of the structural member are known as **normal stresses**. Tension and compression are both examples of normal stress.

Shear stresses tend to divide the material into layers. A shear stress applies the load parallel to the surface, or plane, affected.

When a stress is applied to a piece of material, there is always some deformation of the material, even though it may be very small. This deformation is called **strain**. If the applied stress does not exceed the elastic limit, the material will return to its original shape as soon as the stress is removed. A tension stress will result in the material being

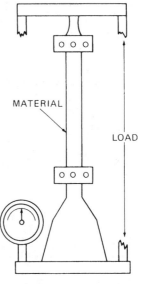

FIGURE 8–2 Tensile testing.

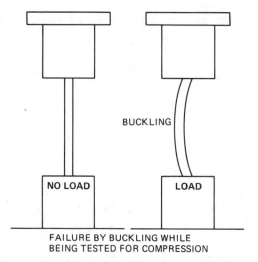

FIGURE 8-3 Compression testing.

horizontal axis is labeled in units of strain. The amount of strain is plotted against the amount of stress which causes it. The resulting curve is called a stress-strain diagram. Much information about the properties of a material can be learned from observing the stress-strain diagram. The diagram begins as a straight line indicating that the stress and the strain are proportional. As stress increases, the elastic limit is reached. This point, also known as the proportional limit, is indicated by the end of the straight line portion of the diagram (*A*). Exceeding the elastic limit causes permanent deformation to take place. The point at which permanent deformation begins is known as the **yield point**, or **yield stress** (*B*). For some materials, when the yield stress is reached, the strain will continue to increase with little if any increase in stress. This is illustrated by the line in Figure 8-4 having a small "hook" at the yield point. Increasing stress beyond the yield point causes a rapid rate of strain or deformation to occur until the material ruptures, or fails (*X*).

The point just prior to failure is known as the ultimate stress point, or the ultimate tensile strength (*C*). The portion of the diagram from the origin (*O*) to the elastic limit (*A*) is known as the **elastic**, or **proportional, range** of the material. From the yield point (*B*) to the ultimate tensile strength (*C*) the material is said to be in the **plastic range**.

Some more ductile materials will have a curve similar to that shown in Figure 8-5. The material does not exhibit an abrupt increase in strain at the yield point. The stress-strain diagram does not have a definite end to the straight-line portion. For a material of this type the yield point has been arbitrarily determined to be the point where a permanent strain of 0.002 occurs.

The angle formed by the straight-line portion of the diagram and the horizontal axis indicates the **modulus of elasticity**. The modulus of elasticity is the value obtained by dividing the stress by the strain. Values for the modulus of elasticity are quite large. For example, the modulus for aluminum is 10×10^6, and for steel it is 30×10^6.

Stress-strain diagrams can be used to provide information about a specific material or to compare several materials. They may also be used to graphically demonstrate mechanical properties covered in this section. Diagrams for four materials with widely varying properties are combined in Figure 8-6.

The plasticity of a material is graphically represented by the portion of the diagram to the right of the yield point. Material *A* would have no plasticity. The abrupt failure of the material with little strain would classify it as brittle. Materials *B* and *C* have virtually the same elastic range and limits. However, material *C* fails shortly after exceeding the yield point and has little plasticity. Material *D* has a lower elastic limit than *B* but has a similar value of plasticity.

Material *B* could be described as **stiffer** than material *D* because of a higher modulus of elasticity. Material *B* also can withstand a higher stress before exceeding its elastic

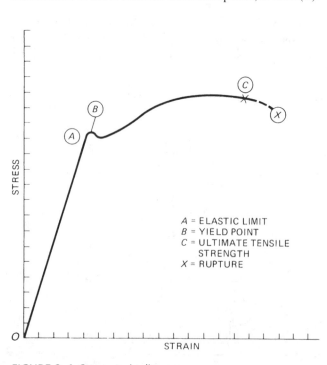

A = ELASTIC LIMIT
B = YIELD POINT
C = ULTIMATE TENSILE
     STRENGTH
X = RUPTURE

FIGURE 8-4 Stress-strain diagram.

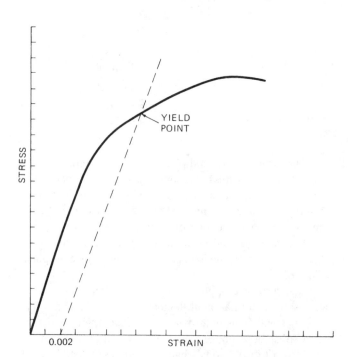

FIGURE 8-5 Stress-strain diagram for a ductile material.

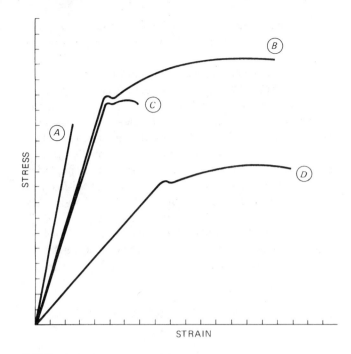

FIGURE 8–6 Comparison of four materials.

The strength of a material is usually reported in terms of the tensile yield stress or ultimate tensile stress. In Figure 8–6, material *B* has the highest yield strength and thus could be described as the strongest material. However, many properties, and the type of load to be applied, need to be considered before any material is said to be the "strongest," "toughest," and so forth.

Strengths of Joints

Many materials are fastened together with bolts or rivets (fasteners) in a simple lap joint. Regardless of the properties of a material, the strength of the assembly can be no greater than that of the joint. A joint will fail for one of four reasons: tensile load, shear load on the fastener, bearing failure of the material, or tear-out of the material.

A **tensile failure** is related to the tensile strength of the material. A hole must be drilled to insert the fastener. The drilled hole will effectively reduce the cross-sectional area of the material, as shown in Figure 8–8*a*. For this reason the tensile load of a joint will never equal 100% of the tensile load of the material.

$$P_{\text{tensile failure}} = f_t \times [A - N(t \times D)]$$

where

$$P_{\text{tensile failure}} = \text{load leading to failure, lb}$$
$$f_t = \text{tensile strength of the material, psi}$$
$$A = \text{cross-sectional area of material, in}^2$$
$$N = \text{number of fasteners}$$
$$t = \text{thickness of the material, in}$$
$$D = \text{diameter of the fastener, in}$$

A joint will experience **shear failure** when a load, applied as shown in Figure 8–8*b*, is sufficient to cause the fastener to fail by shearing. The shear strength of a joint will depend upon the material, number, and size of the fasteners.

limit. Material *D* can have a higher strain, or more stretch, before exceeding the elastic limit and would generally be described as more elastic than *B*.

Toughness is sometimes defined as the total area under the curve of a stress-strain diagram. Using this definition, material *B* would have more toughness than material *D*. Even though they have similar plasticity, material *B* takes a higher stress. Because of the low plasticity of material *C*, it could be considered as having less toughness than material *D*.

The stress-strain diagrams shown are typical of structural materials. Figure 8–7 shows a stress-strain diagram for a highly elastic material, such as rubber.

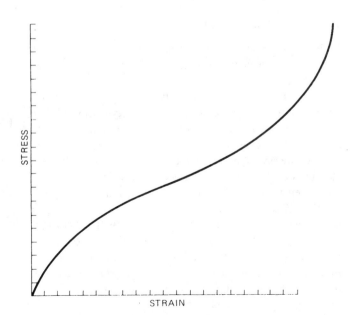

FIGURE 8–7 Stress-strain diagram for an elastic material.

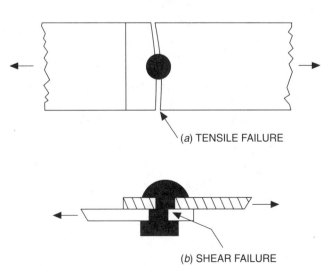

(a) TENSILE FAILURE

(b) SHEAR FAILURE

FIGURE 8–8 Tensile and shear failure of a joint.

$$P_{\text{shear failure}} = f_s \times A \times N$$

where

$P_{\text{shear failure}}$ = load leading to failure, lb
f_s = shear strength of the fastener material, psi
A = cross-sectional area of the fastener, in
N = number of fasteners

A **bearing failure** is a type of compressive failure from the fastener pushing, or bearing, against the sheet. A bearing failure, shown in Figure 8–9a, will show up as an elongated hole. Bearing failure depends upon the bearing strength, the thickness of the material, and the size of the fastener.

$$P_{\text{bearing failure}} = f_b \times t \times D$$

where

$P_{\text{bearing failure}}$ = load leading to failure, lb
f_b = bearing strength of the material, psi
t = thickness of the material, in
D = diameter of fastener, in

Placing a fastener too close to the edge of the sheet will result in **tear-out failure**. As shown in Figure 8–9b, a tear-out is a shear failure of the material.

$$P_{\text{tear-out}} = 2 \times (f_s \times t \times ed)$$

where

$P_{\text{tear-out}}$ = load leading to failure, lb
f_s = shear strength of the material, psi
t = thickness of material, in
ed = edge distance (distance from the center of the hole to the edge of the sheet), in

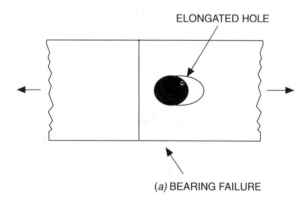

(a) BEARING FAILURE

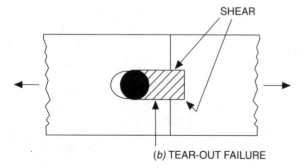

(b) TEAR-OUT FAILURE

FIGURE 8–9 Bearing and tear-out failure of a joint.

It is common practice to use the edge distance to calculate tear-out loads. In actuality, the value will be slightly less than that shown.

The aviation maintenance technician will frequently deal with joint failures when making sheet-metal repairs on aircraft structure. More detailed information can be found in the companion text *Aircraft Maintenance and Repair*. The following example is used here to illustrate the use and interaction of the preceding information: Assume that you are given two sheets of material to be joined in a simple lap joint with a single fastener. The fastener has a 0.25-in diameter with a f_s of 32 000 psi. The material is 0.050 in thick (each piece) and 1 in wide. The material has a f_t of 55 000 lb and a f_b of 85 000 lb. The center of the fastener hole is 0.75 in from the edge of the material. Find the load at which each type of failure will occur.

Tensile failure:

$$55\,000 \times [(0.050 \times 1.0) - (1 \times 0.25 \times 0.050)] =$$
$$55\,000 \times (0.050 - 0.0125) =$$
$$2062.5 \text{ lb}$$

Shear failure:

$$32\,000 \times (0.7857 \times 0.25 \times 0.25) \times 1 =$$
$$32\,000 \times 0.049 \times 1 =$$
$$1517.2 \text{ lb}$$

Bearing failure:

$$85\,000 \times 0.050 \times 0.25 =$$
$$1062.5 \text{ lb}$$

Tear-out failure:

$$2 \times (32\,000 \times .050 \times .75) =$$
$$2400 \text{ lb}$$

In this example the lowest value is that for bearing failure. Therefore, this joint could not support a load of more than 1062.5 lb.

Physical Properties

Physical properties of materials that are of interest to the aircraft technician include density, conductivity, and thermal expansion.

The **density** of a material is its weight per unit volume, such as pounds per cubic inch. The density combined with the strength characteristics of a material produces what is known as the strength/weight ratio of a material. For example, say we are given two materials, A with a tensile yield strength of 120 000 psi and a density of 0.28 lb/in³, and B with a tensile yield strength of 60 000 psi and a density of 0.10 lb/in³. We are then asked to provide a rod 8 in long of each material capable of supporting a load of 10 000 lb. Previously, we used the formula $f = P/A$ to find stress. In this problem, stress and load are known. To find area, the formula will be expressed as $A = P/f$. Carrying out the calculations for the two materials finds the cross-sectional area required for each material to be

$$A = 0.083 \text{ in}^2 \qquad B = 0.167 \text{ in}^2$$

Multiplying the area times the length (8 in) times the density of each material will give the weight of the required rod:

$$A = 0.186 \text{ lb} \qquad B = 0.134 \text{ lb}$$

Even though rod B will be larger in cross section, it will still weigh less than rod A because it has a better strength/weight ratio. Dividing the tensile strength by the density for each material provides the following values:

$$A = 4.3 \times 10^5 \qquad B = 6.0 \times 10^5$$

As illustrated in this example, material B has a better strength-to-weight ratio.

A material's **specific gravity** is used to compare the weights of materials. The specific gravity of a substance is its weight divided by the weight of an equal volume of water. Since equal volumes are involved, specific gravity also is a ratio of densities.

Conductivity of a material may refer to either electricity or heat. **Thermal conductivity** is the property of a material to conduct heat. Thermal conductivity is desirable in a material used to conduct away excess heat. Insulating materials should have a low thermal conductivity. **Electrical conductivity** is a measure of the material's ability to have an electron flow.

The term **thermal expansion** refers to the dimensional change that occurs as materials become hotter. Thermal expansion is of interest in aircraft because of the temperature extremes—seasonal, performance, and altitude-related—in which an aircraft operates.

Chemical Properties

The chemical properties of a material refer to its atomic structure and basic elements. While they may not appear to be of interest or of significance to the aircraft technician, the behavior of metals is greatly affected by chemical properties. Mechanical properties, such as ductility and hardness, as well as response to thermal treatment are related to chemical properties. The chemical properties of a metal also determine its susceptibility to corrosion.

GENERAL PROPERTIES OF METALS

Chemical elements may be roughly categorized into three groups: metals, nonmetals, and inert gases. Metals, when in a solid state, are characterized by the following properties:

1. Crystalline structure
2. High thermal and electrical conductivity
3. Ability to be deformed plastically
4. High reflectivity

All elements are made up from various combinations of atoms. The atoms are bonded together by various forces. Temperature affects the energy levels within the bonds so that at cool temperatures the atoms are closely packed and form a solid material. As the temperature rises, the energy level between the atoms increases and the solid becomes a liquid and eventually a gas.

The arrangement of the atoms affects the properties of a metal. During the process of solidification, the atoms will arrange themselves in an orderly manner called a **space lattice**. The smallest unit having the same arrangement as this crystal is called the unit cell. The unit cell adds atoms in an orderly fashion and grows into a large crystal. The number of atoms in a small crystal may number in the billions. The number of atoms in the unit cell are few, depending on the crystalline system. Seven systems of atom arrangement are known to exist, but the widely used metals form either a cubic or hexagonal system. The cubic shape can be either **face-centered cubic (FCC)** or **body-centered cubic (BCC)**. The hexagonal shape is closed-packed and abbreviated CPH. Figure 8–10 shows the atom arrangement of the three common systems and typical metals for each structure. Some metals will crystallize in one form and upon further cooling change to another form. This type of change is called an **allotropic change** and affects the processing of some metals. Metals with the face-centered lattice lend themselves to a ductile, plastic, workable state. Metals with the hexagonal lattice exhibit a general lack of plasticity and

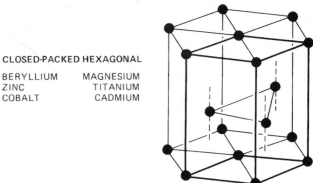

FACE-CENTERED CUBIC

IRON   CHROMIUM
VANADIUM   MOLYBDENUM
LITHIUM

BODY-CENTERED CUBIC

ALUMINUM   NICKEL
COPPER   SILVER
GOLD   LEAD

CLOSED-PACKED HEXAGONAL

BERYLLIUM   MAGNESIUM
ZINC   TITANIUM
COBALT   CADMIUM

FIGURE 8–10 Common unit-cell types for metals.

rapidly lose what they do have upon shaping or cold forming. Metals with body-centered lattice have properties between these two groups.

Figures 8–11a and 8–11b illustrate the formation of a metal crystal. From random unit cells (a), crystal growth takes place in three dimensions and continues until it is stopped by surrounding crystals (b). Crystals in a newly solidified metal will have random shapes and sizes, as shown in Figure 8–11c. Crystals found in commercial metals are commonly called grains. Although each grain has a random external shape, its internal structure is composed of an orderly space lattice.

Within the crystal, there are planes called **slip planes**. When an external force is applied to the crystal, the atoms along a slip plane will move in relation to one another. If the force is less than the elastic limit of the material, the atoms will return to their original position after the force is removed. If the force exceeds the elastic limit, plastic deformation will occur and the shape of the crystal will be permanently altered. As plastic deformation takes place, the slip plane is "used up," or reaches a point where no more slippage occurs. If the force continues, other slip planes must be brought into action or the metal will fail by rupture. Figure 8–12 illustrates grain (crystal) changes that take place as a metal is deformed by being rolled to a thinner size.

The control of **grain size** is important to commercial metals. A fine-grain metal is usually tougher and stronger than one with a coarse grain. A coarse-grain metal will normally be easier to form to shape than one with a fine grain. The size of grain will obviously depend upon the use of the material. The grain size of a metal is affected by the rate of cooling of the molten metal. Slow cooling results in the formation of a small number of large grains. Fast cooling causes the formation of a large number of small grains.

As a metal is formed, or cold worked, slip planes are used up in the crystals. As the slip planes are used up, the material becomes harder and stronger. This is known as **work hardening**, **strain hardening**, or **cold working** of the metal. If greater strength is desired, the effect of work hardening is good. If more forming is to be done on the part, work hardening may not be good. The effects of work hardening can be removed by a process known as **recrystallization**, or **annealing**. Annealing is performed by heating the material, followed by slow cooling. The recrystallization

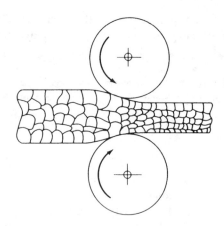

FIGURE 8–12 Grain size and shape being altered by rolling.

process takes place with the metal in a solid state. During the heating and cooling cycle, new crystals are formed with a new set of slip planes. The metal is once again capable of being easily formed or shaped.

All metals are subject to cold working and annealing in varying degrees. In the preceding paragraphs, the discussion has been on pure, or homogeneous, metals. Other than for decorative and low-strength applications, most pure metals are not practical for structural applications. A metal that has been fully cold worked, or work hardened, has a moderate amount of strength. On the other hand, the hardness that increases the metal's strength also makes it more brittle and susceptible to other types of failure. For these reasons, modern civilization could not have developed without metal alloys.

ALLOYS

An alloy is a mixture of two or more metals. Alloys consist of a base metal with small percentages of other materials. Alloys are used to enhance the properties of the base metal, such as its corrosion resistance, tensile strength, or workability. The elements of an alloy and a base metal combine as solid solutions, compounds, or mixtures. An in-depth study of alloy types and behavior is far beyond the scope of this text and the knowledge needed by the aviation maintenance

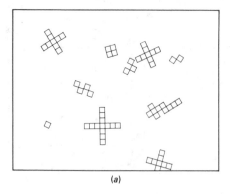
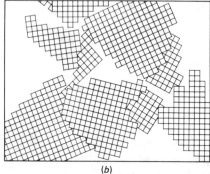
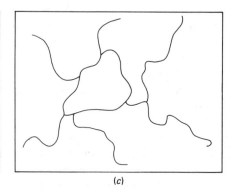

(a) (b) (c)

FIGURE 8–11 Crystal growth as metal solidifies.

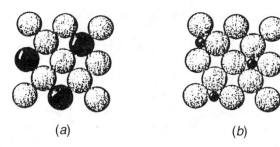

FIGURE 8–13 Atom arrangement for substitutional solid-solution alloy (a) and interstitial solid-solution alloy (b).

technician. However, a basic knowledge of the common alloys used for aircraft will enhance the technician's performance.

Solid-Solution Alloys

The term *solution* brings to mind liquids, but solutions exist in solid materials as well. A solution consists of the **solvent**—the base metal—and the **solute**—the alloying element. A solid-solution alloy forms when the atoms of two elements are combined. A solid-solution alloy may be one of two types: substitutional or interstitial.

A **substitutional solid-solution alloy** occurs when the solute atoms are very close in size to those of the solvent. The solute atom takes the place of, or substitutes for, a solvent atom in the lattice structure (see Figure 8–13a). An **interstitial solid-solution alloy** occurs when the solute atom is much smaller than the solvent atoms. Instead of replacing another atom in the lattice structure, the solute atom wedges into the open space between the solvent atoms (see Figure 8–13b). In both cases the lattice structure will be slightly distorted by the changes, providing different properties of the material.

Some combinations of elements are completely soluble at all times; that is, regardless of the concentration of either element, 100% solubility of one into the other occurs. For other combinations, there may be 100% solubility at elevated temperatures and only limited solubility at room temperature. Because of this, a number of different actions will occur depending upon the elements that are combined. We will look at two types of alloys in this section: Type I, solid-solution alloys, and Type II, eutectic alloys.

A **Type I alloy** occurs when complete solubility exists in both the liquid and solid states. Two metals that exhibit this characteristic are copper (Cu) and nickel (Ni). The chart in Figure 8–14 shows both metals against the melting, or freezing, point for each material. Nickel, plotted on the right, becomes a solid at about 2646°F [1452°C]. Copper, plotted on the left, becomes a solid at 1981°F [1083°C]. It would appear logical that a combination of the two elements would become solid somewhere between these two values, and in this case that is true. The actual solidification will occur over a small range of temperature; thus the two lines on the graph. The upper line is called the **liquidus**; everything above this line will be a liquid solution. The lower line is called the **solidus**; everything below this line will be a solid solution of copper and nickel. The area between the lines represents an area in which the metals are solidifying and crystal growth is taking place.

The Type I alloy will produce crystals that are all a solid solution of copper and nickel. There will be no pure copper or nickel crystals. A common nickel-copper alloy called monel is 67% copper and 33% nickel.

When two elements form a solid-solution alloy, the alloy will be stronger and harder, as shown in Figure 8–15, and have less electrical conductivity than would be predicted from the properties of the individual metals. For example, nickel, with a tensile strength of 50 000 psi [345 MPa], and

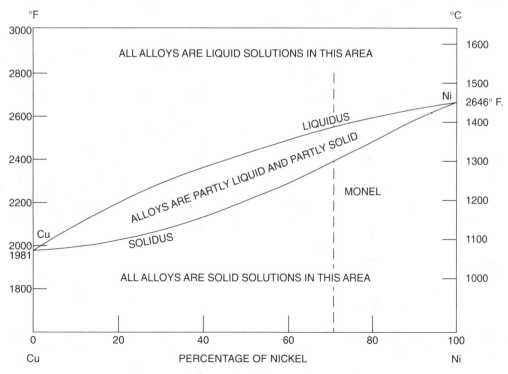

FIGURE 8–14 Copper-nickel alloy concentration diagram.

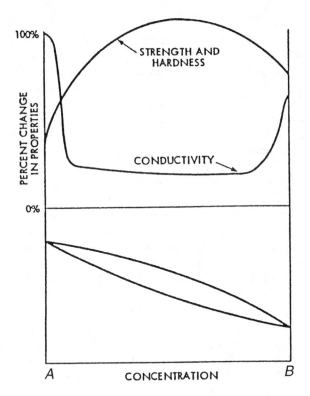

FIGURE 8–15 Mechanical properties for Type I alloy.

copper, with a tensile strength of 32 000 psi [221 MPa], combine to form monel, with a tensile strength of 100 000 psi [670 MPa]. Color and magnetic properties cannot be predicted for solid-solution alloys.

A **Type II alloy** is the result of the combination of two elements that have complete solubility in the liquid state and only limited solubility in the solid state. A cadmium (Cd) and bismuth (Bi) combination, shown in Figure 8–16, is frequently used to illustrate the behavior of this type of alloy. As in the Type I alloy, the diagram shows one material and its freezing point on each side. The liquidus is represented by line *AEB*, and the solidus by line *ACEDB*. Point *E* is referred to as the **eutectic point**. From looking at Figure 8–16 it can be seen that adding a small amount of one metal to the other causes the freezing point to be lowered in both cases. The eutectic point is where the lowest possible solidification temperature occurs for the two-element alloy. In the example illustrated, this occurs at a point where there is a combi-

nation of 60% bismuth and 40% cadmium, called the **eutectic alloy**.

A eutectic alloy will always become a solid solution at a single temperature value; 284°F [120°C] in the example shown in Figure 8–16. Unlike Type I alloys, the crystals formed are not a solid solution of bismuth and cadmium. Instead, a combination of nearly pure bismuth and nearly pure cadmium crystals is formed, 60% bismuth crystals and 40% cadmium crystals. If a liquid solution that contains less than 60% bismuth is cooled, nearly pure cadmium crystals separate from the solution when line *AE* is reached. This, in effect, increases the bismuth concentration, causing the freezing temperature to move down line *AE* with yet more cadmium crystals forming. When the remaining liquid solution reaches the eutectic point (60% Bi, 40% Cd), it will immediately crystallize as a eutectic alloy. Type II alloys will therefore always be composed of a combination of nearly pure crystals of the two elements. The solidification process will depend upon the actual composition of the two elements in relation to the eutectic point.

Type II alloys will have properties intermediate between the two elements, as shown in Figure 8–17. This applies to hardness, electrical conductivity, color, and magnetic properties.

Compounds

In many cases a combination of elements will cause the formation of compounds. In the formation of a compound, the original identities and properties of the two elements are

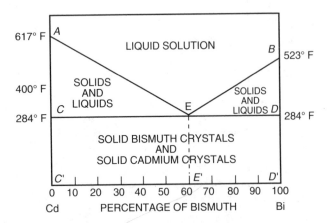

FIGURE 8–16 Cadmium-bismuth alloy diagram.

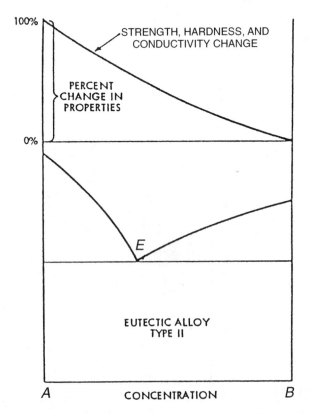

FIGURE 8–17 Mechanical properties for Type II alloys.

lost, replaced by new properties characteristic of that compound. The heat treatment of many commercial metals is made possible by compounds; Fe_3C for ferrous alloys and $CuAl_2$ for aluminum, for example. These compounds, and their effects, are covered in more detail in Chapter 9 of this text.

Commercial Alloys

The examples discussed so far represent simple and basic alloys. In reality, commercial alloys have complex actions taking place as the elements are combined. In most cases more than two elements are involved. Many of the commercial alloys are allotropic; that is, the lattice structure changes under certain conditions. An alloying element may be used to cause such phase changes, or to prevent them. Phase changes are critical to some material's ability to be hardened by heat treatment. Although complex, the alloying action in effect is a series of separate actions. For example, an aluminum alloy (2024) may exhibit the effects of both Type I and Type II alloys as well as form a compound.

CORROSION

Corrosion is a problem for all metals. Corrosion is the decomposition of metallic elements into compounds such as oxides, sulfates, hydroxides, and chlorides. Compound formation is caused by direct chemical action and electrolytic action. Chemical corrosion involves reactions with acid, salts, alkalis, or even the oxygen in air. The presence of moisture will usually accelerate the chemical action. Electrolytic action occurs when metals that have different levels of electrochemical activity are in close proximity in the presence of moisture. The dissimilar metals form the poles of a galvanic cell, an electric current flows, and the more active metal is decomposed, or corroded. The susceptibility of a metal to corrosion varies depending upon its chemical composition and properties. Corrosion protection and prevention is a very important part of aircraft maintenance. Processes involving corrosion control are covered in Chapter 9.

FATIGUE

The phenomenon of metal fatigue has long been known, but in recent years it has become a major concern of aircraft maintenance because of aging aircraft.

Metal fatigue refers to the loss of strength, or resistance to load, experienced by a material as the number of load cycles or load reversals increase. The tensile strength of a material was discussed earlier in this chapter. The strength values discussed then were based on a single load application. Load cycles, or load reversals, refer to a material being loaded and then unloaded. If the elastic limit has not been exceeded during a load condition, the material, in theory, returns to its original state once the load is removed. In reality, the application of the load can cause minute cracks to occur in the material. These cracks are so small as to have no significant effect upon the metal at first. As the load cycles continue, additional cracks are formed and old ones get larger. Eventually, the cumulative effect will be such that the strength of the metal will be compromised. A large number of load cycles are required before significant strength loss will be encountered. A load reversal occurs as parts vibrate or flex. Thus components such as the wings of an aircraft or the rotors of a helicopter will encounter a high number of load reversals in a short period of time.

The **fatigue strength** of a metal can be empirically determined and plotted, as shown in Figure 8–18. The strength of the material is plotted against the vertical axis, and the number of load cycles is plotted on the horizontal axis. Using such a chart allows us to determine how many cycles can be encountered before the strength of the material falls below a safe level. This information can be used to determine the number of cycles, or time, that a part may have on an aircraft before replacement or other action is required.

Fatigue is a natural phenomenon and cannot be prevented. The ability to correctly predict its effects and take necessary action is the problem faced by aircraft design and maintenance personnel. Different metals have different fatigue characteristics. Fatigue is also affected by design characteristics, such as changes in cross-sectional area, holes, notches, and so on. Fatigue cracking will expose unprotected metal to the elements, which increases the possibility of corrosion. Once begun, the constant working and growth of the fatigue cracks enhance the continued spread of corrosion. The combination of corrosion and fatigue can generate serious structural problems in a short period.

The maintenance technician has the task of monitoring the aircraft structure and correcting fatigue and corrosion problems before they become serious. Unfortunately, the damage that occurs in these situations is often not visible until it has occurred. With the aging of the airline fleet, the

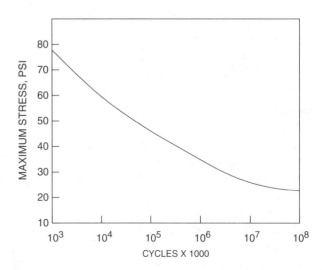

FIGURE 8–18 Strength reduction due to fatigue.

maintenance technician's role in detecting and preventing such problems has taken on added significance. A number of new inspection processes and tools have been developed to meet this concern. Nondestructive inspection processes are discussed in Chapter 9.

AIRCRAFT METALS

An overview of some major metals commonly found in aircraft is given in the following paragraphs. For more complete coverage of aircraft metals, see *Metallic Materials and Elements for Aerospace Vehicle Structures, MIL-HDBK-5*, published by the U.S. Government Printing Office.

Aluminum (Al)

Aluminum is the principal structural metal for aircraft. A unique combination of properties makes aluminum a very versatile engineering and construction material. Light weight is perhaps aluminum's best known characteristic. With a specific gravity of 2.7, the metal weighs only about 0.1 lb/in^3 [2.8 g/cm^3], as compared with 0.28 lb/in^3 [7.8 g/cm^3] for iron and 0.32 lb/in^3 [8.86 g/cm^3] for copper. Commercially pure aluminum has a tensile strength of about 13 000 psi [89.6 MPa]. Its usefulness as a structural metal in this form is somewhat limited, although its strength can be approximately doubled by cold working. Much greater increases in strength can be obtained by alloying with other metals. Aluminum alloys having tensile strengths approaching 100 000 psi [689.6 MPa] are available. Aluminum and its alloys lose strength at elevated temperatures, although some retain good strength at temperatures as high as 400°F [204°C]. At subzero temperatures, aluminum's strength increases without loss of ductility. Aluminum in general is considered as having good corrosion resistance. The ease and versatility with which aluminum is made surpasses that of virtually any other material.

Aluminum products are made in two forms, cast and wrought. **Cast aluminum** is formed into a shape by pouring molten aluminum into a mold of the required shape. **Wrought aluminum** is mechanically worked into the form desired by rolling, drawing, and extruding.

Wrought Aluminum Code. Wrought aluminum and aluminum alloys are designated by a four-digit system, with the first digit of the number indicating the principal alloying element:

1000 Series. When the first digit is 1, the material is 99% or higher pure aluminum.

2000 Series. Copper is the principal alloying element in this group. The addition of copper allows aluminum to be heat-treated to high strengths, but also reduces its corrosion resistance.

3000 Series. Manganese is the principal element in this series. It is added to increase hardness and strength. Alloys of this series are non–heat treatable.

4000 Series. The major alloying element of this group is silicon, which can be added in sufficient quantities to cause substantial lowering of the melting point without producing brittleness. The principle use of 4000-series alloys is for welding wire.

5000 Series. Magnesium is one of the most effective and widely used alloys for aluminum. When it is used as the major alloying element, the result is a moderate- to high-strength, non-heat-treatable alloy. Alloys in this series have good welding characteristics and high corrosion resistance.

6000 Series. Alloys in this group contain silicon and magnesium in proportions to form magnesium-silicide, thus making them heat treatable. Though not as strong as the 2000 and 7000 alloys, the magnesium-silicon alloys possess good formability and corrosion resistance with moderate strength.

7000 Series. Zinc is the major alloying element of this group and, when coupled with a small percentage of magnesium, results in a heat treatable alloy of very high strength. Usually, other elements, such as copper and chromium, are added in small quantities.

The second digit of the code number identifies any modifications made to the original alloy. In the 2000 to the 8000 numbers, the last two digits have little significance other than to identify the alloys and the sequence of development. In the 1000 series, the last two digits indicate the amount of pure aluminum above 99% in hundredths of 1%. For example, aluminum identified with the number 1240 would be 99.40% pure aluminum. In the 1000 series a modification is for the purpose of controlling one or more of the impurities. Examples of the alloy code are shown in Figure 8–19.

Cast Aluminum Code. Cast aluminum also uses a four-digit identification system with the first digit indicating the alloy group. The numbers used are not the same as for wrought alloys. Cast alloy groups are

1 — aluminum, 99% or more
2 — copper
3 — silicon, with copper and/or magnesium
4 — silicon
5 — magnesium
6 — not used
7 — zinc
8 — tin
9 — other elements

The second two digits identify the aluminum alloy or indicate the aluminum purity. The last digit is separated from the other three by a decimal point and indicates the product

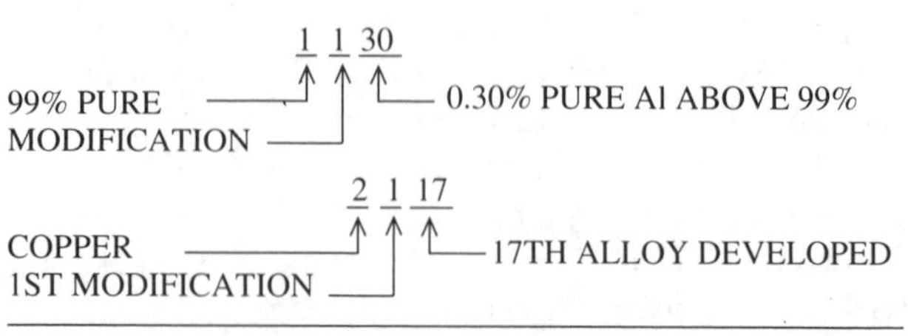

FIGURE 8–19 Aluminum alloy identification.

form; that is, castings or ingots. A modification of the original alloy is indicated by a serial letter before the numerical designation. Alloy A514.0, for example, indicates an aluminum alloy casting with magnesium as the principal alloy. One modification to the original alloy has been made, as indicated by the letter A.

Hardness and Temper Designation. An important factor for aluminum alloy identification is the **temper**, or **hardness** value. A letter and number combination is placed after the alloy code to indicate the processes that have taken place and the degree of hardness. Basic designations are as follows:

F—as fabricated (no treatment)
O—annealed
H—cold worked or strain hardened (wrought products)
W—unstable condition (temporary condition while the material ages after solution heat treatment)
T—solution heat treated

The H designations are used only for the non-heat-treatable alloys. These are generally the alloys in the 1000, 3000, and 5000 series. These materials can only be hardened through the effects of cold working. The T designations are used after alloys capable of being hardened by thermal treatment. Heat-treatable alloys contain elements such as copper, magnesium, silicon, and zinc. Alloys from the 2000, 6000, and 7000 series use T designations. The W designation applies only to those alloys capable of heat treatment. The annealed designation, O, and the as-fabricated designation, F, can apply to any of the alloys.

The H tempers are further subdivided to indicate the specific combination of basic operations. For example,

H1—strain hardened only
H2—strain hardened and partially annealed
H3—strain hardened and stabilized

A number following the H1, H2, or H3 indicates the degree of strain hardening of the alloy. The number 8 indicates that the maximum degree of strain hardening has occurred. The number 2 indicates that the metal is one-quarter hard, 4 indicates one-half hard, and 6 indicates three-quarters hard.

The T designation is followed by a number that indicates specific sequences of basic treatments. Frequently used numbers are

T3—solution heat treated, cold worked
T4—solution heat treated
T6—solution heat treated and artificially aged

The term **cold working**, or **strain hardening** describes any process applied at room temperature that stretches, compresses, bends, draws, or otherwise changes the shape of the metal to any appreciable degree. **Solution heat treating** is a thermal process for hardening aluminum. It involves heating the material to a specified temperature, causing the chemical structure of the material to change. The material is then quickly cooled or quenched. Upon quenching, the material is soft and unstable. Chemical changes continue within the metal at room temperature until a stable condition is reached. This is known as **aging**. Acceleration of the aging

process by additional thermal treatment is known as **artificial aging**. The material will not develop full strength until it has stabilized through aging. Heat treatment of aluminum is covered in more detail in Chapter 9.

Corrosion. Aluminum is considered to be highly corrosion resistant under the majority of service conditions. When aluminum surfaces are exposed to the atmosphere, a thin, invisible oxide (Al_2O_3) skin forms, which protects the metal from further oxidation. Unless this coating is destroyed, the material remains fully protected against corrosion. Aluminum is highly resistant to weathering and is corrosion resistant to many acids. Alkalis are among the few substances that attack the oxide skin and are therefore corrosive to aluminum.

High-strength alloys containing copper are less resistant to corrosion than the other alloys. Alloys of this type often have a thin layer of pure aluminum rolled on each side. The pure aluminum acts as a barrier between the environment and the less resistant alloy. Aluminum of this type is known as **clad aluminum**. (Alclad is a trade name used by ALCOA for this product.)

While highly corrosion resistant by itself, aluminum is susceptible to galvanic corrosion resulting from contact with other materials. Among the structural metals, aluminum is second only to magnesium on the electromotive series. Galvanic corrosion with any structural metal other than magnesium results in aluminum being the material corroded or decomposed.

Workability. Aluminum can be cast by any method known. It can be rolled to any desired thickness, including a foil thinner than paper. It can be stamped, drawn, spun, or roll formed. The metal can be hammered or forged, and there is almost no limit to the different shapes into which it may be extruded. Aluminum can be turned, milled, bored, or machined in other manners at the maximum speed at which the majority of the machines are capable. Almost any method of joining is applicable to aluminum; riveting, welding, brazing, soldering, or adhesive bonding.

Alloys Used for Aircraft. The majority of the aluminum products that the technician encounters will consist of the following alloys:

Non-Heat-Treatable Alloys

1100. Pure aluminum that is soft, ductile, and low strength. Its use on aircraft is limited to nonstructural application. It is also used for making low-strength rivets.

3003. This alloy is similar to 1100 but has about 20% greater strength. It is used on aircraft for fluid lines, with limited application for fairings or cowlings.

5052. The highest strength of the non-heat-treatable alloys, 5052 has high corrosion resistance and high fatigue strength. Having excellent workability, it is widely used for aircraft cowlings, fairings, and other nonstructural parts requiring forming. Fluid lines are frequently made of this alloy.

5056. This alloy is used to make rivets for riveting magnesium sheet.

Heat-Treatable Alloys

2017. Used more frequently on older aircraft, 2017 alloy is seen today, if at all, only in aluminum rivets.

2117. A modification of alloy 2017, this alloy is used exclusively for the manufacture of aluminum rivets. The 2117 rivet is not as strong as a 2024 rivet but can be driven with no special treatment.

2024. 2024 is probably the "standard" structural metal as well as the most-used metal for aircraft. It is found in virtually every form available, including sheet, extrusion, bar stock, standard hardware, and tubing. Heat treatable to high strengths, its ability to be cold worked is good to excellent for everything except rivets. Like most heat-treatable alloys, 2024 is not recommended for welding. To improve corrosion resistance, 2024 sheet stock is available as a clad material. 2024 ages rapidly after heat treatment, and most forms will be in the T3 condition. Alloy 2024 is highly susceptible to **intergranular corrosion** if heat treatment is done improperly.

6061. An easily worked metal, 6061 has a strength only two-thirds that of 2024. A good general-purpose material, it can be welded, offers high corrosion resistance, and can be worked by almost any means. To develop full strength after heat treatment normally requires artificial aging; such metal will have a T6 designation.

7075. This is one of the highest strength aluminum alloys available and also one of the more difficult aluminum alloys to work. Parts should be formed in the annealed state or at an elevated temperature. Arc and gas welding is not recommended. 7075 is available in a clad form to improve corrosion resistance. This alloy is normally artificially aged and carries the T6 designation.

Aluminum-Lithium Alloys. During the 1980s, development work was begun on a series of aluminum alloys with lithium (Li) as the major element. An early alloy, **2090,** used copper as the major alloy (2.4 to 3.0%) with 1.9 to 2.6% lithium. Later alloys have a lower copper content (less than 1.5%) and 1 to 1.5% lithium. The developmental alloys are numbered as 8000.

These alloys were developed partially as a result of competition from lightweight composite materials for the future aircraft market. Aluminum-lithium alloys are 10% lighter than the 2024 alloy: $0.091 - 0.093$ lb/in^3 [$2.51 - 2.59$ g/cm^3] versus 0.10 lb/in^3 [2.8 gm/cm^3]. It is estimated that the weight savings of this alloy would amount to 12 000 lb [5443 kg] on a large transport aircraft. In addition to the lighter weight, these alloys also exhibit 15% greater stiffness, which allows significant design changes.

While products are being manufactured from these alloys, their use is not necessarily widespread. The fact that lithium is a very light and active element requires that special handling techniques be developed, both to ensure the integrity of the material and the safety of those working with it. Widespread use is predicted by 1995.

Ferrous-Based Metals—Iron (Fe) and Steel

Ferrous metals are those whose principal content is iron (Latin, *ferrum*), such as cast iron, steel, and similar products. Most general-purpose steels used for aircraft work are wrought steel products, designated as shown in Table 8–1. In addition to the standard carbon and alloy steels, a substantial number of heat- and corrosion-resistant steels are used in aerospace vehicles. The principal designations for these steels are shown in Table 8–2.

In Table 8–1, the first digit of each number indicates the general classification of the steel, that is, carbon, nickel, and so on. The number 1 indicates a carbon steel. The second digit of the number indicates the approximate percentage of the principal alloying element; for example, a 2330 steel contains more than 3% nickel. The last two digits of the number indicate the approximate amount of carbon in one-hundredths of 1%.

One of the most important considerations for ordinary carbon steel is the quantity of carbon it contains. A low-carbon steel contains 0.10 to 0.15% carbon. Medium-carbon steels contain 0.20 to 0.30% carbon. The higher the carbon content of steel, the greater its hardness and also its brittleness. High-carbon steels are used for cutting tools, springs,

TABLE 8–1 SAE Identification for Wrought Steels

Carbon Steels	
10xx	Nonsulfurized carbon steel (plain carbon)
11xx	Resulfurized carbon steel (free machining)
12xx	Resulfurized and rephosphorized carbon steel
Alloy Steels	
13xx	Manganese 1.75% (1.60–1.90%)
23xx	Nickel 3.50%
25xx	Nickel 5.00%
31xx	Nickel-chromium (Ni 1.25%, Cr 0.65%)
32xx	Nickel-chromium (Ni 1.75%, Cr 1.00%)
33xx	Nickel-chromium (Ni 3.50%, Cr 1.50%)
40xx	Molybdenum 0.25%
41xx	Chromium-molybdenum (Cr 0.50 or 0.95%, Mo 0.12 or 0.20%)
43xx	Nickel-chromium-molybdenum (Ni 1.80%, Cr 0.50 or 0.80%, Mo 0.25%)
46xx	Nickel-molybdenum (Ni 1.75%, Mo 0.25%)
47xx	Nickel-chromium-molybdenum (Ni 1.05%, Cr 0.45%, Mo 0.20%)
48xx	Nickel-molybdenum (Ni 3.50%, Mo 0.25%)
50xx	Chromium 0.28 or 0.40%
51xx	Chromium 0.80, 0.90, 0.95, 1.00, or 1.05%
5xxxx	Chromium 0.50, 1.00, or 1.45%, Carbon 1.00%
61xx	Chromium-vanadium (Cr 0.80 or 0.95%, V 0.10 or 0.15%)
86xx	Nickel-chromium-molybdenum (Ni 0.55 or 0.05 or 0.65%, Mo 0.20%)
87xx	Nickel-chromium-molybdenum (Ni 0.55%, Cr 0.50%, Mo 0.25%)
92xx	Manganese-silicon (Mn 0.85%, Si 2.00%)
93xx	Nickel-chromium-molybdenum (Ni 3.25%, Cr 1.20%, Mo 0.12%)
98xx	Nickel-chromium-molybdenum (Ni 1.00%, Cr 0.80%, Mo 0.25%)

2xx	Chromium-nickel-manganese (nonhardenable, austenitic, nonmagnetic)
3xx	Chromium-nickel (nonhardenable, austenitic, non-magnetic)
4xx	Chromium (hardenable, martensitic, magnetic)
4xx	Chromium (hardenable, ferritic, magnetic)
5xx	Chromium (low chromium, heat resisting)

and other objects. For general purposes, low- or medium-carbon steels are best because they are more easily worked, are tougher, and have a greater impact resistance. Such steels were used for many years in the manufacture of aircraft structures and fittings.

Probably the most commonly used steel for aircraft structural purposes is **SAE 4130 chromium-molybdenum (chrome-moly) steel**. When properly heat treated, it is approximately four times as strong as 1025 mild-carbon steel. The tensile strength of 4130 steel will range from 90 000 psi [620 MPa] to more than 180 000 psi [1241 MPa], depending upon heat treatment. SAE 4130 chrome-moly steel is easily worked, readily weldable by any method, hardenable, heat treatable, easily machined, and well adapted to high-temperature conditions of service.

The **nickel steel**s, SAE 23xx and 25xx, contain from 3.5 to 5% nickel and a small percentage of carbon. The nickel increases the strength, hardness, and elasticity of the steel without appreciably affecting the ductility. Nickel steel is used for making various aircraft hardware, including nuts, bolts, clevis pins, and screws.

Nickel-chromium and chromium-vanadium steels are used where still greater strength, hardness, and toughness are required. Such steels are often found in highly stressed machine parts, such as gears, shafts, springs, and bearings.

Corrosion-Resistant (Stainless) Steels

Since the 1940s the term **stainless steel**, also designated as **corrosion-resistant steel (CRES)**, has become a household word because of its many applications in consumer items as well as in aerospace products. The most important characteristics of stainless steels are corrosion resistance, strength, toughness, and resistance to high temperatures. Stainless steels can be divided into three general groups based on their chemical structure: austenitic, ferritic, and martensitic.

The **austenitic steels** are chromium-nickel (Cr-Ni) and chromium-nickel-manganese (Cr-Ni-Mn) alloys. They can be hardened only by cold working, and heat treatment serves only to anneal them. They are nonmagnetic in the annealed condition, although some may be slightly magnetic after cold working. Austenitic steels are formed by heating the steel mixture above its "critical range" and holding it to allow formation of an internal structure called austenite. A controlled period of partial cooling is allowed, followed by a rapid quench or cooling.

The **ferritic steels** contain no carbon; hence they do not respond to heat treatment. They contain a substantial amount of chromium and may have a small amount of aluminum. They are always magnetic.

The **martensitic steels** are straight chromium alloys that harden intensely if allowed to cool rapidly from high temperatures. They differ from the two preceding groups because they can be hardened by heat treatment.

The most widely used stainless steels for general use are those in the 300 series, called 18-8 because they contain approximately 18% chromium and 8% nickel. Typical of these types are 301, 302, 321, and 347.

Although CRESs have many advantages, there are certain disadvantages that must be faced by the fabricator and designer:

1. Stainless steels are more difficult to cut and form than many materials.
2. Stainless steels have a much greater expansion coefficient than other steels, and they conduct heat at a lower rate, making welding more difficult.
3. Many of the stainless steels lose their corrosion resistance under high temperatures.

When welding CRES, inert-gas-shielded arc welding is preferred because this process causes less deformation due to heat expansion and prevents oxidation. The expansion of stainless steel due to temperature increases may be more than twice that of ordinary carbon steels. Because of its toughness, stainless steel is more difficult to cut, form, shear, machine, or drill than ordinary steel.

Magnesium (Mg)

Magnesium alloys are used frequently in aircraft structures in cast, forged, and sheet form. The greatest advantage of magnesium is that it is one of the lightest metals for its strength. A typical alloy for aircraft use might be AZ31B. This material, which is available in sheet, plate, and extruded forms, has a density about two-thirds that of aluminum: 0.067 lb/in^3 versus 0.100 lb/in^3 [1.85 gm/cm^3 versus 2.8 gm/cm^3]. AZ31B exhibits tensile strength of 30 000 to 40 000 psi [207 to 276 MPa]. The A and Z indicate that aluminum and zinc are the major alloys.

The disadvantages of magnesium are that it is more subject to corrosion than many metals, it is not easily worked at room temperatures, and if it ignites, it is extremely difficult to extinguish. Magnesium cannot be cut easily and has a tendency to tear.

When magnesium is used in an aircraft structure, it can often be recognized by the fact that it has a yellowish surface due to a chromate treatment used to prevent corrosion and furnish a suitable paint base.

Because of its tendency to corrode, it is very important that the correct hardware items, such as the proper rivets, bolts, and screws, be used. For example, rivets used with magnesium should be made of 5056-H aluminum alloy. Any metal part used with magnesium should be of a compatible metal or be properly insulated from the magnesium.

Titanium (Ti)

The use of titanium as a structural material has become widespread only during the last half of this century. Although discovered in 1790, the first isolation of pure metallic titanium in sufficient quantity for practical study was not accomplished until 1906. The Kroll process, widely used for extracting titanium metal, was developed in 1932. This process was improved upon by the United States Bureau of Mines, which in 1946 began to produce titanium sponge in 100-lb batches. Today, titanium is produced in relatively large quantities in rod, bar, sheet, and other forms for use in the manufacture of a wide variety of metal products.

Titanium and its alloys are widely used in the aerospace industry because of their high strength, light weight, temperature resistance, and corrosion resistance. Titanium is equal to iron in strength, at approximately 56% of the weight ($Ti = .167$ lb/in^3 [4.6 gm/cm^3], $Fe = .28$ lb/in^3 [7.8 gm/cm^3]).

The strength of titanium is maintained to temperatures of more than 800°F [427°C], making it useful in the cooler sections of gas-turbine engines, for the cowling and the baffling around engines, and for the skin of aircraft subjected to elevated temperatures. When titanium is exposed to temperatures of 1000°F [538°C] and above, it must be protected from the atmosphere, because at these temperatures it combines rapidly with oxygen. The melting point of titanium is approximately 3100°F [1704°C]. Titanium has a very low coefficient of thermal expansion, much lower than that of other structural metals such as corrosion-resistant steel. Thermal conductivity is approximately the same as that of corrosion-resistant steel.

An outstanding property of titanium is its resistance to corrosive substances and industrial chemicals. It is uniquely resistant to inorganic chloride solutions, chlorinated organic compounds, chlorine solutions, and moist chlorine gas. It has excellent resistance to oxidizing acids, such as nitric or chromic acids, but will be attacked by strong reducing acids. The resistance of titanium to corrosion by natural environmental substances is unequaled by other structural metals.

Titanium may be worked by many of the methods employed for steel and corrosion-resistant steel. It can be sheared, drawn, pressed, machined, routed, sawed, and nibbled. The operator handling titanium must be familiar with its peculiarities and special characteristics in order to obtain good results. The cutting dies and shear blades used in cutting titanium must be of good-quality steel and must be kept very sharp. Because of its affinity for oxygen and nitrogen at elevated temperatures, the metal should be protected by a surrounding atmosphere of inert gas for "hot" working. The usual method of protection is to heat the metal in an atmosphere of argon or helium gas. Welding may be done using a gas-shielded process.

At about 1950°F [1065°C], titanium will ignite in the presence of oxygen and burn with an incandescent flame. A nitrogen atmosphere will result in ignition and flame at about 1500°F [815°C]. When titanium is being machined in quantity, fire-extinguishing materials should be immediately available. Liquid coolants of the proper type should be used during machining to reduce the possibility of fire.

Titanium is an allotropic metal having CPH crystalline structure (see Figure 8–10) at room temperature and changing to a BCC arrangement at elevated temperatures. When in the CPH phase it is referred to as *alpha titanium*, and as *beta titanium* when in the BCC phase. These are sometimes referred to as Type A and Type B titanium. Titanium is receptive to many other metals as alloying elements. The use of alloys has made possible the stabilization of the beta phase at room temperatures. As a result, a number of types of titanium metals are available: pure titanium, alpha titanium, alpha-beta titanium, and beta titanium. Different properties are exhibited by each type.

Commercially **pure (unalloyed) titanium** is available in all familiar product forms and is noted for its excellent formability. A designation of pure titanium alloys usually takes the form of Ti-75. The number (75) indicates that the minimum yield strength of the product is 75 000 psi [517 MPa]. Pure titanium cannot be hardened by heat treatment, but it can be annealed. It can be cold worked to above 100 000 psi [689 MPa] yield strength.

An example of **alpha-titanium** uses the commercial designation of 8Al-1Mo-1V-Ti, sometimes referred to as Ti-8-1-1. The designation indicates that the major alloy elements are 8% aluminum (Al), 1% molybdenum (Mo), and 1% vanadium (V). The addition of an alloying element creates a tendency for the material to change to the beta phase. One of the functions of the alloying elements in this category is to stabilize the alpha structure. Alpha titaniums are stronger but less workable than their pure counterparts. They can be annealed but not strengthened by heat treatment. Annealed Ti-8-1-1 will exhibit a yield strength in excess of 120 000 psi [827 MPa] at room temperature.

The **alpha-beta titanium** has both forms of internal structure. A widely used commercial product is Ti-6Al-4V. This metal is heat treatable to more than 140 000 psi [965 MPa] yield strength. As with many metals, as the alloy becomes more complex the working properties become more difficult. Ti-6Al-4V can have most processes performed, but some are highly developed and require careful attention to detail.

Ti-13V-11Cr-3Al is a heat-treatable, high-strength **beta-titanium** alloy. Solution-heat-treated and aged Ti-13-11-3 will develop yield strengths in the range of 170 000 to 200 000 psi [1172 to 1379 KPa]. As in the case of the alpha-beta alloy just discussed, this alloy can be worked, but the ease of workability varies and special processes are required for some operations.

Copper (Cu)

Copper is one of the comparatively plentiful metals and has been used by human beings for thousands of years. It is easily identified by its reddish color and by the green and blue colors of its oxides and salts. It is very ductile in the annealed state but hardens with cold working. A primary use for copper is as an electrical conductor. Before the development of aluminum as a practical metal, copper was used for tubing and other applications where aluminum is used today.

The principal alloys with a copper base are bronze, brass, and beryllium copper. Bronze is primarily a blend of copper and tin, the tin content being from 10 to 25%. Brass is an alloy of copper with 30 to 45% zinc plus small amounts of other metals. Beryllium copper is approximately 97% copper, 2% beryllium, and 1% other metals.

Bronze and brass are used for bushings, bearings, valve seats, fuel-metering valves, and numerous other applications.

Beryllium copper is heat treatable and can be brought up to a tensile strength of 200 000 psi [1379 MPa]. Beryllium copper is used for precision bearings, bushings, spring washers, diaphragms, ball cages, and other applications where its qualities of wear resistance, toughness, strength, and elasticity are desirable.

Copper is alloyed with aluminum, manganese, silicon, iron, nickel, and other metals to make a variety of "bronzes." These are not true bronzes in the original sense of the word because they do not contain tin. Among these bronzes are aluminum bronze, silicon bronze, and manganese bronze. These alloys are available in sheet, bar, rod, plate, and other standard shapes.

High-Temperature Alloys

Because of the need for metals that can withstand the extremely high temperatures found in gas-turbine engines, afterburners, thrust reversers, and other modern equipment, and because of the high temperatures generated by air friction at supersonic speeds, it has become necessary to develop metal alloys which retain their strength under elevated temperature conditions. The products of high-temperature metal research have led to the development of alloys that use a wide variety of metal elements to produce the desired results.

High-temperature alloys contain high percentages of nickel, cobalt, chromium, molybdenum, titanium, and other alloying elements that make them particularly resistant to heat and corrosion and retain high tensile strength at elevated temperatures such as 1000 to 2200°F [538 to 1205°C]. Other qualities considered in the development of these alloys include thermal stability, tensile strength at elevated temperatures, low-cycle fatigue strength, stress-rupture properties, hot-corrosion resistance, and oxidation resistance. Some examples of the major groups and representative alloys are discussed in the following sections.

Iron-Chromium-Nickel-Based Alloys. The alloys in this group generally fall between the austenitic stainless steels and the nickel- and cobalt-based alloys, both in cost and in maximum service temperature. They are used in airframes, principally in the temperature range of 1000 to 1200°F [538 to 649°C], in those applications in which the stainless steels are inadequate and the service requirements do not justify the use of the more costly nickel or cobalt alloys.

One common alloy in this group, **A-286**, is a precipitation-hardening iron-based alloy designed for parts requiring high strength up to 1300°F [704°C] and oxidation resistance up to 1500°F [816°C]. It is used in turbine engines for parts such as turbine buckets, bolts, and discs. It is also used for sheet-metal assemblies and is available in the usual mill forms. In addition to the iron base, the major alloying elements are 25% nickel and 15% chromium.

Nickel-Based Alloys. Nickel is the base element for most of the higher temperature heat-resistant alloys. While it is much more expensive than iron, nickel provides an austenitic structure that has greater toughness and workability than ferrous alloys of the same strength level. The common alloying elements for nickel are cobalt, iron, chromium, molybdenum, titanium, and aluminum.

One such alloy, **Inconel 702**, is a heat-treatable nickel-based alloy containing chromium and aluminum for oxidation resistance and aluminum and titanium as hardeners. It is used primarily for parts and assemblies requiring oxidation resistance to about 2000°F [1093°C] (and higher under some conditions) rather than high strength and where parts may require welding during fabrication. Although available in strip, sheet, bar, and tubing forms, it is used primarily in sheet form. Inconel has 15% chromium, 3% aluminum, and 0.5% titanium as alloying elements.

Another alloy, **Rene 41**, is a vacuum-melted precipitation-hardening nickel-base alloy designed for highly stressed parts operating between 1200 and 1800°F [649 to 982.7°C]. Its applications include afterburner parts, turbine castings, wheels, buckets, and high-temperature bolts and fasteners. Rene 41 is available in sheets, bars, and forgings. Alloying elements include 19% chromium, 11% cobalt, 10% molybdenum, and 3% titanium.

Cobalt-Based Alloys. The use of cobalt in wrought heat-resistant alloys is usually limited to the addition of cobalt to alloys of other bases. In very few alloys is cobalt the base element. These alloys are designed to have very high strength at very high temperatures. Working properties are extremely limited.

One such alloy, **L-605**, is a cobalt-based alloy also known as Haynes Alloy 25. L-605 contains 20% chromium, 15% tungsten, and 10% nickel as alloying elements. This alloy is used for moderately stressed parts operating between 1000 and 1200°F [649 and 1038°C]. L-605 is not hardenable except by cold working and is usually used in the annealed condition.

PLASTICS

The word *plastic* is derived from the Greek word *plastikos*, meaning "to form." In a broad sense, any material that can be formed into various shapes can be called plastic; however, common usage of the term limits it to synthetic materials developed for industrial use and consumer products. Most of the materials called plastics are often termed *synthetic resins*. Natural resins are usually produced from the sap or pitch of plants or from certain insects. Natural resins, once used for the manufacture of varnishes and lacquers, have been largely replaced by synthetic resins.

Plastics are classified as thermosetting resins or thermoplastic resins. **Thermosetting resins** harden or set when heat of the correct value is applied. This type of plastic cannot be softened and reshaped after having been solidified. **Thermoplastic resins** can be softened by heat and reshaped or reformed many times without changing composition, provided that the heat applied is held within proper limits.

Thermosetting Resins

Common thermosetting resins include phenolics, epoxies, polyurethanes, polyesters, and silicones.

The **phenolic resins** are based on phenol and formaldehyde. These resins are resistant to heat, moisture, chemicals, and oils, and they are excellent insulators. They are therefore used extensively for various parts and insulators in electrical devices.

The **epoxy resins** find many uses in aircraft. They have excellent heat resistance, insulation qualities, dimensional stability, chemical resistance, and moisture resistance. When applied in a liquid state, they have outstanding adhesive qualities. Epoxy resins are used for potting, encapsulating, casting, reinforced laminates, adhesives, and protective coatings. Epoxies are usually supplied in two parts. The resin component is in a syrupy liquid state, and the curing agent may be a liquid or a powder. The curing agent is mixed with the resin just before the material is used. The time allowable between mixing and solidification is stated on the container. In some aircraft, epoxy resin is used as an adhesive agent for metal bonding. The resulting bond has excellent strength and durability. Another common use is in coatings for aircraft. Epoxy primers provide an effective base for other finishes, such as polyurethane.

The **polyurethane plastics** may be used for rigid or flexible structures. They are commonly used as a foam that, when solidified, makes very light heat-resistant and thermal-insulating materials. Polyurethane enamels make superior finish coatings for aircraft. When properly applied, the finish has a high gloss and there is no need for sanding, rubbing, polishing, or waxing. These finishes are weather resistant and retain good appearance and quality for several years.

The **polyester resins** are commonly used as a matrix material for glass-fiber laminates. In aircraft these are found as fairings, tail cones, antenna housings, radomes, cowling, wheel pants, and similar items.

The **silicone resins** can be used for reinforced laminates with glass fiber, fibrous graphite, and other materials. The silicones are superior in heat resistance and for this reason are used in products exposed to high temperatures. Among high-temperature silicone materials are oils, greases, rubbers, and reinforced sheet.

Thermoplastic Resins

Among the thermoplastic resins that may be encountered by the aviation technician are cellulose acetate, polyethylene, polypropylene, vinyls, polymethyl methacrylate (acrylic resin), polytetrafluoroethylene (Teflon), and nylon.

The resin **cellulose acetate** is used for transparent film and sheet. In aircraft, cellulose acetate was once used for windows and windshields but has been replaced by acrylics.

The resin **polyethylene** is made in low-density and high-density qualities. Low-density polyethylene is made in thin, flexible sheet or film and is used for plastic bags, protective sheeting, and electrical insulation. High-density polyethylene is used for containers such as fuel tanks, large drums, and bottles.

The **vinyls** are manufactured in a variety of types and have a wide range of applications. Their use in aircraft includes seat coverings, electrical insulation, moldings, and tubing. They are flexible and resistant to most chemicals and moisture.

Another resin, **acrylic resin**, is a water-clear plastic that has a light transmission of 92%. This property, together with its weather and moisture resistance, makes it an excellent product for aircraft windows and windshields.

The resin **polytetrafluoroethylene (Teflon)** is encountered in nonlubricated bearings, tubing, electrical devices, and other applications. It is extremely tough and almost frictionless. It has good resistance to temperatures as high as 500°F [260°C] and remains flexible at low temperatures.

COMPOSITE MATERIALS

Structural materials known as composites are made of many different materials and in a variety of forms. The use of plastic resins has made possible the development of nonmetallic materials that are often superior to metals in strength/weight ratio, corrosion resistance, ease of fabrication, and cost. Newer aircraft use composite materials for many structural parts, and the use of composite materials for wingtips, cowlings, fairings, flaps, spoilers, and ailerons is common.

For this text, composite materials will be divided into two categories, **laminated** and **cored, or sandwich**, construction. Both types of construction produce lightweight parts and components with high tensile strength. The need for adequate stiffness to handle compression and bending loads has resulted in a large variety of material and design configurations.

An in-depth coverage of the various materials and designs of composite materials exceeds the scope of this text. The objective of this section is to give you a basic knowledge of the common design principles and materials used by the aircraft industry. Detailed information on material application, construction techniques, and repair procedures is available in another text of this series, *Aircraft Maintenance and Repair*.

Laminate Materials

The **laminates** are made by laying a resin-saturated fabric over, or within, a mold to produce a desired shape. The laminate will be made up of a number of individual layers, called **plies**, of the fabric. In laminated construction, much of the tensile strength comes from the fabric, which is known as the

reinforcing material. The resin, which bonds the plies together and provides stiffness, is usually referred to as the **matrix**. The term **lay-up** is often used to describe the process of adding the plies of fabric and resin to the form.

For full strength to be developed, the reinforcing fibers must be completely encapsulated by the resin. It is also essential that individual plies be in contact with the next layers, with no air pockets or areas of excess resin.

For individual parts or those with complex shapes, lay-ups are usually done manually. On laminated items such as tubing, piping, pressure vessels, and cylindrical containers, fabrication may be done by machine, using continuous bundles of fibers called rovings or tows. The roving is drawn through a reservoir of resin and machine-wound onto the form.

The orientation of the reinforcing fibers in the matrix is very important. A component may be designed to have unidirectional or bidirectional strength, depending upon the fiber orientation. Much of the stiffness, or compressive strength, in a laminated part is derived from the shape of the part and is not necessarily an inherent property of the materials used. An advantage of laminated construction is that complex shapes can be made in one piece.

It should be obvious that laminated construction for aircraft components is a highly developed science. The fabrication and repair techniques for laminates require close attention to detail and must be specifically followed to obtain a product with the desired characteristics. Equally important is the use of the correct materials. A number of fibers and resins are used with widely varying properties. Care must be exercised to obtain the correct combination of materials and to follow the correct procedures for construction or repair.

Reinforcing Materials. The most commonly used form of reinforcing material in laminates is woven fabric, although continuous bundles, rovings or tows, and chopped mats are also widely used.

Fabric consists of threads woven at right angles to each other. The **warp** threads are those running parallel to the length of the fabric. Threads running across the warp are called the **fill (weft)**. As it comes from the mill, the warp edges are bound by the fill so that the fabric will not unravel. These edges are called the **selvage edges** and can be used to identify the warp threads. The selvage edges should be removed before use in a laminate because the weave is different from the rest of the fabric. In addition to the base material, fabrics used for laminating will vary by thread count (warp and fill), tensile strength, and weight, usually expressed in ounces per square yard (oz/yd^2).

The weave patterns, or methods in which the warp and fill threads interlace, also vary and determine many of the properties of a fabric. A major advantage of composite construction is that the strength can be tailored to the application. If a component needs strength in only one direction, such as the cap strip of a spar, unidirectional fabric may be used. A **unidirectional fabric** has most of the fibers running parallel to the warp and thus its strength lies in the same direction. Some unidirectional fabrics, such as those shown in Figure 8–20, do not have any fill threads, but use

CROWFOOT WEAVE   PLAIN WEAVE

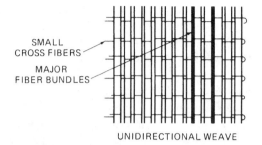

SMALL CROSS FIBERS
MAJOR FIBER BUNDLES

UNIDIRECTIONAL WEAVE

FIGURE 8–20 Bidirectional plain and crowfoot weaves and unidirectional weave.

small cross threads to hold the warp threads in place. A unidirectional material will provide the needed strength at a lighter weight, due to the absence of fill threads. **Bidirectional fabric** is woven to provide strength in both directions, warp and fill. However, the strength will not necessarily be equal in both directions. The actual thread count in each direction will determine the respective strengths. If not equal in both directions, the warp will have the greatest strength.

A number of bidirectional weaves are available, including the two shown in Figure 8–20. The **plain weave** consists of warp threads woven over and under alternate fill threads. The plain weave is characterized by fabric stability, with minimum pliability except at low thread counts. The **crowfoot weave** has warp threads going over three fill threads and then under one thread. The crowfoot weave is one of a number of weaves designed to be more pliable than a plain weave and better able to conform to complex shapes or compound curves. Because of a difference in characteristics, care must be exercised in choosing the fabric with the specified weave.

Fabric composite materials are of several types, including two kinds of fiberglass, aramid, graphite (carbon), and ceramics. Table 8–3 compares these materials in terms of cost, weight, stiffness, heat resistance, toughness, and im-

TABLE 8–3   Comparison of Composite Fabrics (Best = 1, Worst = 5)

Material	E-Glass	S-Glass	Kevlar	Graphite	Ceramic
Cost	1	2	3	4	5
Weight (density)	4	3	1	2	5
Stiffness	5	3	2	1	4
Heat	3	2	4	5	1
Toughness	3	2	1	5	4
Impact resistance	3	2	1	5	4

pact resistance. As with most materials, the one selected for a task will often be a compromise. These materials can be briefly described as follows:

Fiberglass. Fiberglass is made from small strands of glass which are spun together and woven into cloth. Fiberglass is one of the older reinforcing materials. Although inexpensive, it is lower in strength and heavier than other materials. Fiberglass has been widely used in aircraft for nonstructural parts. Two types of fiberglass fabric are used. **E-glass** is the more commonly used fabric and is inexpensive. **S-glass** has a different chemical formulation and is 30% stronger and 15% stiffer. E-glass will retain its properties up to 1500°F [815°C].

Aramid fibers. Kevlar is the registered trademark of Dupont for an aramid fiber. Aramid material ranks high in stiffness, toughness, and impact resistance. It is also lightweight. These materials can usually be identified by their yellow color. An unusual benefit is that cowlings made of aramid fibers transmit less noise and vibration than ones made of glass or graphite.

Graphite/carbon fiber. Made primarily of carbon, the term *graphite* is also used for this material. High strength, high stiffness, and low density are characteristics of these fabrics.

Ceramic. The main advantage of fabric made from ceramic fibers is its ability to withstand temperatures of almost 3000°F [1650°C]. Otherwise it is heavy, very expensive, and comparable with S-glass in strength.

Hybrid materials. Combining different materials in a component is yet another way in which the composite material can be designed for specific applications. A hybrid material may have plies of different material to use the "strengths" of each type. In some cases, fabrics are woven from a combination of thread materials. Such a hybrid fabric might be used to gain the high stiffness of one material combined with the vibration-dampening characteristics of another.

The fabrics just listed are made in many different weaves, weights, and strengths. Table 8–4 shows values for an example of each type of fabric.

Matrix Materials. The matrix bonds the fibers together and transfers the stresses among the fibers. Early composite materials made use of polyester resin. As a matrix material, polyester resin was low strength and brittle. Many newer **matrix systems** have been developed which provide for a variety of needs.

Many of the matrix systems are epoxy resins. Epoxy is a thermosetting resin that has outstanding adhesion and is very good at bonding nonporous and dissimilar materials. Epoxy resins are well known for their strength, resistance to moisture, and compatibility with most chemicals. Epoxy should be thought of as a family of matrix systems, as there are a number of formulations for specific applications. A system may be designed for a specific application or for requirements related to temperature, rigidity/flexibility, cure rate, or a specific fiber.

Virtually all of the resins require the addition of a catalyst, or **hardener**, to set up and develop their full properties. Once the hardener has been added, the resin must be used before the material begins to cure. The rate of cure may vary with the amount of hardener to be added. The mixing of the parts of a matrix system are critical and the manufacturer's directions should be closely followed.

Preimpregnated Materials. Preimpregnated fabrics, commonly call **prepregs**, are those that have the resin system already impregnated into the fabric by the manufacturer. Advantages of using prepregs include the following:

1. The prepreg material will contain the proper amount of the matrix.

2. The reinforcing fibers are completely encapsulated with the matrix. Resin-rich or resin-lean areas are avoided.

3. The matrix will have the correct proportions of resin and hardener.

These advantages help ensure that the component to be made will be successful. The disadvantages in the use of prepregs include the following:

1. The materials must be kept in a freezer to prevent the chemical action of the hardener from beginning.

2. The materials are only produced in large quantities.

3. The materials are usually more expensive than that encountered when doing manual lay-ups.

Sandwich Materials

In most cases a material's tensile strength will be more than adequate for a designed task. The problem lies in the lack of stiffness, or the inability of the material to take a compressive load without buckling. One way in which a material may be "stiffened-up" for higher compressive loads requires increasing the thickness of the material. The stiffness will be increased, but so will the weight. A second method is to fasten structural shapes, such as angles or channels, to the

TABLE 8–4   Typical Composite Fabrics

Material	Weight, oz/yd^2	Thickness, in	Thread count, W × F	Weave	Tensile, warp	Strength, fill
E-glass	3.70	0.0055	24 × 22	Plain	160	135
S-glass	3.70	0.0050	24 × 22	Plain	205	175
Kevlar	5.00	0.0100	17 × 17	Crow	630	650
Graphite	5.70	0.0070	12.5 × 12.5	Plain	1704	1704
Ceramic	7.50	0.0090	48 × 47	Crow	>200	>200

part to increase rigidity. This method not only adds weight but increases the complexity and cost of fabrication. With the development of modern adhesive systems, it is possible to solve this problem with composite, **sandwich construction**. In this method, a "**sandwich**" is constructed using at least two layers of high-strength material for facings and a lightweight, lower strength material for a core. By bonding the facings to the core, a composite material is formed with the tensile strength of the facings and a high level of rigidity from the increased thickness. For example, if two layers of fiberglass laminates are placed on each side of a foam core, the stiffness (resistance to buckling) will be many times greater than when four fiberglass layers form a solid laminate. The foam core, being lightweight, produces an almost negligible increase in weight.

Facing Materials. Virtually any type of material can be used for a facing in a sandwich-type material. In areas of high temperature or high abrasion, metal may be a choice. Because the facing can be thin to allow the overall sandwich material to be relatively light, many materials, such as titanium and stainless steel, can be used as facings. For general-purpose uses or where weight is critical, laminated composite materials are commonly used for facings.

Core Materials. Core materials used for sandwich construction may be either solid materials or honeycomb materials. A section of honeycomb sandwich material is shown in Figure 8–21. The core is made up of corrugated material assembled in a way that resembles a honeycomb. The face, or skin, is bonded to the honeycomb core. Honeycomb core can be made of fiber laminates, paper, or metal. Honeycomb core made of stainless steel and titanium is used for high-temperature applications. The honeycomb core is specified by the type of material, the thickness of the material, and the size of the hexagon-shaped cell. Honeycomb material is manufactured in blocks, allowing the core material to be cut to the desired thickness or shape. The facings can be metal or solid laminate materials. The combination of materials chosen for a honeycomb sandwich will depend upon the conditions in which the component will be used.

A solid core is a low-density solid material, such as balsa wood or expanded foam materials. Foam materials used include styrofoam and urethane. As with most of the materials used in composites, foams come in a variety of densities and working properties. Care must be exercised to acquire the right material for the task being done.

An example of the use of a sandwich composite for an aircraft is shown in Figure 8–22, a wing made entirely of laminate and foam-core sandwich construction. The wing consists of molded upper and lower halves. Each half is

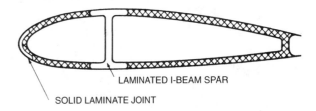

= FOAM-CORE SANDWICH STRUCTURE
(MIN. 3 PLIES FIBERGLASS EACH SIDE)

LAMINATED I-BEAM SPAR

SOLID LAMINATE JOINT

FIGURE 8–22 Wing design with sandwich construction.

made of a sandwich of foam with laminate facings. An I-beam-shaped main spar is built of solid laminate. A rectangular rear spar and ribs are all made with the same foam-core sandwich construction. When all the parts are bonded in place, the wing becomes one piece with no fasteners to wear or loosen.

Figure 8–23 shows the cross section of a wing built of similar materials but in what may be called a "moldless" method of construction. The whole wing has a solid foam core cut to the desired airfoil shape. All lines, wires, cables, and so forth that will pass through the wing are placed in the foam, or a passage is made for them. The entire core is then covered with layers of fabric, impregnated with resin, and finished.

Metal-Matrix Composites

Although most work is still in the developmental stage, the potential for metal-matrix composites is promising. This composite is formed by adding the reinforcing materials, in the form of chopped fibers or strands of fibers, into the metal while in a molten state. When cooled, the solid metal matrix is strengthened by the fibers. The metal can be worked with conventional metalworking processes, with allowance for the changed properties of the composite.

Metal Bonding

Although not necessarily a composite material, the bonding of metals, as a means of joining them, uses many of the same adhesive materials, fabrication methods, and inspection techniques as in composite construction. Bonded metal joints are usually higher strength and have a longer fatigue life than conventional riveted joints. Once a process is set up, bonded joints are usually more economical in terms of manufacturing cost than riveted joints. The disadvantages of bonded joints include the need for special inspection processes and specialized repair considerations.

FIGURE 8–21 Honeycomb core material.

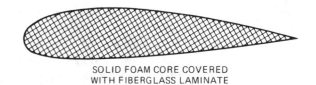

SOLID FOAM CORE COVERED
WITH FIBERGLASS LAMINATE

FIGURE 8–23 Moldless wing design with solid foam core.

AIRCRAFT WOOD

Three forms of wood are commonly used in aircraft: solid wood, laminated wood, and plywood. To be used in an aircraft, the wood must be of **aircraft quality**.

Solid Wood

The standard species of solid wood for aircraft use is Sitka spruce. Other species of wood that may be substituted for spruce include Douglas fir, noble fir, western hemlock, northern white pine, white cedar, and yellow poplar. The substitution of these materials for spruce should only be done in accordance with the guidelines established by the FAA and published in *Advisory Circular AC43.13-1A*.

To be aircraft quality, a piece of wood must meet a number of criteria. Many of these are determined by the original growth conditions encountered by the tree from which the wood has been cut. A tree grows by developing new fibers around its circumference each year. During the spring, the growth is more rapid and is distinguished by fibers of a larger size and thinner walls. The wood formed during this period is called **spring wood** and is lighter in color than the smaller, thicker walled fibers formed during summer growth. Spring wood is also weaker than summer wood. The dark-colored layers of summer growth form what are called **annual rings**. The **grain pattern** of a wood refers to the directional orientation of the fibers. Since annual rings are layers of fibers, they also show the grain pattern. The grain pattern and annual rings are both important criteria for the evaluation of wood as aircraft quality.

Specification AN-W-2 for Sitka Spruce specifies that the slope of the grain shall not be steeper than 1 in 15 and that the wood must be sawn vertical grain and shall have no fewer than six annular rings per inch. The determination of the slope of the grain is shown in Figure 8–24. *Sawn vertical grain* refers to the log being sawn in such a manner that the annual rings form an angle of 45 to 90° with the face of the board. This is accomplished by a process known as **radial sawing**, as compared to **plain sawing**. Examples of radial- and plain-sawn wood are shown in Figure 8–24. Having no less than six annual rings per inch ensures that the piece does not have an excess of the weaker spring wood.

The strength of wood will vary directly with the density. Aircraft-quality spruce must have a specific gravity of at least 0.36. A defect called *compression wood* has the appearance of an excessive growth of summer wood. Wood with this defect has a high specific gravity and should be rejected.

Wood must be kiln dried to be aircraft quality. As the moisture of a wood increases, the strength decreases. The use of kiln drying ensures that the wood has been brought to an acceptable moisture content. Wood fibers continually exchange moisture with the surrounding air, depending upon the humidity. As moisture is absorbed or released, the fibers expand and contract. Provisions must be made for the dimensional change which occurs. The dimensional change of a board will be the greatest across the fibers and parallel to the growth rings, somewhat less across the fibers and per-

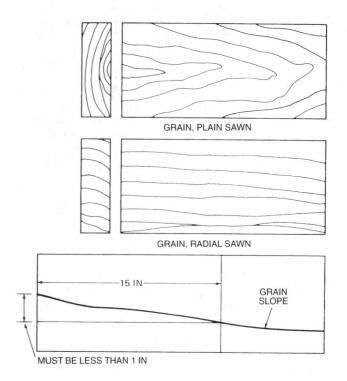

FIGURE 8–24 Radial- and plain-sawn wood, grain slope.

pendicular to the growth rings, and negligible in a longitudinal direction. In essence, the greatest dimensional change (due to moisture) will occur to the thickness of a radial-sawn board.

Certain defects in wood will affect its use for aircraft. **Checks** are longitudinal cracks extending across annual rings. **Shakes** are longitudinal cracks between annual rings. Checks and shakes are formed during tree growth, as shown in Figure 8–25. **Splits** are longitudinal cracks caused by an artificially induced stress. Wood containing checks, shakes, or splits is not aircraft quality. A **spike knot**, shown in Figure 8–26, runs perpendicular to the annual rings. Spike knots are not allowed in aircraft quality wood.

Certain other defects may exist if they are within limitations as listed in *AC43.13-1A*. These include wavy grain, hard knots, pin-knot clusters, pitch pockets, and mineral streaks. Any form or evidence of decay is cause for the wood to be rejected.

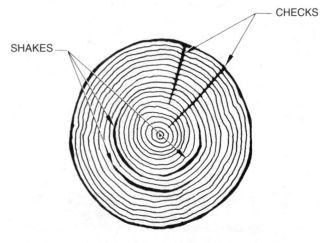

FIGURE 8–25 Shakes and checks in wood.

FIGURE 8–26 Spike knot.

Laminated Wood

Laminated wood is made up of a number of layers glued together. Unlike plywood, laminated wood has the grain running the same direction in all layers. The wood may have been laminated in order to form it to a shape, to use materials that have had defects removed, or to get a specific size. The FAA allows solid spruce spars to be replaced with laminated spars or vice versa, provided the material is of the same quality.

Plywood

Aircraft-grade plywood is made of imported African mahogany or American birch veneers laminated to cores of poplar or basswood with waterproof glue. Plywood made to Specification MIL-P-6070 calls for shear testing of the plywood after immersion in boiling water for three hours. Conventional plywood has the grain of alternate layers running at 90°. Aircraft plywood is made in two styles, with one having grain at 45° angles in alternate plies. Aircraft plywood is used for skin or for reinforcing plates.

Further details on aircraft-quality wood as well as the maintenance and repair of wood structures may be found in the FAA's *Advisory Circular AC43.13-1A.*

AIRCRAFT FABRICS

The fabric discussed in this section is used for covering aircraft wings, control surfaces, and fuselages. As was discussed in the section on the reinforcing material for laminates, fabric is a material of various types of threads woven at right angles to each other. The **warp** threads are those running parallel to the length of the cloth. Threads running across the warp are called the **fill**. The fabric as it comes from the mill has each edge bound by the fill so that it will not unravel. These edges are called the **selvage edges** and can be used to identify which are warp threads. The manufacturer will normally print identification information on the fabric along the selvage edge.

Specifications for fabric will normally reference the thread count, or the number of threads per inch. The thread count will be specified for both warp and fill, as they may be different. The fabric may have a specified weight stated in ounces per square yard.

Aircraft-covering materials were originally made from cotton. Three specifications exist for the manufacture of cotton fabric. **Grade A cloth** (AMS3806) has a minimum of 80, and a maximum of 84, threads warp and fill. It is required on fabric-covered airplanes with wing loading in excess of 9 lb/ft^2 or a never-exceed-speed in excess of 160 mph. Grade A cloth must have a tensile strength of 80 lb/in, warp and fill, when new. It is allowed to deteriorate to 56 lb/in (70%) in use before being replaced. Fabric is tested by putting a 1-in-wide strip in a tensile tester and pulling it to failure. Note that the figures are given as units of pounds per inches of width and not pounds per square inch.

Aircraft other than those requiring Grade A fabric may use AMS3804 **airplane cloth**, sometimes called intermediate grade. Fabric made to this specification has a minimum of 80 and a maximum of 94 threads, warp and fill. New strength for intermediate fabric is 65 lb/in. Gliders with a wing loading of less than 8 psi and a never-exceed speed of 135 mph or less can use AMS3802 cotton cloth. This **glider cloth** has a maximum of 110 threads per inch, warp and fill, with a new strength of 50 lb/in. **Aircraft linen** is manufactured to British specifications and may be substituted for Grade A cloth.

The installation of a fabric cover requires several other types of materials. A waxed cord, normally called rib-lacing cord, is used to lace the fabric to the ribs and other parts of the aircraft structure. **Reinforcing tape** is a heavy cotton tape used over ribs between the fabric and the rib-lacing cord. Its purpose is to keep the cord from cutting through the fabric. **Surface tape**, also called **finishing tape**, is usually made in various widths from the same material as that used for the cover. The purpose of surface tape is to cover seams and rib-lacing cords.

The cover is not complete until the required finishes have been applied. The finishes for fabric covers include nitrate dope, cellulose-nitrate dope, and cellulose-acetate-butyrate dope. Both clear and pigmented dope are used. The cover must be installed and finished in accordance with set standards. The FAA's *Advisory Circular AC43.13-1A* provides detailed information on fabric-covering installation and maintenance.

Fabrics made from synthetic fibers have been developed as a replacement for cotton fabrics. The most common synthetic is a **Dacron cloth** sold under several trade names. Dacron cloth has the advantage of being longer lasting than cotton. It can be heat shrunk to size and bonded seams can be used, making the application of the cover easier than cotton. Fiberglass has also been used as a cotton replacement. Virtually all fabric-covered aircraft were originally built with cotton fabric. Replacement of the cover with another material is a major alteration. The manufacturer of a synthetic material may have obtained a supplemental type certificate (STC) to allow the installation of his or her product. If a synthetic material is being used, the instructions supplied by the manufacturer, with the STC, must be strictly followed.

REVIEW QUESTIONS

1. Discuss the five types of stress. Give two common examples of each type.
2. Why are materials normally not rated in terms of compressive strength?
3. What is the difference between the elasticity and the plasticity of a material?
4. How can you estimate the modulus of elasticity from looking at a stress-strain diagram?
5. What is the difference between a substitutional solid-solution alloy and an interstitial solid-solution alloy?
6. What type of crystals do you find in a Type II alloy?
7. Why is aluminum alloy a good structural material for aircraft?
8. Describe the composition and properties of aluminum alloy 7075-T6.
9. What is the SAE number for chrome-molybdenum steel?
10. What alloying elements are used in 18-8 stainless steel?
11. What type of CRES may be hardened by heat treatment?
12. What are the advantages of magnesium for aircraft parts?
13. What are the disadvantages of magnesium?
14. What precautions must be taken when titanium is heated to a temperature above 1000°F [538°C] and when welding titanium?
15. For what purpose is beryllium copper used?
16. What is the difference between thermosetting and thermoplastic resins?
17. What are the advantages of polyurethane enamels?
18. For what purpose are polyester resins commonly used?
19. What materials are used for aircraft windshields and windows?
20. What is the function of the matrix system in a laminate?
21. Describe honeycomb core material.
22. What materials may be used in honeycomb sandwich construction?
23. What types of wood are considered standard for aircraft structures?
24. What defects in aircraft woods make them unsuitable for use?
25. Describe the criteria used to determine if an aircraft fabric is Grade A.

Metal Fabrication Techniques and Processes 9

INTRODUCTION

This chapter describes various processes used in the production and fabrication of metal parts and components for aircraft. The intent is not to provide detailed instruction on the processes that the technician is expected to perform. That material is covered in the other texts of this series. The objective of this chapter is to give you a better understanding of the techniques used to make the materials with which you will be working. As mentioned in the preceding chapter, the aircraft maintenance technician of the twenty-first century will be required not only to perform a number of skilled tasks but to understand what he or she is doing. The technician must understand the new materials as well as the new and improved processes for working with them.

MILL PRODUCTS

Metal materials are manufactured in a number of shapes and sizes. The technician should be able to use the correct terminology to specify such products, including the following:

Bar. A solid product that is long in relation to its cross-sectional dimensions and that is square or rectangular (excluding plate or flattened wire), with sharp or rounded corners or edges, or hexagonal or octagonal, with regular edges. At least one perpendicular distance between parallel faces must be $\frac{3}{8}$ in [9.5 mm] or greater.

Foil. A rolled product, rectangular in cross section, with a thickness less than 0.006 in [0.15 mm].

Forging. A metal part mechanically worked to a predetermined shape by one or more processes such as hammering, upsetting, pressing, rolling, and so forth.

Pipe. A tube in standardized combinations of outside diameter and wall thickness, commonly designated by nominal pipe sizes and ANSI schedule numbers.

Plate. A rolled product rectangular in cross section and form, with a thickness of 0.250 in [6.35 mm] or more and sheared or sawed edges.

Rod. A solid product that is long in relation to its cross-sectional dimensions and $\frac{3}{8}$ in [9.55 mm] or greater in diameter.

Shape. A wrought product that is long in relation to its cross-sectional dimensions and has a cross section other than that of sheet, plate, rod, bar, tube, or wire.

Sheet. A rolled product that is rectangular in cross section and form, with a thickness of 0.006 through 0.249 in [0.15 through 6.32 mm] and sheared, slit, or sawed edges.

Tube. A hollow, wrought product that is long in relation to its cross-sectional dimension and that is round, regularly hexagonal, regularly octagonal, elliptical, square, or rectangular in cross section, with sharp or rounded corners and a uniform wall thickness except as affected by corner radii.

Wire. A solid, wrought product that is long in relation to its cross-sectional area and that is square or rectangular (with sharp or rounded corners or edges), round, regularly hexagonal, or regularly octagonal in cross section; it also has a diameter, or the greatest perpendicular distance between parallel faces (except for flattened wire), less than $\frac{3}{8}$ in [9.5 mm].

Identification of Mill Products

Various standards have been developed for marking materials for identification. Those used for aluminum are described here; standards for other metals are similar.

Aluminum-alloy sheet, as it comes from the manufacturer, is marked with letters and numbers in rows about 6 in [15.24 cm] apart. The identification symbols may include a federal specification number, the alloy number with temper designation, and the thickness of the material in thousandths of an inch. The standard methods for marking aluminum sheet are shown in Figure 9–1. The rows of letters and figures are parallel to the grain of the metal. The color used for the marking may also be used to identify the alloy material.

Methods used for marking foil sheet and other shapes are shown in Figure 9–2. Items too small for conventional markings, like rivets, are identified by symbols and numbers that are an integral part of the metal.

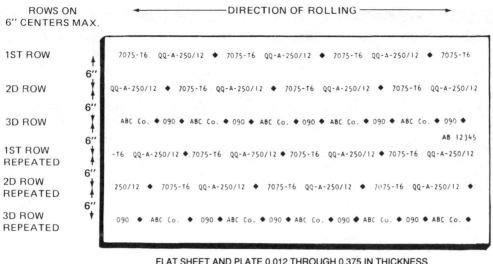

FLAT SHEET AND PLATE 0.012 THROUGH 0.375 IN THICKNESS
6 THROUGH 60 IN WIDTH, 36 THROUGH
200 IN LENGTH

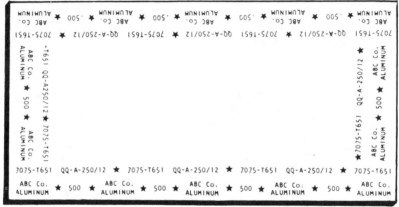

FLAT SHEET AND PLATE, GREATER IN THICKNESS AND SIZE
THAN THAT SHOWN ABOVE

FIGURE 9–1 Identification markings for aluminum-alloy sheet. *(Aluminum Association, Inc.)*

FABRICATION OF METAL COMPONENTS

Most metal products begin as **ingots**, formed by pouring molten metal into large molds and allowing it to cool. At a later time the ingot is further processed by either casting or plastically forming the material into a wrought shape. Both cast and wrought materials are subjected to a variety of processes, which might include heat treatment, machining, welding, and numerous types of inspection.

Those products in which the metal has been mechanically worked or plastically formed to a desired shape are called **wrought products**. Forming of the metal may be by cold working or hot working. **Hot working** involves plastic deformation of the metal at a temperature which results in the metal being continually recrystallized (annealed); therefore, strain hardening, or work hardening, does not occur. For aluminum, hot working is performed at 650 to 900°F [343 to 482°C], depending on the alloy. This is above the temperature required for recrystallization, but well below the melting point or other heat-treatment temperatures.

The term **cold working** refers to work done at a temperature below that required for recrystallization. Cold working causes a permanent change in the grain shape and orientation. The strength and hardness that is gained from the cold working is usually desirable. However, in cases of severe deformation, the cold-working effects may need to be removed by annealing. Since a metal is limited as to how much deformation can occur before failure or rupture, it may be necessary to periodically anneal the material during the forming process. The alteration of the grain structure can also strengthen a product by providing a directional orientation for the flow of stresses within the metal.

The term **cast products** describes those products formed by melting the metal and pouring it into a mold of the desired shape. Since plastic deformation of the metal does not occur, no alteration of the grain shape or orientation is possible. The grain size of the metal can be controlled by the cooling rate, the alloys of the metal, and the thermal treatment. Castings are usually considered as being lower in strength and more brittle than a wrought item of the same material. For a product with an intricate shape or internal passages, casting may be the most economical process.

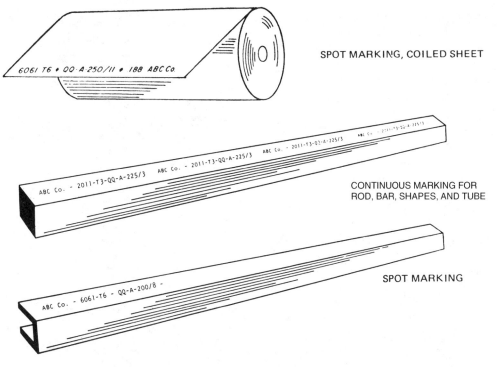

SPOT MARKING, COILED SHEET

6061 T6 • QQ-A-250/11 • 188 ABC Co.

ABC Co. - 2011-T3-QQ-A-225/3

CONTINUOUS MARKING FOR
ROD, BAR, SHAPES, AND TUBE

SPOT MARKING

ABC Co. - 6061-T6 - QQ-A-200/8 -

FIGURE 9–2 Markings for coiled metal and other shapes. *(Aluminum Association, Inc.)*

Except for some engine parts, most metal components found on an aircraft will be wrought.

It should be noted that all metal products start in a cast form. Wrought metals are converted from cast ingots by plastic deformation. For high-strength aluminum alloys, an 80 to 90% reduction (i.e., dimensional change in thickness) of the material is required to obtain the high mechanical properties of a fully wrought structure.

Plastic Deformation of Metals

Many different processes are used to shape metal into products usable by the aircraft technician. Some of these processes are described in the following paragraphs.

The term **rolling** describes a widely used process of plastic deformation that reduces the thickness of metal by subjecting it to compressive forces by passing it between driven cylindrical rollers. Many ingots are first rolled into plates or bars to provide easier handling of the material for future processing. Rolling can provide finished materials such as plate, sheet, bars, and some shapes. Both hot- and cold-rolling processes are used. Hot rolling requires less energy and is used when alteration of the grain structure is either not desired or not important. Many hot-rolled materials may be identified by the presence of a hard, oxide scale formed by the oxygen in the atmosphere combining with the surface of the hot metal. While the scale does not appreciably affect the strength, it may affect later machining and forming operations. In some cases the process specification for a particular product may call for the scale to be removed.

Cold rolling alters the grain structure and mechanically changes the properties of the material. Severe reduction by the rolling process may require thermal treatment to relieve work-induced, internal stresses. Cold-rolled metals will usually have a very smooth, natural color and finish.

In another process, **forging**, the metal is forced to flow (deform) under high compressive stresses. Forging will usually cause the metal to flow into the shape of a die cavity, or it may be an upsetting operation, as shown in Figure 9–3. Upsetting is usually limited to a cylindrical workpiece whose length is not more than three times its diameter, because of the tendency to buckle. The aircraft maintenance technician uses a forging process to drive or upset rivets. Forging can also be done freehand, such as a blacksmith hand-forging a chisel. Such methods are limited to craft-type or special-purpose activities. Forging is usually done hot, but thinner and more malleable materials may be cold worked.

The **extrusion process** uses compressive force to cause the metal to flow through a die. The shape of the die will be the cross section of an angle, channel, tube, or some other

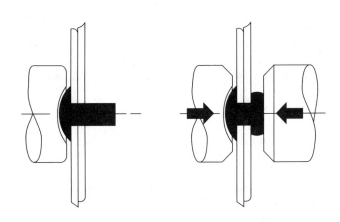

FIGURE 9–3 Upsetting a rivet, a forging process.

intricate shape. Many aluminum shapes have been extruded. Aluminum is extruded at a temperature of 700 to 900°F [371 to 482°C] and requires pressure of up to 80 000 psi [552 MPa]. After extrusion, the product frequently will be subjected to both thermal and mechanical processes to obtain the desired properties. Extrusion processes are limited to the more ductile materials.

A process very similar to extrusion is **die drawing**. A die is used with the desired cross-sectional shape, but the metal is pulled, or drawn, through the die rather than pushed. Die drawing is commonly used for wires, tubes, and some other shapes.

Metal **forming** refers to plastically forming a flat sheet to a desired shape. **Bending** is a forming process in which plastic deformation only occurs in a relatively small area along a bend line. Most of the metal is not subjected to any change in shape or internal structure when a **straight-line bend** is formed. Other forming operations form **compound curves**, in which the metal is stretched or drawn to a new shape. These processes may involve considerable metal flow and thickness changes accompanied by the expected changes in mechanical properties. Straight-line bends may be formed in a **press brake**, which incorporates forming dies to form the bend in the metal. Complex shapes made up of a number of straight-line bends may be formed. Most press-brake work is done cold. Some "harder" materials (e.g., 7075 aluminum or magnesium) may require the material to be heated above room temperature but below that for hot working. Straight-line bends may also be made around a radius bar in a cornice (or similar) brake. The aviation maintenance technician is expected to develop proficiency in making bends by this method. More details on this process may be found in the text *Aircraft Maintenance and Repair*.

The process of **continuous-roll forming** involves forming shapes, such as angles, channels, and tubes, by passing a strip of sheet metal through a series of rollers. Each pair of rollers produces a small amount of plastic deformation, with the cumulative effect being the desired shape. This method involves making a number of straight-line bends, similar to the press-brake process. The press brake is limited by size as to the length of material that can be formed. Continuous-roll forming is limited only by the length of the stock. Even a simple shape, such as an angle, requires several pairs of rolls to form. The amount of deformation through any one pair must be limited to avoid excessive work hardening. Aluminum, for example, should not be bent more than 22.5° at any one pair of rolls. This method of forming requires expensive equipment and extensive setup time. Continuous-roll forming is not economical unless large quantities of the shape are to be produced.

The **drawing** process is used in the production of sheet-metal parts having compound curves. Drawing usually requires a movable punch and some sort of die (see Figure 9–4) to form the desired shape. Deep or severe drawing may require several steps, with annealing occurring between steps. **Flexible-die forming** will use either a punch or a die but not both. Figure 9–5 shows a laminated rubber pad substituted for the die. The workpiece and a mold of the desired shape are on the platen, or punch, of the press. Under pres-

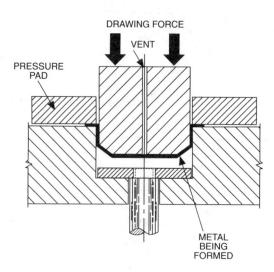

FIGURE 9–4 Drawing a metal part.

sures of 1000 to 2000 psi [6.89 to 13.70 MPa], the rubber will flow and form to the shape of the mold, causing the metal to also form to the mold. Several variations of this process exist, including the use of a liquid and hydraulic pressure in place of the rubber pad.

The **spinning** process involves the forming of a turning, or spinning, part over a mandrel with a tool or roller. Figure 9–6 shows a part similar to a propeller spinner being formed by spinning. Most parts made by spinning can also be made by drawing, with economic reasons being the determining factor for a given part.

The **shearing** process is used for separating (cutting) a metal by forcing two opposing and slightly offset blades against it with sufficient force to cause fracture, as shown in Figure 9–7. During initial stages of shearing, the metal is elastically, then plastically, deformed. During plastic deformation there will be a reduction in the area of the metal, followed by fracture. Sheared edges are not perfectly square but may show burnished and rough or torn edges. The clearance between the blades is very important in providing clean-sheared surfaces. Soft, ductile materials require greater plastic deformation before fracture occurs and re-

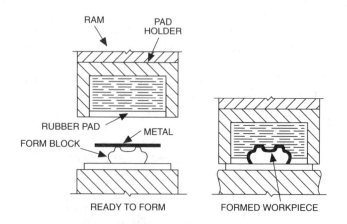

FIGURE 9–5 Forming metal with a flexible die.

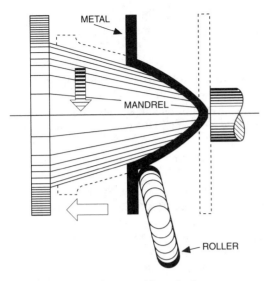

FIGURE 9–6 Forming metal by spinning.

quire a greater clearance between blades than a harder, more brittle material.

HEAT TREATMENT OF METALS

Heat treatment can be any one of several thermal processes used on metals in order to develop the desired hardness, softness, ductility, tensile strength, or grain structure. For the remainder of this chapter, the term **heat treatment** refers to the processes used to harden or strengthen metal. *Annealing* will be used to refer to processes used for recrystallization. A variety of other processes—normalizing, tempering, stress relieving, and aging—will be defined as they are discussed.

The effects of cold working on the internal structure of the metal were covered in Chapter 8. Figure 9–8 graphically depicts the changes that occur with work hardening. The effects of the work hardening can be removed. Annealing, or recrystallization, is accomplished by heating the metal to a temperature that allows new grain formation to occur. A pe-

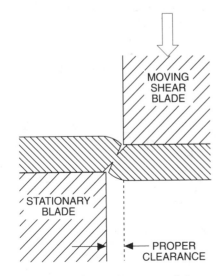

FIGURE 9–7 Shearing action on a metal.

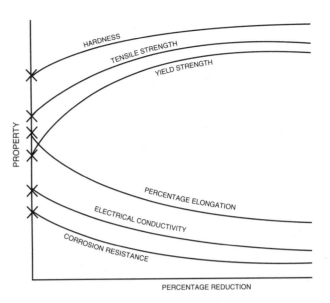

FIGURE 9–8 Effects of cold working.

riod of slow cooling enhances the new grain, or crystal, formation along with a new set of slip-planes, therefore restoring the pre-cold-work properties. All metals are affected by cold working, and all metals will respond to annealing treatments.

All metals and alloys do not respond to heat treatment. Ferrous metals, iron and steel, can usually be heat treated; many corrosion-resistant (stainless) steels cannot. Some alloys of aluminum are strengthened by heat treatment, but others must be hardened by cold working. The high-temperature nickel-base alloys can be heat treated in some cases, depending upon their composition. The temperatures and processes of heat treatment vary considerably among the different metals. It is essential that the process specification for a particular product be followed if the desired properties are to be attained.

The hardening of metal by heat treatment is usually the result of one of two phenomena. Some metals are allotropic, that is, their lattice structure will change at elevated temperatures. By using certain alloys and thermal treatment, a stronger internal structure can be stabilized for use at normal operating temperatures. Steel is hardened through this process.

Aluminum is not allotropic. The hardening of aluminum is accomplished by alloying an element that is soluble only at higher temperatures. At lower temperatures, the alloy precipitates as a metallic compound, producing hardening effects.

Solution Heat Treatment of Aluminum

To be hardened by heat treatment, aluminum must have copper as an alloy. Figure 9–9 shows the aluminum end of an aluminum-copper diagram similar to those discussed in Chapter 8. From the diagram it can be seen that aluminum will only dissolve approximately 0.5% copper from room temperature to about 600°F [316°C]. As the temperature is increased, the percentage of copper in solution increases. At 1018°F [548°C] the alloy will have slightly over 5% copper in solution. This is the maximum concentration of copper

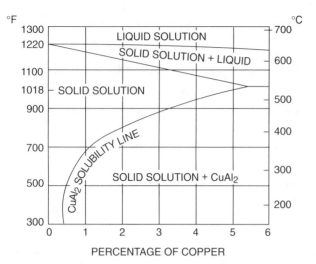

FIGURE 9–9 Aluminum-copper alloy chart.

that will enter a solid solution with aluminum. Higher concentrations will cause the alloyed combination to become a Type II, eutectic alloy. This temperature versus concentration point is on the solidus line and is the temperature at which some crystals in this combination begin to melt. The diagram also shows that if the copper content exceeds 0.5%, the copper that is not in solution forms a compound, $CuAl_2$.

For illustrative purposes, let us assume we have an aluminum with a 3% copper concentration. An aluminum alloy with 3% copper at room temperature would have 0.5% copper in solid solution. The remainder of the copper would be in the form of the compound $CuAl_2$. As the temperature of the alloy is increased, copper dissolves from the compound and enters a solid solution with the aluminum. If the alloy is heated to approximately 890°F [477°C], 100% of the copper would be in solution. The material must be kept at this temperature for sufficient time for the copper to be dissolved. This is referred to as the **soaking time**. *The first step of heat treatment is heating the aluminum to allow the alloying element, copper, to enter into a solid solution.*

The next step, cooling the metal, is critical if the hardening effects are to occur. As the metal cools, the excess copper again forms the $CuAl_2$ compound. Initially, the compound will consist of a large number of very small particles. If the solution cools slowly, the energy levels present in the atomic structure allow the $CuAl_2$ particles to move and congregate into a few relatively large masses scattered throughout the aluminum. Other than some distortion of the

lattice structure, there will be little effective hardening. In short, no lasting effects have been achieved. Cooling the metal very quickly, almost instantly, brings about different results. If cooling occurs very quickly, virtually none of the copper has time to **precipitate** out of solution. Using the preceding example, we would find a 100% solid solution containing 3% copper at room temperature. This is a temporary, or unstable, condition and the material is soft and ductile. As time passes, the excess copper precipitates out in the form of $CuAl_2$. At room temperature the energy levels in the atomic structure are much lower, resulting in the $CuAl_2$ particles not congregating, but remaining stable as a large number of small particles. The large number of particles are interspersed among the atoms in the grain structure, causing distortion and "locking" the slip planes, in effect producing a metal that is harder to deform, thus stronger. *The second step of heat treatment for aluminum is to cool, or quench, the material as quickly as possible.* These two steps are the process known as **solution heat treatment**.

Hardening has not occurred with the quench because the compound has not yet precipitated out of solution. The precipitation process is known as **aging** and may take minutes to months, depending upon the alloy. The process can be accelerated by slightly reheating the material and allowing it to soak for a specified time. This is known as **artificial aging**, or more technically as **precipitation heat treatment** or **precipitation hardening**. To ensure that a metal has reached full hardness, many metals will be aged by this process.

Table 9–1 provides data for the solution heat treating and the aging of aluminum alloys. Table 9–2 lists soaking times.

There are several areas of special concern to consider when heat treating aluminum. The first is that the solution-heat-treating temperatures are very close to the melting point of the metal. Care must be taken to ensure that the correct temperature is achieved.

Copper and aluminum are not friendly metals in terms of corrosion. As long as the material is quickly quenched, the formation of the $CuAl_2$ compound will be in minute enough particles to avoid problems. The formation of larger particles, through slow cooling, presents the potential of dissimilar metals causing **intergranular corrosion**.

Clad aluminum has a layer of pure aluminum on both faces. The copper alloying element cannot distinguish between the pure cladding and the alloyed base. After several heat-treating cycles, the cladding will no longer be pure aluminum.

TABLE 9–1 Aluminum-Alloy Heat-Treating Data

Alloy	Temperature		Quench	Aging Temperature		Time of Aging
	°F	°C		°F	°C	
2014T	930–950	499–510	Hot water	335–345	168–174	10 h
2017T	930–950	499–510	Cold water	Room	Room	4 days
2117T	930–950	499–510	Cold water	Room	Room	4 days
2024T	910–930	488–499	Cold water	Room	Room	4 days
5053T	960–980	515–527	Water	312–325	155–163	18 h
6061T	960–980	515–527	Water	315–325	157–163	18 h
7075T	860–930	460–499	Cold water	345–355	174–179	6–10 h

TABLE 9–2   Aluminum-Alloy Soaking Time

Alloy	Soaking Time, min			
	Less Than 0.032 in [0.081 cm] Thick	0.032– 0.125 in [0.317 cm] Thick	0.125– 0.250 in [0.635 cm] Thick	More Than 0.250 in [0.635 cm] Thick
2014T	20	20	30	60
2017T	20	30	30	60
2117T	20	20	30	60
2024T	30	30	40	60
2024T (clad)	20	30	40	60
5053T	20	30	40	60
6061T	20	30	40	60
7075T	25	30	40	60
7075T (clad)	20	30	40	60

Some materials, such as 2024 aluminum rivets, are too hard to work in the hardened condition. It is common practice to solution heat-treat the rivets and drive them while they are in the soft, unstable stage before precipitation occurs. The 2024 alloy ages very quickly, and the unstable "workable" phase lasts only a short time. The precipitation process may be retarded by cold storage. By keeping the rivets at a low temperature, the hardening effects of precipitation can be delayed for days—thus the term *icebox rivets*.

Temper, or condition, designations for aluminum alloy were discussed in Chapter 8 and are also shown in Table 9–3. Note that the **T** designation indicates that the material has been solution heat-treated. The remainder of the designation tells whether artificial or natural aging has occurred and whether or not the material has been cold-worked during the aging process.

Annealing Aluminum

Aluminum is annealed by heating it to the required temperature, allowing it to soak for the time necessary for recrystallization to occur, and cooling the metal slowly. Table 9–4 provides specific data for annealing aluminum.

TABLE 9–3   Aluminum-Alloy Temperature Designations

T1	Cooled from an elevated-temperature shaping process and naturally aged to a substantially stable condition.
T2	Cooled from an elevated-temperature shaping process, cold worked, and naturally aged to a substantially stable condition.
T3	Solution heat treated, cold worked, and naturally aged to a substantially stable condition.
T4	Solution heat treated and naturally aged to a substantially stable condition.
T5	Cooled from an elevated-temperature shaping process and then artificially aged.
T6	Solution heat treated and then artificially aged.
T7	Solution heat treated and stabilized.
T8	Solution heat treated, cold worked, and then artificially aged.
T9	Solution heat treated, artificially aged, and then cold worked.
T10	Cooled from an elevated-temperature shaping process, cold worked, and then artificially aged.

TABLE 9–4   Annealing Treatment for Aluminum

Alloy	Metal Temperature, °F	Approx. Time of Heating, hours
1100	650	[1]
3003	775	[1]
2024	775[2]	2–3
5052	650	[1]
6061	775[2]	2–3
7075	775[3]	2–3

[1] Time in furnace need not be longer than time necessary to bring all parts to annealing temperature. Cooling rate is unimportant.

[2] This treatment is intended to remove the effect of heat treatment and includes cooling at a rate of about 50°F per hour from the annealing temperature to 500°F. The rate of subsequent cooling is unimportant. To remove the effects of cold working only, the temperatures for 1100 may be used.

[3] Should be followed by heating for about six hours at about 450°F if material is to be stored an extended period of time.

Heat Treatment of Ferrous Metals

Ferrous metals are those that are iron based or contain large percentages of iron. Carbon is the element that makes it possible to harden iron and make steel. Pure iron cannot be heat treated. The presence of carbon makes possible several different combinations of iron and carbon that affect the properties of the metal. Heat-treating processes rearrange the atoms of the carbon-and-iron alloy to alter the strength, toughness, and hardness of the steel. The chemical action that occurs in iron-based alloys is very complex. Unlike aluminum, steel can be hardened and tempered to a range of values. A detailed explanation of these actions is beyond the scope of this text. Figure 9–10 is a simplified iron-carbon diagram.

When carbon steel is heated to a temperature of 1341°F [727°C], a solid solution of iron and carbon forms that consists of 0.76% carbon. This material is called **austenite**. If the steel contains less than 0.76% carbon, the result will be a mixture of **ferrite** and austenite. Ferrite is pure iron of relatively low tensile strength with a maximum of 0.025% carbon in a dissolved state.

When austenite is cooled below 1341°F [727°C], a compound of iron carbide (Fe_3), called **cementite**, precipitates from the solution at the crystal boundaries. As cooling continues, the austenite continues to decompose into alternate platelets of ferrite and cementite in a form called **pearlite**. Pearlite is a layered structure of platelets of iron carbide in a

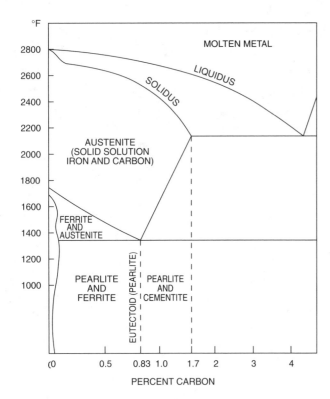

°F

MOLTEN METAL

LIQUIDUS

SOLIDUS

AUSTENITE
(SOLID SOLUTION
IRON AND CARBON)

FERRITE
AND
AUSTENITE

PEARLITE
AND
FERRITE

EUTECTOID (PEARLITE)

PEARLITE
AND
CEMENTITE

PERCENT CARBON

FIGURE 9–10 Simplified iron-carbon diagram.

ferrite matrix. The amount of pearlite formed and the nature of its structure will depend upon the rate of cooling and the carbon content of the steel.

Very rapid cooling of austenite precludes the formation of pearlite, and it changes immediately to a structure called **martensite**. In this form, the steel attains a maximum of hardness, particularly when the carbon content is high.

The hardness, toughness, ductility, and brittleness of steel will depend upon the mixture and arrangement of ferrite, cementite, pearlite, and martensite in the steel, as brought about by the specific heat-treating process.

Alloy steels can develop greater strength and toughness than carbon steels because of the effects of the alloying elements, such as chromium, molybdenum, tungsten, nickel, and vanadium. Heat treatment of the alloy steels must be tailored to the particular type of steel being treated, in accordance with the elements it contains.

The **hardening process** is performed by heating the metal slightly above its critical temperature and then rapidly cooling it by quenching in oil, water, or brine. This produces a fine-grained structure, great hardness, maximum tensile strength, and minimum ductility. Material in this condition is usually too brittle for most uses, but this treatment is the first step in the production of high-strength steel.

The hardening of steel in industry is a carefully controlled process. The rate of heating, the time of soaking, and the rate of cooling are all regulated to achieve the desired results. The more rapidly a steel is cooled from the critical temperature, the harder the steel will be. It is usually necessary to temper the steel to achieve the necessary toughness and to reduce brittleness. Table 9–5 gives the temperatures for the heat treating and the tempering of steel.

The process of **tempering** (**drawing**) is generally applied to steel to relieve the strains induced during the hard-

ening process. It consists of heating the hardened steel to a temperature below the critical range, holding this temperature for a sufficient period, then cooling in water, oil, air, or brine. The degree of strength, hardness, and ductility obtained depends directly upon the temperatures to which the steel is heated. When high temperatures are used for tempering, ductility is improved at the expense of hardness, tensile strength, and yield strength.

The term **normalizing** describes the process of heating iron-based metals above their critical temperature to obtain better solubility of the carbon in the iron, followed by cooling in still air. The normalizing process reduces stresses in the metal that have been caused by forging, machining, cold working, or other processes. It improves the grain structure, toughness, and ductility of the metal. Welded steel parts should always be normalized to eliminate stresses caused by uneven heating and to improve the grain structure of the weld.

Annealing of Steel

Steel products can be fully annealed by heating to the appropriate temperatures (see Table 9–5) and cooling slowly.

Case-Hardening Processes

Case-hardening processes produce a hard, wear-resisting surface, while leaving the core of the metal tough and resilient. Three common methods of case hardening are carburizing, nitriding, and cyaniding.

The process of **carburizing** consists of holding the metal at an elevated temperature while it is in contact with a solid, liquid, or gaseous material rich in carbon. Time is allowed for the surface metal to absorb enough carbon to become high-carbon steel.

The **nitriding** process is accomplished with special alloy steels containing small amounts of chromium, molybdenum, and aluminum by holding at temperatures below the critical point in anhydrous ammonia. Nitrogen from the ammonia is absorbed into the surface of the steel as iron nitride. This produces a greater hardness than carburizing, but the hardened area does not reach as great a depth.

The process of **cyaniding** is a fast method of producing surface hardness on iron-based alloys of low carbon content. The metal may be immersed in a molten bath of cyanide salt, or powdered cyanide may be applied to the surface of the heated metal. During this process, the temperature of the steel must range from 1300 to 1600°F [538 to 871°C]. The exact temperature depends upon the type of steel, the depth of the case hardening desired, the type of cyanide compound used, and the time that the steel is exposed to the cyanide. In using sodium cyanide or potassium cyanide, great care must be taken to avoid getting any of the cyanide into the mouth, eyes, or any other part of the body. These materials are deadly poisons.

Heat Treatment of Stainless Steels

The chrome-nickel stainless steels, types 302 through 347 (18-8 steels), cannot be hardened by heat treating. They can

Steel No.	Temperatures, °F				Tempering (Drawing) Temperatures for Tensile Strength (psi), °F				
	Normalizing Air Cool	Annealing	Hardening	Quenching Medium[n]	100 000	125 000	150 000	180 000	200 000
1020	1650–1750	1600–1700	1575–1675	Water	—	—	—	—	—
1022 (x1020)	1650–1750	1600–1700	1575–1675	Water	—	—	—	—	—
1025	1600–1700	1575–1650	1575–1675	Water	a	—	—	—	—
1035	1575–1650	1575–1625	1525–1600	Water	875	—	—	—	—
1045	1550–1600	1550–1600	1475–1550	Oil or water	1150	—	—	n	—
1095	1475–1550	1450–1500	1425–1500	Oil	b	—	1100	850	750
2330	1475–1525	1425–1475	1450–1500	Oil or water	1100	950	800	—	—
3135	1600–1650	1500–1550	1475–1525	Oil	1250	1050	900	750	650
3140	1600–1650	1500–1550	1475–1525	Oil	1325	1075	925	775	700
4037	1600	1525–1575	1525–1575	Oil or water	1225	1100	975	—	—
4130 (x4130)	1600–1700	1525–1575	1575–1625	Oil[c]	d	1050	900	700	575
4140	1600–1650	1525–1575	1525–1575	Oil	1350	1100	1025	825	675
4150	1550–1600	1475–1525	1500–1550	Oil	—	1275	1175	1050	950
4340 (x4340)	1550–1625	1525–1575	1475–1550	Oil	—	1200	1050	950	850
4640	1675–1700	1525–1575	1500–1550	Oil	—	1200	1050	750	625
6135	1600–1700	1550–1600	1575–1625	Oil	1300	1075	950	800	750
6150	1600–1650	1525–1575	1550–1625	Oil	d, e	1200	1000	900	800
6195	1600–1650	1525–1575	1500–1550	Oil	f	—	—	—	—
NE8620	—	—	1525–1575	Oil	—	1000	—	—	—
NE8630	1650	1525–1575	1525–1575	Oil	—	1125	975	775	675
NE8735	1650	1525–1575	1525–1575	Oil	—	1175	1025	875	775
NE8740	1625	1500–1550	1500–1550	Oil	—	1200	1075	925	850
30905	—	g, h	i	—	—	—	—	—	—
51210	1525–1575	1525–1575	1775–1825 / j	Oil	1200	1100	k	750	—
51335	—	1525–1575	1775–1850	Oil	—	—	—	—	—
52100	1625–1700	1400–1450	1525–1550	Oil	f	—	—	—	—
Corrosion resisting (16–2)	—	—	—	—	m	—	—	—	—
Silicon chromium (for springs)	—	—	1700–1725	Oil					

a Draw at 1150°F for tensile strength of 70 000 psi.
b For spring temper draw at 800 to 900°F Rockwell hardness C-40-45.
c Bars or forgings may be quenched in water from 1500–1600°F.
d Air-cooling from the normalizing temperature will produce a tensile strength of approximately 90 000 psi.
e For spring temper draw at 850 to 950°F Rockwell hardness C-40-45.
f Draw at 350 to 450°F to remove quenching strains.
f Draw at 350 to 450°F to remove quenching strains. Rockwell hardness C-60-65.
g Anneal at 1600 to 1700°F to remove residual stresses due to welding or cold work. May be applied only to steel containing titanium or columbium.
h Anneal at 1900 to 2100°F to produce maximum softness and corrosion resistance. Cool in air or quench in water.
i Harden by cold work only.
j Lower side of range for sheet 0.06 in and under. Middle of range for sheet and wire 0.125 in. Upper side of range for forgings.
k Not recommended for intermediate tensile strengths because of low impact.
l AN-QQ-S-770.—It is recommended that, prior to tempering, corrosion-resisting (16 Cr-2 Ni) steel be quenched in oil from a temperature of 1875 to 1900°F, after a soaking period of ½ h at this temperature. To obtain a tensile strength at 115 000 psi, the tempering temperature should be approximately 525°F. A holding time at these temperatures of about 2 h is recommended. Tempering temperatures between 700 and 1100°F will not be approved.
m Draw at approximately 800°F and cool in air for Rockwell hardness of C-50.
n Water used for quenching shall not exceed 65°F. Oil used for quenching shall be within the temperature range of 80–150°F.

be annealed at temperatures of 1850 to 2050°F [1010 to 1121°C] and can be work hardened.

The chromium stainless steels, types 410, 416, 420, and 431, can be heat treated to increase their hardness and strength. Hardening is accomplished by heating to a range of 1750 to 1900°F [954 to 1038°C] and quenching in oil.

Heat Treatment of Magnesium Alloys

Magnesium alloys may be heat treated in much the same manner as that employed for aluminum alloys; however, the heating times, soaking times, and cooling rates will vary in accordance with the type of alloy.

Magnesium-alloy castings are solution heat treated to improve such characteristics as tensile strength, ductility, and shock resistance. The temperatures to which magnesium alloys are heated for heat treatment are less than those used for aluminum alloys, being in a range of 730 to 780°F [388 to 416°C]. The actual temperature for heat treatment depends upon the particular alloy involved.

Precipitation heat treatment (artificial aging) is used for magnesium alloys after solution heat treatment. This treatment improves hardness and yield strength and also increases corrosion resistance. Temperatures employed for precipitation heat treating are in the range of 325 to 500°F [163 to 260°C].

The heating of magnesium alloys must be undertaken with careful control to avoid overheating. At high temperatures, magnesium oxidizes rapidly and will ignite if the temperature is too high. Magnesium fires are very difficult to extinguish, particularly where large quantities of chips and shavings may have accumulated. To avoid oxidation and the danger of fire at high temperatures, magnesium should be heated in an inert-gas atmosphere.

Heat Treatment of Titanium

Heat treatment for titanium is possible for some alloys and not for others. The temperatures used for heat treatment of titanium range from 1450 to 1850°F [788 to 1010°C]. At these temperatures the atmosphere in the furnace should be inert to prevent the material from combining with oxygen and nitrogen. The temperature and soaking times vary according to the material. For example, Ti-8-1-1 should be soaked for 8 hours at 1450°F [788°C], then air cooled.

Annealing of titanium alloys must be accomplished according to the specifications for the particular alloy being treated. A typical annealing procedure is to heat the metal to 1350°F [732°C] and soak it at this temperature for 1 hour. The metal is then removed from the furnace and allowed to air cool. Titanium alloys may also be stress relieved to remove stress concentrations developed during cold working. The stress-relieving temperature is usually under 1000°F [538°C], and the alloy is soaked at this temperature for about 30 minutes.

In stress relieving and annealing, a scale may form on the metal. This can be removed by pickling the metal in a bath consisting of 15% nitric acid, 2% hydrofluoric acid, and water.

HARDNESS TESTING

Materials used for the structural parts of an aircraft are frequently examined to determine their hardness. Hardness is used as an indication of strength for many materials and it is based on the desired value given in the material specifications.

Hardness testing is accomplished by means of various types of instruments, all of which enable the operator to estimate the tensile strength of the material.

Brinell Hardness Test

The Brinell hardness tester is shown in Figure 9–11. The Brinell test is performed by forcing a steel ball of known diameter into the material being tested, as shown in Figure 9–12. Three different pressures are used to force the ball into ferrous, nonferrous, and soft materials, depending upon the material being tested. The reading is taken by measuring the width of the impression made by the steel ball in the material. A microscope is used to compare the size of the impression with an established comparison chart. The chart will give a tensile-strength value for each reading. The surface of the material must be clean, free from scale, flat, and

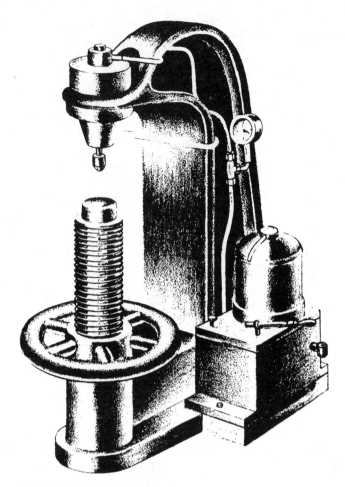

FIGURE 9–11 Brinell hardness tester.

fairly smooth in order to obtain an accurate reading. The work must be adequately supported to avoid twist or movement when applying the test load. The Brinell tester should never be used on thin material where the indentation will show through.

The Rockwell Hardness Test

The Rockwell hardness tester shown in Figure 9–13 is another instrument used in examining the hardness of metals. The Rockwell test is made by using either a diamond point

FIGURE 9–12 Forcing a steel ball into the surface of a metal.

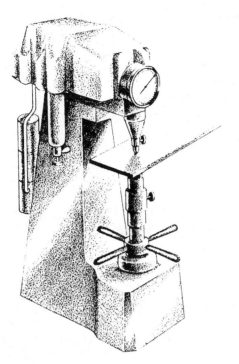

FIGURE 9–13 Rockwell hardness tester.

or a ball of fixed size under a predetermined load. A reading is made directly from a calibrated dial showing a given number for the depth of penetration. This number may then be compared with the comparison chart to estimate the tensile strength. Table 9–6 shows sample values for Brinell, Rockwell, and Vickers instruments.

The diamond point, or brale, penetrator is used for harder materials. Predetermined pressures allow three ranges of pressure to be measured. For softer material, a given-sized steel ball and a predetermined pressure are used. Each range, for both ball and brale, is covered by an alphabetically designated scale, such as the *Rockwell C scale*.

When using the Rockwell instrument, the operator must be certain that the surface of the material is clean and smooth. The material should be held square, as shown in Figure 9–14, and in cases where the material is large or heavy it should be adequately supported to avoid any movement of the material while the reading is taken.

Using too small of a sample of metal may result in the problem shown in Figure 9–15. If the metal can flow away from the ball or brale as shown, a false reading will be obtained. False readings may also result from taking two readings too close together, as illustrated in Figure 9–16.

TABLE 9–6   Rockwell, Brinell and Vickers Hardness Values

| Rockwell Hardness Number | | | | Vickers Hardness Number, Diamond-Pyramid Penetrator | Brinell Hardness Number, 3000-kg Load, 10-mm Standard Ball | Approx. Tensile Strength 1000 psi |
A Scale, 60-kg Load, Brale Penetrator	B Scale, 100-kg load, $\frac{1}{16}$-in [0.16 cm] Diam. Ball	C Scale, 150-kg Load, Brale Penetrator	D Scale, 100-kg Load, Brale Penetrator			
78.5		55	66.9	595		287
78.0		54	66.1	577		278
77.4		53	65.4	560		269
76.8		52	64.6	544	500	262
76.3		51	63.8	528	487	
75.9		50	63.1	513	475	245
75.2		49	62.1	498	464	239
74.7		48	61.4	484	451	232
74.1		47	60.8	471	442	225
73.6		46	60.0	458	432	219
73.1		45	59.2	446	421	212
72.5		44	58.5	434	409	206
72.0		43	57.7	423	400	201
71.5		42	56.9	412	390	196
70.9		41	56.2	402	381	191
70.4		40	55.4	392	371	186
69.9		39	54.6	382	362	181
69.4		38	53.8	372	353	176
68.9		37	53.1	363	344	172
68.4	(109.0)*	36	52.3	354	336	168
67.9	(108.5)	35	51.5	345	327	163
67.4	(108.0)	34	50.8	336	319	159
66.8	(107.5)	33	50.0	327	311	154
66.3	(107.0)	32	49.2	318	301	150
65.8	(106.0)	31	48.4	310	294	146
65.3	(105.5)	30	47.7	302	286	142
64.7	(104.5)	29	47.0	294	279	138
64.3	(104.0)	28	46.1	286	271	134
63.8	(103.0)	27	45.2	279	264	131
63.3	(102.5)	26	44.6	272	258	127
62.8	(101.5)	25	43.8	266	253	124
62.4	(101.0)	24	43.1	260	247	121
62.0	100.0	23	42.1	254	243	118
61.5	99.0	22	41.6	248	237	115
61.0	98.5	21	40.9	243	231	113

*Values in parentheses are beyond normal range and are given for information only.

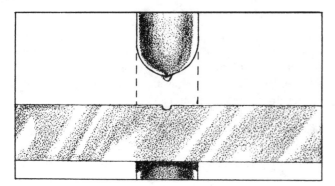

FIGURE 9–14 Position of material being tested for hardness.

FIGURE 9–15 Result of testing too small a part.

The Rockwell hardness tester is a precision instrument. Before attempting to use it, a technician should carefully study the manufacturer's instructions.

Vickers Hardness Test

The Vickers tester operates on a principle similar to that of the Brinell and Rockwell machines; however, the penetrator is a diamond pyramid. The Vickers tester is particularly useful for testing the hardness of very hard steels.

Calibration of Hardness Testers

The accuracy of the hardness-tester readings should be checked frequently. Checking for accuracy, or calibration, is done by means of test blocks supplied by the manufacturers of the instruments.

NONDESTRUCTIVE INSPECTION

Because of the need to save weight, aircraft components are usually designed to operate at high stress levels. This is no problem if a component is able to carry its design load. Flaws, or discontinuities of the metal's structure, may develop during the manufacturing process or during operation

of the aircraft. Even a small defect may cause a part not to carry its design load or even to fail. Both the manufacturer and the aircraft maintenance agency must have a means of determining the integrity of the component or part. **Nondestructive inspection (NDI)** refers to a number of methods that have been developed to detect and measure the extent of a discontinuity without damaging or destroying the part or component. Most NDI equipment is not highly complex, but it does require a knowledgeable and skilled operator that can interpret the test results. To do so requires that the technician not only have knowledge of the NDI process but also of the properties of the materials being tested, the types and effect of flaws that might be present, and what constitutes an unacceptable level of defects.

Types of Defects

A number of defects may develop in a piece of metal during the manufacturing processes. Gas pockets developing in molten metal may cause **porosity** or **voids** in the metal. During forming, these pockets may be shaped into a **fold**, a **lap**, or a **seam**. All of these terms are used to describe a break, or discontinuity, in the grain structure of the metal. Many of these defects are subsurface and thus not visible.

Materials may also be overstressed during forming, causing cracks to occur. Heat and stress buildup during the machining of a part may also result in cracks.

In operation, cracks may develop from overstress or high-temperature operations. Such defects would be quite likely to occur on certain types of components, such as wheels, brakes, and certain engine parts. Perhaps the major defects of interest to the aircraft maintenance technician are those resulting from corrosion or fatigue.

Acceptable Levels of Defects

Very few materials are perfect. If examined closely, virtually any component will show some flaws. Specifications for the inspection of a part will set the levels of acceptability. These may take the form of the number or size of defects, or they may be lists of the settings used for the recommended testing equipment. For this reason it is very important that the test specifications be understood and followed.

FIGURE 9–16 Incorrect reading from tests too close together.

Types of Nondestructive Inspection

With a variety of different metals and numerous types of defects, there is no one "best" method of NDI. Several methods have been developed, and each has unique applications. The ability to select the correct method of NDI is a skill that the aviation maintenance technician is expected to have.

One requirement common to all methods is that the part to be tested must be clean and free of foreign objects. Failure to properly clean and prepare the part for inspection will result in invalid testing.

Magnetic-Particle Inspection. Magnetic-particle inspection is a method of detecting cracks or other flaws on the surface or subsurface of materials that are readily magnetized. Magnetic fields are induced in the material by an electric current. Any flaws or defects in the structure of the material will cause variations in the magnetic field. Parts may be magnetized by either longitudinal or circular magnetization. Longitudinal magnetization, shown in Figure 9–17, is normally accomplished by placing the part in a coil. Electric current flow through the coil will cause the magnetic lines to flow longitudinally through the metal. The magnetic lines of force will be most affected by defects at right angles. Thus a defect at a right angle to the magnetic lines will show clearly. Defects parallel to the magnetic lines will show poorly, if at all.

Circular magnetization is accomplished by passing the electric current directly through the part, causing the magnetic lines of force to encircle it. As can be seen in Figure 9–18, circular magnetism will show longitudinal defects the

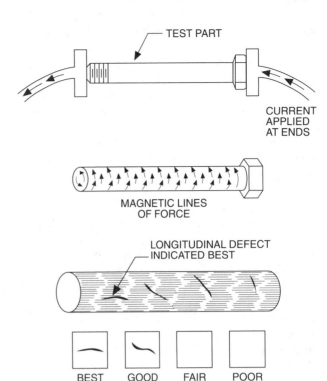

FIGURE 9–18 Circular magnetism.

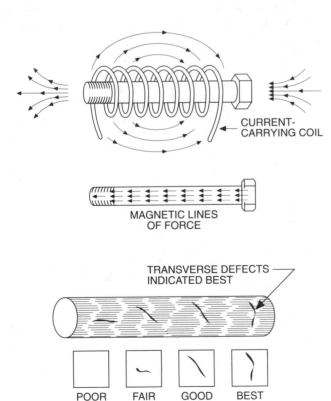

FIGURE 9–17 Longitudinal magnetism.

best. By testing with both methods, all possible defects should be found.

The disruptions of the magnetic lines of force are invisible. To make them visible, small magnetic particles are used. **Wet particles** consist of a suspension of magnetic particles in a light oil. The particles are carefully made for a specific process and may be colored to make them more visible. Other particles may have a fluorescent coating to enable them to be better seen with a black light. Wet particles may be applied by immersing the part in the solution or flowing the solution over the part. **Dry particles** are in powder form and may be applied by hand shakers, spray bulbs, or similar instruments.

The actual inspection of a part will be done in accordance with the specifications found in manufacturers' manuals or similar technical documents. Variations in the magnetic-particle process include the type of current used (AC or DC), the amount of current (the degree of magnetization), and the type of particles.

After inspecting a part by this method, the part must be demagnetized before being put back in service.

Most magnetic-particle inspection is done on stationary machines and requires that the part be disassembled. A number of portable units and processes have been developed allowing some inspection to be done on the aircraft.

The word **Magnaflux** is a copyrighted name used to denote the magnetic process and equipment developed by the Magnaflux Corporation.

Penetrant Inspection. The penetrant process involves the use of a penetrating **dye** which seeps into cracks or other defects. Before applying the dye, the part must be thoroughly cleaned and dried to remove oil, rust, scale, and other foreign materials. The dye is applied by dipping the part in the dye, by brushing, or by spraying. The dye is then allowed to "soak" on the part to ensure that it has penetrated all possible cracks or fissures. The part is then washed with a special cleaner or with water to remove all dye left on the surface, which is then dried. The **developer**, a fine white powder suspended in a solvent, is sprayed or brushed on the part, or the part can be dipped in the developer. The developer is allowed to dry, and the penetrant will flow from any cracks or fissures, revealing their presence, as shown in Figure 9–19. After inspection, the developer is washed off with solvent.

Penetrant inspection can be used on any material but will only detect surface defects or subsurface defects if they are open to the surface. Porous or rough surface materials are not commonly tested by this method, as the penetrant would enter the pores and other surface irregularities and give false readings.

In the use of any particular penetrant process, the technician should follow carefully the instructions provided by the manufacturer. These instructions are usually found on the container for the penetrant and the developer.

Fluorescent-penetrant inspection is a penetrant inspection that uses a fluorescent penetrant. When viewed under a black light (ultraviolet light), the indications will be fluorescent and thus easier to see.

Eddy-Current Inspection. When an electrically conductive material is subjected to an alternating magnetic field, small circulating electric currents, called **eddy currents**, are generated in the material. The magnitude of these currents is affected by the conductivity, homogeneity, mag-

netic permeability, and mass of the material. When performing an eddy-current test, a coil carrying an alternating current is used to induce eddy currents in a material. The eddy currents generate their own magnetic field which interacts with the magnetic field of the coil, influencing the impedance, or opposition, to current flow in the coil. By monitoring or measuring the impedance of the coil (see Figure 9–20), tests can be made to locate flaws in metal; to sort metals on the basis of alloy, temper, or other metallurgical conditions; and to gage metals according to size, shape, thickness, and so forth.

The coil is housed in a probe, which generates the magnetic field to induce the eddy currents and also acts as the sensing device. Probes are made in a number of specialized styles, differing in shape and basic impedance, for use on different types and shapes of materials.

The impedance changes in the coil are measured with electronic circuits. The readout of the instrument may be by the movement of a needle on an analog scale or by a display on a cathode-ray tube.

The eddy-current tester is a comparison tool. A reading will be compared to a standard of known value. Without a standard for comparison, the readings will not have a meaningful value.

Eddy-current testing is a relatively inexpensive, very versatile, and easy-to-use means of testing metals. As with all types of testing, it is important to learn the process and follow the manufacturer's specifications if the test is to be valid.

Ultrasonic Inspection. Ultrasonic inspection uses high-frequency sound waves to reveal flaws in metal parts. The element transmitting the waves is placed on the part, and a reflected wave is received and registered on a cathode-ray tube. By examining the variations of a given response pattern, the material's discontinuities, flaws, or boundary conditions may be detected.

Ultrasonic inspection can be accomplished satisfactorily by a well-trained and experienced technician. The technician must be familiar with the responses of the equipment and be able to interpret all of the indications observed accurately.

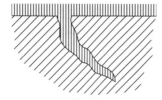

LIQUID PENETRANT, APPLIED TO SURFACE, SEEPS INTO DEFECT

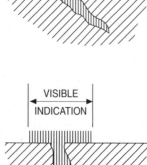

PENETRANT REMOVED FROM SURFACE, BUT DEFECT REMAINS FULL

VISIBLE INDICATION

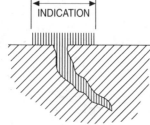

DEVELOPER ACTS AS A BLOTTER TO DRAW PENETRANT OUT OF SURFACE DEFECT AND PRODUCE VISIBLE INDICATION

FIGURE 9–19 Penetrant and developer action.

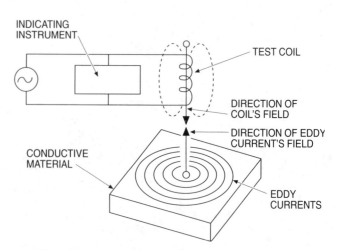

INDICATING INSTRUMENT

TEST COIL

DIRECTION OF COIL'S FIELD

DIRECTION OF EDDY CURRENT'S FIELD

CONDUCTIVE MATERIAL

EDDY CURRENTS

FIGURE 9–20 Basic eddy-current test setup.

X-Ray Inspection. X-ray inspection, or radiography, is often used for the inspection of metal parts. A very powerful X-ray machine is used to produce the rays necessary to penetrate metal. The rays pass through the metal and impinge upon a photographic plate. Flaws in the metal are revealed as shadows in the picture of the part.

X-ray inspection is performed by persons certificated for this type of work. There is danger from the high voltages required and from inadvertent exposure to the powerful radiation. Furthermore, the technician must know what type of photographic film to use and what voltages to apply, depending upon the type of material being tested and the thickness of the material.

Radiographic inspection may also be accomplished by means of a small cobalt "bomb" that is placed inside parts where it would be impossible to use an X-ray machine. The cobalt is radioactive and continuously emits gamma radiation. This radiation is the same as X-ray radiation and penetrates the material being tested. The photographic film is placed on the side of the material opposite the cobalt and reveals the nature of the interior of the material by the shadows on the developed film.

The use of radioactive cobalt for nondestructive testing is subject to numerous regulations to assure that no one is exposed to the radioactive material.

CORROSION CONTROL

Essentially, corrosion is the decomposition of metallic elements into compounds such as oxides, sulfates, hydroxides, and chlorides. It is brought about by direct chemical action and by electrolytic action. Chemical corrosion is caused by an acid, salt, or alkali in the presence of moisture. Water provides the vehicle through which the chemical action takes place. Electrolytic corrosion takes place when metals that have a different level of chemical activity are touching or are in close proximity in the presence of moisture. The two dissimilar metals form the poles of a galvanic cell, an electric current flows, and the more active metal is decomposed. This is a process similar to electroplating.

Types of Corrosion

The types of corrosion most commonly encountered in aircraft are surface corrosion, dissimilar-metals (electrolytic) corrosion, fretting corrosion, stress corrosion, and intergranular corrosion. Corrosion is noted as discoloration on a metal surface, as rust on steel parts, as greenish deposits on brass and copper, as white or gray powder or deposits on aluminum and magnesium, or as some other color or deposit depending upon the chemical combination involved.

On aluminum, **surface corrosion** occurs most often on bare aluminum alloy that is not painted or otherwise treated to prevent corrosion. Clad aluminum is much more corrosion resistant than bare aluminum; however, clad aluminum will also suffer corrosion if allowed to become dirty, especially in a humid climate or when exposed to sea air. Surface corrosion on aluminum generally appears as white

blotches if on the surface and as small dark-gray lumps if it has penetrated below the clad surface and is attacking the interior of the metal. Surface corrosion on metal that has been painted is evidenced by peeling or blistering of the paint. In the case of steel, the reddish color of rust will often show up on the outer surface of the paint.

Another type of corrosion, **dissimilar-metals corrosion**, is caused when metals with different chemical activity are in contact in the presence of moisture. Table 9–7 shows the electrochemical series for metals. The higher the metal is on the list, the more active it is and the more susceptible to corrosive activity. When dissimilar metals are in contact, the more active metal will be destroyed. FAA *Advisory Circular AC 43.13-1A* has divided the metals into four groups, as shown in Table 9–8. The most active metals are in group I, and the least active are in group IV. The aluminum alloys in subgroups A and B should be considered as dissimilar metals with respect to corrosion prevention. This is particularly true when a large area of an alloy in subgroup B is in contact with a small area of an alloy in subgroup A. Severe corrosion of the alloy from subgroup A may be expected.

In the design of aircraft metal structures, every effort is made to prevent contact between metals of different groups. This is the reason that aircraft bolts are usually cadmium plated; note that cadmium is in group II along with the alu-

TABLE 9–7 Electrochemical Series for Metals

Arranged in order of electrode potential
Most Anodic (gives up electrons most easily)
Magnesium
Zinc
Commercially pure aluminum (1100)
Clad 2024 aluminum
Cadmium
7075-T6 aluminum alloy
2024-T3 aluminum alloy
Mild steel
Lead
Tin
Copper
Stainless steel
Silver
Nickel
Chromium
Gold
Most Cathodic (least corrosive)

Any metal appearing before another in this series is anodic to any metal which follows it and will be the one corroded when they are subject to galvanic action.

TABLE 9–8 Dissimilar Metals

Group I	Magnesium and its alloys	
Group II	All aluminum alloys, cadmium, and zinc	
	Subgroup A	Aluminum alloys 1100, 3003, 5052, 6061, 220, 355, 356, and all clad alloys
	Subgroup B'	Aluminum alloys 2014, 2017, 2024, 7075, 7079, 7178, and 195
Group III	Iron, lead, tin, and their alloys except stainless steels	
Group IV	Stainless steels, titanium, chromium, nickel, copper, and their alloys, plus graphite	

minum alloys. Copper electric terminals are cadmium plated for the same reason.

Improper heat treatment causes **intergranular corrosion**. If an alloy is not quenched rapidly enough following precipitation heat treatment, alloying elements will precipitate along grain boundaries and form minute galvanic cells. The aluminum alloy 2024 is particularly vulnerable to intergranular corrosion because of its copper content. The zinc in the 7000-series alloys leads to intergranular corrosion unless the heat treatment is accurately controlled. Intergranular corrosion also occurs in some corrosion-resistant steels.

Intergranular corrosion shows up as raised surface flaws unless it is deep inside a casting or forging. In this case, ultrasonic or eddy-current inspection techniques are effective means of checking for intergranular corrosion. If apparent blisters are noted under the surface of bare or painted aluminum alloy, intergranular corrosion may be present. Scraping or piercing the raised area will reveal the nature of the damage. Intergranular corrosion also shows up under the surface of the metal, and the outer surface may not show any corrosion. Under the surface will be white powder mixed with aluminum particles. When the aluminum breaks up or delaminates, the condition is called **exfoliation**.

When a metal part is highly stressed over a long period of time under corrosive conditions, **stress corrosion** results. Parts that are susceptible to stress corrosion include overtightened nuts in plumbing fittings, parts joined by taper pins that are overtorqued, fittings with pressed-in bearings, and any other assembly where the metal is stressed almost to the yield point. Stress corrosion is not easy to detect until cracks begin to appear.

One type of stress corrosion is known as **corrosion fatigue**. This occurs where cyclic stresses are applied to a part or an assembly. These stresses not only affect the metal but also produce minute cracks in the surface coating, thus removing any corrosion protection.

Another type of corrosion, **fretting corrosion**, occurs when there is slight movement between close-fitting metal parts. The movement prevents the formation of oxides that inhibit corrosion and produces fine particles of metal and oxide that tend to absorb and retain moisture. This further aggravates the corrosion. The particles also act as an abrasive that keeps the bare metal exposed to the corrosive conditions. Fretting corrosion is sometimes called *false Brinelling* because of the appearance of parts that are affected.

Repair of Corroded Parts

The repair of parts that are affected by corrosion but are not rendered unserviceable is simply a matter of removing the existing corrosion products and applying a coating or finish that will prevent further corrosion. Care must be taken to assure that the parts are not damaged by the removal process and that the correct type of refinish is applied.

Parts made of carbon or alloy steel and not highly stressed can be cleaned with buffers, wire brushes, sandblasting, steel wool, or abrasive papers. If corrosion remains in pits and crevices, it may be necessary to use chemical inhibitors to arrest further corrosion. Unpainted steel parts are often coated with rust-inhibiting oil or greases. Steel parts

that are highly stressed must not have appreciable material removed during cleaning and must not have pits or appreciable surface damage. Parts that are satisfactory after corrosion removal can be primed and painted in accordance with the original finish.

When removing corrosion from aluminum parts, the use of steel or wire brushes, steel wool, emery cloth, or other harsh abrasives should be avoided. Steel brushes and steel wool are likely to leave metal particles embedded in the surface, and these will lead to electrolytic corrosion. Mild abrasives or polish meeting U.S. Military Specification MIL-P-6888 can be used on clad aluminum alloy but must not be used for cleaning anodized alloys. The anodized film is so thin that it can be removed by polishing. When it is necessary to remove severe corrosion that cannot be removed by polishing, a number of chemical cleaners are available. One such process uses a 10% solution of chromic acid to which a small amount of sulfuric acid has been added. This solution can be brushed on with a stiff bristle brush and allowed to remain for more than 5 minutes. If the existing corrosion is removed, the solution can be rinsed off with water. The cleaned surface should be repainted within a few hours after cleaning.

If an anodized film has been partially removed from an aluminum-alloy surface, a protective coating can be partially restored by treating the surface with a chromic-acid solution. After treatment, the part should be primed and painted as soon as possible.

Structural aluminum-alloy parts that have suffered severe intergranular corrosion must usually be replaced because of the loss of strength in the parts. Sometimes a small amount of intergranular corrosion can be removed from the outer surface of a part, and if there is no evidence of additional penetration, the part can be treated chemically and then refinished.

Since magnesium is the most chemically active of the metals used in aircraft, it is also the most critical with respect to corrosion. Magnesium parts are chemically treated to produce a film coating that prevents corrosion; however, this film can be scratched or worn off with the result that corrosion starts immediately if any moisture is present. When the protective coating on magnesium is damaged, it should immediately be repaired by chemical treating with a 10% chromic-acid solution to which has been added a small amount of sulfuric acid. This acid can be obtained in the form of lead-acid battery electrolyte. The recommended amount for the chromic-acid solution is 20 drops of the electrolyte to 1 gal of the solution. The solution should be brushed on the cleaned magnesium part and allowed to remain for 10 to 20 minutes.

FINISH AND SURFACE-ROUGHNESS SYMBOLS

The surface of a metal part is finished by a machining, coating, or hand-finishing operation. Scraping, file fitting, reaming, lathe turning, shaping, and grinding are some additional finishing operations.

On many existing blueprints the symbol for a finished surface is a letter *V* with its point touching the surface to be finished, drawn with an angle of 60° between the sides of the *V*. Numbers may be placed within the angle formed by the sides of the *V* to represent the type of finish to be applied to that particular surface. Where a part is to be finished on all surfaces, the abbreviation FAO is sometimes used to represent "finish all over."

Some factories use symbols on shop orders but not necessarily on drawings, according to what they call a *finish code*. Examples taken at random are as follows: 1, zinc chromate primer; 2, darkened primer; 3, aluminized primer; and 9, dope. However, it is a more common practice to write a finish specification to show how each material and portion of a particular assembly are finished.

Many manufacturers in the aerospace industry have adopted the **root-mean-square (rms) microinch system of surface-roughness designation**. This system has been standardized by the National Aerospace Standards Committee in Specification NAS30 and is also set forth in U.S. Military Specification MIL-STD10A. All new drawings of machined castings, machined forgings, and other machined parts will use this method of specification of surface finishes. **Surface roughness** is a term used to designate recurrent or random irregularities that may be considered as being superimposed upon a plane or a wavy surface. On smooth-machined surfaces, these irregularities generally have a maximum crest-to-crest distance of not greater than 0.010 in and a height that may vary from 0.000 001 to 0.000 05 in. *Waviness* should not be confused with *roughness*, as the crest distances with waviness are much greater, generally running from 0.04 to 1.00 in [0.10 to 2.54 cm], and the height distances can be as much as several thousandths of an inch.

The need for a simple control of the surface quality of a machined part by means of production drawings has long been apparent. Dimension tolerances as well as process notes such as "rough machine," "smooth-machine finish," "grind," and "polish" limit the surface characteristics in a general way but are not sufficiently specific to describe the desired result. Certain machining processes, such as grinding, could produce several degrees of smoothness; hence, the decision as to which degree was intended generally was made in the past by the shop or the vendor.

By means of the rms system of surface-roughness designation, it is possible for the engineering department of any company to specify precisely the degree of finish required and for the shop to produce the specified finish without resorting to judgment. The rms average is a unit of measurement of surface roughness and is expressed in microinches. The microinch is one-millionth (0.000 001) part of an inch. The rms average is chiefly affected by the highest and lowest deviations from a mean surface and is a mathematical indication of average surface roughness.

The National Aerospace Standards Committee has selected a series of preferred roughness numbers that cover the range of aircraft requirements. These numbers are 1, 2, 5, 10, 20, 40, 100, 250, 500, and 1000. All these numbers, with the exception of number 1, are used by manufacturers. They indicate the maximum allowable or the acceptable

roughness of the surface on which they are specified, in rms microinches.

Engineers have a fairly clear conception of finish characteristics when expressed in terms of the machining process used to produce a surface finish. It is therefore necessary for the designer to designate this surface condition as an rms finish number. Figure 9–21 lists roughness numbers for surface-finish designation.

In order to allow the machine shop latitude in developing the specified finish, drawings specify the maximum roughness allowable. Any smoother finish will be acceptable as long as economy is not sacrificed. The shop may develop the specific finish by whatever method is the most practical. Surface defects, flaws, irregularities, and waves that generally occur in only a few places will be handled by the inspection department.

The shop and engineering departments must use the standard reference samples for comparison with machined parts. By visual inspection and by feel, they can check the completed surfaces. In borderline cases the surface can be checked by two different instruments, the profilometer and the Bush analyzer. These are tracer-point instruments that measure the surface roughness when drawn along the surface to be measured.

The rms numbers on drawings always are called out by the use of the standard symbol illustrated in Figure 9–22. The roughness number must always be on the left side of the long leg close to the horizontal bar, as indicated by *xx* in the

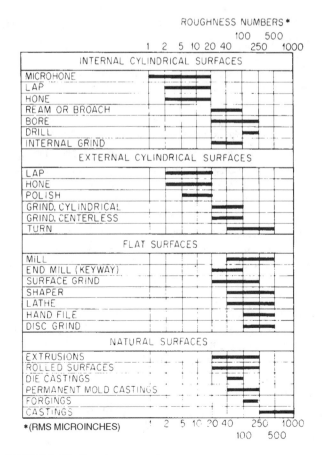

FIGURE 9–21 Roughness numbers.

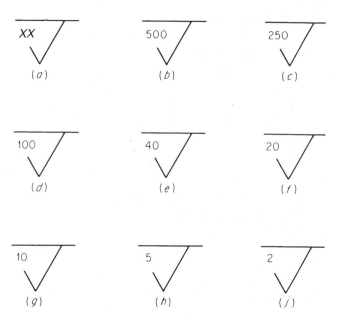

FIGURE 9–22 Root-mean-square (rms) numbers and symbols.

Root-mean-square 20, a very fine finish, is indicated by the symbol in Figure 9–22f. This finish is produced by a fine cylindrical or surface grind, a very smooth ream, a smooth emery buff, or a coarse to medium lap or hone. The extremely smooth finishes are indicated by Figures 9–22g, h, and j. These finishes are produced by honing, lapping, microhoning, polishing, or buffing.

Root-mean-square 10 and the finer finishes may have either a dull or a bright appearance, depending upon the method used to produce them. The surface appearance must not be considered in judging quality; the degree of smoothness must be determined by feel or by roughness-measuring instruments.

METAL SURFACE TREATMENTS

A number of surface treatments for metals have been developed to reduce or eliminate corrosion. Treatments for aluminum alloys include anodizing, alodizing, and processes involving chromic acid and sodium dichromate. All these processes provide a protective film, and parts so treated must be handled carefully to avoid damaging the film. The aviation maintenance technician is not usually required to apply the more complicated treatments because they require equipment not generally available in the field. Usually, such treatments are applied by manufacturers and large overhaul bases.

Magnesium is usually treated with a chrome-pickle or sodium-dichromate process. These processes produce an excellent protective film that can be repaired by brushing chromic-acid solution on damaged areas.

Steel parts are protected by cadmium plating, chrome plating, or phosphate processes such as parkerizing, bonderizing, parco lubrizing, or granodizing. These processes effectively protect the surface from oxidation, and the parts may be painted over after preparation as specified by the manufacturer of the product applied.

Cadmium Plating

Cadmium plating is a nonporous, electrolytically deposited layer of cadmium that offers high corrosion resistance for steel. Plating is done per U.S. Military Specification MIL-P-416A. Three types of cadmium plating are considered in this specification:

Type I. This type is pure silver-colored cadmium plate without supplementary treatment. This type of cadmium coating was used on all steel aircraft hardware in the past.

Type II. This type consists of Type I plating followed by a chromate treatment. Type II plating is a light to dark gold color. It has improved corrosion resistance. Procurement specifications for most aircraft now specify Type II plating.

Type III. This is Type I coating followed by a phosphate treatment. It is used mainly as a paint base.

illustration. The roughness of natural surfaces is not specified unless such surfaces are critical because of functional or manufacturing requirements.

The finish of such items as drilled holes, reamed holes, and spot faces is not usually specified if the maximum roughness to be produced will be acceptable. The roughness of fillets and chamfers conforms to the rougher of the two connected or adjacent surfaces, unless otherwise indicated.

Unless specified, a symbol used on a plated or a coated surface always signifies that a control applies to the parent-metal surface before plating or coating.

Lay may be defined, for the purpose of this discussion, as the direction of tool marks or the grain of the surface roughness. Waviness and tool-lay designations also covered in the National Aerospace Standards Committee Specification NAS30 have not yet been adopted by all manufacturers.

The symbol in Figure 9–22b indicates rms 500. This is a very rough, low-grade machine surface resulting from heavy cuts and coarse feeds in milling, turning, shaping, and boring as well as from rough filing and rough disk grinding. This is also the natural finish of some forgings and sand castings. It is not used for aluminum alloys or other soft metals, for surfaces in tension, or where notch sensitivity is a factor. It may be used on secondary items but is generally not called out as a finish for aircraft or missile parts.

Figure 9–22c is the symbol for rms 250. This is a medium machine finish and is fairly inexpensive to produce. Figure 9–22d is the symbol for rms 100. This finish is generally known as a smooth machine finish and is the product of high-grade machine work in which relatively high speed and fine feeds are used in taking light cuts with well-sharpened cutters.

Figure 9–22e is the symbol for rms 40. This is a fine machine finish produced by a carbide or diamond bore, a medium surface or cylindrical grind, a rough emery buff, a ream, a burnish, or a similar tool.

Anodizing

A finish produced by anodizing, applied to aluminum by an acid-plating process, hardens the surface, reduces porosity, increases abrasion resistance, and has high dielectric strength. Anodized aluminum can be dyed almost any color. U.S. Military Specification MIL-A-8625R covers three types of anodizing:

Type I. Chromic anodize coating will vary from a light to a dark gray color depending on the alloy. This coating is given a chromate treatment to seal the surface.

Type II. Sulfuric anodize coating is the best coating for dying. Nondyed coating will have a dull yellow-green (gold) appearance when sealed with a chromate treatment.

Type III. Hard anodize coating can be used as an electrical insulation coating or as an abrasion-resisting coating on devices such as hydraulic cylinders and actuating cams.

REVIEW QUESTIONS

1. How is aluminum alloy marked for identification?
2. What is the difference between sheet metal and metal plate?
3. How is a cast material converted to a wrought material?
4. What are the disadvantages to continuous-roll forming?
5. What is the difference between die drawing and extrusion?
6. What is meant by *heat treating*?
7. What is meant by *solution heat treatment of aluminum alloy*?
8. What is artificial aging?
9. How is maximum hardness attained in carbon steel?
10. What is meant by *tempering of steel*?
11. Describe how case hardening of steel is accomplished.
12. Explain the purpose of precipitation heat treatment.
13. Compare the heat treatment of magnesium with that of aluminum alloy.
14. What precautions must be observed in heating magnesium?
15. Explain the principles of hardness testing with typical testers.
16. How are hardness testers calibrated?
17. What is meant by *nondestructive testing*?
18. If a part is suspected of having cracks approximately perpendicular to the longitudinal axis, what type of magnetization should be employed?
19. How is penetrant inspection accomplished?
20. What is the principle of ultrasonic inspection?
21. How can dissimilar-metal corrosion be prevented?
22. What is the cause of intergranular corrosion?
23. What should be done to repair corrosion where no appreciable damage has occurred?
24. How should corrosion be removed from aluminum alloys?
25. What precaution must be taken when cleaning aluminum alloys that have been anodized?

10 Standard Aircraft Hardware

INTRODUCTION

In the design, production, and maintenance of aircraft, space vehicles, or any other device for which specific quality and performance requirements are established, it is necessary to select standards and develop specifications to assure that the aircraft or other device will meet its requirements. Generally speaking, standards and specifications establish quality, size, shape, performance, strength, finish, materials used, and numerous other conditions for aircraft and their components. Because of the almost infinite number of sizes, shapes, and materials involved in mechanical devices, a wide variety of standards and specifications have been developed covering hardware, metals, plastics, coatings, nonmetallics, and manufactured components.

In order to provide measures of uniformity, the military services, technical societies, manufacturers, and other agencies have attempted to establish uniform standards that are universally acceptable for particular materials, products, dimensions, and other items.

This chapter is primarily concerned with the use of such standards for parts known as **hardware.** Standard aircraft hardware includes such items as fastener assemblies, control-cable fittings, fluid-line fittings, and electrical-wiring components. Fluid-line fittings are covered in Chapter 12 of this text, and wiring components are covered in a separate text of this series.

A listing of all the aircraft hardware for fasteners, cables, and miscellaneous applications would require a publication much larger than this text. The hardware covered in this chapter has been chosen as being representative of the types of hardware that you will encounter and that will also be found on a large number of aircraft. Much of the information in this chapter relates to standards and designation codes. Learning this information will enable you to use the technical information available for hardware not covered in this text.

The emphasis in this chapter will be on the identification and the function of various types of hardware items. It is not practical to provide you with specific instructions relating to the installation of all types of fasteners and other items of hardware that you may

encounter. In various sections of the texts of this series, specific information is given relating to the items of hardware involved in particular operations. For example, the installation of rivets is discussed in the structural-repair section of *Aircraft Maintenance and Repair.*

It is very important that all items be installed or applied in accordance with the manufacturer's instructions. Numerous items of hardware apply only to certain makes and models of aircraft and should be handled as specified by the manufacturer of the product.

STANDARDS

A **standard** is variously defined as (1) something established for use as a rule or basis of comparison in measuring or judging capacity, quantity, content, extent, value, quality, and so forth; (2) a level or grade of excellence; and (3) any measure of extent, quality, or value established by law or by general use or consent.

In the normal performance of their duties, technicians encounter an extensive array of standards establishing the characteristics of the materials and components that they will use from day to day in repair and maintenance work. Among these standards are **AN** (Air Force and Navy), **AND** (Air Force-Navy Aeronautical Design), **MS** (Military Standard), **NAS** (National Aerospace Standard), **NAF** (Naval Aircraft Factory), and **AS** (Aeronautical Standard).

Fluid-line (hydraulic) fittings manufactured before World War II were manufactured to **AC** (Air Corps) standards. Some of these fittings may still be in use on old aircraft. All fittings are currently manufactured to meet AN or MS standards.

The most widely used standards for aircraft hardware are AN and MS. These standards have been established by the military to ensure the quality and uniformity of hardware acquisitions. Items manufactured in accordance with these standards are not limited to military use and are found in all classifications of aircraft. Examples of AN parts are shown in Figure 10–1, and MS parts are shown in Figure 10–2.

In recent years, all items approved for military applications have been manufactured according to MS drawings and procedures. Many former AN parts are now produced as MS parts; for example, the universal-head rivet listed under AN470 is now manufactured under MS20470.

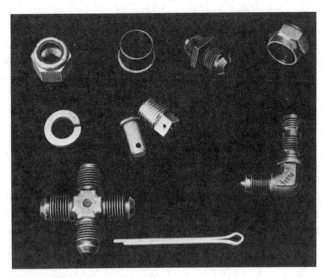

FIGURE 10–1 Some AN parts.

Hardware items not used for military aviation but which have been proven satisfactory by the aerospace industry may be given a NAS designation.

In addition to items produced under these standards, manufacturers often design their own hardware items, which are given manufacturers' part numbers. Many such parts may be identical, or almost identical, to standard parts. The technician must be careful to ensure that only the specified approved parts are installed. If the manufacturer specifies the item by a "standard" part number, then any part meeting that standard can usually be installed.

The NAF standards are those developed and approved for use by the Naval Aircraft Factory. Items or parts manufactured under these standards are almost all superseded by AN or MS standards.

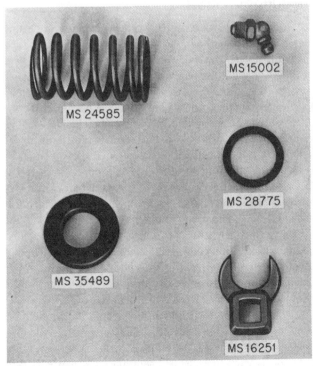

FIGURE 10–2 Typical MS parts.

AND standards are generally of more interest to the engineer than the technician because they are primarily standards for design.

AS standards have been established by the Society of Automotive Engineers (SAE). They include design standards, parts standards, and specifications that have not been assigned **AMS** (Aeronautical Materials Specifications) numbers. Materials manufactured under AMS specifications are often required for use on civil aircraft by the FAA; for example, Grade A fabric used for covering aircraft is designated AMS3806.

Table 10–1 lists a few of the AN, MS, and NAS numbers and the items to which they apply. This listing is only a sample, and it should be understood that there are hundreds of items covered by these standards. The agency for which the aircraft maintenance technician works should have complete listings among its technical publications.

Industry Standards

In addition to the standards previously described, industry organizations have also developed standards and specifications that are not necessarily covered by other standards. The American Society for Testing and Materials (ASTM) is one of the more active organizations in the establishment of material standards.

The American National Standards Institute (ANSI) is a federation of other "standard-setting" organizations. Its function is to serve as a clearinghouse for standards. When a standard is established by the ANSI, the item standardized is generally accepted by all groups concerned with the particular item.

One of the leading organizations concerned with standards for iron and steel products is the American Iron and Steel Institute (AISI). Standards established for most iron and steel products are designated by AISI numbers. Many iron and steel products are also covered by standards issued by the SAE and ASTM. Other steel and iron products are standardized by the Alloy Casting Institute (ACI), the Invest-

TABLE 10–1 Sample AN, MS, and NAS Numbers

AN214	Pulley, control, plain bearing
AN253	Hinge
AN501	Screw, machine, fillister head
AN175	Bolt, close tolerance
AN960	Washer
AN815	Union
MS20365	Nut
MS20667	Fork end, cable
MS20995	Safety wire
MS20470	Rivet
MS24693	Screw, flat head
NAS1134	Bolt, pan head, close tolerance
NAS697	Nut, plate
NAS501	Bolt, hex head
NAS2007	Lock bolt, pan head
NAS1291	Nut, lightweight

ment Casting Institute (ICI), the Gray Iron Founders' Society (GIFS), the Steel Founders' Society of America (SFSA), and the Metal Powder Association (MPA).

The SAE, ANSI, and ASTM are concerned with almost all the materials used in the aircraft field. Other organizations concern themselves with limited fields, such as iron and steel, nonferrous metals, plastics and rubber, non-metallics, and finishes and coatings.

SPECIFICATIONS

A **specification** may be defined as a particular and detailed account or description of an item; specifically, it is a statement of particulars describing the dimensions, details, or peculiarities of any work about to be undertaken, as in architecture, building, engineering, or manufacturing. A specification also sets forth the standards of quality and performance that a particular aircraft or aircraft component must meet to be acceptable for the purpose intended. Products of many types are manufactured according to specifications and then tested to assure that the specifications are met.

Military Specifications

Of concern to persons working in the aircraft field are the frequently encountered military specifications for materials and products. Such specifications are designated MIL followed by a letter and a number. For example, the specification for standard aircraft cable, 7 by 19, carbon steel, is MIL-C-5424. *MIL* is the abbreviation for military, *C* is the first letter of the first word in the title of the specification (cable), and *5424* is the basic serial number. The number could be followed by a letter, which would indicate a revision in the basic specification.

A few of the products produced to an MIL specification that the technician may encounter are listed in Table 10–2.

FAA Specifications

The FAA requires all aircraft, aircraft engines, and propellers to be certificated. As part of the certification process, specifications covering all parts and components of the aircraft are developed. These specifications may include reference to parts built under AN, MS, or NAS standards.

Replacement or modification parts for a certificated aircraft must be one of several types:

1. Parts produced under the aircraft type or production certificate, which are primarily those produced by the manufacturer or its licensee.
2. Parts produced under a Technical Standard Order (TSO) of the FAA. A TSO is an FAA-generated specification for products to be used on an aircraft (tires, seat belts, radios, and so on).
3. Parts produced by the holder of FAA Parts Manufacturer Approval (PMA). PMA parts are usually components or parts other than hardware. To receive a PMA, the manu-

TABLE 10–2 Sample MIL Specifications

MIL-R-5521	Ring, hydraulic fitting
MIL-P-5315A	Packing, O-ring
MIL-P-7034	Pulley
MIL-R-5674	Rivet
MIL-B-6946	Bronze bar stock
MIL-S-5000	Steel, annealed

facturer will have to demonstrate to the FAA that it has developed specifications that will produce a part equal to the original.

4. Standard parts (such as hardware) conforming to established industry or government specifications, that is, AN, MS, NAS parts.

FAA certification procedures for products and parts are covered in Part 21 of the *Federal Aviation Regulations*.

THREADED FASTENERS

Hardware items used to fasten two or more parts together are called **fasteners**. Fasteners are essential to the structural integrity of the aircraft. It is essential that the aircraft technician be able to determine that the proper fastener is or has been used. **Threaded fasteners** allow parts to be taken apart and reassembled as necessary. In most cases, threaded fasteners are reusable.

Threaded fasteners, for the purposes of this chapter, consist of bolts, nuts, and other associated items, such as washers or locking devices.

Designation Codes

All standard hardware will have a designation code that identifies the basic item and its dimensions, materials, and special features. The code will usually relate to an AN, MS, or NAS standard. In some cases the manufacturer may use its own code to supplement the standard. While a code may look quite complex, most consist of a meaningful assortment of letters and numbers. The technician must have some knowledge of the basic part (bolt, screw, rivet) to develop a full understanding of the code. The code structure for a few common items should be committed to memory. Because of the number of items of standard hardware, it will be necessary in many cases for the technician to make reference to technical literature or distributor's catalogs.

The term **dash number** is frequently used when discussing designation codes. This comes from the practice of separating the basic parts of the code with dashes (-). For example, in a code such as AN4-16, the *AN4* designates the item and the first size, usually the diameter of the item. The *-16* would give an indication of the length. It is possible for some hardware to have several dash numbers. In some designations, the dash may be replaced with a letter or a number to designate some special condition, such as AN4H16 or AN4H-16.

Machine Screw Thread

The winding groove around a bolt or a screw or in the hole of a nut forms what is called a **screw thread**. Screw threads can take a variety of shapes, such as those on a wood screw, a sheet-metal screw, or a bolt. The thread on bolts or screws designed for use with nuts is referred to as the **machine screw thread**.

A number of machine-screw-thread designs are in use. Aircraft hardware uses a design based upon a 60° thread. Figure 10–3 shows that the angles cut into the bolt or nut form a series of equilateral triangles when viewed in cross section. Sharp points on the crest and at the root of the thread created several problems when these items were first made. These problems were overcome by a slight flattening of the crest and filling of the root. This modification reduces the thread depth by about 25%. This thread form was standardized around the time of World War I as the **American (national) form**. In 1948 the United States, Canada, and Great Britain agreed to adopt the **unified national form thread** to provide for interchangeability of threaded parts. This form is essentially the same as the national form except that the roots must be rounded and the crests may be either flat or rounded. Fasteners using national and unified national thread forms are interchangeable except for some of the close-tolerance fits. The European metric thread also uses a similar 60° design, but the difference in the systems of measurement prevents interchangeability of metric and unified national threads.

Figure 10–4 shows some of the dimensions that are applied to screw threads. Threads are classified as internal and external depending upon their application. The **major diam-**eter of an external thread is the diameter measured across the thread crest. The **minor diameter** is the diameter at the root of the thread. For internal threads, the major diameter applies to the root and the minor diameter to the crest. For a given size, the major and minor diameters will be the same for both the internal and external threads. **Pitch diameter** is a standard value for a given thread size and is approximately halfway between the major and minor diameter.

The **lead** of a screw refers to the distance that the screw will advance into another threaded object with one revolution. This is the same as the **pitch**, or the distance between crests, for most threads. Most threads are described in terms of the number of **threads per inch (tpi)** or the number of crests in a length of one inch. It should be apparent that tpi and pitch are related, in that the reciprocal of tpi (1/tpi) equals pitch.

Unified national threads are made in two series, **unified national fine (UNF)** and **unified national coarse (UNC)**. The basic difference for these two series is in the number of threads per inch for a given diameter. For example, a $\frac{1}{4}$-in-diameter coarse-thread bolt (UNC) will have 20 threads per inch, while a $\frac{1}{4}$-in fine-thread bolt (UNF) will have 28. A discussion of the advantages and disadvantages of *coarse* or *fine* threads is beyond the scope of this chapter.

Threaded aircraft fasteners larger than $\frac{1}{4}$-in diameter are dimensioned by fractions of an inch. Those smaller than $\frac{1}{4}$ in are dimensioned by screw sizes. The AN3 ($\frac{3}{16}$-in diameter) bolt used for aircraft is an exception to this practice. Machine-screw sizes, shown in Table 10–3, range from 0, the smallest, to 12, the largest. It should be noted that the dimension of a No. 10 machine screw is 0.190 in, or very close to $\frac{3}{16}$ in (0.1875 in). The screw threads for a bolt that has a $\frac{3}{16}$-in diameter will use the same threads specified for a No. 10 machine screw. Screw-thread sizes may be better visualized by remembering that a No. 10 screw has approximately a $\frac{3}{16}$-in diameter and that a No. 5 screw has a $\frac{1}{8}$-in diameter. The technician will encounter extensive use of No. 8- and No. 10-sized screws in aircraft maintenance.

In Table 10–3, the number of threads per inch for both UNC and UNF is shown for the various sizes. Technicians

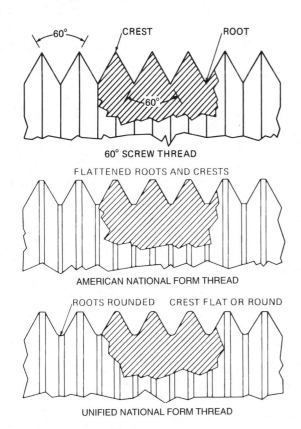

FIGURE 10–3 Thread forms.

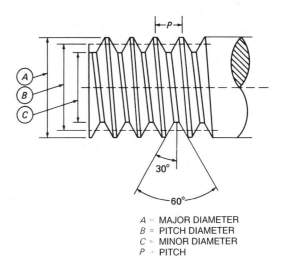

FIGURE 10–4 Thread dimensions.

will find that their work is enhanced by a knowledge of the threads per inch for the common sizes.

External threads are usually cut with a **die**. Some dies are adjustable so that the depth of the thread can be slightly varied. This will allow the tightness, or fit, of the thread to be varied. Internal threads are cut with a **tap**. Before using the tap, a hole must be drilled for the tap to go into. The drill used for this is called the **tap drill**. The tap drill must be at least as large as the minor diameter. Tap drill sizes are given in Table 10–3. Threads may also be cut on machine tools or formed by rolling.

The fit between internal and external threads has been standardized into five classes ranging from Class 1, loose, to Class 5, tight. Aircraft bolts use a Class 3 thread, while aircraft screws may use either a Class 2 or 3 thread.

Figure 10–5 shows a code used to specify screw threads. The letters designate it as unified national coarse or unified national fine, followed by the diameter expressed as either a screw size or a fraction of an inch. Following a dash is the number of threads per inch. A second dash number, from one to five, specifies the class of thread fit. A letter *A* following the class of fit indicates an external thread, and a *B* indicates an internal thread.

Bolts

A bolt is designed to hold two or more parts together. It may be loaded in shear, in tension, or both. Bolts are designed to be used with a nut and to have a portion of the shank that is not threaded, which is called the **grip**. Machine screws and cap screws have the entire length of the shank threaded.

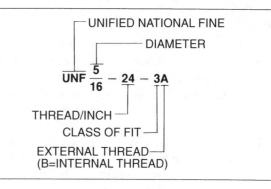

FIGURE 10–5 Machine-screw thread designation.

The dimensions used for a bolt are shown in Figure 10–6. Bolt sizes are expressed in terms of diameter and length. The diameter is the diameter of the shank, and the length is the distance from the bottom of the head to the end of the bolt. The grip length should be the same as the thickness of the material being held together. The grip length can be determined by using a reference chart for AN bolts.

Bolt heads are made in a variety of shapes, with the hexagon-shaped, or hex-shaped, head being the most common.

General-Purpose Bolts. An all-purpose structural bolt used for both tension and shear loading is made under AN standards 3 through 20. The bolt diameter is specified by the AN number in sixteenths of an inch. For example,

AN3 $\frac{3}{16}$ in diameter

AN11 $\frac{11}{16}$ in diameter

TABLE 10–3   Thread Chart

Unified National Coarse-Thread Series, Medium Fit, Class 3 (UNC)					Unified National Fine-Thread Series, Medium Fit, Class 3 (UNF)				
Size and Threads	Dia. of Body for Thread	Body Drill	Pref'd Dia. of Hole	Nearest Stand'd Drill Size	Size and Threads	Dia. of Body for Thread	Body Drill	Pref'd Dia. of Hole	Nearest Stand'd Drill Size
					0–80	0.0600	52	0.0472	$\frac{3}{64}$
1–64	0.0730	47	0.0575	No. 53	1–72	0.0730	47	0.0591	No. 53
2–56	0.0860	42	0.0682	No. 51	2–64	0.0860	42	0.0700	No. 50
						0.0990			
3–48	0.0990	37	0.0780	$\frac{5}{64}$	3–56		37	0.0810	No. 46
4–40	0.1120	31	0.0866	No. 44	4–48	0.1120	31	0.0911	No. 42
5–40	0.1250	29	0.0995	No. 39	5–44	0.1250	25	0.1024	No. 38
6–32	0.1380	27	0.1063	No. 36	6–40	0.1380	27	0.1130	No. 33
8–32	0.1640	18	0.1324	No. 29	8–36	0.1640	18	0.1360	No. 29
10–24	0.1900	10	0.1472	No. 26	10–32	0.1900	10	0.1590	No. 21
12–24	0.2160	2	0.1732	No. 17	12–28	0.2160	2	0.1800	No. 15
$\frac{1}{4}$–20	0.2500	$\frac{1}{4}$	0.1990	No. 8	$\frac{1}{4}$–28	0.2500	F	0.2130	No. 3
$\frac{5}{16}$–18	0.3125	$\frac{5}{16}$	0.2559	No. F	$\frac{5}{16}$–24	0.3125	$\frac{5}{16}$	0.2703	I
$\frac{3}{8}$–16	0.3750	$\frac{3}{8}$	0.3110	$\frac{5}{16}$ in	$\frac{3}{8}$–24	0.3750	$\frac{3}{8}$	0.3320	Q
$\frac{7}{16}$–14	0.4375	$\frac{7}{16}$	0.3642	U	$\frac{7}{16}$–20	0.4375	$\frac{7}{16}$	0.3860	W
$\frac{1}{2}$–13	0.5000	$\frac{1}{2}$	0.4219	$\frac{27}{64}$ in	$\frac{1}{2}$–20	0.5000	$\frac{1}{2}$	0.4490	$\frac{7}{16}$ in
$\frac{9}{16}$–12	0.5625	$\frac{9}{16}$	0.4776	$\frac{31}{64}$ in	$\frac{9}{16}$–18	0.5625	$\frac{9}{16}$	0.5060	$\frac{1}{2}$ in
$\frac{5}{8}$–11	0.6250	$\frac{5}{8}$	0.5315	$\frac{17}{32}$ in	$\frac{5}{8}$–18	0.6250	$\frac{5}{8}$	0.5680	$\frac{9}{16}$ in
$\frac{3}{4}$–10	0.7500	$\frac{3}{4}$	0.6480	$\frac{41}{64}$ in	$\frac{3}{4}$–16	0.7500	$\frac{3}{4}$	0.6688	$\frac{11}{16}$ in
$\frac{7}{8}$–9	0.8750	$\frac{7}{8}$	0.7307	$\frac{49}{64}$ in	$\frac{7}{8}$–14	0.8750	$\frac{7}{8}$	0.7822	$\frac{51}{64}$ in
1–8	1.0000	1.0	0.8376	$\frac{7}{8}$ in	1–14	1.0000	1.0	0.9072	$\frac{49}{64}$ in

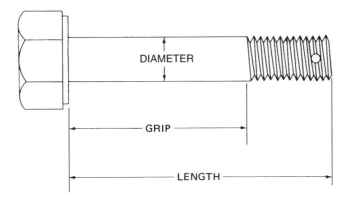

FIGURE 10–6 Bolt dimensions.

These bolts range in size from having a $\frac{3}{16}$-in diameter (AN3) to a $1\frac{1}{4}$-in diameter (AN20). AN 3 through 20 bolts have hex heads with drilled shanks, are made from alloy steel, and have UNF (fine) threads.

The length of the bolt is expressed by a dash number. Bolts increase in lengths by eighths of an inch. The dash number will be one or two digits. If there is one digit, it will state the length in eighths of an inch. A two-digit dash number will always be 1 in or longer. The first number gives inches of length, and the second gives the eighths of an inch. For example,

AN3-7	$\frac{7}{8}$ in long
AN3-15	$1\frac{5}{8}$ in long

The lengths stated will be nominal; the actual length may be from $\frac{1}{32}$ to $\frac{3}{32}$ in longer than shown.

The bolts in the AN3 to 20 series were designed for use with a castellated nut and cotter pin. The standard bolt has a hole drilled in the shank for a cotter pin. If self-locking nuts are used, a plain or nondrilled shank should be used. To specify a bolt with a plain shank, the letter *A* is placed after the dash number. For example,

AN3-7	Drilled shank
AN3-7A	Plain shank

In some installations it may be necessary to safety, or secure, the bolt with safety wire through the head. A drilled head is indicated with the letter *H* placed before the length designation. For example,

AN3-7	Nondrilled head
AN3H7	Drilled head

The standard bolt is made of alloy steel (2330). AN all-purpose bolts are also available made from aluminum alloy and corrosion-resistant steel. To specify aluminum alloy, the letters *DD* are inserted in front of the length designation. Corrosion-resistant steel is specified by the letter *C*. For example,

AN3-7	Alloy steel
AN3DD7	Aluminum alloy
AN3C7	Corrosion-resistant steel

The bolt's material can be identified by head markings, as shown in Figure 10–7. Alloy-steel bolts will have a cross (or asterisk), aluminum-alloy bolts will have two raised dashes, and corrosion-resistant-steel bolts will have one raised dash. Alloy-steel bolts will be cadmium plated, and aluminum-alloy bolts will be anodized.

The use of alloy-steel bolts with diameters smaller than $\frac{3}{16}$ in and aluminum-alloy bolts with diameters smaller than $\frac{1}{4}$ in is prohibited for use in aircraft primary structure by the FAA. Aluminum-alloy bolts should not be used where they will be repeatedly removed for purposes of maintenance and inspection.

Close-Tolerance Bolts. AN4 bolts have a specified diameter of 0.249 (+0.000/-0.003) in. This tolerance is adequate for all applications except when the joint is subject to frequent load reversal or severe vibration. A close-tolerance bolt is made for this type of application under AN standards 173 through 186. AN174 bolts have a specified diameter of 0.2492 (+0.0000/-0.0005) in. The close-tolerance bolt may be identified by a triangle marking (see Figure 10–7) on the head in addition to the standard material markings. The designation code and options are the same for close-tolerance bolts as for the general-purpose bolts, with the exception of the AN number. The AN173 bolt has a $\frac{3}{16}$-in diameter, and the diameter increases by $\frac{1}{16}$ in for each AN number increase. Therefore, an AN181 bolt would have an $\frac{11}{16}$-in diameter. For example,

AN5-7A	General-purpose bolt, $\frac{5}{16}$-in diameter, $\frac{7}{8}$-in length, plain shank
AN175-7A	Close-tolerance bolt, $\frac{5}{16}$-in diameter, $\frac{7}{8}$-in length, plain shank

FAA *Advisory Circular 4313-1A* allows a general-purpose bolt to be used in place of the close-tolerance bolt if the bolt has a light-drive fit in the hole. One definition of a light-drive fit is a maximum clearance between the bolt and the hole of 0.0015 in. A second definition is a fit that requires

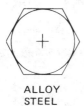

ALLOY STEEL

ALUMINUM ALLOY

CORROSION-RESISTANT STEEL

CLOSE TOLERANCE

FIGURE 10–7 Code marks for aircraft bolts.

the bolt to be pushed into the hole using the handle of a hammer. A tight-drive fit, in contrast, would require a sharp blow from a 12- to 14-oz hammer.

Drilled-Head Engine Bolts. Drilled-head bolts have a deeper, or thicker, head than general-purpose AN bolts. The extra material is required because there are three holes drilled through the head for safety wire (see Figure 10–8). Drilled-head bolts are designed to be installed into threads tapped in an engine crankcase. In this installation, the bolt can only be secured by wires through the head. Drilled-head bolts are identical to AN3 through 20 bolts in terms of shear and tensile strength. A major difference is that these bolts are available with either fine or coarse threads. Drilled-head bolts were originally manufactured under AN73 through AN81 specifications, with a code similar to that for the bolts described previously. A coarse-thread bolt is identified by a letter *A* in front of the dash number. The shank is not drilled for a cotter pin. For example,

AN74A7	Drilled-head bolt, $\frac{1}{4}$-in diameter, $\frac{7}{8}$-in length, UNC thread
AN74-7	Drilled-head bolt, $\frac{1}{4}$-in diameter, $\frac{7}{8}$-in length, UNF thread

AN standards 73 through 81 have been superseded by MS standards. Drilled-head bolts are produced under MS20073 for fine thread and MS20074 for coarse threads. Two dash numbers are used to indicate diameter and length.

Clevis Bolts. Clevis bolts (AN21 through 36) are designed only for shear-load applications. The slotted, domed head results in this bolt often being mistaken for a machine screw. Closer examination (see Figure 10–9) reveals that, unlike a machine screw, it has only a relatively short portion of the shank threaded. Since it will only be loaded in shear, a thin "shear" nut can be used. Use of this design provides more clearance for moving parts at each end than with a regular bolt-and-nut combination. The clevis bolt is made of alloy steel with an identifying cross on the head. The threaded shank is drilled for a cotter pin unless the letter *A* appears after the dash number in the designation code.

Designation of diameter follows the method used for other AN bolts. AN23 clevis bolts have a $\frac{3}{16}$-in diameter. Unlike other AN bolts, clevis bolts are made in diameters smaller than $\frac{3}{16}$ in. An AN21 has the diameter of a No. 6 machine screw. Another difference is that the clevis bolt's length increases in increments of $\frac{1}{16}$ in. The dash number of the code states the nominal length in sixteenths of an inch. For example, AN26-14A designates a clevis bolt with a $\frac{3}{8}$-in diameter ($\frac{6}{16}$), a $\frac{7}{8}$-in ($\frac{14}{16}$) length, and a plain shank (*A*).

MS and NAS Bolts. A large number of bolts are produced under MS and NAS numbers. These include high-strength bolts, high-temperature bolts, and internal-wrenching bolts (shown in Figure 10–10). The standards for these bolts will have a designated code for identification. The particulars of these codes as well as the bolt applications may be found in various technical publications relating to MS and NAS standards and in hardware-distributor catalogs.

Nuts

Nuts are used to hold bolts in place and to provide the necessary clamping force to make a strong joint. The strength of a bolted joint depends upon the bolt and the nut being tightened to a specified torque. To ensure that the torque is maintained, various means are used to "lock" the bolt and the nut together. Nuts are locked, or safetied, to the bolt by several means. One method involves mechanically locking the two together with safety wire or a cotter pin. A second method is to cause adequate friction between the threads of the nut and the bolt to, in effect, lock them together. The amount of friction can be increased in several ways. One way is by the use of a spring-type lock washer. A second way is the use of a second nut on the bolt, called a check nut. A third way is to use a nut with special design features that cause an increase in friction.

The designation code for nuts includes information similar to that for bolts: the basic part, the size, the material, and the thread size.

The nut designed for use with AN bolts is the **AN310 castellated nut**, shown in Figure 10–11. The AN310 nut is shaped for a cotter pin to be used to mechanically lock the nut and bolt together. The standard material for the AN310 nut is alloy steel; aluminum and corrosion-resistant steel are also used. The size of the AN310 nut is specified with a

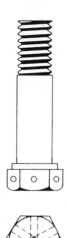

FIGURE 10–8 Drilled-head bolt.

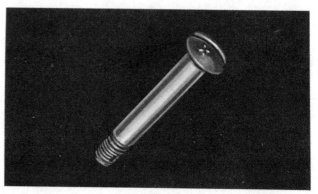

FIGURE 10–9 Clevis bolt.

FIGURE 10–10 AN internal-wrenching NAS bolt.

dash number that is the same as the number of the bolt it fits. The type of material is designated with the letter *D* before the dash number for aluminum, the letter *C* for corrosion-resistant steel, and no letter for steel. For example, AN310-5 is a steel-castellated nut for a $\frac{5}{16}$-in-diameter bolt. All AN310 nuts have UNF Class 3 threads.

Similar to the AN310 nut is the **AN320 shear nut**. The AN320 shear nut is designed for use with a cotter pin but is only about one-half to two-thirds as thick as the AN310. It is designed for use with the clevis bolt. The code for the AN320 is the same as for the AN310. The shear nut is made in dash numbers from 1 to 20 (-1 for No. 6 and -2 for No. 8 machine-screw sizes). AN320 nuts are made only with fine threads.

The **AN315 plain nut** is shown in Figure 10–11. The name is very descriptive, as the AN315 has no special features. Plain nuts are available with either left-hand (counterclockwise to tighten) or right-hand (clockwise to tighten) threads. The code for size and material is the same as for the AN310. Right-hand or left-hand thread is designated with an *L* or an *R* following the dash number. For example,

AN315-6R   Plain nut, right-hand thread for a $\frac{3}{8}$-in bolt
AN315C6L   Plain nut, left-hand thread for a $\frac{3}{8}$-in bolt from stainless steel

Standard aircraft hardware normally has right-hand threads. Plain nuts are often used on control rods and wires, which have right-hand threads on one end and left-hand threads on the other. This arrangement of threads allows the rod or wire to be adjustable in length.

The plain nut can be locked with a lock washer or a check nut. **The AN316 check nut** is a thinner version of the plain nut. To lock the plain nut, the check nut is tightened against it. This loads the plain nut so that the bolt and nut threads are pushed tightly together. The identification code for the check nut is the same as for the plain nut, including the designation of right-hand and left-hand threads.

Self-Locking Nuts. Self-locking nuts are made with a nonmetallic insert or with the top two or three threads distorted. As the bolt screws into the nut and encounters the insert or the distorted threads, a downward force is placed on the nut. The force removes all the axial play between the threads of the nut and the bolt and creates adequate friction to "lock" the nut.

Self-locking nuts are made in both coarse- and fine-thread types. The type of thread is designated in the identification code by listing the size and number of threads per inch as the dash number. A $\frac{1}{4}$-in bolt (-4) and a No. 4 machine screw will both have dash numbers starting with 4. The technician must use the number of threads per inch to designate the correct size (see Table 10–3). For example,

-440   No. 4 machine screw (UNC)
-448   No. 4 machine screw (UNF)
-420 $\frac{1}{4}$-in bolt (UNC)
-428 $\frac{1}{4}$-in bolt (UNF)

Earlier in this chapter it was stated that a $\frac{3}{16}$-in bolt uses a No. 10 machine screw thread. The self-locking nut used with an AN3 bolt will have a dash number of 1032 (No. 10 screw, 32 threads per inch).

One of the first self-locking nuts was the **AN365** (see Figure 10–11), which uses a nonmetallic insert to provide the locking force. Original inserts were of an elastic, fibrous material, giving the nut the name of *fiber locknut* or *elastic-stop nut*. Current nonmetallic inserts are made from nylon. In turning through the insert, the bolt does not cut threads but forces its way into the elastic material. As previously mentioned, the force required to "push" through the insert removes all axial play and locks the nut and bolt together. As long as the insert material retains its elasticity, the nut may be reused. The **AN364 nut** is a shear-nut version of the AN365. The AN364 is used with clevis pins or in other applications where the bolt is loaded primarily in shear. The nonmetallic insert will soften and melt under high temperatures. Both the AN364 and the AN365 are limited to operating temperatures below 250°F. Materials for locknuts include steel, aluminum, and brass. The letters *D* and *B* before the dash number designate aluminum and brass as the material. For example,

AN365-1032   Steel locknut for AN3 bolt
AN365B1032   Brass locknut for AN3 bolt
AN365D1032   Aluminum locknut for AN3 bolt

Standards for AN364 and AN365 nuts have been superseded by MS20364 and MS20365. The material and size designations of the code remain the same.

The need for a nut to operate in higher temperatures resulted in the development of the AN363 metallic locknut. The locking action in these nuts is caused by slightly distorting the top threads of the nut. As the bolt is screwed into these threads, a downward force is placed upon the nut, pushing it against the bolt threads and holding it tight. The AN363 nut is usable to 550°F [288°C]. The AN363C, made of corrosion-resistant steel, can be used up to 800°F [427°C]. The standard for AN363 has been superseded by MS20363. The size designation is the same as that used for AN365 nuts.

FIGURE 10–11 Common AN-standard nuts.

Two types of lightweight, self-locking nuts usable to 450°F [232°C] have been developed under NAS1291 (MS21042) and NAS679A (MS21040). The NAS1291 is an all-metal hexagon design. An identifying feature is that the wrench size is much smaller than that used on conventional nuts. For example, an AN310-4 uses a $\frac{7}{16}$-in wrench; an NAS1291 for a $\frac{1}{4}$-in bolt will use a $\frac{5}{16}$-in wrench. The NAS679A is of conventional size but has been stamped to shape. Both of these nuts have the threads slightly distorted to provide the self-locking force.

Federal Aviation Regulations prohibit the use of self-locking nuts on any bolt subject to rotation in operation, unless a nonfriction locking device is used in addition to the friction lock.

Plate Nuts. Nuts which are made to be riveted in place in the aircraft are called **plate nuts**. Their purpose is to allow bolts and screws to be inserted without having to hold the nut. Self-locking plate nuts are made under a number of standards and in a variety of shapes and sizes. Figure 10–12 shows an AN366 two-lug plate nut with a nonmetallic insert. Also shown is an NAS680A lightweight, all-metal, 450°F [232°C] plate nut.

Washers

Washers serve up to three functions when used with a bolt and a nut. One is to protect the material being fastened from being marred or crushed. A second is to take up any excess grip length on the bolt and to allow the nut to tighten before reaching the ends of the threads. The grip length of the bolt should be equal to but never less than the thickness of the metal. Since bolts vary in length by eighths of an inch, the grip length will usually be slightly longer than the thickness of the material. This extra length is compensated for by adding washers. A third function of a washer is to provide a locking force between the nut and the bolt.

The **AN960 flat washer** is a general-purpose washer for use under the heads of bolts or nuts. The AN960 is available in cadmium-plated steel, corrosion-resistant steel, or aluminum. This washer is made in two thicknesses, regular and light, to provide more variation in the positioning of a nut on the threads. The thin washer is one-half the thickness of the regular washer. The AN960 washer is sized by the screw or the bolt that it fits. The dash number for a washer to be used with a screw is the same as the screw size. For use with

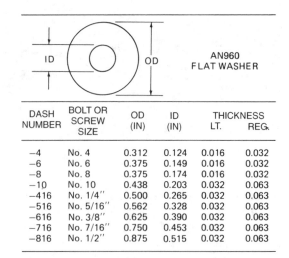

DASH NUMBER	BOLT OR SCREW SIZE	OD (IN)	ID (IN)	THICKNESS LT.	THICKNESS REG.
−4	No. 4	0.312	0.124	0.016	0.032
−6	No. 6	0.375	0.149	0.016	0.032
−8	No. 8	0.375	0.174	0.016	0.032
−10	No. 10	0.438	0.203	0.032	0.063
−416	No. 1/4″	0.500	0.265	0.032	0.063
−516	No. 5/16″	0.562	0.328	0.032	0.063
−616	No. 3/8″	0.625	0.390	0.032	0.063
−716	No. 7/16″	0.750	0.453	0.032	0.063
−816	No. 1/2″	0.875	0.515	0.032	0.063

FIGURE 10–13 Dimensions of AN960 washers.

an AN bolt, the dash number for the diameter of the bolt followed by 16 (e.g., 416, 516) is used. The exception is the AN3 bolt, which uses a -10 washer. Material designation is similar to that for nuts. A washer from the light, or thin, series is designated by placing the letter *L* after the dash number. For example:

AN960-4	Steel flat washer for a No. 4 screw
AN960-416	Steel flat washer for an AN4 bolt
AN960D416L	Aluminum flat washer for an AN4 bolt, one-half the regular thickness

Figure 10–13 shows the dimensions for an AN960 washer.

The **AN970 large-area flat washer** was designed to be used with bolts in wood structures. Figure 10–14 shows an AN970 washer for an AN4 bolt compared to an AN960 washer for an AN8 bolt. The large area of the washer spreads the clamping force over a larger area and keeps the relatively soft wood fibers from being crushed. Although originally designed for wood, the AN970 will work equally well in any installation of a similar nature. Since the AN970 is only made in sizes for AN bolts, the dash number of the size code will be the same as that for the bolt. For example, AN970-5 is a steel large-area washer for an AN5 bolt.

Lock washers are made in two designs. The **AN935 split-ring lock washer** is made of a twisted piece of steel. As the nut turns on the thread, the steel is flattened. The spring action of the steel provides the friction force to keep the nut tight. The size code for the AN935 standard uses the same dash number scheme as for AN960 washers. The

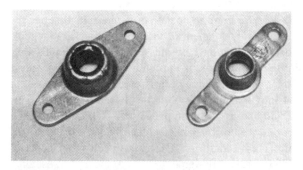

FIGURE 10–12 Self-locking plate nuts.

FIGURE 10–14 General-purpose and large-area washers.

AN935 standard has been superseded by an MS35338 standard. The size in the MS35338 code is given by a sequential number. For example,

AN935-416   Split-ring lock washer for an AN4 bolt
MS35338-40   Split-ring lock washer for a No. 4 machine screw
MS35338-44   Split-ring lock washer for an AN4 bolt

A second design of a lock washer is the **AN936 shakeproof lock washer**. The AN936 is a relatively thin washer with a number of twisted teeth that are flattened as the nut is tightened. As shown in Figure 10–15, there are two design variations. The *Type A* design has internal teeth, and the *Type B* design has external teeth. The locking forces generated by this washer are less than those generated by the AN935 washer. The size code for the AN936 follows that of the AN935, with the letter *A* or *B* being inserted before the dash number to identify the teeth location. The AN936 has been superseded by the MS35333 for the Type A washer and by the MS35335 for the Type B. As with the split-ring washer, the size is designated by a sequential number under the MS standard. For example,

AN936A416   Shakeproof washer for an AN4 bolt, internal teeth
AN936B416   Shakeproof washer for an AN4 bolt, external teeth
MS35333-40   Shakeproof washer for an AN4 bolt, internal teeth
MS35335-33   Shakeproof washer for an AN4 bolt, external teeth

Cotter Pins

Cotter pins are used to lock castellated nuts onto drilled bolts or to secure plain-shank pins in a hole. Figure 10–16 illustrates the proper use of cotter pins. AN380 is the standard for cadmium-plated steel cotter pins, with AN381 being the standard for those made of corrosion-resistant steel. The size of cotter pins under the two standards are designated by two dash numbers. The diameter of both types is expressed in thirty-seconds of an inch. The second dash number for the AN380 gives the length in quarters of an inch. The length for AN381 cotter pins is given in sixteenths of an inch. Both standards for cotter pins have been superseded by a combined standard, MS24665. Diameter, length, and type of material are represented by a single series of sequential numbers under MS24665 specifications. For example,

FIGURE 10–15 Shakeproof washers.

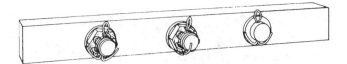

FIGURE 10–16 Applications of cotter pins.

AN380-2-2   Steel cotter pin, $\frac{1}{16}$ by $\frac{1}{2}$ in
AN381-2-8   Corrosion-resistant-steel cotter pin, $\frac{1}{16}$ by $\frac{1}{2}$ in
MS24665-132   Steel cotter pin, $\frac{1}{16}$ by $\frac{1}{2}$ in
MS24665-151   Corrosion-resistant-steel cotter pin, $\frac{1}{16}$ by $\frac{1}{2}$ in

The size of cotter pin to use with AN3, 4, and 5 bolts is the AN380-2-2 ($\frac{1}{16}$ by $\frac{1}{2}$ in). AN380-3-3 cotter pins ($\frac{3}{32}$ by $\frac{3}{4}$ in) are used for AN6, 7, and 8 bolts.

Safety Wire

Safety wire is used in some cases to lock castellated nuts to drilled bolts or to secure a bolt with a drilled head. Stainless-steel safety wire is made in accordance with standard MS20995. The size of the safety wire is specified by a dash number representing the diameter of the wire in one-thousandths of an inch. Safety wire is available in diameters of 0.021, 0.025, 0.035, 0.041, and 0.051 in. The designation code contains the letter *C* to indicate that the material is corrosion-resistant steel. For example, MS20995-C41 is 0.041 stainless-steel safety wire. Safety wire is usually purchased in 5-lb spools or 1-lb dispensing packages.

Aircraft Screws

Aircraft use a large number of machine screws and self-tapping screws. Screws are used to fasten inspection panels, cowling, fairings, and similar components not requiring high-strength fasteners. Screws are designed to be installed into threaded objects. On aircraft, they are used with plate nuts or regular AN nuts.

A screw has the shank threaded all the way to the head, as can be seen in Figure 10–17. Most machine screws specified for aircraft have coarse threads with a Class 2 fit. An exception is the No. 10 screw, which has a fine thread compatible with nuts used for AN3 bolts. The common screws used for aircraft are the 6-32, 8-32, and 10-32, all three having the same 32 tpi. AN standards exist for Nos. 6 and 8 fine-thread screws, and some of these may be encountered by the technician.

The head of the screw is designed to accept the installation or driving tool. The slotted head for a plain screwdriver is the original design. A head made for use with a Phillips screwdriver is called a **recessed head** and is available on all AN screws. Because the Phillips design provides a better grip, the recessed head has, in effect, become the standard head used for aircraft.

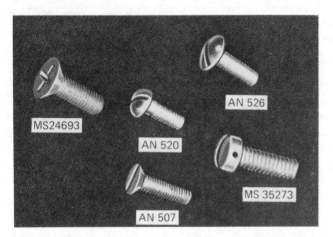

FIGURE 10–17 Typical machine screws.

Machine screws used for aircraft are made with a number of head shapes, which determine the screw type. Figure 10–17 shows five head styles, which are as follows:

MS24693	Flat head (recessed), 100°
AN507	Flat head (slotted), 100°
AN520	Round head (slotted)
AN526	Truss head (slotted)
MS35273	Fillister head (slotted)

Two additional, commonly used machine-screw head styles are shown in Figure 10–18, the AN525 washer head and the MS35206 pan head.

The **AN507 flat-head screw** (superseded by MS24694) requires that the material in which it is placed have a 100° countersunk area. Many flat-head screws are made for an 82° countersink and should not be used in place of the AN507. The flat-head screw is used where flush surfaces are desired.

The **AN526 truss-head screw** is a widely used protruding-head machine screw. Its large head area provides a good clamping force on the thin sheet-metal parts with which it is commonly used. AN526 and 507 are the two most common screws encountered in general-aviation aircraft.

The **AN520 round-head screw** has been replaced in most cases by truss- or pan-head screws.

The **pan-head screw** (MS35206, coarse thread; MS35207, fine thread) has a large head area similar to the

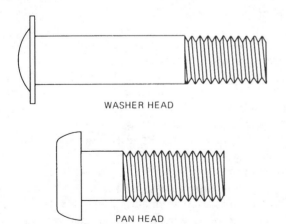

WASHER HEAD

PAN HEAD

FIGURE 10–18 Washer-head and pan-head screw.

truss head, but with a deeper, or thicker, head, allowing for more tension loads.

The **fillister-head screw** has a deep head of a small diameter when compared to other screws. This screw is widely used for non-sheet-metal applications, such as carburetors and magnetos. The deep head design allows a large clamping force to be applied. The head is usually drilled for a safety wire. The fillister-head screw was originally manufactured to one of four standards. The AN500 screw has a coarse thread, a Class 2 fit, and is made of plain-carbon steel. The AN501 is identical to the AN500 except for fine threads. The AN502 is made of alloy steel with a Class 3 fit and fine threads. The AN503 is identical to the AN502 except for coarse threads. The AN502 and 503 screw heads are marked with a cross to indicate that they are made from alloy steel. The AN500 and 501 standards have been replaced with several different MS standards.

With the exception of structural screws and some fillister-head screws, most screws are made from plain-carbon steel. Specifications also provide for screws made from aluminum alloy, brass, or corrosion-resistant steel.

Structural Screws. Certain machine screws are made from alloy steel and have a portion of the shank with no threads, thus providing a grip length. These screws are known as **structural screws** and are used in a manner similar to a bolt. Structural screws include the AN509 (100° flat head), the AN525 (washer head), and the MS27039 (pan head). The AN525 is called a washer-head screw because the head shape appears to have a washer on it. The AN509 standard has been superseded by MS24694. Structural screws have a grip length that must be considered in choosing screw length and that is part of the designation code.

Screw Designations. AN screws use two dash numbers to indicate screw size. The first number is the screw diameter (6, 8, 10), and the second gives the length in sixteenths of an inch. The letter *R* before the second dash number designates a recessed (Phillips) head. The letter *B*, *C*, or *D* before the first dash number reveals the material to be brass, corrosion-resistant steel, or aluminum.

Some AN standards have been superseded by MS standards. The MS standards use sequential numbers to designate size, similar to the system described earlier for cotter pins. Because of the simplicity of the AN code, it is still in widespread use. For example,

AN507-8R6	Flat head, No. 8 × $\frac{3}{8}$-in screw with Phillips head
MS24693-S48	Flat head, No. 8 × $\frac{3}{8}$-in screw with Phillips head
MS24693-C48	Flat head, No. 8 × $\frac{3}{8}$-in screw with Phillips head made of corrosion-resistant steel
AN526-10-8	Truss head, No. 10 × $\frac{1}{2}$-in screw with slotted head
AN526C8R6	Truss head, No. 8 × $\frac{3}{8}$-in screw with slotted head made of corrosion-resistant steel

FIGURE 10–19 Self-tapping screws.

Self-Tapping Screws. Self-tapping screws are also called *sheet-metal screws* and *PK screws*. At one time, a commonly used self-tapping screw was the Parker-Kalon screw, hence the use of the term *PK screw*. The term *sheet-metal screw* is derived from the fact that the self-tapping screw is designed to fasten sheet-metal parts together. The large, coarse threads pull their way through sheet-metal parts that have not been tapped, or threaded. AN standards exist for self-tapping screws, but those in common use are made to industry standards. Self-tapping screws are only used for nonstructural applications.

Self-tapping screws are made in two styles: Type A, with sharp points (see Figure 10–19), and Type B, with blunt points. The sharp-pointed style can start threading into a smaller hole. Head styles for self-tapping screws include flat-head (both 82° and 100°), truss-head, and pan-head styles, similar to the shapes found in machine screws. The heads can be slotted or have a recessed design for a Phillips screwdriver. Self-tapping screws are made of cadmium- or nickel-plated steel or stainless steel. The diameters are specified in terms of machine-screw sizes. Lengths are available from $\frac{1}{4}$ in up, depending upon the diameter. To properly specify a self-tapping screw the technician should designate the head style, screwdriver style, material, type of point, diameter, and length.

NONTHREADED FASTENERS

With the exception of pins, the fasteners covered in this section are not designed to be taken apart. To disassemble parts held together with these systems requires the destruction of the fastener.

Pins

Metal pins of various shapes are used in certain locations in aircraft where their characteristics make their use beneficial. A flat-head pin, taper pin, and roll pin are shown in Figure 10–20.

A **flat-head pin**, also called a **clevis pin**, for use in aircraft is covered by MS20392. Pins of this type are frequently used to join rod ends to bellcranks, to link secondary-control-cable terminals to control arms or levers,

and to perform similar tasks where the control is not in continuous operation. This type of pin is installed with the head up to reduce the possibility of its dropping out if the cotter pin comes loose through wear. The pin is safetied with a cotter pin or safety wire.

A **taper pin** is designed to carry shear loads in a situation where a rod and a tube are telescoped or where two tubular members of different diameters are telescoped to form a rigid joint. Since the pin is tapered, it will eliminate all play in the joint when properly installed. The most satisfactory type of taper pin is threaded on one end so it can be secured with a taper-pin washer and a shear nut. The length and diameter of a taper pin are critical for any particular installation because a pin of the wrong size will not secure the joint in a rigid condition.

The **roll pin** is a split tube made of spring steel and chamfered at the ends. The split extends the full length of the pin on one side. When the pin is driven into an undersized hole, the pin reduces in diameter just enough to enter the hole. Since the pin is normally larger than the hole, when compressed in the hole it will maintain strong pressure against the sides of the hole, thus keeping it securely in place. It can best be removed with a pin punch.

Rivets

Rivets are metal pin-type fasteners designed primarily for shear-type loads. Thousands of rivets are used in aircraft to join sheet-metal skins and to fasten the skin to the aircraft structure. Rivets are fastened in place by forming, or **upsetting**, a second head on the end of the shank. As the rivet's upset head is being formed, the entire shank of the rivet swells in diameter and completely fills the hole. With the

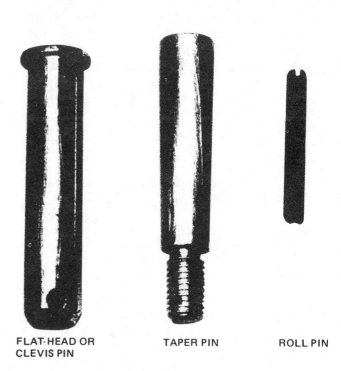

FLAT-HEAD OR CLEVIS PIN   TAPER PIN   ROLL PIN

FIGURE 10–20 Flat-head, taper, and roll pins.

hole filled, all play between the rivet and the sheet metal is eliminated, forming a very tight joint.

The majority of the rivets for aircraft are made of aluminum-alloy material. Aluminum rivets manufactured to AN standards (see Figure 10–21) have the following head shapes:

AN426 Flush head, 100° countersink
AN430 Round head
AN442 Flat head
AN455 Brazier head
AN456 Modified brazier head
AN470 Universal head

The AN standards have been superseded by two MS standards. The MS20470 universal-head rivet can replace any other protruding-head rivet. The MS20426 flush-head rivet supersedes the AN426.

Aircraft rivets are made from five aluminum alloys. The material from which a rivet has been made can be determined by the head marking on the rivet. A one- or two-letter code in the rivet-designation code identifies the alloy used. The alloys used, their head marking, and letter code (see Figure 10–21) are as follows:

Alloy 1100 A Plain head
Alloy 2017 D Raised dot
Alloy 2117 AD Recessed dot
Alloy 2024 DD Raised double dash
Alloy 5056 B Raised cross

Alloy 1100 Type A rivets are made from pure aluminum. The rivets are soft and of low strength and are used only for nonstructural purposes.

Alloy 2024 Type DD rivets are the strongest of the aluminum rivets. Alloy 2024 is too hard to drive, or form, in its normal state. Forming an upset head creates internal stresses, causing the rivet to crack. The rivets must be heat treated and quenched immediately prior to driving. Immediately after they have been quenched, the material is relatively soft

and can be formed (driven) without adverse effects. The material age-hardens very rapidly after heat treatment and quenching. In a very short time the material will again become too hard to drive. If the rivets are refrigerated after quenching, the cold temperature will slow down the aging process, allowing more time to elapse before the rivets have to be re–heat treated. For this reason, 2024 rivets are called **icebox rivets**.

Alloy 2017 Type D rivets have approximately 85% of the strength of 2024 rivets. Although not as much of a problem as 2024 rivets, the Type D rivets are difficult to drive and age-harden rapidly. Larger sizes of these rivets must be kept refrigerated or must be heat treated immediately prior to use.

Alloy 2117 Type AD rivets are made from a modification of the 2017 alloy that allows the rivets to be driven "off the shelf," that is, at any time. The AD rivet has 77% of the strength of a 2024 rivet. The 2117 rivet, identified by a recessed dot, is the "standard" rivet for aluminum aircraft structures.

Alloy 5056 Type B rivets are used for riveting magnesium sheets. Magnesium is too hard of a material to be used for rivets. The 2117 rivet, used for aluminum, has copper as a major alloy, which presents a potential for corrosion in the proximity of magnesium. Aluminum alloy 5056 has magnesium as a major alloying element and provides adequate strength for rivets in magnesium structure.

Some other aircraft-rivet materials and the letter used to identify the material are

Copper   C
Stainless steel   F
Monel   M

Rivets are sized by the diameter of their shanks and by their length. Rivets are made in diameters that increase in $\frac{1}{32}$-in increments. Common sizes of rivets used for general-aviation aircraft have a $\frac{3}{32}$- to $\frac{3}{16}$-in shank diameter. A protruding-head-rivet length is the distance from the bottom of the head to the end of the shank. The length of a flush-head

MATERIAL	HEAD MARKING		AN MATERIAL CODE
1100	PLAIN	◯	A
2117T	RECESSED DOT	⊙	AD
2017T	RAISED DOT	⊙	D
2024T	RAISED DOUBLE DASH	⊖	DD
5056T	RAISED CROSS	⊕	B

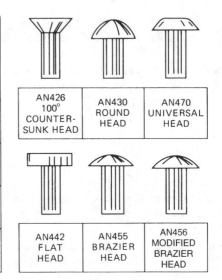

FIGURE 10–21 Rivet heads and markings.

rivet is the overall length of the rivet. The designation code for rivets uses two dash numbers for the size. The first dash number gives the diameter of the rivet in thirty-seconds of an inch. The length of the rivet is stated in sixteenths of an inch by the second dash number.

The complete designation code for rivets includes the basic AN number, which gives the head shape. This is followed by the letter code indicating the material. Finally, the size is given by diameter and length. For example, MS20470AD3-7 is a universal-head rivet of 2117 aluminum alloy, $\frac{3}{32}$-in diameter, and $\frac{7}{16}$-in length. Most rivets are anodized or given a zinc-chromate treatment by the manufacturer. The color of the rivet provides no information in terms of material.

Detailed instructions for rivet installation are given in the associated text *Aircraft Maintenance and Repair*. Information is also available in the FAA's *Advisory Circular 43.13-1A*.

Special Fasteners

In addition to the standard nuts, bolts, screws, and rivets, there are many specialized fasteners that have been developed to join parts or structures where the more common fasteners cannot meet all of the requirements. Special fasteners are used for installations where only one side of the material is accessible. Other specialized fasteners are used because of the need for high strength. Still others are used because they provide high strength with less weight than conventional hardware.

Space does not permit descriptions of all the special fasteners available. A few of the typical fasteners will be described so that you may better understand the principles behind their design and installation.

Many of the special fasteners use manufacturers' part numbers even though they are made to MS or NAS standards. Before installing a special fastener, the manufacturer's technical data should be checked to make sure the fastener is approved for that application. The installation of many special fasteners requires special tooling and precise handling. It is essential that the manufacturer's instructions be followed when working with this type of hardware. The information on installation in this text is for general knowledge and should not be used for specific applications.

Blind Rivets. Blind rivets, or fasteners, are designed to be installed where access to both sides of a sheet assembly or structure is not possible or practical. The blind rivet usually consists of a tubular sleeve, or rivet shell, in which a stem having an enlarged end is installed. The heads of such rivets are made in standard configurations such as brazier, universal, and flush. The rivet and stem are inserted in a correctly sized hole, and the stem is drawn into the sleeve by means of a special tool. The bulb or other enlargement on the end of the stem expands the end of the rivet and locks it into the hole. The rivet shell will usually be of a soft, low-strength material that can easily be formed. The stem, which does not become deformed, will be made of a harder, stronger material. Most of the strength of the rivet comes from the stem.

Hole size is critical for blind-rivet installation. Unlike conventional rivets, a blind rivet will not fill the hole while it is being driven. A hole that is slightly oversized will often cause the installation process to fail. Many of the blind rivets are made in standard and oversized diameters to help alleviate problems caused by enlarged holes.

In early blind rivets, the stem was retained in the shell only by friction. Vibration and other forces would often overcome the friction forces, causing the stem to fall out. Since the majority of the strength is in the stem, the rivet, in effect, had failed. Current designs of blind fasteners incorporate a method of mechanically locking the stem into the shell.

Typical blind rivets are shown in Figures 10–22, 10–23, and 10–24. The rivets shown are Cherrylock and Cherrymax rivets, which are registered tradenames of Cherry Textron, Inc. Many other companies produce rivets similar to these rivets.

The Cherrylock rivet shown in Figure 10–22 is a typical blind rivet. A collar is used to lock the stem into the rivet shell. Cherrylock rivets can be installed with a hand gun or with a pneumatic power gun. The guns are manufactured with a provision for changing the pulling heads to accommodate different sizes and types of rivets. The technician must carefully follow the specifications given by the manufacturer for the installation of the rivets. The *Cherry Rivet Process Manual*, supplied by the manufacturer, is a good source of detailed and specific information.

The **standard Cherrylock rivet** is installed as shown in Figure 10–22. The rivet is inserted into the correctly drilled hole, as shown in view *a*. View *b* shows the action as the pulling head (not shown in the illustration) begins to pull the stem, which begins the formation of a blind head. At this time the two sheets are also being drawn together. In view *c* the rivet head is firmly seated and the sheets are tightly clamped together. In view *d* the stem continues to be pulled through the shell. When the stem reaches the point illustrated in view *e*, the installation tool stops pulling on the stem and pushes the locking collar into place. Once the collar is in place, the stem is pulled, causing it to fracture flush with the rivet head, as shown in view *f*.

The **bulbed Cherrylock rivet** in Figure 10–23 is similar to the standard Cherrylock but forms a larger, bulbed head. The larger head provides more clamping force and works better for thinner sheet metals.

An improved blind rivet, the **Cherrymax rivet**, is shown in Figure 10–24. This rivet is similar to the Cherrylock except, as can be seen in the illustration, the locking collar is inside the shell before installation begins. During installation, the collar is drawn up with the stem until it contacts the driving anvil. Further movement of the stem forces the collar to form into a lock. The installation tooling is much less expensive and easier to use for this rivet.

Cherry rivets are manufactured in a variety of shell and stem materials, including aluminum alloy, stainless steel, and monel. Cherry rivets use the manufacturer's code for identification. For example, CR2163-6-4 indicates a Cherrylock universal-head rivet made from 2017 aluminum alloy that has a $\frac{6}{32}$-in diameter (-6) and a maximum grip length of $\frac{4}{16}$ in (-4).

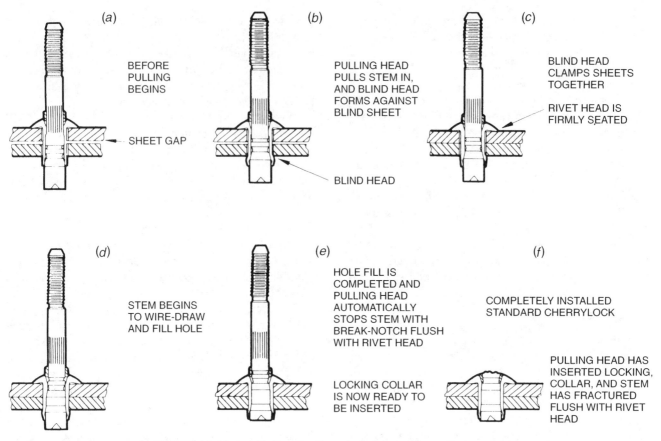

(a) BEFORE PULLING BEGINS — SHEET GAP

(b) PULLING HEAD PULLS STEM IN, AND BLIND HEAD FORMS AGAINST BLIND SHEET — BLIND HEAD

(c) BLIND HEAD CLAMPS SHEETS TOGETHER — RIVET HEAD IS FIRMLY SEATED

(d) STEM BEGINS TO WIRE-DRAW AND FILL HOLE

(e) HOLE FILL IS COMPLETED AND PULLING HEAD AUTOMATICALLY STOPS STEM WITH BREAK-NOTCH FLUSH WITH RIVET HEAD

LOCKING COLLAR IS NOW READY TO BE INSERTED

(f) COMPLETELY INSTALLED STANDARD CHERRYLOCK

PULLING HEAD HAS INSERTED LOCKING, COLLAR, AND STEM HAS FRACTURED FLUSH WITH RIVET HEAD

FIGURE 10–22 Installation of a Cherrylock rivet. *(Cherry Textron, Inc.)*

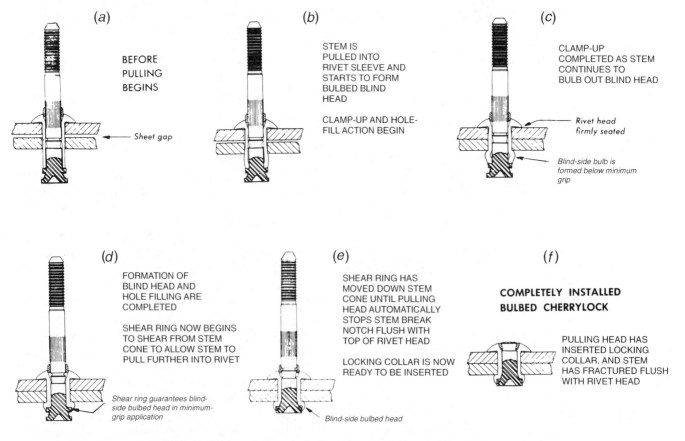

(a) BEFORE PULLING BEGINS — *Sheet gap*

(b) STEM IS PULLED INTO RIVET SLEEVE AND STARTS TO FORM BULBED BLIND HEAD

CLAMP-UP AND HOLE-FILL ACTION BEGIN

(c) CLAMP-UP COMPLETED AS STEM CONTINUES TO BULB OUT BLIND HEAD — *Rivet head firmly seated* — *Blind-side bulb is formed below minimum grip*

(d) FORMATION OF BLIND HEAD AND HOLE FILLING ARE COMPLETED

SHEAR RING NOW BEGINS TO SHEAR FROM STEM CONE TO ALLOW STEM TO PULL FURTHER INTO RIVET

Shear ring guarantees blind-side bulbed head in minimum-grip application

(e) SHEAR RING HAS MOVED DOWN STEM CONE UNTIL PULLING HEAD AUTOMATICALLY STOPS STEM BREAK NOTCH FLUSH WITH TOP OF RIVET HEAD

LOCKING COLLAR IS NOW READY TO BE INSERTED

Blind-side bulbed head

(f) COMPLETELY INSTALLED BULBED CHERRYLOCK

PULLING HEAD HAS INSERTED LOCKING COLLAR, AND STEM HAS FRACTURED FLUSH WITH RIVET HEAD

FIGURE 10–23 Installation of a bulbed Cherrylock Rivet. *(Cherry Textron, Inc.)*

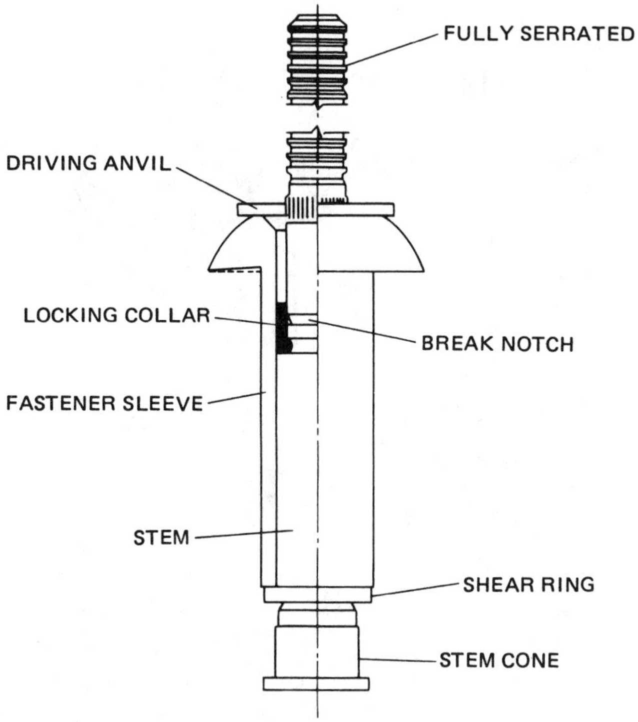

FULLY SERRATED
DRIVING ANVIL
LOCKING COLLAR
BREAK NOTCH
FASTENER SLEEVE
STEM
SHEAR RING
STEM CONE

FIGURE 10–24 Drawing of a Cherrymax rivet. *(Cherry Textron, Inc.)*

Because of the large number of material combinations, it is recommended that the manufacturer's technical manuals be used when working with these rivets.

Products similar to those just described are produced by a number of manufacturers. Many but not all of the blind fasteners are produced under NAS and MS standards. As mentioned previously, the technician must be very careful to use only fasteners approved for the application with which she or he is working.

A **rivnut** is a hollow blind rivet that serves as a nut. It is manufactured by the B. F. Goodrich Company and was originally designed for the attachment of deicer boots to aircraft wings. Rivnuts are now used in many places where threads are required but the metal is too thin to be threaded. The installation of a rivnut is shown in Figure 10–25. Because of their low strength, rivnuts are used primarily to receive threaded fasteners. As with other fasteners, a standard designation code is used to identify the material, size, and configuration of rivnuts.

Blind Bolts. A blind bolt, like a blind rivet, is one that can be completely installed from only one side of a structure or assembly. The blind bolt is used when it is necessary to use a fastener with high shear strength. The bolt is usually made of alloy steel, titanium, or other high-strength material. A typical blind lock bolt is shown in Figure 10–26, consisting of a nut, sleeve, and screw. Selecting the bolt of the correct length and diameter for the particular installation is very important, and the specification sheet for the product used should be followed closely.

Installation consists of inserting the sleeve in the hole and pulling the nut toward the sleeve, causing the nut to form a collar against the workpiece. A core bolt is inserted in the sleeve and torqued as specified in the installation instructions. There are many sizes, types, and designs of blind bolts, and the technician must make certain to use the size, type, and material that has been approved for the repair or replacement that is being made.

A number of high-strength fasteners have been developed to use for permanent installations where it is desired to reduce weight and installation time. Such rivets or bolts can be used only where their use has been approved by the appropriate authority.

Swaged-Collar Fasteners. The **Hi-Shear rivet** has a pin made of steel or other high-strength material and is held in place by a swaged collar of aluminum or other soft material. The collar is driven onto the end of the rivet by means of a special tool in a conventional pneumatic rivet gun. The installation of a Hi-Shear rivet is shown in Figure 10–27. Because there is no change in the pin diameter during installation, the hole size is very critical and may require reaming to size in many cases. Although the pin is relatively expensive, the installation tools are not, and the process can be easily learned. As with all fasteners, it is very important that the correct size be selected for the particular application.

The **lock bolt** is similar to the Hi-Shear rivet in function and is produced by several manufacturers. These fasteners are made in two types. The **pull-type lock bolt**, shown in Figure 10–28, includes a grooved pintail by which the bolt is pulled into place by means of a pneumatic installation gun. Once the pin has pulled the sheets together, the collar is swaged into grooves on the lock bolt. When the collar is in place, the force of the gun fractures the pintail at a breakneck groove (see Figure 10–29). The installation of this fastener requires expensive specialized equipment.

The **stump-type lock bolt** is installed in locations where the clearance is such that the installation tool for the pull-type bolt cannot be used. Stump-type lock bolt installation, shown in Figure 10–30 on page 236, uses tools similar to those used for the Hi-Shear rivet.

Lock bolts are used where permanent assemblies are made because they save weight, provide strength equivalent to standard bolts, and are easy to install. They provide excellent shear strength, and the tension lock bolts provide good tension strength. The tension lock bolts have four collar-locking grooves, while the shear-type lock bolts have only two locking grooves. In selecting a lock bolt, the shear and tension loads must be known.

A **Hi-Lok bolt**, illustrated in Figure 10–31 on page 236, is a threaded fastener but has features of the swaged-collar type. The installation of the Hi-Lok fastener is completed on one side of the assembly after the bolt has been inserted through the hole from the other side. The hexagonal wrench tip of the installing tool is inserted into a recess in the bolt, which holds the pin (bolt) while the tool turns the collar (nut). As the collar is tightened to the designed torque level built into the collar, the hex portion of the collar is sheared off automatically by the driving tool. This leaves the installation with the correct amount of torque and preload. A permanent collar has, in effect, been swaged onto the pin threads without subjecting the structure to pneumatic pounding, as

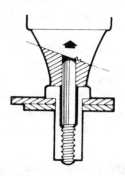 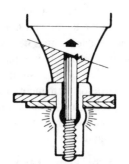 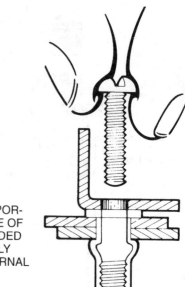

RIVNUT ON HEADER TOOL MANDREL, INSERTED IN DRILLED HOLE READY FOR INSTALLATION. ARROW INDICATES DIRECTION OF MANDREL MOVEMENT AS TOOL IS OPERATED.

MANDREL RETRACTS, PULLING THREADED POR-TION OF RIVNUT SHANK TOWARD BLIND SIDE OF WORK, FORMING BULGE AROUND UNTHREADED SHANK AREA. RIVNUT IS CLINCHED SECURELY IN PLACE. THE TOOL MANDREL LEAVES INTERNAL RIVNUT THREADS INTACT, UNHARMED.

INSTALLED RIVNUTS ALSO SERVE AS BLIND-NUT PLATES FOR SIMPLE SCREW ATTACH-MENTS. WHEN IMPERATIVE THAT ATTACHED PART BE FLUSH FIT, COUNTERSUNK RIVNUT HEADS ARE USED INSTEAD OF FLAT HEADS.

FIGURE 10–25 Installation of a rivnut.

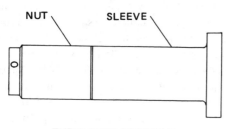

THE SLEEVE ASSEMBLY

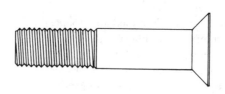

CORE BOLT

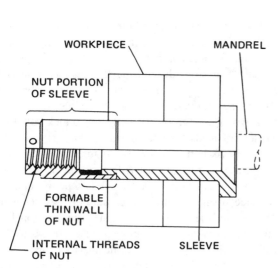

BEFORE PULL-UP

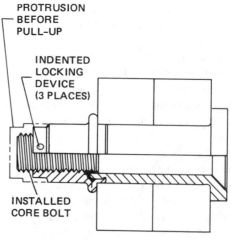

AFTER PULL-UP
(COMPLETED INSTALLATION
WITH CORE BOLT INSTALLED)

FIGURE 10–26 Blind lock-bolt.

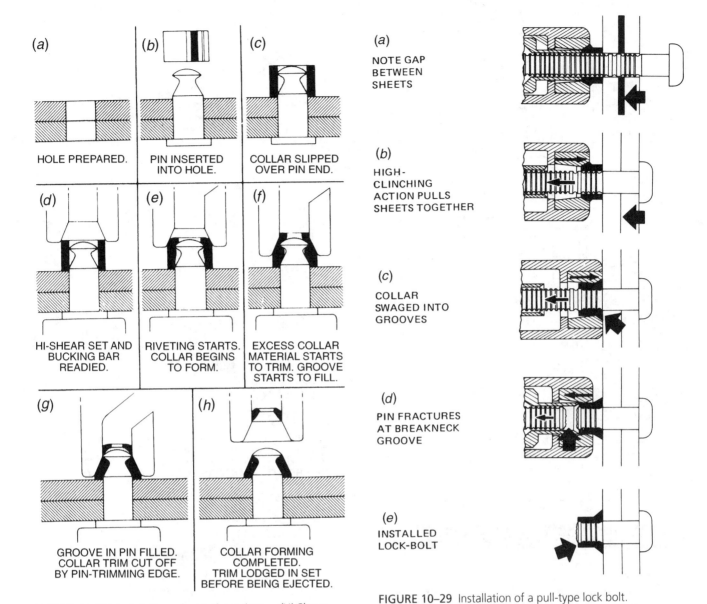

(a) HOLE PREPARED.

(b) PIN INSERTED INTO HOLE.

(c) COLLAR SLIPPED OVER PIN END.

(d) HI-SHEAR SET AND BUCKING BAR READIED.

(e) RIVETING STARTS. COLLAR BEGINS TO FORM.

(f) EXCESS COLLAR MATERIAL STARTS TO TRIM. GROOVE STARTS TO FILL.

(g) GROOVE IN PIN FILLED. COLLAR TRIM CUT OFF BY PIN-TRIMMING EDGE.

(h) COLLAR FORMING COMPLETED. TRIM LODGED IN SET BEFORE BEING EJECTED.

FIGURE 10–27 Installation of a Hi-Shear rivet. *(Hi-Shear Corp.)*

(a) NOTE GAP BETWEEN SHEETS

(b) HIGH-CLINCHING ACTION PULLS SHEETS TOGETHER

(c) COLLAR SWAGED INTO GROOVES

(d) PIN FRACTURES AT BREAKNECK GROOVE

(e) INSTALLED LOCK-BOLT

FIGURE 10–29 Installation of a pull-type lock bolt.

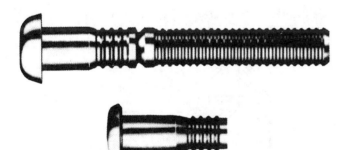

FIGURE 10–28 Pull-type and stump-type lock bolts.

in the installation of the Hi-Shear rivet. Installation of the Hi-Lok fastener is much less complex than the lock bolts just described. The Hi-Lok fastener is made in several styles and material combinations. While relatively expensive, the installation procedures are easy.

The high-strength fasteners described in this section are made in a large number of configurations, material combi-nations, and sizes. All will have designation codes devel-oped by their manufacturers for identification and selection. Most will also have been manufactured under an MS or NAS standard. The use of these fasteners virtually requires that the manufacturer's technical material be available.

PANEL AND COWLING FASTENERS

Panel and cowling fasteners that may be disengaged quickly are important in the inspection and servicing of an aircraft. For small aircraft, fasteners such as the Dzus, Camloc, and Airloc are suitable; these are illustrated in Figure 10–32. For high-performance aircraft, a panel fastener is needed that not only holds the panel in place but will pull the panel into a se-cure fit so that the panel becomes a part of the load-bearing structure. This type of fastener is called a structural-panel fastener and is illustrated in Figure 10–33 on page 238. With this type of fastener, the sleeve bolt is retained in the panel by means of a retainer ring. The receptacle assembly is riveted to the structure.

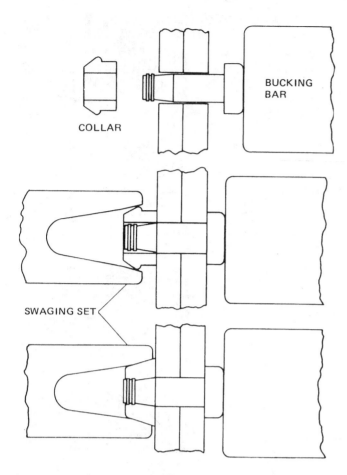

COLLAR

BUCKING BAR

SWAGING SET

FIGURE 10–30 Installation of a stump-type lock bolt.

CABLE FITTINGS

Cable fittings are required to connect a cable to control arms, to other fittings, to turnbuckles, and to other sections of cable. When it is necessary to attach a cable to a turnbuckle or some other device and swage-type fittings are not available, the AN100 cable thimble is used. A thimble is illustrated in Figure 10–34. It is attached by a swaged Nicopress sleeve.

The Nicopress sleeve and a spliced cable fitting are shown in Figure 10–35. The Nicopress sleeve is composed of copper and is pressed (swaged) on the cable by means of a special tool. The specifications for a properly installed sleeve are established by the manufacturer.

When swaging equipment is available, it is desirable to use swaged fittings because these fittings develop the full cable strength when properly installed. The AN664 ball end, AN666 stud end, AN667 fork end, and AN668 eye end, in Figure 10–36 on page 239, are typical swaged fittings.

To install a swaged fitting, the cable is inserted to the full depth of the barrel in the fitting and is held firmly in this position while the swaging operation is completed. The swaging can be done either with a hand machine or with a power swaging machine. The machine presses the metal of the fitting barrel into the cable to the extent that there is no visible division between the fitting and the cable if a cross-sectional cut is made through the swaged portion. After the swaging

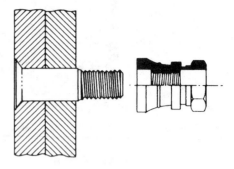

THE INSTALLATION OF THE HI-LOK FASTENER IS COMPLETED ON ONE SIDE OF THE ASSEMBLY AFTER THE PIN HAS BEEN INSERTED THROUGH THE HOLE FROM THE OTHER SIDE.

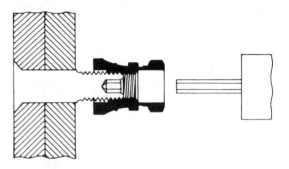

FOR NON INTERFERENCE-FIT APPLICATIONS, THE HEX WRENCH TIP OF THE POWER DRIVER IS INSERTED INTO THE HEX RECESS OF THE PIN. THIS KEEPS THE PIN FROM ROTATING AS THE COLLAR IS DRIVEN.

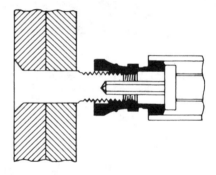

PROGRESSIVE TIGHTENING TAKES PLACE AS TORQUE IS APPLIED.

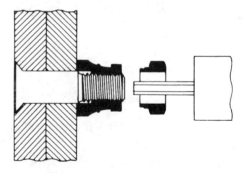

AT THE DESIGNED TORQUE LEVEL BUILT INTO THE HI-LOK COLLAR, THE HEX PORTION OF THE COLLAR IS SHEARED OFF AUTOMATICALLY BY THE DRIVING TOOL. REMOVAL OF THE INSTALLATION TOOL FROM THE HI-LOK PIN COMPLETES THE INSTALLATION.

FIGURE 10–31 Installation of a Hi-Lok fastener. *(VOI-SHAN)*

operation is completed, the swaged barrel should be checked with a go/no-go gage to make sure that the proper degree of swaging has been accomplished. It is also advisable to mark the cable when it is inserted into the barrel of the fitting to make sure that it does not slip during the swaging operation. It is also good practice to mark the junction of the terminal or fitting and the cable with paint to provide a means of detecting slipped cable at later inspections.

TURNBUCKLES

Turnbuckles are used for adjusting the tension of control cables. A standard turnbuckle consists of a **barrel** and two steel ends, one end having a right-hand thread and the other having a left-hand thread. When the barrel is rotated, the ends are moved together or away from each other. The end

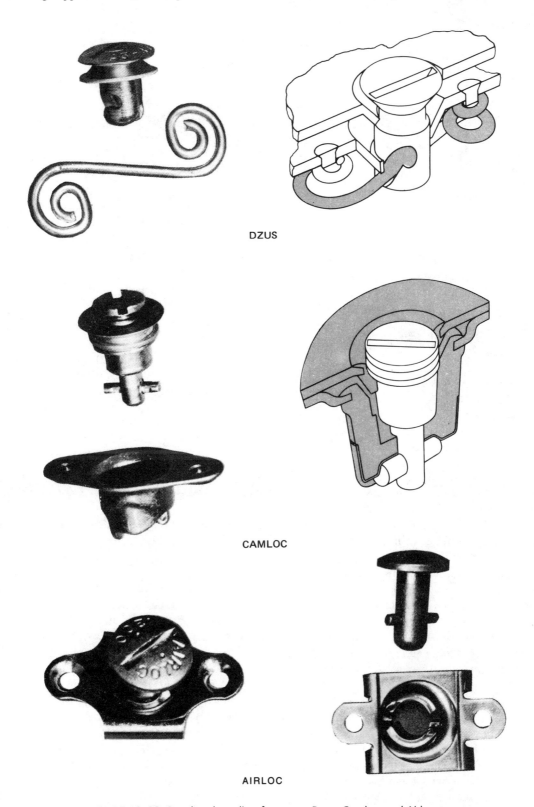

FIGURE 10–32 Panel and cowling fasteners: Dzus, Camloc, and Airloc.

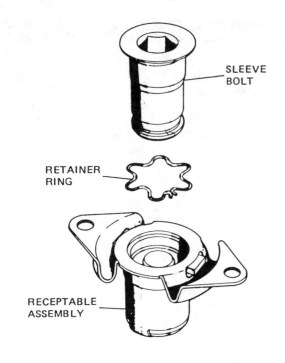

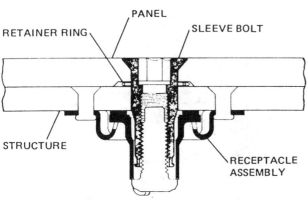

FIGURE 10–33 Structural-panel fastener. *(VOI-SHAN)*

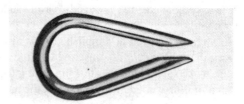

FIGURE 10–34 Cable thimble.

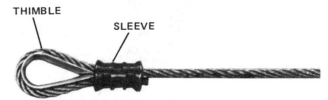

FIGURE 10–35 Nicopress cable splice.

Two principal factors must be considered when installing and adjusting turnbuckles: (1) when the turnbuckle is tightened, not more than three threads must show outside the barrel at each end; and (2) the turnbuckle must be properly safetied.

Typical turnbuckle parts are designated by MS and AN standards as follows:

Barrel	MS21251	AN155
Fork	MS21252	AN161
Pin eye	MS21254	AN165
Cable eye	MS21255	AN170
Swaging terminal		AN669
Safety clip	MS21256	

of the barrel having the left-hand threads is marked with a groove completely around the end of the barrel.

Typical turnbuckles are illustrated in Figure 10–37. The AN-type turnbuckle is safetied with safety wire, and the MS-type turnbuckle is safetied with an MS21256 safety clip. The barrel (MS21251) of this turnbuckle has a groove in the threads on both ends to receive the straight end of the safety clip. The end fittings also have the threads grooved for the clip. The clip locks the threads of the barrel and fittings together to prevent any rotation.

Turnbuckles may be supplied with several different types of ends. Some of these are illustrated in Figure 10–38. Figure 10–38*a* is a cable eye for use with a cable thimble; Figure 10–38*b* is a fork by which the turnbuckle can be attached to a flat fitting; Figure 10–38*c* is a pin eye to be inserted into a forked or double-sided fitting; and Figure 10–38*d* is a swage fitting by which the cable can be attached to the turnbuckle after having been swaged into a sleeve. The barrel of a turnbuckle is made of brass and, as stated previously, is grooved on one end to indicate the left-hand thread. The hole through the center of the turnbuckle is used to turn the barrel and to safety the assembly.

SAFETY BELTS

Although safety belts for civil aircraft may not be considered as aircraft hardware, they do involve hardware and must meet rigid standards. For this reason, they are included in this section.

All seats that may be occupied during takeoff or landing must be equipped with approved safety (seat) belts. The requirements for safety belts are set forth in Technical Standard Order C22d, issued by the FAA. Models of safety belts manufactured for installation on civil aircraft since November 30, 1960, have had to meet the standards of National Aerospace Standards (NAS) Specification 802, with certain exceptions. Any safety belt meeting all the requirements of NAS802 may be approved for use.

The principal difference between the requirements of TSO C22d and NAS802 is in the strength of the safety-belt assembly. TSO C22d states that the safety-belt strength need be only 1500 lb [680 kg] for a single-person belt and 3000 lb [1360 kg] for a two-person belt. NAS802 requires 3000 and 6000 lb [1360 and 2720 kg], respectively. Any

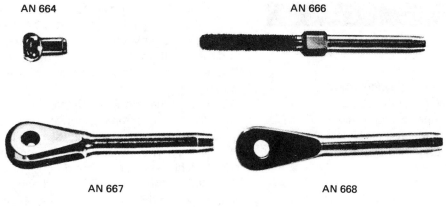

FIGURE 10–36 Swaged cable fittings.

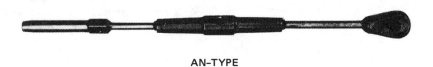

AN–TYPE

MS–TYPE

FIGURE 10–37 AN- and MS-type turnbuckles.

safety belt meeting the requirements of TSO C22d may be used on civil aircraft.

Safety belts must be designed so as to be easily adjustable. Each belt must be at least $1\frac{15}{16}$ in [4.92 cm] wide and equipped with a quick-release mechanism designed so that it cannot be released accidentally. A safety belt may be approved for one person or for two adjacent persons, depending upon its strength. A belt for one person must be capable of withstanding a load of 1500 lb [680 kg] applied in alignment with the anchored belt. The quick-release mechanism must be capable of withstanding this load without undue distortion and must be easily releasable under a load simulating a person hanging on the belt.

A safety belt approved for two adjacent persons must be capable of withstanding a load of 3000 lb [1360 kg] applied in alignment with the anchored belt. The quick-release mechanism must be easily releasable, as described previously. After a test under extreme load, the release mechanism must be releasable with a pull of not more than 45 lb [20 kg].

The strength of a safety belt is determined by a test as specified in NAS802. The static testing of the belt and its attachments must be accomplished under conditions simulating the belt pulling against a human body. Each half of an approved safety-belt assembly must have a legible and permanent nameplate or identification label with the following information: (1) the manufacturer's name and address; (2) the

equipment name or type or model designation; (3) the serial number and/or date of manufacture; and (4) the applicable TSO or NAS number.

KEEP CURRENT

It must be emphasized that new standards are constantly being issued and that older standards become obsolete. Nevertheless, many of the older standards are still effective, and it is up to the maintenance technician to use only approved parts and materials. It is always safe to select materials and parts that are specified in the manufacturer's overhaul or maintenance manual for a particular airplane, power plant, or accessory.

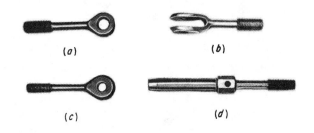

FIGURE 10–38 End fittings for turnbuckles.

REVIEW QUESTIONS

1. Define *standard*.
2. List the meanings of the following standard designations: AC, AF, AN, AS, NAS, and NAF.
3. What standard has superseded the AN standard?
4. In case of doubt, should a technician install a standard part or a part designated by the manufacturer's part number?
5. What does AMS stand for?
6. What is the purpose of the ASTM?
7. Who establishes standards for iron and steel?
8. What does SAE stand for?
9. What is an MIL specification?
10. What agency establishes aircraft, engine, and propeller specifications?
11. What is the purpose of a close-tolerance bolt?
12. What is meant by a UNF Class 3 thread?
13. How is the material of an aircraft-standard nut indicated in the standard identification number?
14. Describe a castellated nut and how it may be safetied.
15. Explain the principle of a fiber locknut.
16. What is the difference between a standard machine screw and a structural screw?
17. How are rivet heads marked to indicate material?
18. Describe and explain the purpose of a blind rivet.
19. What is the purpose of a blind bolt?
20. What are the advantages of lock bolts?
21. What is a structural-panel fastener?
22. Explain the purpose of a turnbuckle.
23. Describe a Nicopress sleeve.
24. Give the requirements for a safety belt in an aircraft.

Hand Tools and Their Application 11

INTRODUCTION

Technicians are often judged by their knowledge of tools and by the manner in which they care for their tools. A variety of hand tools are necessary for the maintenance of aircraft. It is essential that the aviation maintenance technician be well informed and skilled in the use of these tools.

Common hand tools have very broad application and are used by people with a wide variety of training and experience. As a result, hand tools are often taken for granted and misused. Many users are surprised to find that there are many details affecting the design, construction, and use of the various tools. Tools are designed for specific purposes and for working with specific types of materials. Using a tool in a manner for which it was not designed can result in damage to the part or the tool as well as serious injury to the user. The proper and safe use of hand tools is very important to the aircraft technician.

Because of the number of tools in existence, this chapter concentrates on the frequently used hand tools that all technicians would have in their personal tool box. Specialized tools for use in areas such as sheet-metal fabrication are covered in the appropriate sections of other textbooks in this series. The emphasis of this chapter is on the ability to identify different tools and understand their function.

MEASUREMENT AND LAYOUT

One of the most important considerations in the manufacture, maintenance, and overhaul of machinery is measurement. The concept of interchangeable parts depends upon precision measurement. The repair of an aircraft structure requires the measurement of damaged and replacement material.

The term **layout** refers to the process of transferring measurements from a drawing to the materials from which parts or repairs will be made. The tools used for measurement and layout are closely related.

Rules and Scales

The most commonly used measuring instrument is the rule. A rule is a straight-edged piece of material, either steel, wood, or plastic, that is marked off in units of length. The **scale** of the rule depends on how the units of length are marked off. The markings of the scale are referred to as **graduations**. The graduations may divide the rule into fractional parts of an inch ($\frac{1}{2}$, $\frac{1}{4}$, $\frac{1}{8}$, $\frac{1}{16}$, $\frac{1}{32}$, and $\frac{1}{64}$), decimal parts of an inch (0.1 and 0.01), or metric graduations of centimeters and millimeters.

Rules exist in a wide variety of sizes and shapes. One of the most useful rules for an aviation technician is the 6-in **flexible steel rule** illustrated in Figure 11–1. Rules of this size are available with a number of different graduations. The one shown here has scales graduated in tenths and one-hundredths of an inch on one side and thirty-seconds and sixty-fourths of an inch on the other. A similar rule that has widespread use is pictured in Figure 11–2 and is a part of a combination-square set. The rule of a combination square is usually 12 in long and has four scales. The types of scales available for this rule are similar to those described for the 6-in rule.

An instrument for taking measurements up to several feet in length is the flexible steel tape shown in Figure 11–3. The **flexible steel tape** is equipped with a hook on one end so that it will hold to a corner or ledge, making it possible for the rule to be used by one individual.

Examples of specialized rules are shown in Figures 11–4 and 11–5. The triangular-shaped **engineering rule** has six scales graduated in tenths, twentieths, thirtieths, fortieths, fiftieths, and sixtieths of an inch. The graduations on the six scales are numbered so that the instrument may be used to scale down a drawing to any desired proportion of an actual measurement. The **shrink rule** shown in Figure 11–5 is used by those making patterns for use in casting. The pat-

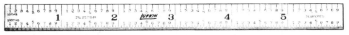

FIGURE 11–1 Six-inch flexible rule.

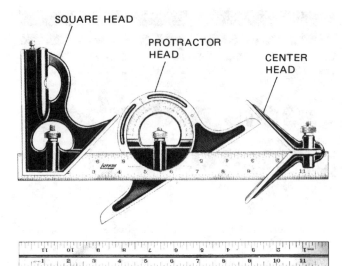

FIGURE 11–2 Rule from a combination-square set.

FIGURE 11–3 Flexible steel tape.

FIGURE 11–4 Engineering rule.

FIGURE 11–5 Shrink rule.

tern for the part must always be made a certain proportion larger than the actual part, depending upon the metal to be cast. This is due to the metal shrinking a small amount as it solidifies in the mold. A shrink rule made to conform to the shrinkage of the metal is used in laying out and measuring the dimensions of the pattern. Many other types of specialized rules exist.

Reading a Rule

The first step in reading a rule is to know the value of the graduations on the rule. A careful study of the rule prior to taking measurements will make it possible for the technician to read the measurements quickly and accurately. Figure 11–6 shows a measurement being taken with a steel rule graduated in thirty-seconds of an inch. Examination of the drawing will show that the measurement being taken is $\frac{23}{32}$ in. Observe that the measurement is taken from the 1-in point on the scale rather than from the end. This method is more

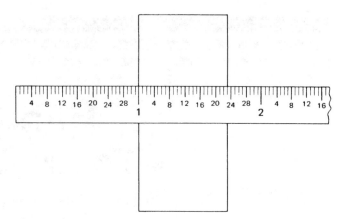

FIGURE 11–6 Reading a measurement.

accurate because it is difficult to align and easy to damage the end of the rule.

The reading of a rule is not difficult; however, the technician must know the units into which the rule is divided and also must know the subdivisions of these units. Figure 11–7 shows four different edges of rules to illustrate given measurements. The rules shown are subdivided into sixteenths of an inch. The rule in Figure 11–7a shows a measurement at the arrowhead of $1\frac{1}{4}$ in. This could also be read $1\frac{2}{8}$ in or $1\frac{4}{16}$ in. In b, the rule shows a measurement of $2\frac{5}{8}$ in at the arrowhead, in c, $\frac{3}{16}$ in, and in d, $1\frac{13}{16}$ in.

A decimal rule is shown in Figure 11–8. The user of this rule should note that the divisions of the inches are in tenths rather than eighths. If a person were not careful, it would be easy to read a measurement as $1\frac{5}{8}$ instead of $1\frac{5}{10}$. Decimal rules are marked with divisions as small as $\frac{1}{100}$ in.

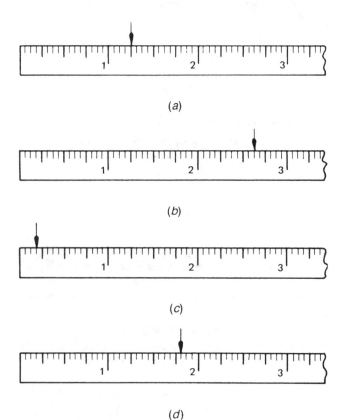

(a)

(b)

(c)

(d)

FIGURE 11–7 Scale readings.

FIGURE 11–8 Decimal rule.

To obtain accurate measurements with a rule, the readings should be taken very carefully, and the rule should be kept in good condition. A rule should not be thrown into a box with other tools because it may become scratched, nicked, or otherwise damaged. It also may become so dirty and worn that it is impossible to read the scales accurately. A steel rule should be wiped clean and dry after each use and from time to time should be wiped with a cloth soaked with a light oil.

Calipers

In some cases it is not possible or convenient to take measurements with a rule, such as when measuring inside and outside diameters or the width of a slot. A caliper can be used for these measurements. A basic caliper has two parts that can be moved in relation to one another, allowing different dimensions to be set. There are many types of calipers. Figure 11–9 shows two **spring calipers** commonly used for inside or outside measurement. The distance between the ends of the legs are adjusted with a screw working against a spring. The calipers shown require the use of a rule or other direct reading device to determine the units of measurement. Calipers can be used as a gage when machining a part to size by first setting the caliper, with the aid of a rule, to the desired dimension and then comparing it to the part as work progresses.

Similar in appearance to the caliper is the **divider** shown in Figure 11–10. The legs of the divider are straight and have sharp points. Dividers are designed to transfer a desired dimension from a scale to a piece of material. One set-

FIGURE 11–10 Standard divider.

ting will allow a line to be divided into a number of segments of equal length. The divider can be used as a compass for drawing arcs and circles for layout work. It can also be used in a manner similar to a caliper to measure thickness and other characteristics. Calipers and dividers are sized by the length of the legs. Common sizes used by technicians are 3 in and 6 in.

Figure 11–11 shows a **slide caliper**. This device has a scale located on a beam with one fixed jaw. A second jaw is free to move up and down the beam. The ends of the jaws are formed so that either inside or outside measurement can be made. The sliding jaw has two reference lines marked *in* and *out*. The distance between the jaws will be the value on the scale opposite the appropriate reference line.

Precision Measurement

The ability of the human eye to accurately read the graduations of a scale is limited. The need for precision measurement as small as ten-thousandths of an inch has necessitated the development of equipment capable of this function. Two tools used to obtain more precise measurements are the micrometer and the vernier scale.

Micrometers. The **micrometer** is a device used to make small measurements. The term comes from *micro* plus *meter*, or one-millionth of a meter. In terms of English units,

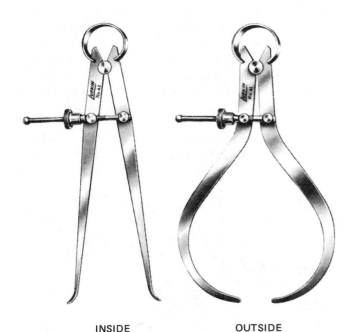

INSIDE   OUTSIDE

FIGURE 11–9 Spring calipers.

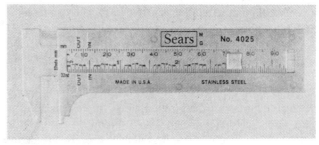

FIGURE 11–11 Slide caliper.

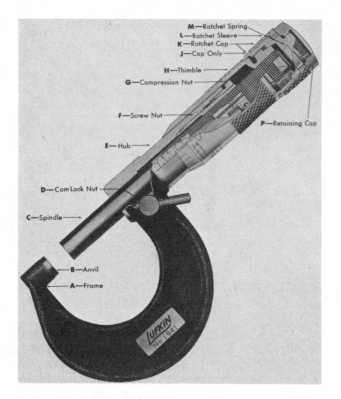

FIGURE 11–12 Standard micrometer caliper.

FIGURE 11–13 Graduations on a micrometer.

the basic micrometer is designed to accurately measure to one-thousandth of an inch. A micrometer caliper is shown in Figure 11–12. The basic micrometer unit is made up of three parts, a **hub**, or barrel, a **spindle**, and a **thimble**. Adding a C-shaped frame and an anvil completes the caliper. The anvil is attached by the frame to the hub. The primary scale is also on the hub. The spindle serves the same function as the sliding jaw on the slide caliper. Although the basic micrometer mechanism is used on a number of different devices, the micrometer caliper is the most common. Micrometers are designed for a range of measurement of 1 in. The size of the frame determines if the micrometer will measure from 0 to 1 in, 1 to 2 in, or another range. The ability of the micrometer to measure to 0.001 in is based on the fact that it has a spindle threaded with forty threads per inch. The spindle screws into internal threads in the hub. The thimble is attached to the spindle and moves with it. By rotating the thimble, the spindle is moved toward or away from the anvil at the rate of $\frac{1}{40}$ in per turn of the thimble. Since $\frac{1}{40}$ in is equal to 0.025 in, each full turn of the thimble will move the spindle 0.025 in either in or out, depending on the direction of rotation. The hub or barrel of the micrometer is marked in divisions of 0.025 in each. The beveled edge of the thimble, which surrounds the barrel, is marked in divisions of one twenty-fifth of a revolution. Therefore, each division on the thimble represents a distance of 0.001 in of spindle travel.

The graduations on the barrel and on the thimble of a micrometer are shown in Figure 11–13. The reading on the micrometer shown is 0.975 in. The figures on the barrel of the micrometer are spaced 0.10 in apart; hence the figure 1 represents a space of 0.10 in, the figure 2 represents 0.20 in, and so on. The micrometer scale in Figure 11–14 is set to read 0.203 in. Observe that the thimble is barely outside the graduation at the figure 2 on the barrel. Then note that the 0

mark on the thimble is three spaces below the 0 line of the barrel. If the 0 on the thimble were aligned with the 0 line on the barrel, the reading would be 0.200 in. However, since the 0 is three graduations below the 0 line, 0.003 in must be added to the reading to make a total of 0.203 in.

In Figure 11–15 the scales for four different micrometer readings are shown. Examine these settings and see if your readings show that a is 0.235, b is 0.036, c is 0.121, and d is 0.762. To develop proficiency in the use of a micrometer, you should obtain one and practice taking readings on a variety of small objects. These readings should be compared for accuracy with similar readings taken by an experienced technician.

The micrometer shown in Figure 11–12 has a part called the cam locknut on the frame and a ratchet at the end of the thimble. The purpose of the cam locknut is to lock the spindle in place after a setting has been made so that there will be no change in the reading when the micrometer is removed from the material. The purpose of the ratchet is to provide a uniform pressure for taking readings. Figure 11–16 shows the micrometer being used to take a reading on a small part; the pressure of the spindle is applied through the ratchet. The ratchet and spindle are turned to the left by means of the ratchet itself, and when the proper pressure is reached, the ratchet will slip, preventing any further pressure from being applied to the spindle, providing a uniform pressure for each reading. Special micrometers are made to take readings which cannot be made with a standard micrometer. Figure 11–17 shows a micrometer designed to measure the distance between a concave and a convex surface, such as the wall thickness of a tube. Figure 11–18 shows a micrometer which has a 60° angle on the spindle and anvil to make it suitable for measuring screw-thread diameters.

An inside-measuring micrometer is shown in Figure 11–19. The **inside micrometer** consists of a basic microme-

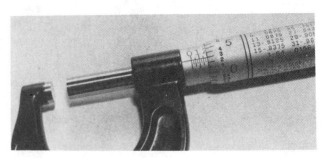

FIGURE 11–14 Reading a micrometer scale.

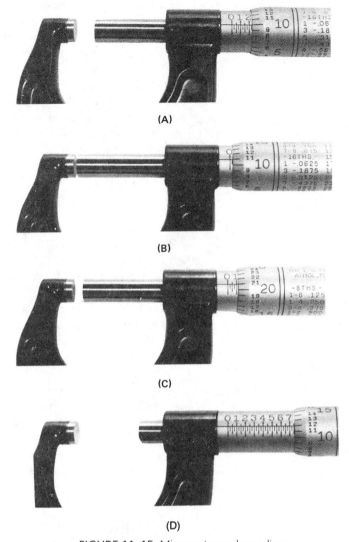

(A)

(B)

(C)

(D)

FIGURE 11–15 Micrometer scale readings.

ter mechanism with extension rods of various lengths to provide a range of measurements. The micrometer unit shown measures 0 to 0.5 in. Adding the extension rods shown will allow it to measure up to 6 in. For example, the addition of a 2-in rod will provide measurements of 2.0 to 2.5 in. Inside micrometers are used to measure inside diameters and widths of slots.

A **micrometer depth gage** can be used to measure the depth of a hole or slot or the range of travel between two parts. As shown in Figure 11–20, the depth gage has a mi-

crometer unit mounted on a base. As the thimble is turned, the spindle will advance or retract into the base. The distance it has extended from the base can be read on the micrometer scale. Depth micrometers have interchangeable spindles of various lengths to provide for measurements greater than 1 in.

Vernier Scales. When it is necessary to make measurements smaller than 0.001 in a technician must use a

FIGURE 11–16 Taking a measurement with a micrometer.

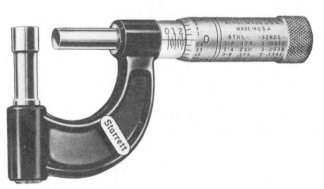

FIGURE 11–17 Special micrometer for measuring tubing-wall thickness.

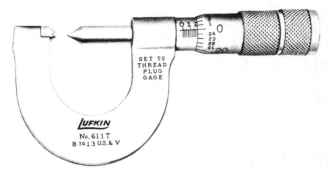

FIGURE 11–18 Micrometer for measuring screw threads.

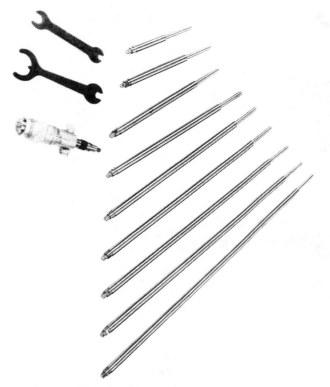

FIGURE 11–19 Inside micrometer set.

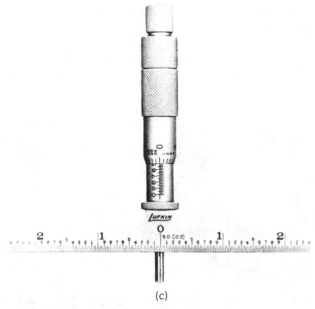

(c)

FIGURE 11–20 Depth-gage micrometer.

micrometer with a vernier scale. A **vernier scale** is an auxiliary scale on a measurement tool that divides the smallest graduation on the main scale into even smaller increments. On the micrometer, a vernier scale will divide each unit of 0.001 into 10 parts, or into units of 0.0001. On other tools, it may divide an increment of 0.025 into 25 parts (0.001). A fractional scale with $\frac{1}{16}$-in graduations may be subdivided into increments of $\frac{1}{128}$ in.

An instrument with a vernier scale will have two scales, the main scale and the vernier scale, located side by side but not connected. The vernier scale will have slightly smaller divisions than the main scale. Looking at Figure 11–21, it can be seen that the scale of the vernier has been divided into 4 parts. It can also be seen that the length of 4 divisions on the vernier scale is equal to 3 divisions on the main scale. The vernier scale will always have one more division than an equal length on the main scale. The number of divisions on the vernier scale determines how much the main-scale increment is subdivided. With 4 vernier divisions, each main-scale division can be divided by 4. To divide a scale increment into 25 parts would require a vernier scale having 25 divisions equal in length to 24 divisions on the main scale.

Figure 11–22 shows a micrometer with a vernier scale. The scale is located on the barrel adjacent to the scale on the thimble. The scale on the barrel has 10 divisions compared to 9 on the thimble. Each thimble division (0.001) can be broken down into units of 0.0001. To read the vernier micrometer, you first read the barrel and thimble, as with a standard micrometer. In Figure 11–22, the reading appears to be between 0.215 and 0.216. The vernier scale will tell exactly how far it is between these two values.

Figure 11–23 illustrates an enlarged view of a vernier scale (*A*) and a thimble scale (*B*) from a micrometer. In looking at Figure 11–23, it can be seen that only one of the marks (0 through 9) on the vernier scale will line up with a mark on the thimble. In the left-hand scales, the 0 mark of the vernier scale is the one that lines up. This would indicate that the reading in this case is an even one-thousandth measurement. In the right-hand scales, it is the 5 mark that lines up, indicating that the measurement is five-tenths of the way between the two one-thousandths marks. Using the reading from Figure 11–22, we add 0.0005 to the 0.215 value, with a final reading of 0.2155.

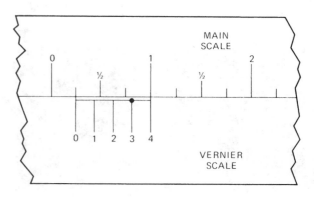

FIGURE 11–21 Vernier scale.

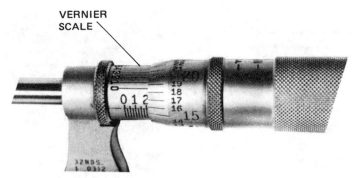

FIGURE 11–22 Micrometer with a vernier scale.

To summarize, in reading a vernier scale, look for the line on the vernier scale that lines up with a line on the main scale. The value of the vernier-scale line will be the amount to add to your base value. Figure 11–24 represents a micrometer barrel scale (*A*), a thimble scale (*B*), and a vernier scale (*C*). Determine what value is shown. Your answer should be 0.1724.

Care of Micrometers. The micrometer is a delicate precision instrument and must therefore be handled with great care. The technician should always use a light touch and avoid excessive clamping pressures. The micrometer should be kept in its own case or wrapped in a soft cloth and put in a special compartment. It should not be dropped or thrown into the tool box. The technician should check the micrometer periodically for accuracy. Better quality micrometers are adjustable and will have instructions for calibration.

Vernier Calipers. A slide caliper with a vernier scale is commonly called a **vernier caliper**. The vernier caliper shown in Figure 11–25 has two separate scales, each with its own vernier. One scale will normally have the capability of measuring to 0.001 in. The second scale may be decimal inches, fractions, or metric. A decimal-scale area is illustrated in Figure 11–26. Most vernier calipers will measure up to

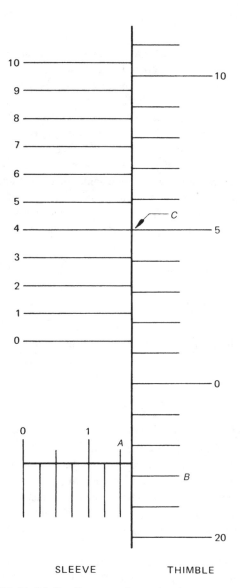

FIGURE 11–23 Principle of the vernier scale.

FIGURE 11–24 Reading a vernier-scale micrometer.

FIGURE 11–25 Vernier caliper.

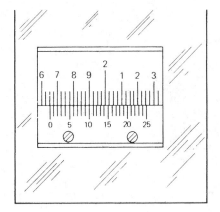

FIGURE 11–26 Vernier-caliper scale.

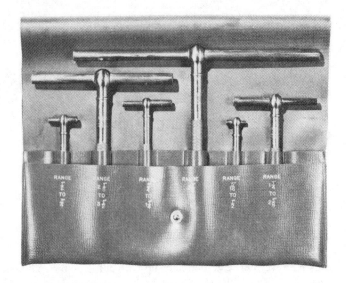

FIGURE 11–27 Telescoping gages.

6 in, although other sizes are available. The graduations on the main scale are similar to those on the barrel of the micrometer. The largest number represents inches. The smaller number is tenths of an inch. Each tenth is divided into four parts, or 0.025 per graduation. The vernier scale has 25 divisions and thus will divide each graduation on the main scale in 0.001. The vernier value is read the same as with the micrometer. The jaws of the vernier caliper are designed so that the scale reading will be the same for both inside and outside measurements. Therefore, only a single reference mark is needed to read the scale. The reference mark is the 0 mark on the vernier scale. The scale in Figure 11–26 has a reading of

main scale	1.600
plus	.050
vernier scale	.004
	1.654

The vernier caliper is a precision tool that should be cared for in the same manner as the micrometer.

Gages

The inside micrometer and the vernier caliper can make inside measurements. However, there are many occasions when they are difficult to use. Various gages have been developed to be used with micrometer calipers that in general provide more convenient and accurate inside measurements.

A set of telescoping gages is shown in Figure 11–27. The telescoping gage is shaped like a T, with one fixed arm and one telescoping arm. The fixed arm is hollow and contains a coiled spring which extends into the telescoping arm. The gage has a screw-stop arrangement by which the telescoping arm may be held at any desired point of extension.

To use a telescoping gage, the gage with the correct size range must be chosen. For example, if a technician wants to measure a hole having a diameter of 1 in, the $\frac{3}{4}$ to $1\frac{1}{4}$ in gage should be selected. The telescoping arm is depressed, and the knurled spindle is turned to the holding position. The gage is then inserted into the hole to be measured, and the telescoping arm is released. When it seems certain that the gage is in the maximum extended position which the hole will allow, the operator tightens the knurled spindle to hold the gage in a fixed position. The gage is then removed carefully and measured with a micrometer. It is best to take more than one measurement for each dimension, because there is a possibility of error if the gage is not exactly in the right position when the telescoping arm of the gage is set.

In using the telescoping gage, the gage *must not be forced* into any position, and the holding mechanism must not be screwed in too tightly. As is true with any precision measuring tool, the telescoping gage must be handled with care at all times.

To accurately measure holes smaller than $\frac{1}{2}$ in, **small-hole gages** are recommended. A set of small-hole gages is shown in Figure 11–28. Each gage consists of a small split ball or half ball with an inner shaft upon which is mounted a cone. When the knurled knob on the end of the gage is turned, the shaft moves axially and changes the position of the cone in the split ball. This, in turn, changes the outer dimension of the split ball. When a technician wants to measure the inside of a small bearing or any other small hole requiring exact dimensions, a gage is selected which will fit easily into the hole when the gage is at its smallest dimension. The gage is inserted into the hole and expanded until it fits the hole firmly but not tightly. It is then removed, and its largest dimension is measured with a micrometer. This procedure will provide an accurate measurement of the hole's diameter.

For certain measurements, manufacturers will sometimes specify that **go and no-go gages** be used. For example, if the dimension of a particular hole is specified as 0.525 ± 0.0005 in, the go gage will have a diameter of 0.5245 in, and

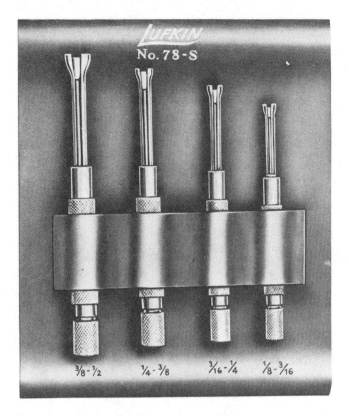

FIGURE 11–28 Small-hole gages.

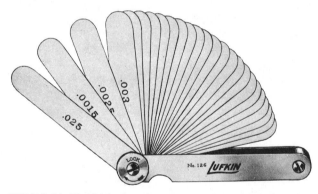

FIGURE 11–29 Thickness gage.

the no-go gage will have a diameter of 0.5256 in. If the go gage will fit into the hole, it is known that the hole is large enough, and if the no-go gage will not fit into the hole, the hole is not too large.

A **thickness gage** is pictured in Figure 11–29. This gage consists of a number of metal leaves ranging from as thin as 0.001 in to as thick as .060 in. The thickness gage is used to determine the dimension of a gap or the clearance between two parts, such as a set of breaker points.

The **radius and fillet gage** shown in Figure 11–30 consists of a number of metal leaves which have been shaped with inside and outside radii of specific sizes. This gage can be used to check the radius of a part being fabricated or to check for stretch or other deformations of a part in use. It can also be used for laying out radii for new parts.

The **screw-pitch gage**, or **thread gage**, is shown in Figure 11–31. It is used to measure the number of threads-per-inch on a threaded fastener.

Angular Measurement

The measurement of angles is involved in many aircraft maintenance operations. Many of these, such as setting propeller angles or checking control-surface travel, require precision equipment beyond the scope of this chapter. The general-purpose measurement of angles can be done with a protractor or a combination square.

A **protractor** is an instrument for drawing or measuring angles. One type of protractor used by many technicians is shown in Figure 11–32. The **combination square**, shown in Figure 11–2, consists of a steel rule used with one of three heads. The **square head** is used to measure, layout, or check 90 and 45° angles. This head can also be positioned on the scale to assist in making a number of repetitive measurements, or it can be used as a marking gage. The **protractor head** provides for measuring or drawing angles. The **center head** can be used to locate the center of round stock.

A number of other types of fixed and adjustable squares are available to the technician. Because many are of a specialized nature, they are not covered in this chapter.

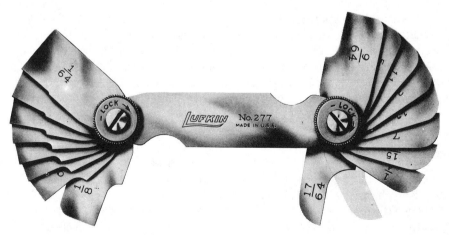

FIGURE 11–30 Radius and fillet gage.

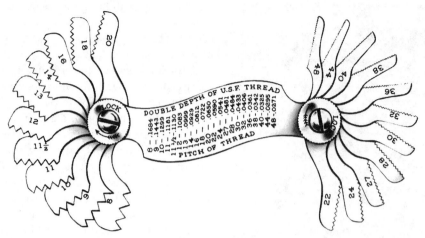

FIGURE 11–31 Screw-pitch gage.

FIGURE 11–32 Protractor.

TABLE 11–1   Wrench Sizes

Wrench Size	Fits*
$\frac{11}{32}$	Nuts for 8-32 screws
$\frac{3}{8}$	Nuts for 10-32 screws
	AN3 bolts
$\frac{7}{16}$	AN4 bolts and all nuts
$\frac{1}{2}$	AN 5 bolts and all nuts
$\frac{9}{16}$	AN6 bolts and all nuts
$\frac{5}{8}$	AN7 bolts and all nuts
	AN8 bolts and AN363 and AN365 nuts
$\frac{1}{2}$	AN310, 315, 316, and 320 -8 nuts

*The bolt and nut sizes shown are for standard AN hardware. Other types of hardware may be different sizes.

TABLE 11-2 Metric Wrench and Socket Sizes, mm

4.0	10	17
5.0	11	18
5.5	12	19
6.0	13	20
7.0	14	21
8.0	15	22
9.0	16	23

WRENCHES

Wrenches are used for the installation and tightening of threaded fasteners. They are made in a variety of shapes to meet the requirements of the fastener, the installation design, or the physical location of the fastener assembly. The technician should always use a wrench designed for the operation to be performed. If the wrong wrench is used, the material may be damaged.

Basic aircraft hardware uses the hexagonal, or six-point, shape for bolt heads and nuts. Hardware is dimensioned by the distance across flats. Wrenches are sized by the hardware they fit and are made in fractional and metric sizes. Table 11–1 lists commonly used fractional wrench sizes. The bolt and nut sizes shown are for standard AN hardware. Other types of hardware may use different sizes of wrenches. Metric-sized wrenches and sockets are shown in Table 11–2.

The wrenches covered in this chapter are designed as hand tools and should be used in that manner. Specially designed wrenches for use with power or impact tools are available. Wrenches may be purchased individually or in sets.

Open-End Wrenches

Open-end wrenches are commonly used for general mechanical work requiring the tightening or loosening of threaded fasteners. The **open-end wrench** is used for both square and hexagonal-headed bolts and nuts. The illustration in Figure 11–33 shows that an open-end wrench will make contact only on two faces of the bolt or nut. For this reason it is essential that the open-end wrench be precisely made to fit the head. Clearance between the wrench and the head may result in the wrench slipping and rounding off the corners of the head. An open-end wrench of good quality will have the proper fit but can become worn or deformed with use. A wrench that does not fit tightly should be rejected by the technician.

The jaws of an open-end wrench are offset, as shown in Figure 11–34. A 15° offset allows the wrench to operate in a space as small as 30° when turning a hexagonal nut. Other degrees of offset are used to allow the wrench to work in tight locations. The open-end wrench is usually double ended, as shown in the figure. This allows two different wrench sizes to be on the same handle.

In using open-end wrenches, it is better to pull the wrench towards you than to push it. There is less likelihood

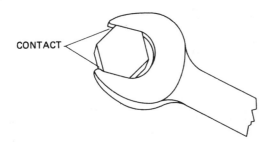

FIGURE 11–33 How open-end wrenches apply force to a nut.

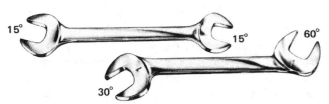

FIGURE 11–34 Open-end wrenches. *(Snap-on Tool Corp.)*

of slipping when the wrench is pulled. Slipping of the wrench will damage the head of the bolt or nut as well as expose the hands of the technician to injury.

Box-End Wrenches

The most effective type of wrench to use in turning a nut or the head of a bolt is the **box-end wrench**, illustrated in Figure 11–35. In the illustration, the 12-point, or double-hex box wrench, is shown. Box-end wrenches are the most effective wrenches because they apply pressure to six points on the nut or bolt head. Figure 11–36 shows how two differ-

FIGURE 11–35 Box-end wrench. *(Proto Tools)*

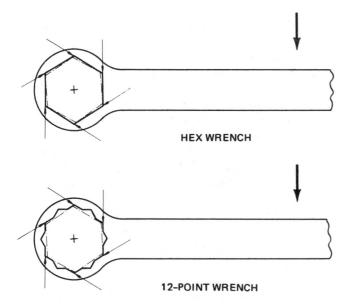

HEX WRENCH

12-POINT WRENCH

FIGURE 11–36 How box-end wrenches apply force to a nut.

ent types of box-end wrenches apply force to the head of a bolt. Note that both wrenches apply force at six points equally spaced around the head. For this reason, the 12-point wrench may be used just as effectively as the 6-point wrench; however, the 12-point wrench may operate through an angle of 30°, whereas the hex wrench requires an angle of 60° through which to operate. This makes the 12-point wrench more useful in close or restricted areas. The 12-point wrench is required by some specialized aircraft hardware using 12-point bolt heads and nuts.

Box-end wrenches are among the most useful tools used by the technician. The box-end wrench should be the preferred wrench to use whenever possible. Like open-end wrenches, most box-end wrenches will have two different size ends. Box-end wrenches are made with a variety of angular offsets between the handle and the end, as shown in Figure 11–37. A ratcheting box-end wrench is shown in Figure 11–38. This wrench makes it possible to tighten a nut or bolt completely without having to remove the wrench.

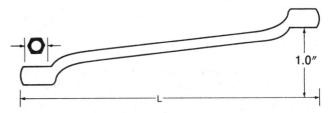

FIGURE 11–37 Box-end wrench offsets.

FIGURE 11–38 Ratcheting box-end wrench. *(Snap-on Tool Corp.)*

Combination Wrenches

The **combination wrench** is manufactured with a box-end wrench on one end and an open-end wrench on the other end. Such a wrench is illustrated in Figure 11–39. These wrenches are quite popular with technicians because they provide the versatility necessary in many mechanical operations.

FIGURE 11–39 Combination wrench. *(Proto Tools)*

Flare-Nut Wrenches

The **flare-nut wrench** (see Figure 11–40) looks like a box-end wrench with one side removed, forming a gap. The gap allows the wrench to be slipped over a tube and placed on a flare nut. The flare-nut wrench provides the advantages of a

Wrenches **251**

FIGURE 11–40 Flare-nut wrench.

FIGURE 11–41 Open-end adjustable wrench.

box-end wrench in tightening a flare nut. Flare nuts are of lower strength than standard hardware nuts and are easily damaged by tightening with an open-end wrench. Flare-nut wrenches are manufactured in both 6- and 12-point designs.

Adjustable Wrenches

An open-end **adjustable wrench** is shown in Figure 11–41. This wrench should be considered as an "emergency" tool and should not be used unless other types of wrenches are not available. Because of the open-end design, the wrench makes contact on only two sides of the bolt head. The adjustment mechanism makes slippage of the wrench, with resulting damage to the bolt or nut, more likely. When using a wrench of this type, care should be taken to see that the jaws tightly fit the nut. Placing the thumb on the adjusting screw and applying pressure in the direction to tighten the jaws will help prevent slippage. The wrench should be turned in a direction so that the maximum stress is applied toward the inner end of the fixed jaw, as shown in Figure 11–42. Adjustable wrenches are manufactured in sizes designated by their overall length. Table 11–3 shows the sizes and the maximum openings of some common adjustable wrenches.

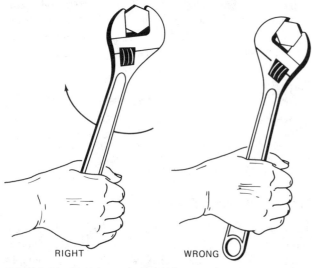

FIGURE 11–42 Using an adjustable wrench.

TABLE 11–3 Adjustable Wrench Sizes

Size, in	Jaw Capacity, in
4	$\frac{1}{2}$
6	$\frac{3}{4}$
8	$\frac{15}{16}$
10	$1\frac{1}{8}$
12	$1\frac{5}{16}$

Allen Wrenches

Set screws and certain aircraft bolts use an internal wrenching head. These fasteners do not have a hex head but rather a hex shape recessed into the shank. Tightening these fasteners requires the use of a hex-shaped tool called an **Allen wrench**. Most of these wrenches are shaped as an *L*, as shown in Figure 11–43, although *T* shapes are also available. Allen wrenches are made in a variety of sizes from 0.028 in to $\frac{3}{4}$ in, as measured across the flats.

FIGURE 11–43 Allen wrenches.   *(Proto Tools)*

Socket Wrenches

One of the tools most useful to the technician is the **socket wrench**. A set of socket wrenches is shown in Figure 11–44. Socket sets are made of a variety of types and sizes of drivers and sockets.

Socket Drivers. Socket wrenches have a square opening which accepts the driving tool. The basic size of a socket set is determined by the size of the drive. Sets are made in $\frac{1}{4}$-, $\frac{3}{8}$-, $\frac{1}{2}$-, and $\frac{3}{4}$-in drives. Adapters are made to allow different sizes of sockets and drivers to be used together. The aircraft technician will make extensive use of the $\frac{1}{4}$- and $\frac{3}{8}$-in drive sets. Metric-sized socket wrenches use standard fractional-sized drivers.

Sockets can be turned, or driven, by a number of tools. These include the **speed handle**, the **breaker bar**, the **sliding T-handle**, and the **ratchet handle**, as shown in Figure 11–45. The speed handle is used with a socket to make the rotation of the bolt or nut more rapid than with other types of handles. The breaker bar and T-handle are used where it is necessary to get more torque on a bolt, such as to "break loose" a tight nut. The ratchet handle is used to rotate the socket wrench in a very restricted area where only a small amount of handle movement is available. The ratchet handle allows the socket to be left on the head of the bolt or on the

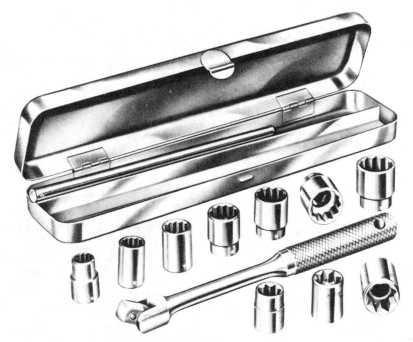

FIGURE 11–44 Set of socket wrenches.

nut until it is completely tightened. Not pictured is a handle best described as a **screwdriver-type handle**. This handle is widely used for socket wrenches in low-torque applications.

When it is necessary to reach a nut that is deeply recessed or in an extremely tight area, an extension may be used. The extension attaches between the driver and the socket wrench. Extensions are available in lengths ranging from 2 to 24 in.

Sockets. Sockets are made in a variety of shapes and sizes for use with a given bolt head or nut size. Figure 11–46 shows a **standard socket** and a **deep socket**. The standard socket and the deep socket are similar in design

and use except for the amount of bolt clearance provided. Figure 11–47 shows the dimensional difference between the two types of sockets. Both types pictured are available as either a 6-point or 12-point socket.

A variety of special tools are made to be used with the socket-wrench drivers. Figure 11–48 shows two examples, which are called **crowfoot wrenches**. Crowfoot wrenches fit on the extension of a ratchet handle and can be used on bolt heads, nuts, and fittings that would be inaccessible to normal socket wrenches. Socket wrenches are also made with Allen-wrench bits for use with internal-wrenching bolts. In addition, all types of screwdriver ends are available as bits for socket sets.

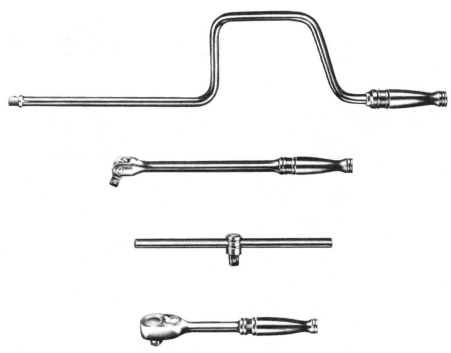

FIGURE 11–45 Socket-wrench drivers.

FIGURE 11–46 Sockets.

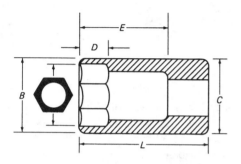

DIMENSION	STD SOCKET	DEEP SOCKET
B	23/32	23/32
C	21/32	21/32
D	5/16	13/32
E	1/2	1 1/16
L	15/16	2 1/8

FIGURE 11–47 Standard- and deep-socket dimensions.

FIGURE 11–48 Crowfoot wrenches.

Special socket wrenches are made for use with electric or pneumatic impact drivers. These are of heavier construction than socket wrenches designed for use with hand tools.

Torque Wrenches

A **torque wrench** includes a measuring device designed to indicate the twisting force being applied to a nut or a bolt in inch-pounds (in-lb), foot-pounds (ft-lb), or newton-meters (N-m). Figure 11–49 illustrates three torque wrenches with

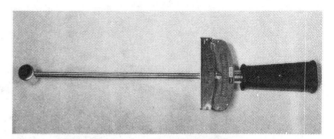

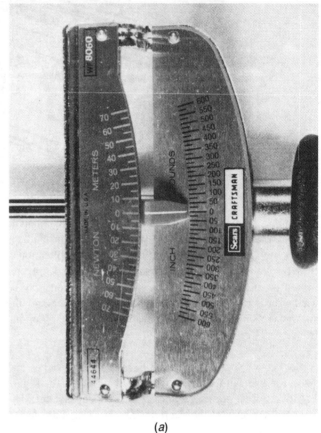

(a)

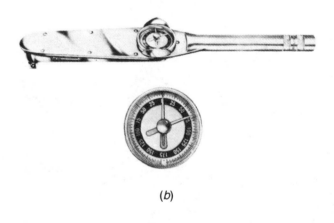

(b)

(c)

FIGURE 11–49 Torque wrenches. *(Sears Roebuck & Co., Snap-on Tool Corp.)*

three different methods for indicating torque. Each of the torque wrenches is designed to be used with a socket wrench. The wrench in Figure 11–49*a* is a **beam-type torque wrench**. It has a pointer which moves across the indicating scale in an amount proportional to the force exerted by the wrench. The movement of the pointer is caused by deflection of the spring-steel handle. A **torsion-type torque wrench** is shown in Figure 11–49*b*. With this wrench the torque is indicated by a small dial gage attached to the wrench. The gage is actuated by a spring mechanism in the handle. An adjustable **toggle-type torque wrench** is shown in Figure 11–49*c*. The operator sets the wrench at the desired torque value with a micrometer-type adjustment on the handle. When the desired torque is applied to the wrench, the wrench clicks and the handle *momentarily* releases. The wrench is capable of applying increased torque once this point has been reached. Technicians using this type of wrench must be sure they recognize the release action. This toggle-type wrench does not have the direct reading capability of the other two torque wrenches.

Torque wrenches are available in a number of ranges, such as 30 to 200 in-lb, 150 to 1000 in-lb, 5 to 75 ft-lb, and 30 to 250 ft-lb. The technician should choose a wrench that has a range of torque values compatible with the work he or she is doing. Torque wrenches are precision instruments and should be handled and stored with care. Torque wrenches must be calibrated on a periodic basis.

Torque wrenches should be used in every case where the force applied to a nut or bolt is critical. This is especially true in the assembly of components in the aircraft engine. Extensions and special attachments are required for use of the torque wrenches in certain applications. These will change the effective arm of the wrench, and a correction value must be calculated to get the correct torque setting. Methods for calculating this correction are explained in the FAA's *Advisory Circular 4313-1A.*

Pipe Wrenches

The **pipe wrench** is used for assembling and disassembling pipes and fittings. A pipe wrench, illustrated in Figure 11–50, has jaws fitted with sharp hardened-steel teeth designed to grip pipes or rods. The wrench is designed so that when the jaws are adjusted to a pipe and the handle is pulled forward, the jaw pressure will increase, causing the teeth to grip the pipe very firmly. The pipe wrench must not be used on any pipe or rod where the finish must be preserved, because the teeth of the wrench will mark the surface of the

FIGURE 11–51 Strap wrench.

object being gripped. Furthermore, a pipe wrench will crush thin-walled tubing. It is unlikely that the technician will ever use a pipe wrench on an aircraft, but she or he will find it useful for other jobs around the shop. When it is necessary to use a pipe wrench on a surface that must not be marred, a **strap wrench**, shown in Figure 11–51, should be used.

Spanner Wrenches

Several different types of **spanner wrenches** are illustrated in Figure 11–52. The first is an **adjustable-hook spanner wrench**. This wrench is designed for use with large nuts having notches cut in the periphery into which the hook may be inserted. With the adjustable hook on the wrench, it is possible to accommodate several different sizes of nuts. The second wrench is called a **pin-hook spanner wrench**. This wrench is designed for use on nuts having holes drilled in the periphery and works in a similar manner as the just-discussed wrench. The third wrench is an **adjustable pin-face spanner wrench**. This wrench is used to turn large nuts which are recessed flush with the part in which they are installed. Holes are drilled on the face of such nuts so that an adjustable pin-face spanner wrench may be used for installing or removing the nut.

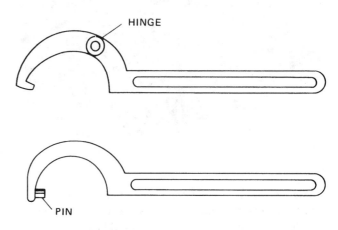
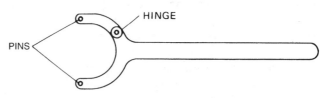

FIGURE 11–52 Spanner wrenches.

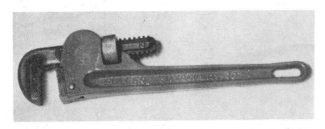

FIGURE 11–50 Pipe wrench.

SCREWDRIVERS

The most frequently used tool for aircraft maintenance is probably the screwdriver. Hundreds of machine screws and sheet-metal screws are used in aircraft to hold cowling, fairings, access panels, and inspection plates in place. Screwdrivers are made in a wide range of sizes and blade design.

Plain Screwdrivers

A **plain screwdriver** is shown in Figure 11–53. When driving a screw, the width of the screwdriver blade should not exceed the diameter of the screw head. If the screwdriver tip is too wide, it will mar the material into which the screw is being driven. A screwdriver on which the tip has been rounded or beveled should not be used to drive a screw because there is a danger that it may slip out of the slot in the screwhead, thus damaging both the screw and the material into which the screw is being driven. If the screwdriver tip is found to be worn, it should be replaced or ground to the shape shown in Figure 11–54.

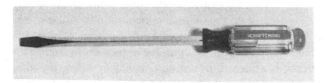

FIGURE 11–53 Plain screwdriver.

FIGURE 11–54 Tip of a correctly ground screwdriver.

Plain screwdrivers are manufactured in many sizes. Size criteria include the width and thickness of the tip and the length of the blade. The length of the blade is measured from the end of the handle to the tip. A large number of different-sized screwdrivers are manufactured. The technician has excellent catalog and reference material available from the manufacturers to provide guidance in the selection of screwdrivers. The screwdriver blade may have a round or a square shank. The square shank permits a wrench to be used (see Figure 11–55) to give more turning force to the screw. A provision for the use of a wrench on some round shanks is made with a hex shape, as shown in Figure 11–56.

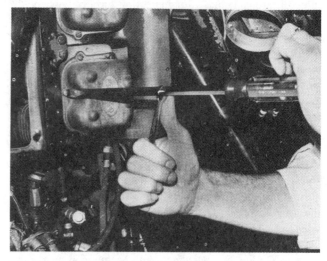

FIGURE 11–55 Using a wrench with a heavy-duty screwdriver.

FIGURE 11–56 Hex shape on a shank for a wrench.

Cross-Point Screwdrivers

Screws are manufactured with a number of designs for the driving tool in addition to the straight slot. A recessed-cross design, known as a **Phillips head**, provides a more positive drive than is possible with a straight-slot screwdriver. The Phillips-head screw has become a virtual standard for aircraft. The Phillips-screwdriver tip is made with four flutes, as shown in Figure 11–57, which fit the recess in the Phillips-head screw. Phillips-screwdriver tips are made in four sizes, with No. 1 being the smallest and No. 4 the largest. It is important when using this screwdriver to have the size which matches the recess in the screw head. Phillips screwdrivers will wear on the ends, as shown in Figure 11–58. If this condition exists, the screwdriver should be replaced.

A different cross-point-head design was used extensively with older aircraft. This design, used in the **Reed & Prince** screwdriver, is shown in Figure 11–59. By comparing the pictures of the Phillips and the Reed & Prince tips, it can be seen that the Reed & Prince has a sharper point. Each screwdriver will only work correctly with the screw head of its design.

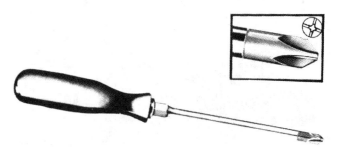

FIGURE 11–57 Phillips (cross-point) screwdriver. *(Snap-on Tool Corp.)*

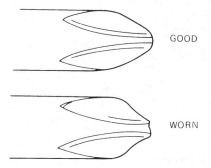

FIGURE 11–58 Phillips screwdriver with a worn tip.

GOOD

WORN

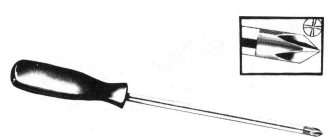

FIGURE 11–59 Reed & Prince screwdriver. *(Snap-on Tool Corp.)*

The majority of the screws used on aircraft use the Phillips design. A number of other screw-head designs have been developed and may be encountered on an aircraft. An example is the clutch-head design shown in Figure 11–60. Some of these designs appear similar to the Phillips head. The technician must make sure that a screwdriver with the proper tip is used.

Special Screwdrivers

To compensate for a variety of tip shapes and sizes, screwdriver sets have been developed. The set consists of a screwdriver-handled driver with hex-shaped bits. In addition to the variety of shapes and sizes available in such sets, the replacement of worn tips is more economical.

The **stubby screwdriver** is designed for use where longer screwdrivers cannot be used conveniently. These screwdrivers have different bit designs, such as standard and Phillips, as shown in Figure 11–61. In many installations, screws are placed in a position where they cannot be

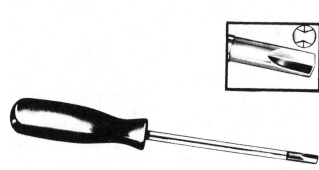

FIGURE 11–60 Clutch-head screwdriver.

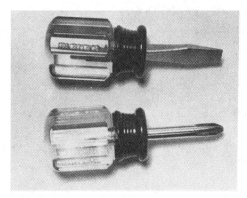

FIGURE 11–61 Stubby screwdrivers.

reached conveniently with a regular screwdriver. In order to get on a screw placed in such a restricted position, it is necessary to use an **offset screwdriver**, such as the ones illustrated in Figure 11–62.

To provide for rapid, semiautomatic rotation of a screw, the **spiral-ratchet screwdriver** (see Figure 11–63) was developed. This type of screwdriver may be adjusted to rotate either to the right or to the left in order to install or remove screws. Pressing down on the handle of the screwdriver causes the spiral grooves in the spindle to rotate the screwdriver blade. This screwdriver is also provided with a locking ring to lock the spindle so the screwdriver can be used in a conventional manner. It is also provided with a ratchet to allow continuous rotation without removing the blade from the screw head.

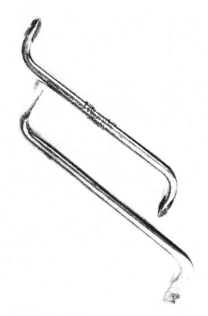

FIGURE 11–62 Offset screwdrivers.

FIGURE 11–63 Spiral-ratchet screwdriver. *(Snap-on Tool Corp.)*

Power Screwdrivers

The use of power screwdrivers has been limited by the cost of electric and pneumatic units and the need for a power source. Lightweight battery-powered power screwdrivers have been developed that are an economical addition to the technician's tool box. The speed that a powered unit provides will make the technician's work more time-efficient. The torque that a power screwdriver can provide should be controllable to prevent damage to screw heads from excessive force.

Use of Screwdrivers

It must be emphasized that screwdrivers are designed only for driving screws and should not be used as pry bars, as chisels, or for any other purpose which may damage the tool.

PLIERS

The technician will need an assortment of pliers to perform aircraft maintenance. As with the other tools covered in this chapter, pliers exist in many different designs for specific functions.

A pair of **combination slip-joint pliers** is shown in Figure 11–64. Combination pliers have jaws capable of gripping flat or round objects. Most combination pliers have light cutting capability. A slip-joint makes it possible to enlarge the jaw width. Combination slip-joint pliers are used for general-purpose holding and gripping. Pliers are sized by their overall length. Combination pliers are made in sizes ranging from $4\frac{1}{2}$ in to almost 10 in.

Another type of pliers, **adjustable slip-joint pliers**, is shown in Figure 11–65. These pliers are often called "water-pump pliers" because an early use of the design was to tighten the packing nut on an automotive water pump. The jaws of these pliers are designed for gripping. The

FIGURE 11–64 Combination slip-joint pliers. *(Snap-on Tool Corp.)*

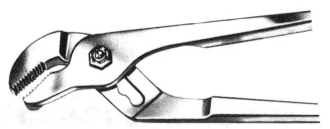

FIGURE 11–65 Adjustable slip-joint pliers. *(Snap-on Tool Corp.)*

FIGURE 11–66 Channel-lock pliers. *(Snap-on Tool Corp.)*

FIGURE 11–67 Adjustable lever-wrench pliers.

shape and length of the handles are made to provide a heavy gripping force on the jaws. A slip-joint allows the jaws to be set to several widths but still stay parallel to each other.

The **channel-lock pliers** shown in Figure 11–66 are similar in appearance and function to the adjustable slip-joint set. The major difference is in the design of the adjustable joint. Curved grooves make up a series of interlocking joints. Considerable force may be exerted by these pliers, with virtually no danger that the adjustment may slip. The adjustable slip-joint and the channel-lock pliers are made in sizes ranging from $4\frac{1}{2}$ to 20 in.

Figure 11–67 illustrates **adjustable lever-wrench pliers**. *VISE-GRIP*, a common name for these pliers, is a registered manufacturer's trade name. These pliers provide a powerful clamping force through the action of a cam and lever mechanism. The amount of jaw opening and clamping force is varied by turning the knurled screw on the handle. The pliers shown have standard jaws. A variety of other jaw styles are made for gripping, cutting, and clamping. Standard lever-wrench pliers are made in sizes from 4 to 10 in. Under normal conditions, adjustable pliers should never be used as a substitute for a wrench.

Another type of pliers, **diagonal-cutting pliers**, are so named because the cutting face is at an angle of approximately 15° to the plane of the handles. Diagonal-cutting pliers are designed exclusively for the cutting of wire, cotter pins, nails, and other comparatively small soft-metal pins. Diagonal-cutting pliers are made in a number of lengths (4 to $7\frac{1}{2}$ in) and cutting capabilities. They should not be abused by cutting hard steel or objects greater than their capacity. A pair of diagonal-cutting pliers is shown in Figure 11–68.

Figure 11–69 illustrates **needle-nose pliers**, which are also known as long-nose pliers. They are designed to reach into restricted areas. Needle-nose pliers are made in overall lengths of 4 to 7 in and jaw lengths of 1 to 3 in. The jaws of

FIGURE 11–68 Diagonal-cutting pliers. *(Snap-on Tool Corp.)*

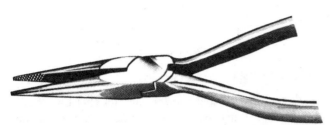

FIGURE 11–69 Needle-nose pliers. *(Snap-on Tool Corp.)*

some needle-nose pliers are bent at an angle, while others may have wire-cutting capability.

The **duckbill pliers** shown in Figure 11–70 are similar in function to needle-nose pliers. The jaws are wide and flat, resembling the bill of a duck. The design of the jaws provides a greater gripping area than that provided by the needle nose. They can be used for many applications where a heavy clamping force is not required. The major use by the aircraft technician is for the safety wiring of fasteners.

The **safety-wire twisters** shown in Figure 11–71 are specially designed pliers for the safety wiring of aircraft fasteners. The jaws of these pliers are designed to both cut and grip the safety wire. The handles have a lock that clamps the jaws on the wire. A helical screw arrangement then provides a twisting action to tighten the wire. The technician must understand the principles of safety wiring to be able to properly use these pliers.

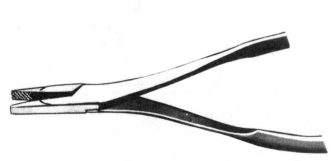

FIGURE 11–70 Duckbill pliers. *(Snap-on Tool Corp.)*

FIGURE 11–71 Safety-wire twisters.

HAMMERS

As a group of tools, hammers are a strong candidate for the title of most misused. While primarily designed as a pounding tool, hammers have a number of designs for different types of pounding as well as different types of material. The use of the wrong hammer can cause damage to the material and personal injury to the user.

One method of classifying hammers is by the shape of the **peen end**. *Peen* may be spelled *pein* or *peen*; in this text we shall use the spelling *peen*. The peen end is the end of the hammer head opposite the face. The aircraft maintenance technician will use the **ball-peen hammer**, shown in Figure 11–72, frequently. The ball-peen hammer is also called a toolmaker's or machinist's hammer because of its all-around usefulness when working with metals. The head is made of a tough steel to permit its use in pounding against other steel parts, such as punches and chisels. The combination of the flat face and the rounded peen make the ball-peen hammer useful for a variety of metal-forming tasks. Ball-peen hammers are sized by the weight of the head. The weights range from 4 oz, for light work, to as much as 3 lb. A common size used by technicians would be 8 oz.

Figure 11–73 shows two other types of metal-working hammers, the **straight peen** and the **cross peen**. These hammers are used where the shape of the ball peen will not reach into a sharp corner or other restricted area.

FIGURE 11–72 Ball-peen hammer. *(Proto Tools)*

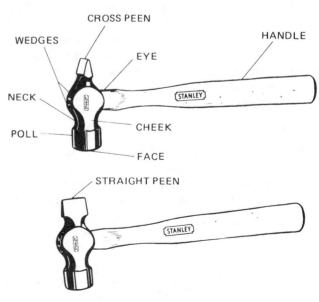

FIGURE 11–73 Straight- and cross-peen hammers. *(Stanley Tool Co.)*

FIGURE 11–74 Soft hammer with replaceable tips.

FIGURE 11–75 Rubber mallet. *(Snap-on Tool Corp.)*

Soft hammers are made with heads or faces of plastic, rawhide, wood, brass, copper, lead, or rubber. Soft hammers are designed to apply moderately heavy blows to various materials, including metals, without causing damage. For many jobs, a soft hammer will serve as well as a steel hammer. Soft hammers are used for forming metal, separating tightly fitted parts, assembling parts where a tight fit is required, removing tight-fit bolts, and a variety of other jobs. The soft hammer may have replaceable tips, as shown in Figure 11–74, or be of construction similar to the rubber mallet in Figure 11–75.

CUTTING TOOLS

Cutting is the process of shearing away or removing portions of a material. The material removed can be a large piece, such as a piece of sheet metal cut into two parts, or it can take the form of a number of small particles or chips, such as those made by a hacksaw. To cut a material, the tool must be harder than the material to be cut. Materials are cut both to size and to shape.

Chisels

The **chisel** is the basic metal-cutting tool. The chisel consists of a steel tool with a hardened point. The shearing force is usually supplied by a hammer. The term *cold chisel* refers to the act of cutting a metal without having to heat it up. Cutting with a chisel is called **chipping**.

Chisels are made with four different shapes of cutting edges, as shown in Figure 11–76. The **cape chisel** is normally used for cutting grooves, such as keyways or slots. The point of the chisel is narrow and has a very short cutting edge. The **diamond-point chisel** is ground to a sharp V-edge. It is used for chipping metal in sharp corners or for cutting V-grooves. The **round-nose chisel** is used for cut-

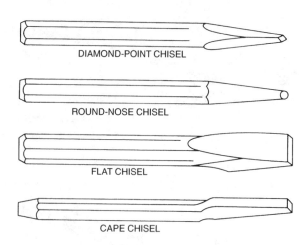

FIGURE 11–76 Types of chisels.

ting rounded or filleted corners. The **flat chisel** has a wide cutting edge. It may be used for cutting sheet metal or comparatively thin bar stock. It may also be used to chip a surface to a desired shape.

The chisel most commonly used by the technician will be the flat chisel. The wedge-type cutting action is shown in Figure 11–77. The cutting angle of a flat chisel should be sharpened to between 50 and 75°, depending upon the type of metal. In general, as the material gets softer, the angle should become sharper (closer to 50°). An angle of 60° is adequate for most work. The head of the chisel is the end opposite the point. The head is not hardened and may become "mushroomed," as shown in Figure 11–78, during use. If mushrooming develops, the head should be ground to eliminate the possibility of injury from flying chips.

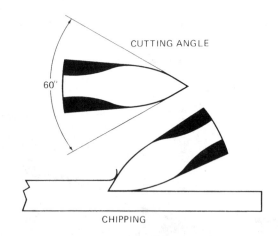

FIGURE 11–77 Chisel cutting action.

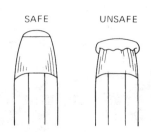

FIGURE 11–78 Mushroomed chisel head.

The aviation technician's use of a chisel will be limited, with removal of rivets a possible use. Flat chisels are sized by the width of the tip. A chisel of $\frac{3}{8}$ in will be adequate for most aircraft maintenance work.

Hacksaws

Hand sawing of metal is done with a **hacksaw**. The hacksaw, as shown in Figure 11–79, consists of the **frame** and the **blade**. The frame is designed to hold the blade. It is adjustable for different-length blades and is designed so that the blade can be installed in any of four positions. The blade can be compared to a number of small chisels, each cutting or chipping the metal. The hacksaw blade varies in length, number of teeth, and type of material.

FIGURE 11–79 A hacksaw and frame.   *(Snap-on Tool Corp.)*

As with all cutting tools, the blade must be harder than the material it is cutting. Blades are made in various degrees of hardness. A harder blade will usually cut longer and straighter than a softer, more flexible blade. The harder blade will also be more brittle and break more easily if it is twisted while cutting. Some blades are made that have a flexible back with hardened cutting edges.

Hacksaw blades are made with different sizes of teeth, varying from 14 to 32 per inch. The number of teeth to use depends upon the thickness of the metal. The teeth must be far enough apart to provide clearance for the chips to get out. If the clearance is inadequate, the blade will clog with chips and stop cutting. When cutting thin material, there should be at least two teeth in contact with the work at all times to prevent the teeth from breaking. The teeth of hacksaw blades have a **set** (see Figure 11–80) that makes the cut, or kerf, slightly wider than the blade. The purpose of the set is to provide clearance for the blade as it passes through the metal. The width of the cut will decrease as the blade wears out during use. A cut made with a worn blade will be narrower than that required for the teeth of a new blade to pass through. Thus it is recommended that a new blade not be used to continue a cut started with a worn blade.

Hacksaw blades are commonly made in lengths of 10 and 12 in. Common sizes of teeth available are 14, 18, 24, and 32 per inch. A blade marked 1032 would be 10 in long and have 32 teeth per inch. A 1218 would have 18 teeth per inch and be 12 in long.

The hacksaw blade should be installed in the frame with the teeth pointing forward. Cutting action takes place on the forward movement of the saw. To prevent excessive wear, pressure on the blade should be lightened during the backstroke.

Hand Shears

Aircraft sheet metal can be cut with hand shears. The most commonly used shears for sheet-metal work are the **aviation snips** shown in Figure 11–81. The snips shown in the figure are a design intended for making straight-line cuts. Aviation snips are made in two other designs, **right hand** and **left hand**, with curved jaws for curved cuts. The terms *right hand* and *left hand* refer to the relative position of the blades. The design type can be identified by holding the snips with the tip of the blades pointing away from your body. The right-hand snips will have the lower jaw on the right side and vice versa for the left. A common practice among manufacturers is to have handles that are green on right-hand snips, red on left-hand snips, and yellow on straight snips.

FIGURE 11–81 Aviation snips.

Making a curved cut with the snips requires that the metal on one side of the snips be distorted to provide for movement of the tool. In general, the smaller the radius of the curve, the more the metal will be distorted. Aviation snips are designed so that the lower jaw will push the metal, on that side, up and out of the way. Figure 11–82 illustrates the use of both types of snips to cut a round circle. Note that the direction of cut depends upon which portion of the metal is to be used and which is scrap. In Figure 11–82*a*, the object is to end up with a round disk, and the direction of cut would be as shown. In Figure 11–82*b*, the object is to end up with a square piece with a round hole, and the direction of cut is the opposite.

Aviation snips are made with serrated teeth on the jaws to provide better shearing action. If the jaws are damaged or become dull, they should be sharpened by a specialist or replaced.

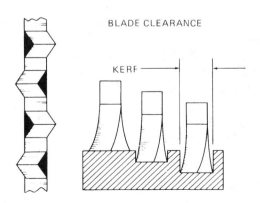

BLADE CLEARANCE

KERF

FIGURE 11–80 Set of a hacksaw blade.

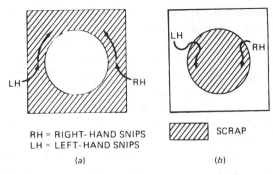

FIGURE 11–82 Directions of cut.

Files

Files are used by technicians to cut and shape metal. Files consist of hardened metal shapes with a number of chisel-like teeth cut into them. While there are hundreds of file types, the aviation technician needs to only be familiar with a few general-purpose types. Files are categorized by size, shape, type of cut, and coarseness of cut. The selection of a file from these criteria will depend upon the material being filed, the nature of the cut to be made, the type of finish desired, and the amount of material to be removed at a given time.

Figure 11–83 shows the names of the parts of a file and how the length is measured. The **tang** of the file is the sharp-pointed end and is designed for a handle. The **heel** of the file is next to the tang and contains the start of the teeth. The **face** of the file is one of the wide, flat sides. The side opposite the face is called the **back**. The **point** of the file is the end opposite the tang. The length, as shown, is from the heel to the point. Files vary in length in 2-in increments from 4 to 20 in. Common sizes are 6, 8, 10, and 12 in.

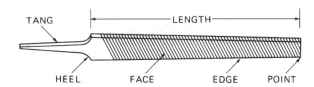

FIGURE 11–83 Parts of a file.

Type and Coarseness of Cut. The type of cut refers to the shape of the teeth on the file. Figure 11–84 shows, from left to right, a single cut, a double cut, a rasp cut, and a curved tooth, or vixen, cut. The **single cut** has a single series of teeth. The **double cut** has a second series of teeth cut at an angle to the first. The first set of teeth, called the **overcut**, is deeper than the second set, called the **upcut**. In general, a double-cut file will remove metal faster but will leave a rougher finish than a single cut.

The **rasp cut** has a series of raised, individual teeth. The rasp makes fast but rough cuts. It is used primarily on soft materials, such as aluminum and lead, where fast removal of material is desired.

The **curved-tooth file** has relatively large spaces between the teeth, which allows the file to be used on soft materials without clogging. Curved-tooth files will produce a very smooth surface on soft materials, such as aluminum.

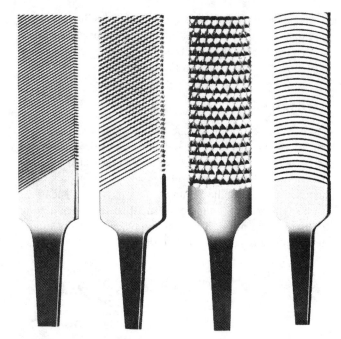

FIGURE 11–84 Types of file-tooth cuts.

The coarseness of cut, for both single- and double-cut files, is related to the spacing between the teeth. Six terms used to describe these spacings are (from most coarse to finest) **rough**, **coarse**, **bastard**, **second cut**, **smooth**, and **dead-smooth**. The spacing for these cuts is not fixed but is dependent upon the length of the file. As the files get shorter, the spacing decreases and the cut becomes finer. For example, a 14-in smooth file may be more coarse than an 8-in bastard file. The aviation technician will have the most need for bastard, second-cut, and smooth files.

File Shape. The shape of a file is its general outline and cross section. Many files are **tapered**, which means that they decrease in width and/or thickness from the heel to the point. Files that do not change in cross section are called **blunt** files. Figure 11–85 shows several shapes of files. The **flat file** (a) has a rectangular cross section that is slightly tapered toward the point in both width and thickness. The flat file has double-cut teeth on both faces and on the edges. A **hand file** is similar to the flat file, with tapered thickness but uniform width. The hand file also has one **safe edge**, which means the edge does not have teeth cut in it. The **mill file** may be tapered or blunt but has single-cut teeth and is used where smooth finishes are desired.

A **triangular file** (b) is also called a **three-square file**. The triangular file is tapered and double cut. It is used to file corners and angles of less than 90°.

A **half-round file** (c) is tapered with one rounded side (the back) and one flat side. The flat side is always double cut. The back side is double cut except for files with a smooth cut. The half-round file is used for concave surfaces. The **round**, (d) or **rat-tail**, **file** is tapered and can be single or double cut. The round file is used to enlarge holes and file curved surfaces.

A **square file** (e) is square in cross section and tapered towards the point. It is principally used for filing slots and keyways. A **pillar file** (f) is rectangular in cross section but

(a)　　(b)　　(c)　　(d)　　(e)　　(f)

FIGURE 11–85 Different shapes of files.

is narrower and thicker than a hand file. It has at least one safe edge, so the file may be used close to a corner without danger of filing on the side. Like the square file, the pillar file is used primarily for keyways and slots.

File Identification. The length of a file and the type of teeth are readily determined by observation. Most of the manufacturers will stamp the file shape and coarseness of cut on the heel of the file.

Use and Care of Files. There are three basic ways in which a file can be put to work: (1) straight filing, (2) draw filing, and (3) lathe filing.

The process of **straight filing** consists of pushing the file lengthwise across the metal and applying sufficient pressure to make the file cut. **Draw filing** consists of grasping the file at each end and pushing or pulling it across the work to produce a fine finish. **Lathe filing** is the application of the file to a piece of metal turning in a lathe.

Work to be filed should usually be held in a vise. The jaws of the vise should be covered with soft sheet metal to prevent damage to the work being filed. The vise should be at approximately elbow height for best results. If the work is of a heavy nature requiring a great amount of filing, the vise should be lower, but if it is to be fine, delicate work, it is better for the vise to be nearer eye level.

Before using a file, a handle should be installed on the tang. A typical handle is turned from wood and has a metal ferrule on the end into which the tang is to be inserted. The metal ferrule prevents the handle from splitting. The handle

is installed simply by inserting the tang into the hole at the end of the handle and tapping the handle to set the tang into place.

When taking a cut with the file, pressure should be applied only on the forward stroke. On the return stroke, the file should be lifted from the work unless the material being filed is soft. In that case the file may be permitted to remain in contact with the material, but no pressure should be applied other than that of the file's weight.

One of the quickest ways to ruin a file is to use too much or too little pressure on the forward stroke. Different materials require different touches; in general, just enough pressure should be applied to keep the file cutting at all times. If allowed to slide over the harder metals, the teeth of the file rapidly become dull, and if they are overloaded by too much pressure, they are likely to chip or clog.

Draw filing is used extensively to produce a perfectly smooth, level surface. A standard-mill bastard file is normally used for draw filing, but where a considerable amount of stock is to be removed, a double-cut flat or hand file will work faster. The double-cut file usually leaves small ridges in the work and consequently does not produce a finished job if a smooth surface is required. In such cases the double-cut file may be used for the "roughing down," and the single-cut mill file is used for finishing.

File life is greatly shortened by improper care as well as by improper use and improper selection. Files should never be thrown into a drawer or tool box containing other tools or objects. They should never be laid on top of each other or stacked together. Such treatment ruins the cutting edges of the teeth and causes nicks in the edges. Files should be kept separate, standing with their tangs in a row of holes or hung in a rack by their handles. They may also be placed in separate grooves made by attaching strips of wood to a baseboard.

It is important to keep the teeth of a file clean of filings and chips, which collect as a result of use. After every few strokes, the end of the file should be lightly tapped on the bench to loosen the metal particles from the teeth. The teeth of the file should be brushed frequently with a file brush or card (see Figure 11–86). To remove stubborn pinnings from the file teeth, a sharp scorer is used. This tool is made of soft iron and is often included with the file card. **Pinnings** are the soft metal particles which clog the teeth of the file and stick in place so they are not easily removed by brushing.

FIGURE 11–86 File card.

Drills

The maintenance technician is often faced with the necessity of boring accurately sized holes in metal parts in order to make attachments and to join parts in an assembly. The tool usually used for boring such holes is the **spiral**, or **twist**, **drill**. This steel drill usually consists of a cylinder into

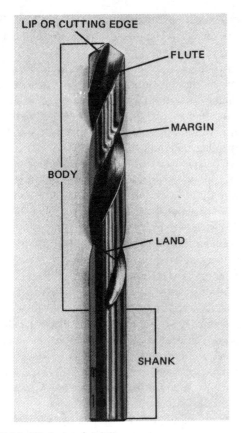

FIGURE 11–87 Parts of a drill.

start with $\frac{1}{64}$ in (0.0156 in) and increase in $\frac{1}{64}$-in increments to 1 in (1.0 in). Table 11–4 presents all the drill sizes and their decimal conversions. A study of Table 11–4 reveals that the three series of drill bits have a total of 138 sizes between No. 80 and $\frac{1}{2}$ in. The only common sizes in the three sets occurs with the letter E and the fractional $\frac{1}{4}$-in drills. Drills are also available in sizes above 1 in. In addition to the three series of drill sizes listed, drills are available in metric sizes.

Drill-Point Angles. In order to perform correctly, the drill must be ground or sharpened to the correct angle. The angles on the drill point will vary depending upon the hardness of the material. Figure 11–88 shows correct angles for three different types of materials. The **drill-point angle** has an effect on the length of the cutting lips. For harder materials, a shorter lip is desirable, and therefore a blunter angle is used. Softer materials can use a sharper point with a longer cutting lip. The general-purpose cutting-lip angle (118°) is

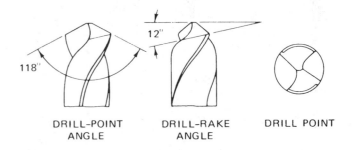

DRILL-POINT ANGLE   DRILL-RAKE ANGLE   DRILL POINT

GENERAL-PURPOSE POINT

which spiral grooves or "flutes" have been cut. One end of the drill is pointed, and the other is shaped to fit a drilling machine.

The principal parts of a drill are illustrated in Figure 11–87. The illustration shows the shank, body, flute, land, margin, and lip, or cutting edge.

The **shank** of a drill is the part designed to fit into the drilling machine. It may be a plain cylinder in shape for use in a drill chuck on a drill motor, drill press, or hand drill. The drill shank may also be tapered for use in drill presses.

The **body** of a drill is the part between the point and shank. It includes the spiral flutes, the lands, and the margin. The body is slightly larger in diameter at the tip than at the shank, thus causing it to bore a hole with clearance to prevent the drill from binding.

The **point** of a drill includes the entire cone-shaped cutting end of the drill. The point includes the **cutting edges**, or **lips**, which are sharpened when the drill is ground.

The **web** is the portion of the drill at the center along the axis. It becomes thicker near the shank. The web may also be defined as the material remaining at the center of the drill after the flutes have been cut. The web forms the dead-center tip at the point of the drill. The **dead center** is in the exact center of the tip and is on the line forming the axis of the drill.

Drill Sizes. Drills are available in three series of sizes. **Number-size drills** are available in sizes from No. 80 (0.0135 in) to No. 1 (0.2280 in). **Letter-size drills** start with A (0.2340 in) and go to Z (0.4130 in). **Fractional-size drills**

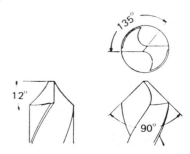

LONG POINT FOR WOOD, BAKELITE, PLASTICS

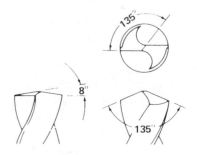

SHORT POINT FOR HARD MATERIAL

FIGURE 11–88 Properly sharpened drill points.

TABLE 11–4 Drill Size Chart

Millimeter	Dec. Equiv.	Fractional	Number	Millimeter	Dec. Equiv.	Fractional	Number	Millimeter	Dec. Equiv.	Fractional	Number	Millimeter	Dec. Equiv.	Fractional	Number	Millimeter	Dec. Equiv.	Fractional
0.10	0.0039			1.75	0.0689			—	0.1570		22	6.80	0.2677			10.72	0.4219	$\frac{27}{64}$
0.15	0.0059			—	0.0700		50	4.00	0.1575			6.90	0.2716			11.00	0.4330	
0.20	0.0079			1.80	0.0709			—	0.1590		21	—	0.2720		I	11.11	0.4375	$\frac{7}{16}$
0.25	0.0098			1.85	0.0728			—	0.1610		20	7.00	0.2756			11.50	0.4528	
0.30	0.0118			—	0.0730		49	4.10	0.1614			—	0.2770		J	11.51	0.4531	$\frac{29}{64}$
—	0.0135		80	1.90	0.0748			4.20	0.1654			7.10	0.2795			11.91	0.4687	$\frac{15}{32}$
0.35	0.0138			—	0.0760		48	—	0.1660		19	7.14	0.2812	$\frac{9}{32}$	—	12.00	0.4724	
—	0.0145		79	1.95	0.0767			4.25	0.1673			7.20	0.2835			12.30	0.4843	$\frac{31}{64}$
0.39	0.0156	$\frac{1}{64}$	—	1.98	0.0781	$\frac{5}{64}$	—	4.30	0.1693			7.25	0.2854			12.50	0.4921	
0.40	0.0157			—	0.0785		47	—	0.1695		18	7.30	0.2874			12.70	0.5000	$\frac{1}{2}$
—	0.0160		78	2.00	0.0787			4.37	0.1719	$\frac{11}{64}$	—	—	0.2900		L	13.00	0.5118	
0.45	0.0177			2.05	0.0807			—	0.1730		17	7.40	0.2913			13.10	0.5156	$\frac{33}{64}$
—	0.0180		77	—	0.0810		46	4.40	0.1732			—	0.2950		M	13.49	0.5312	$\frac{17}{32}$
0.50	0.0197			—	0.0820		45	—	0.1770		16	7.50	0.2953			13.50	0.5315	
—	0.0200		76	2.10	0.0827			4.50	0.1771			7.54	0.2968	$\frac{19}{64}$	—	13.89	0.5469	$\frac{35}{64}$
—	0.0210		75	2.15	0.0846			—	0.1800		15	7.60	0.2992			14.00	0.5512	
0.55	0.0217			—	0.0860		44	4.60	0.1811			—	0.3020		N	14.29	0.5625	$\frac{9}{16}$
—	0.0225		74	2.20	0.0866			—	0.1820		14	7.70	0.3031			14.50	0.5709	
0.6	0.0236			2.25	0.0885			4.70	0.1850		13	7.75	0.3051			14.68	0.5781	$\frac{37}{64}$
—	0.0240		73	—	0.0890		43	4.75	0.1870			7.80	0.3071			15.00	0.5906	
—	0.0250		72	2.30	0.0905			4.76	0.1875	$\frac{3}{16}$	—	7.90	0.3110			15.08	0.5937	$\frac{19}{32}$
0.65	0.0256			2.35	0.0925			4.80	0.1890		12	7.94	0.3125	$\frac{5}{16}$	—	15.48	0.6094	$\frac{39}{64}$
—	0.0260		71	—	0.0935		42	—	0.1910		11	8.00	0.3150			15.50	0.6102	
—	0.0280		70	2.35	0.0937	$\frac{3}{32}$	—	—	0.1935		10	—	0.3160		O	15.88	0.6250	$\frac{5}{8}$
0.7	0.0276			2.40	0.0945			—	0.1960		9	8.10	0.3189			16.00	0.6299	
—	0.0292		69	—	0.0960		41	5.00	0.1968			8.20	0.3228			16.27	0.6406	$\frac{41}{64}$
0.75	0.0295			2.45	0.0964			—	0.1990		8	—	0.3230		P	16.50	0.6496	
—	0.0310		68	—	0.0980		40	5.10	0.2008			8.25	0.3248			16.67	0.6562	$\frac{21}{32}$
0.79	0.0312	$\frac{1}{32}$	—	2.50	0.0984			—	0.2010		7	8.30	0.3268			17.00	0.6693	
0.80	0.0315			—	0.0995		39	5.16	0.2031	$\frac{13}{64}$	—	8.33	0.3281	$\frac{21}{64}$	—	17.06	0.6719	$\frac{43}{64}$
—	0.0320		67	—	0.1015		38	—	0.2040		6	8.40	0.3307			17.46	0.6875	$\frac{11}{16}$
—	0.0330		66	2.60	0.1024			5.20	0.2047			—	0.3320		Q	17.50	0.6890	
0.85	0.0335			—	0.1040		37	—	0.2055		5	8.50	0.3346			17.86	0.7031	$\frac{45}{64}$
—	0.0350		65	2.70	0.1063			5.25	0.2067			8.60	0.3386			18.00	0.7087	
0.90	0.0354			—	0.1065		36	5.30	0.2086			—	0.3390		R	18.26	0.7187	$\frac{23}{32}$
—	0.0360		64	2.75	0.1082			—	0.2090		4	8.70	0.3425			18.50	0.7283	
—	0.0370		63	2.78	0.1094	$\frac{7}{64}$	—	5.40	0.2126			8.73	0.3437	$\frac{11}{32}$	—	18.65	0.7344	$\frac{47}{64}$
0.95	0.0374			—	0.1100		35	—	0.2130		3	8.75	0.3445			19.00	0.7480	
—	0.0380		62	2.80	0.1102			5.50	0.2165			8.80	0.3465			19.05	0.7500	$\frac{3}{4}$
—	0.0390		61	—	0.1110		34	5.56	0.2187	$\frac{7}{32}$	—	—	0.3480		S	19.45	0.7656	$\frac{49}{64}$
1.00	0.0394			—	0.1130		33	5.60	0.2205			8.90	0.3504			19.50	0.7677	
—	0.0400		60	2.90	0.1141			—	0.2210		2	9.00	0.3543			19.84	0.7812	$\frac{25}{32}$
—	0.0410		59	—	0.1160		32	5.70	0.2244			—	0.3580		T	20.00	0.7874	
1.05	0.0413			3.00	0.1181			5.75	0.2263			9.10	0.3583			20.24	0.7969	$\frac{51}{64}$
—	0.0420		58	—	0.1200		31	—	0.2280		1	9.13	0.3594	$\frac{23}{64}$	—	20.50	0.8071	
—	0.0430		57	3.10	0.1220			5.80	0.2283			9.20	0.3622			20.64	0.8125	$\frac{13}{16}$
1.10	0.0433			3.18	0.1250	$\frac{1}{8}$	—	5.90	0.2323			9.25	0.3641			21.00	0.8268	
1.15	0.0452			3.20	0.1260			—	0.2340		A	9.30	0.3661			21.03	0.8281	$\frac{53}{64}$
—	0.0465		56	3.25	0.1279			5.95	0.2344	$\frac{15}{64}$	—	—	0.3680		U	21.43	0.8437	$\frac{27}{32}$
1.19	0.0469	$\frac{3}{64}$	—	—	0.1285		30	6.00	0.2362			9.40	0.3701			21.50	0.8465	
1.20	0.0472			3.30	0.1299			—	0.2380		B	9.50	0.3740			21.83	0.8594	$\frac{55}{64}$
1.25	0.0492			3.40	0.1338			6.10	0.2401			9.53	0.3750	$\frac{3}{8}$	—	22.00	0.8661	
1.30	0.0512			—	0.1360		29	—	0.2420		C	—	0.3770		V	22.23	0.8750	$\frac{7}{8}$
—	0.0520		55	3.50	0.1378			6.20	0.2441			9.60	0.3780			22.50	0.8858	
1.35	0.0531			—	0.1405		28	6.25	0.2460		D	9.70	0.3819			22.62	0.8906	$\frac{57}{64}$
—	0.0550		54	3.57	0.1406	$\frac{9}{64}$	—	6.30	0.2480			9.75	0.3838			23.00	0.9055	
1.40	0.0551			3.60	0.1417			6.35	0.2500	$\frac{1}{4}$	E	9.80	0.3858			23.02	0.9062	$\frac{29}{32}$
1.45	0.0570			—	0.1440		27	6.40	0.2520			—	0.3860		W	23.42	0.9219	$\frac{59}{64}$
1.50	0.0592			3.70	0.1457			6.50	0.2559			9.90	0.3898			23.50	0.9252	
—	0.0595		53	—	0.1470		26	—	0.2570		F	9.92	0.3906	$\frac{25}{64}$	—	23.81	0.9375	$\frac{15}{16}$
1.55	0.0610			3.75	0.1476			6.60	0.2598			10.00	0.3937			24.00	0.9449	
1.59	0.0625	$\frac{1}{16}$	—	—	0.1495		25	—	0.2610		G	—	0.3970		X	24.21	0.9531	$\frac{61}{64}$
1.60	0.0629			3.80	0.1496			6.70	0.2638			—	0.4040		Y	24.50	0.9646	
—	0.0635		52	—	0.1520		24	6.75	0.2657	$\frac{17}{64}$	—	10.32	0.4062	$\frac{13}{32}$	—	24.61	0.9687	$\frac{31}{32}$
1.65	0.0649			3.90	0.1535			6.75	0.2657			—	0.4130		Z	25.00	0.9843	
1.70	0.0669			—	0.1540		23	—	0.2660		H	10.50	0.4134			25.03	0.9844	$\frac{63}{64}$
—	0.0670		51	3.97	0.1562	$\frac{5}{32}$	—									25.40	1.0000	1

also given as 59°, the angle formed by one cutting lip and the centerline of the drill bit.

The **lip-clearance angle** provides clearance between the cutting lip and the rest of the bit. The cutting lips are comparable to chisels. To cut effectively, the area back of the lip must be relieved. Too little clearance results in the drill point rubbing and not penetrating. If the angle is too large, the cutting lip will be weakened. Because of the shape of the bit, lip clearance should be increased slightly as the cutting lip approaches the dead center. If the lip clearance has been correctly ground, the angle formed by the dead center and the cutting lip should be between 120 and 135°.

The angle and lip of each cutting lip must be equal. If they are not equal, the dead center will be displaced from the centerline of the drill. The resulting hole will be over-sized or distorted.

Drill Speeds and Feeds. A properly sharpened drill will not function properly if the correct drill speed and feed are not used. **Drill speed** refers to the speed at which the cutting lips pass through the metal. Cutting speed varies depending upon the metal being drilled and the metal from which the drill is made. Cutting speed is stated in terms of feet per minute (fpm). The speed of drill bits is normally measured in terms of revolutions per minute (rpm). It is necessary therefore to convert the rpm of a drill bit, sized in fractions of an inch, to a lip-cutting speed in fpm. A simplified formula for doing this is

$$CS = \frac{D \times rpm}{4}$$

where

$$CS = \text{cutting speed, fpm}$$
$$D = \text{drill size (diameter), in}$$
$$rpm = \text{speed of drill}$$

The **feed** of a drill bit refers to the advancement of the cutting lips into the work and is given in terms of inches per revolution of the bit. A detailed study of feeds and speeds is beyond the scope of this text. A general rule of thumb is that a slow speed and heavy feed are used for hard materials. Softer materials should have a higher speed but lighter feed. Operating a drill at a speed above its designed speed will usually result in overheating and failure of the drill bit. Table 11–5 gives some cutting speed and feed values for different materials.

Countersinks. A **countersink** is a pointed cutting tool designed to produce a conical-shaped hole in metal or in other materials to fit the head of a rivet or screw. The countersunk hole forms the mouth of a previously drilled straight hole. A countersink is usually provided with a straight shank for use in a hand drill, a drill motor, or a drill press. Two different types of countersinks are shown in Figure 11–89. When a technician wants to drill a hole and a countersink in the same operation, a combination drill and countersink is used. This tool is made in small diameters so the body may also serve as the shank to be inserted in a drill

TABLE 11–5   Speeds and Feeds

Material	Cutting Speed, fpm	Feed per Revolution		
		Less Than $\frac{1}{8}$	$\frac{1}{8}$ to $\frac{1}{4}$	$\frac{1}{4}$ to $\frac{1}{2}$
Aluminum	200–300	0.0015	0.003	0.006
Bronze	50–100	0.0015	0.003	0.006
Cast iron				
Soft	100–150	0.0025	0.005	0.010
Hard	30–100	0.0010	0.002	0.003
Magnesium	200–400	0.0025	0.005	0.010
CRES	20–100	0.0025	0.005	0.010
Steel				
Low carbon	60–100	0.0015	0.003	0.006
Tool	25–50	0.0025	0.005	0.010
Titanium	15–20	0.0015	0.003	0.006

PLAIN COUNTERSINK

COUNTERSINK WITH STOP AND PILOT

FIGURE 11–89 Countersinks.

chuck. Countersinks are made with a variety of point angles. Technicians should make sure that they use the correct countersink for the screw or rivet to be installed.

Counterbores. The **counterbore**, illustrated in Figure 11–90, is a tool designed to bore a second hole that is larger than the first and concentric with it. The pilot on the counterbore fits the first hole drilled and keeps the tool concentric with it. The pilot is interchangeable, so it may be used with different sizes of tools. One of the purposes for counterboring is to make a recess for a bolt head.

FIGURE 11–90 Counterbore.

The counterbore is also used for **spot facing**. Spot facing is the process of cutting a smooth surface around the edge of a hole to provide a square seat for a bolt head. In this case, the counterbore is permitted to cut just deep enough to make the desired seat.

Reamers

A **reamer** is a cutting tool designed to enlarge a hole, to produce an accurately sized hole, or to cut a tapered hole. These tools are made in a wide variety of sizes and styles for a multitude of applications. Figure 11–91 pictures two types of reamers for hand use. The first reamer pictured is a standard straight-fluted reamer, and the second is an adjustable, or expansion, reamer.

ADJUSTING SCREW
FIGURE 11–91 Two types of reamers.

For very accurate work, a fixed reamer of a standard size is employed. Standard-sized reamers may be obtained in increments of $\frac{1}{64}$ in. If other sizes are required, special reamers may be ordered from the manufacturer.

Expansion reamers may be used when a technician wants to size a hole slightly larger than a standard measurement. Expansion reamers are not as accurate as fixed reamers, but for many purposes they are satisfactory.

Reamers must be handled with great care. They should never be placed in a rack or tool box where they will contact other metal tools. Each reamer should be provided with a case of a soft material or should be placed in a separate space provided in a wooden reamer board. The sharp cutting edges are nicked easily and thus may be rendered useless.

Taps and Dies

Taps and dies are used for cutting threads. The **tap** is employed to cut threads inside a drilled hole, and the **die** is used to cut threads on metal-rod stock.

Three common taps are shown in Figure 11–92. These are, in order, a taper tap, a plug tap, and a bottoming tap. The **taper tap** is used more often than the others and is always used to start cutting threads in a hole. If the hole passes through a piece of metal, no other tap is required. The **plug tap** is used when the hole is blind and the user wants to cut threads nearer the bottom than is possible with a taper tap. If it is required to cut threads all the way to the bottom of a hole, the **bottoming tap** is used.

To hold small taps for cutting threads, the T-handle tap wrench is used. This wrench grips the tap firmly and may be

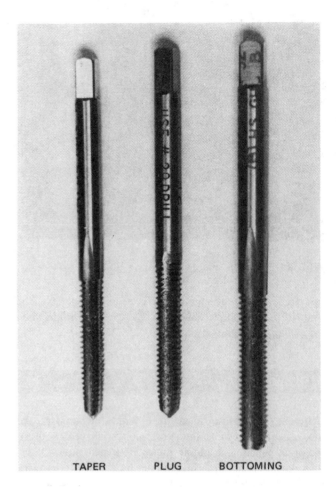

TAPER   PLUG   BOTTOMING

FIGURE 11–92 Taps for cutting internal threads.

turned with one hand. Care must be taken to hold the handle straight so the tap will be in exact line with the hole to be threaded. It is advisable to apply some light cutting oil to the tap when cutting threads in steel or any other hard metal. For large taps, a larger hand-tap wrench, shown in Figure 11–93, is used.

Figure 11–94 shows the two types of dies and a die stock. The die on the left is a solid die. The die on the right is a split die and can have small adjustments made to it. The solid die has no provision for adjustment.

The threading of a metal rod with a die mounted in a die stock is shown in Figure 11–95. The rod to be threaded is held securely in a vise, and a light oil is applied to the end. The starting side of the die is placed over the end of the rod and is turned in a clockwise direction. From time to time the

FIGURE 11–93 Hand-tap wrench.

FIGURE 11–94 Dies and a die stock.

FIGURE 11–95 Threading a metal rod with a die.

direction of the die movement is reversed to clear the threads of metal chips. This same practice is advisable when using a tap to cut threads in a hole.

PUNCHES

Different **punches** are made for a variety of purposes, the most common types being the drift punch, the center punch, the prick punch, and the pin punch. The **drift punch** has a long tapered end with a blunt point. It is normally used for aligning holes for bolts or pins to facilitate installation. The **center punch** has a 90° conical point and is used to make an indentation in metal to mark the location of holes to be drilled and to make the drill start at the correct point. The **prick punch** resembles the center punch in appearance except it has a sharper point. It is used during layout work to precisely mark center locations. Prick-punch marks should be enlarged with a center punch prior to drilling. The **pin punch** has a long, straight cylindrical end and is used to drive and remove various types of pins from shafts or other parts. Pin punches are sized by the diameter of the punch end. Aviation technicians need a minimum set of $\frac{3}{32}$-, $\frac{1}{8}$-, and $\frac{5}{32}$-in pin punches. Punches are illustrated in Figure 11–96.

Scribes

The **scribe**, illustrated in Figure 11–97, is used for marking accurate lines on metal for layout purposes. The point of the scribe is kept very sharp and may be removed from the tool and turned **point in** when not in use. This protects the point and also protects the operator from injury. Scribes should not be used on pieces of metal that will be installed on the aircraft. The scratches may cause stress concentrations and lead to metal failure.

SAFETY EQUIPMENT

The most valuable tool in the tool box is the one that will prevent personal injury to the individual. These pieces of

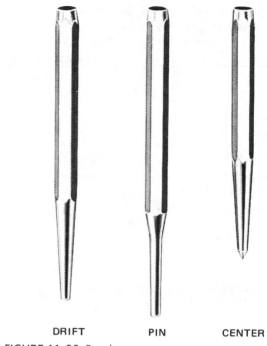

DRIFT   PIN   CENTER

FIGURE 11–96 Punches.

FIGURE 11–97 Scribe.

equipment are also among the most inexpensive items that the tool box can contain. At a bare minimum, the technician should have his or her own personal eye protection, hearing protection, and respirator.

An example of **safety glasses** for eye protection is shown in Figure 11–98. Some personal glasses have hardened, or

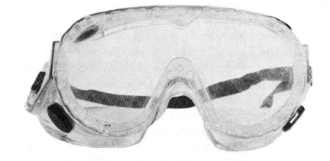

FIGURE 11–98 Safety glasses.

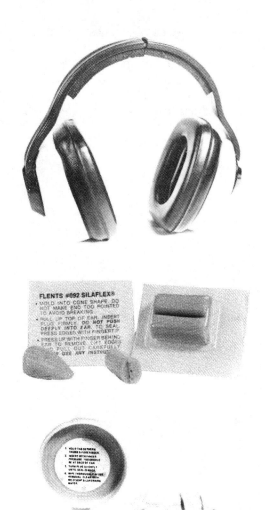

FIGURE 11–99 Ear protectors.

FIGURE 11–100 Respirator.

chemical vapors. The life of the charcoal filter is short when exposed to air. The manufacturer's directions should be followed very closely when using a respirator. The two filters can be used individually or together, as the situation warrants.

safety, lenses. If a technician wants to use personal glasses, it is recommended that clip-on side shields be purchased to prevent injury from the side.

The aircraft technician works in a noisy environment. The sound of turbine engines, rivet guns, air drills, and numerous electric pumps and motors all contribute to excessive noise pollution. Figure 11–99 shows two kinds of hearing protectors. The **ear plugs** are very inexpensive, easy to store, and adequate for most situations. The **ear-muff-style protectors** will usually do a better job and are felt by many to be more comfortable than plugs.

A **respirator** is shown in Figure 11–100. The technician has two areas of air pollution to contend with. The first is dust and other solid particles from cutting and sanding. The second area of concern is chemical pollution from solvents, cleaners, finishes, and resins. Certain substances, such as polyurethane paint, can produce very serious health problems if proper protective measures are not taken. The respirator shown has two separate filtration systems. The first is a **paper**, or **dust**, **filter** that will remove dust particles. The second device contains **activated charcoal** that absorbs

REVIEW QUESTIONS

1. Why is it a good idea to measure from the 1-in point on a scale rather than from the end of the scale?
2. Give a physical description of an outside caliper, an inside caliper, and a set of dividers.
3. Explain the graduations on a micrometer scale.
4. On a micrometer that measures in 0.001 inches, what distance is represented by one turn of the thimble?
5. Explain the use of a vernier scale on a micrometer.
6. A vernier caliper has four divisions on the vernier scale, and the main scale is graduated in $\frac{1}{8}$-in increments. What is the smallest measurement that can be made with this caliper?
7. Explain the use of a telescoping gage.
8. How would you take a measurement with a small-hole gage?
9. Why are box-end wrenches considered the best wrench?
10. Describe the shape of a flare-nut wrench.
11. Describe three types of torque wrenches.
12. Describe three types of spanner wrenches.
13. What precaution must be taken when using cross-point screwdrivers?
14. What are some of the uses of a ball-peen hammer?
15. For what purpose is a chisel used?

16. What is the purpose of *set* in the teeth of a hacksaw blade?
17. What is the difference between a single-cut and a double-cut file?
18. Explain what is meant by *draw filing*.
19. Describe a correctly sharpened drill point.
20. What drill-point angle is best for drilling stainless steel?
21. Which drill in the number series is the largest? the smallest?
22. List three types of taps and the use of each.
23. What is the purpose of a split-thread die?
24. Under what conditions would you use a reamer?
25. What is the difference between a center punch and a prick punch?

Aircraft Fluid Lines and Fittings 12

INTRODUCTION

All aircraft depend upon a number of systems to provide vital functions for operation. Fuel, oxygen, lubrication, hydraulic, instrument, fire extinguishing, air conditioning, heating, and water systems all require fluid lines. The malfunction of these systems due to fluid-line failure seriously jeopardizes the aircraft's safety.

The pressure of the fluids in fluid lines varies from very low, 5 to 10 pounds per square inch (psi) [34 to 69 kPa] for an instrument system, to as much as 5000 psi [34.450 kPa] for a hydraulic system. Low-pressure fluid lines may be made of plastic or rubber hose. High-pressure fluid lines are made in a variety of materials, including aluminum alloy, stainless steel, and reinforced flexible hoses.

Fluid lines are made of rigid, semirigid, or flexible tubes, depending upon the application. A **tube** is defined as a hollow object, long in relation to its cross section, with a uniform wall thickness. The cross section of a tube may be round, hexagonal, octagonal, elliptical, or square. Tubes used for fluid lines are usually round in cross section.

A rigid fluid line would be one that is not normally bent to shape or flared. Directional changes and connections are made by the use of threaded fittings. A **fitting**, in the context of this chapter, is a device used to connect two or more fluid lines. Rigid fluid lines will usually have threads cut into their walls.

Semirigid fluid lines are bent and formed to shape and have a relatively thin wall thickness in comparison to rigid lines. Various types of fittings are used to make connections between semirigid tubes.

Flexible fluid lines are made from rubber or synthetic materials and are usually called **hoses**. Depending upon the pressure they are designed to carry, hoses may have reinforcing material wrapped around them. Various types of fittings are used to attach hoses to each other or to other components.

Fluid-line fittings are made from a variety of materials for many different purposes. New designs, some highly specialized, are being developed on a continuous basis. As in the other chapters of this text, emphasis is placed on basic materials and practices that have widespread use. Specialized technical information on a specific system is available from the product's manufacturer or distributor. As with all aircraft maintenance, the technician has the responsibility of ensuring that the correct materials are used.

TYPES OF FLUID-LINE SYSTEMS

Pipes (Rigid Fluid Lines)

The term **pipe** refers to a rigid fluid line that may be defined as a tube made in standardized combinations of **outside diameter (OD)** and **wall thickness**. The various combinations of diameter and wall thickness are specified as **schedule numbers** and have been standardized by ANSI (American National Standards Institute). Pipe is specified by its nominal diameter ($\frac{1}{4}$ in, $\frac{1}{2}$ in, 1 in, etc.) and the desired schedule number. Table 12–1 shows how the dimensions of a nominal size of pipe will vary in relation to the schedule numbers. The outside diameter of a nominal size is the same for each schedule number, but the wall thickness increases with the schedule numbers. As the schedule number increases, the pressure carried by the pipe can be greater. However, the fluid-carrying capability of the pipe will be reduced due to a smaller inner diameter. A pipe's nominal size cannot be obtained by direct measurement.

Pipes are joined by fittings that use threads cut in the wall of the pipe. To provide a fluid-tight seal, the threads are tapered, unlike machine-screw threads, covered in Chapter 10. As two parts are screwed together, the taper will cause a very tight metal-to-metal seal. Figure 12–1 shows how pipe

TABLE 12–1   ANSI Pipe Schedule Dimensions

Nominal Size, in	Schedule No.	Outside Diameter, in	Wall Thickness, in
$\frac{1}{4}$	40	0.540	0.088
$\frac{1}{4}$	80	0.540	0.119
$\frac{1}{2}$	40	0.840	0.109
$\frac{1}{2}$	80	0.840	0.147
$\frac{1}{2}$	160	0.840	0.187
1	40	1.315	0.133
1	80	1.315	0.179
1	160	1.315	0.250

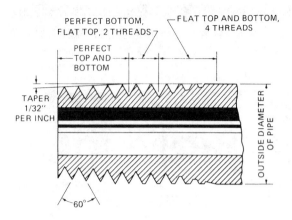

FIGURE 12-1 Tapered pipe threads.

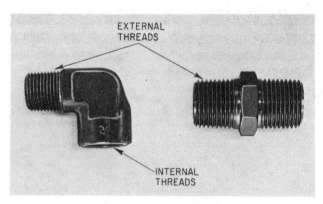

FIGURE 12-2 AN fittings with pipe threads.

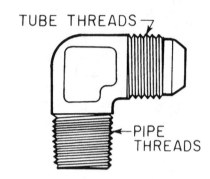

FIGURE 12-3 Tube threads and pipe threads.

threads are tapered to form the metal-to-metal seal. The pipe shown has external threads with the taper occurring on the crests. For internal threads, the taper is reversed and affects the thread root.

Large-scale use of pipe on aircraft is impractical because of weight. However, many aircraft components will use pipe threads. This requires the use of fittings with pipe threads as well as special fittings that connect pipe threads to other types of line connecters. Figures 12-2 and 12-3 show fittings used for these purposes. The aircraft maintenance technician should understand the use of pipe threads and be able to identify them.

Tubes (Semirigid Fluid Lines)

Semirigid fluid lines are usually referred to as tubes or tubing. Tubes can be bent to shape and are often flared for connecters. Tubes used for fluid lines are sized by the outside diameter in inches and the wall thickness. For example, a tube may be $\frac{1}{2}$ in [12.7 mm] OD by 0.035 in [0.89 mm] wall thickness. Tube sizes (OD) increase in $\frac{1}{16}$-in increments. Tube fittings use dash numbers to indicate their size. A -8 fitting is sized to fit an $\frac{8}{16}$- or $\frac{1}{2}$-in tube.

The tube itself may be referred to as a -6 ($\frac{3}{8}$-in) or -8 ($\frac{1}{2}$-in) tube rather than by its fractional size.

Tubes are made from several metals. Table 12-2 shows some of the metals used and their typical applications. Only the materials authorized by the manufacturer, or approved substitutes, may be used on a specific aircraft system.

Flared Tube Connections. Because of the thin wall thickness of semirigid tubes, threads usually cannot be cut in the tube. Special fittings have been designed to allow tubes to be connected to other tubes as well as to aircraft system

TABLE 12-2   Metal Used for Fluid-Line Tubing

Metal	Spec. No.	Application
Aluminum alloy:		
5052	WW-T-787	Used for fuel, oil, instrument, and low-pressure hydraulic lines below 1500 psi where flared fittings are specified.
6061-0	WW-T-789	Used for air ducts where bending is required in diameters up to 6 in. This alloy is weldable.
6061-T6	AMS-4083	Used for hydraulic lines where flareless fittings are specified.
Soft copper	WW-T-799, type N	Used for high-pressure oxygen systems and pressure transmitter lines.
Corrosion-resistant steel (CRES):		
304-$\frac{1}{8}$H	Mil-T-6845	Used for hydraulic lines above 1500 psi and other lines where aluminum alloy is not satisfactory. Temperatures must be below 800° F. This steel is not weldable.
304-1A	Mil-T-8504	Used in place of 304-$\frac{1}{8}$H where greater ductility is necessary and high strength is not required. It is not weldable.
347-1A	WW-T-858	Used in place of 304 steel for welded parts or where operating temperatures are above 800° F.
Titanium	3 AL 2.5V	Used for high-pressure hydraulic systems. Requires swaged fittings.

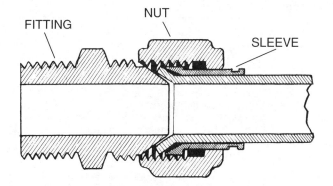

FIGURE 12–4 Position of fitting parts for flared assembly.

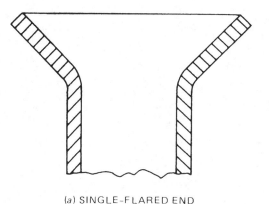

(a) SINGLE-FLARED END

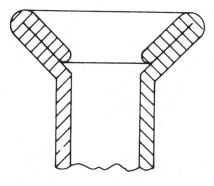

(b) DOUBLE-FLARED END

FIGURE 12–5 Single and double flares.

components. These fittings may be classified as flared, flare-less, swaged, soldered, or brazed. Many of the fittings used for these connections are standard parts and carry AN, AND, or MS specification numbers (see Chapter 10).

Flared fittings require a 37° flare to be formed on the end of the tube. The flare of the tube matches a cone on the fitting. The fitting is threaded with standard machine-screw threads. A special nut and a sleeve are used to pull the flare into contact with the cone and form a fluid-tight metal-to-metal seal (see Figure 12–4). Small-sized or thin-wall tubing may have a double flare, as shown in Figure 12–5. The double flare is used to provide a greater thickness of metal and thus more strength for the seal.

Flared fittings are made to AN or MS standards. Flared fittings made prior to World War II were made to an AC standard. Although AC fittings used a similar flare, they are not 100% compatible with AN fittings. AC fittings may still be found on older aircraft. They can be easily recognized, as shown in Figure 12–6. The AN fitting has a space on each side of the threads, while the AC does not. AC fittings found in a system may be replaced with AN fittings.

The basic components of a flared connection are the AN818 nut, the AN819 sleeve, and one of a number of fittings with a cone to match the tube's flare. Fittings commonly used are classified as tube fittings, universal/bulkhead fittings, and pipe-to-tube (or pipe-to-AN) fittings. Fittings are specified by AN or MS numbers that identify the function of the fitting. Table 12–3 lists some common fittings and their function. Figure 12–7 shows three common tube fittings used to join two or more tubes.

It is often necessary to have fuel, oil, or other tubes pass through structural portions (bulkheads) of an aircraft. This requires a fitting with a long body and provisions for securing the fitting to the bulkhead. Two fittings of this type are shown in Figure 12–8 and are called **universal** and **bulkhead fittings**. The AN924 nut would be used to fasten the fitting to a bulkhead. The term *universal* applies to fittings used to connect a component with internal machine-screw threads to a tube or hose. Machine screws have straight threads and will not provide a fluid-tight seal. A rubber O-ring gasket is required to prevent fluid leakage and is installed with an AN6289 nut, as shown in Figure 12–9. The use of an angle fitting, such as the AN833 in Figure 12–8, allows the fitting to be rotated to any desired orientation

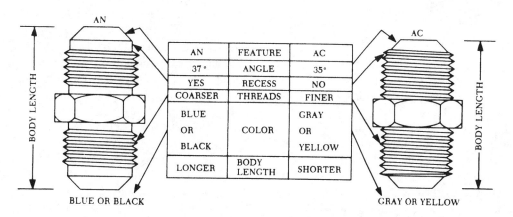

AN	FEATURE	AC
37°	ANGLE	35°
YES	RECESS	NO
COARSER	THREADS	FINER
BLUE OR BLACK	COLOR	GRAY OR YELLOW
LONGER	BODY LENGTH	SHORTER

FIGURE 12–6 AN and AC fittings.

TABLE 12-3 Common Fittings and Their Function

Fitting No.	Name	Function
AN818	Coupling nut	Join flared tube to fitting
AN819	Coupling sleeve	Used with AN818 nut
AN815	Union, flared tube	Join two flared tubes
AN821	Elbow, 90°	Join two tubes at 90° angle
AN824	TEE, flared tube	Join three flared tubes
AN827	Cross, flared tube	Join four flared tubes
AN832	Union, universal or bulkhead	Join two tubes through bulkhead or in boss fitting
AN833	Elbow, 90°, universal or bulkhead	Same as AN821 plus AN832
AN834	TEE, flared tube, universal or bulkhead on tee	Same as AN824 plus AN832
AN816	Nipple, pipe to AN (straight)	Join internal pipe threads to flared tube
AN822	Nipple, pipe to AN (90°)	Join internal pipe threads to flared tube, 90° angle
AN823	Nipple, pipe to AN (45°)	Join internal pipe threads to flared tube, 45° angle

FIGURE 12–7 Typical fittings for aircraft tubing.

FIGURE 12–8 Universal and bulkhead fittings.

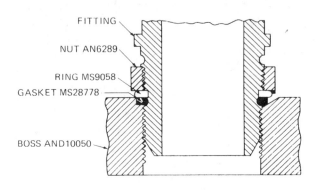

FIGURE 12–9 O-ring seal for boss fitting.

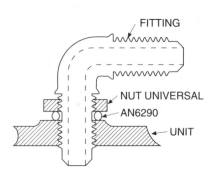

FIGURE 12–10 Fitting can be turned to any angle before nut is tightened.

before tightening. This allows proper tube and component alignment, as illustrated in Figure 12–10.

Another fitting called a **universal fitting** is shown in Figure 12–11. This type fitting is used in installations that require a certain degree of flexibility. They may be used with either stationary or moving units. When used with moving units, such as hydraulic actuating cylinders, they are often called **banjo fittings**. The complete assembly includes a bolt (AN775), with longitudinal and transverse fluid pas-

sages, and the fitting, which is held in place by the bolt. Soft-metal "crush" washers (gaskets) are placed under the head of the bolt and between the fitting and the unit to which it is attached. Serrations on the clamping surfaces provide a satisfactory seal of the joints.

Many aircraft components use internal pipe threads for connections. This requires the use of a special fitting called a **pipe-to-AN nipple**, as shown in Figure 12–12. The fluid-tight seal is provided by pipe threads for the component and a metal-to-metal flare for the tube. Various designs of pipe-to-AN fittings are available, such as 45 and 90° elbows and tees with one leg having pipe threads. The use of pipe-to-AN fittings eliminates the need for rubber seals and nuts. However, directional positioning of the line from the component is not as versatile as with a universal fitting and is primarily a function of tightening the pipe threads.

AN775   AN776   AN778

FIGURE 12–11 Universal fittings.

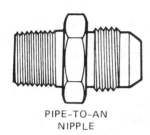

PIPE-TO-AN
NIPPLE

FIGURE 12–12 Pipe-to-AN nipple.

Flared fittings are usually made of aluminum, steel, or stainless steel, with the fitting being the same material as the tube on which it is used. An exception is the use of steel fittings in higher temperature areas, such as in an engine compartment.

Flared-Fittings Designations. Fittings are designated by an AN or MS number, which indicates the function of the fitting. The fitting is sized by the OD of the tube it is used with. The outside diameter of the tube is expressed in sixteenths of an inch. An AN819 fitting for a $\frac{1}{4}$-in tube would have a designation of AN819-4, and for a $\frac{3}{4}$-in tube it would be AN819-12. Aluminum-alloy fittings are indicated by a *D* before the dash. The letter *C* indicates the fitting is made from corrosion-resistant steel (see Figure 12–13). No letter before the dash number indicates that the material is steel. Table 12–4 provides additional information on tube-fitting sizes and dimensions.

Pipe-to-AN nipples are sized by the tube used on the flared end. Common fittings have a specified relationship between the size of the tube and the pipe thread size, as shown in Table 12–5.

Fittings designed to connect or adapt two different sizes of tubes are numbered to show the combinations of sizes. Table 12–6 shows the numbering system used for AN919

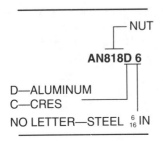

FIGURE 12–13 AN-fitting designation.

adapters. The manufacturer's specifications or distributor's catalog should be consulted to determine the proper size and number of other adapters.

Flareless Tube Connections. High-pressure fluid lines are made of material which is too hard to form satisfactory flares. **Flareless fittings**, made to MS standards, are designed to eliminate the need for a flare on the tube. A sleeve seizes and grooves the tube as the fitting is tightened, producing a fluid-tight seal.

The flareless tube fitting consists of three units: a body, a sleeve, and a nut. The body of the fitting has a counterbored

TABLE 12-4 Thread and Wrench Size for AN Fittings

Fitting Dash no.	Tube Size OD, in	Wrench Size, in	Thread
-3	$\frac{3}{16}$	$\frac{1}{2}$	$\frac{3}{8}$-24
-4	$\frac{1}{4}$	$\frac{9}{16}$	$\frac{7}{16}$-20
-5	$\frac{5}{16}$	$\frac{5}{8}$	$\frac{1}{2}$-20
-6	$\frac{3}{8}$	$\frac{11}{16}$	$\frac{9}{16}$-18
-8	$\frac{1}{2}$	$\frac{7}{8}$	$\frac{3}{4}$-16
-10	$\frac{5}{8}$	1	$\frac{7}{8}$-14
-12	$\frac{3}{4}$	$1\frac{1}{4}$	$1\frac{1}{16}$-12

TABLE 12-5 Pipe-to-AN Nipple Dimensions

AN Dash No.	Tube Size, in	Thread Size, in
-2	$\frac{1}{8}$	$\frac{1}{8}$
-3	$\frac{3}{16}$	$\frac{1}{8}$
-4	$\frac{1}{4}$	$\frac{1}{8}$
-4-4	$\frac{1}{4}$	$\frac{1}{4}$
-5	$\frac{5}{16}$	$\frac{1}{8}$
-6	$\frac{3}{8}$	$\frac{1}{4}$
-6-2	$\frac{3}{8}$	$\frac{1}{8}$
-6-6	$\frac{3}{8}$	$\frac{3}{8}$
-8	$\frac{1}{2}$	$\frac{3}{8}$
-10	$\frac{5}{8}$	$\frac{1}{2}$
-12	$\frac{3}{4}$	$\frac{3}{4}$

TABLE 12-6 Angig Adapters

AN Dash No.	Connects
-0	$\frac{3}{16}$, $\frac{1}{8}$
-1	$\frac{1}{4}$, $\frac{1}{8}$
-2	$\frac{1}{4}$, $\frac{1}{16}$
-3	$\frac{5}{16}$, $\frac{1}{4}$
-4	$\frac{3}{8}$, $\frac{1}{8}$
-5	$\frac{3}{8}$, $\frac{3}{16}$
-6	$\frac{3}{8}$, $\frac{1}{4}$
-7	$\frac{3}{8}$, $\frac{5}{16}$
-8	$\frac{1}{2}$, $\frac{1}{8}$
-9	$\frac{1}{2}$, $\frac{3}{16}$
-10	$\frac{1}{2}$, $\frac{1}{4}$
-11	$\frac{1}{2}$, $\frac{5}{16}$
-12	$\frac{1}{2}$, $\frac{3}{8}$

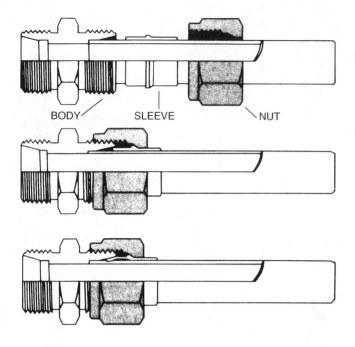

FIGURE 12–14 Flareless tube fitting.

FIGURE 12–15 Cutaway view of a swaged fitting. *(Deutsch Metal Components Div.)*

shoulder against which the end of the tube rests. The counterbore has a cone angle of about 24° which, upon assembly, causes the cutting edge of the sleeve to cut into the outside of the tube. Tightening of the nut forces the sleeve to form the metal-to-metal seal. A cutaway illustration of the MS flareless tube fitting is shown in Figure 12–14. Flareless fittings are easy to identify since there is no flare cone or no space between the threads and the end of the fitting.

The fabrication and installation of flared and flareless fittings are covered in a later section of this chapter.

Swaged Tube Fittings. Many aircraft use swaged fittings to join tubes in areas where routine disconnections are not required. Military specifications require that all tubing be permanently joined, either by swaging or welding, except where it is necessary to make disconnections. Swaged fittings are made by Deutsch Metal Components from aluminum, stainless steel, and titanium and they are used with tubes of the same material. The fittings are attached quickly and easily by means of a hydraulically operated portable swaging tool. A cutaway view of a swaged fitting is shown in Figure 12–15. The advantages of the swaged-type tube fittings are that the original cost is low compared with that of standard AN or MS fittings, the installation takes less time, substantial weight is saved, and repairs can be made on the aircraft without removing complete sections of tubing.

Soldered and Brazed Fittings. Fittings that require soldering or brazing are found on many high-pressure systems. Many applications that have used this process in the past are now using swaged fittings. Brazed or solder fittings require considerably more skill to create and are difficult to make on the aircraft.

Quick-Disconnect Couplings. Quick-disconnect couplings are required at various points in aircraft systems. The purpose of these couplings is to save time in the removal and replacement of components, to prevent the loss of the fluid, and to protect the fluid from contamination. The use of these couplings also reduces the maintenance cost for the system involved. Typical uses are in fuel, oil, hydraulic, and pneumatic systems.

The coupling illustrated in Figure 12–16 consists of a male and a female assembly. Each assembly has a sealing piston (poppet) that prevents the loss of fluid when the coupling is disconnected. Three check points may be used to verify a positive connection. These involve sound, visual observation, and touch. A *click* may be heard at the time the coupling is locked. On some couplers, indicator pins, shown in Figure 12–17, will extend from the outer sleeve upon locking. These pins can be seen and felt by hand.

Hoses (Flexible Fluid Lines)

Because of the need for flexibility in many areas of aerospace vehicle construction, it is often necessary to employ hose instead of semirigid tubing for the transmission of fluids and gases under pressure. A number of standards and specifications, both government and industry, have been developed for the manufacture and application of flexible

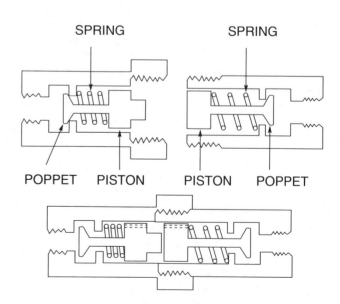

FIGURE 12–16 Quick-disconnect fitting.

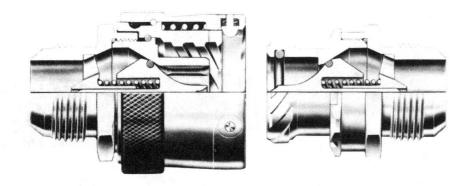

DISCONNECTED—SEALED

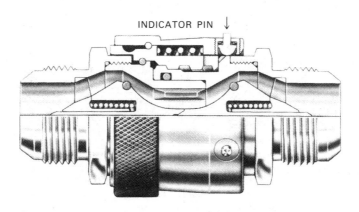

INDICATOR PIN ↓

CONNECTED—OPEN

FIGURE 12–17 Indicator pin in a quick-disconnect fitting. *(Aeroquip Corp.)*

hose. For many years most hoses used on aircraft were made of rubber to one of three MIL-H specifications. These were considered the "universal standard" for aircraft, and the technician was and still is expected to have a good working knowledge of their use. Synthetic materials such as Teflon and several elastomers have proven to be superior in many ways to rubber hoses. The newer products are manufactured under a large number of standards, hose-manufacturers' part numbers, and even aircraft part numbers. Detailed information is available in manufacturers' technical publications. The scope of this text does not permit a detailed study of the many types of hose currently available. Rubber hose is still in use and in fact required on many aircraft. In addition, the technician is still expected to have knowledge of this material. The major emphasis in this section will be on rubber hose manufactured under MIL-H-5593, -8794, and -8788.

Flexible hose for use in aircraft systems is manufactured in four different pressure categories: low-pressure, with a maximum operating pressure of 300 psi [2068 kPa]; medium-pressure, 300 to 1500 psi [2068 to 10 342 kPa]; high-pressure, 1500 to 3000 psi [10 342 to 20 685 kPa]; and extra-high-pressure, 3000 to 6000 psi [20 865 to 41 370 kPa]. It must be noted that the pressure ranges just given are general and that variations will be encountered in specific systems and installations. Hose assemblies, the hose and the installed fittings, are usually pressure tested at a pressure at least twice the maximum operating pressure. The burst pressure is usu-ally required to be at least four times the maximum operating pressure.

Low-Pressure Hoses. The construction of a low-pressure hose is shown in Figure 12–18. This hose conforms to specification MIL-H-5593. The inner tube of the hose consists of synthetic rubber with a braided-cotton reinforcement. The outer cover is synthetic rubber. The hose is identified by a yellow stripe and markings. A linear stripe, called a lay line, is interspersed with the symbol *LP*, the hose manufacturer's code, the hose size, and the quarter/year of manufacture.

Low-pressure hose is used for instrument air or vacuum systems, automatic pilots, and instruments where the maximum pressure for the hose is not exceeded.

Lightweight plastic or rubber tubing has replaced MIL-H-5593 hose in newer installations.

SYNTHETIC RUBBER SYNTHETIC RUBBER

COTTON BRAID

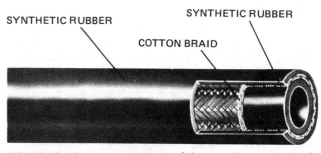

FIGURE 12–18 Low-pressure aircraft hose. *(Aeroquip Corp.)*

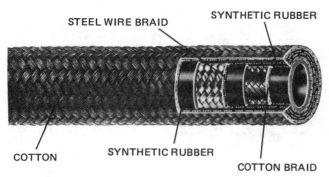

FIGURE 12–19 Medium-pressure hose. *(Aeroquip Corp.)*

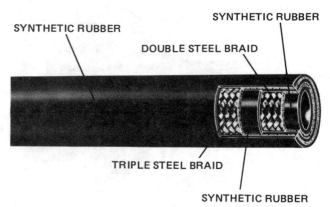

FIGURE 12–20 High-pressure hose. *(Aeroquip Corp.)*

Medium-Pressure Hoses. MIL-H-8794 hose is used for pressures up to 1500 psi [10 342 kPa], and slightly higher in certain sizes. This type of hose, shown in Figure 12–19, has a synthetic rubber tube with one layer of braided cotton and one layer of stainless-steel braid for reinforcement. A rubber-impregnated braided cotton cover makes this type of hose easy to identify. Yellow markings are the same as on the 5593 hose with the exception of the *LP* symbols. MIL-H-8794 hose is approved for aircraft hydraulic (mineral-based), pneumatic, coolant, fuel, and oil systems. There are other hoses that have braided cotton covers that are not compatible with the fluids or pressure that 8794 hose is used for. Only MIL-H-8794-specification hose will have yellow markings.

Many hoses currently being manufactured for this pressure range use inner tubes made of tetrafluoroethylene (TFE, or Teflon) or a variety of new synthetic rubber materials. Teflon hose can be used for practically all fluids that may be encountered on an airplane. Teflon is nonaging, chemically inert, and physically stable, and it can withstand relatively high temperatures. Medium-pressure Teflon hose is produced under specification MIL-H-27267. Most of the newer hoses may be identified by their covers, which are a stainless-steel braid. Identification of these hoses requires that a printed tag from the manufacturer be attached. Technical information on these hoses can be easily obtained from the manufacturer and distributor.

High-Pressure Hoses. The construction of high-pressure MIL-H-8788 hose is shown in Figure 12–20. This hose has a seamless inner tube made of synthetic rubber and is reinforced with high-tensile-strength, carbon-steel wire braid. The smaller sizes will have two layers of wire braid, and the larger sizes (above $\frac{3}{4}$ in [1.905-cm]) have a triple-wire-braid reinforcement. The application of this hose, except for handling higher pressure, is similar to the MIL-H-8794 hose.

Another high-pressure hose complies with specification MIL-H-38360. This hose has a Teflon-type inner tube with reinforcements of stainless-steel wire and braid, the amount of reinforcement depending upon the diameter of the hose. Because of the TFE inner tubing, this hose can be used for most of the fluids employed in aircraft systems.

Extra-High-Pressure Hoses. The MIL-H-8788 hose has proven capable of handling most system pressures in the past. As new, high-performance aircraft are developed, a need has been generated for higher pressures in the operating systems. A hose developed for systems with pressures up to 6000 psi [41 370 kPa] is called extra-high-pressure hose. This type of hose uses spiral stainless-steel wire layers cushioned with a high-temperature elastomer between layers. The synthetic inner tube allows this hose to be used for practically all aircraft fluids, including phosphate esters. The usable temperature range for hose in this category will be up to 400°F [204°C].

Size of Hoses. A hose is sized in accordance with the size of a tube with similar fluid-carrying capabilities. A $\frac{1}{2}$-in hose will carry the same amount of fluid as a $\frac{1}{2}$-in tube. The inner diameter (ID) of the hose inner tube would therefore be approximately the same as the ID of a tube. The hose will use the same dash number to indicate size as that used on a comparable-sized tube. Thus a $\frac{1}{2}$-in hose will be identified as a -8 hose (e.g., MIL-H-8794-8 indicates a $\frac{1}{2}$-in medium-pressure hose).

Identification of Hoses. Care must be taken in the selection of hose for a particular application. Hose is selected on the basis of size, pressure rating, temperature rating, and material. The technician must be very careful that a replacement hose meets the necessary requirements. For rubber hose, the pertinent information must be printed on the hose. The color of the printing is significant, with yellow being used on general-purpose rubber hose. A color other than yellow identifies a hose with different properties. Newer hoses with braided stainless-steel covers should have metal tags with the necessary identification material. If markings are not legible or are missing, that hose should not be installed on an aircraft.

Hose Fittings. Hose-end fittings may be of a permanent, factory-assembled design, or they may be designed to be removed and reused on new hose. Reusable fittings consist of three parts: the socket, the nipple, and the coupling nut. A reusable fitting is pictured in Figure 12–21. Each type of hose (high-, medium-, and low-pressure) will have its own end fittings. These are not interchangeable. Reusable fittings are easily installed in the field using common hand tools. The installation of hose ends is covered in a later section of this chapter.

The permanent-type fitting requires replacement of the entire hose assembly, both the hose and the fittings. A per-

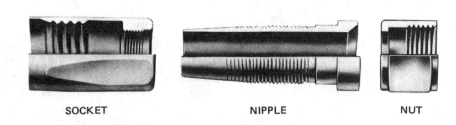

SOCKET   NIPPLE   NUT

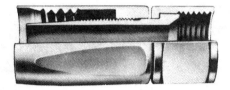

FITTING ASSEMBLED

FIGURE 12–21 Reusable hose-end fitting.

manent fitting for a high-pressure hose is shown in Figure 12–22. During assembly, the socket is swaged circumferentially to compress the hose between the socket wall and the nipple. The serrations inside the socket and on the nipple grip the hose to prevent slippage. The socket is welded to the nipple at the forward projection and is also provided with an interlock that prevents a blow-off under maximum pressure. The skirt of the socket is flared to allow for the hose to bend at the fitting without creating concentrated stress at the end of the socket.

A fitting used for extra-high-pressure hose is shown in Figure 12–23. This fitting is designed with an internal locking and sealing member called a lip seal. The inner tube of the hose is inside the lip seal, with the reinforcing layers outside of the lip seal but inside the socket wall. After the fitting is attached to the hose, the socket is swaged, causing the circumferential serrations to grip both the inner tubing and the reinforcing layers, inside and outside, producing a blow-off-proof attachment.

Hose fittings are made in a variety of configurations, such as straight, 45°, and 90°. They are made to mate with either flared tube fittings (AN and MS types) and flareless fittings (MS types). Typical fitting configurations used for hose assemblies are shown in Figure 12–24. The fittings illustrated are designed to mate with MS33514 flareless fittings. Hose

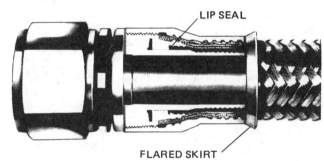

FIGURE 12–23 Fitting for extra-high-pressure hose.
(Resistoflex Corp.)

fittings designed to mate with flared tube fittings incorporate a 37° bevel to match the cone of the fitting.

Protective Sleeves for Hose. In certain areas of an aircraft, it is advisable to protect hoses from heat and wear. For these purposes, protective sleeves of various types have been developed. Figure 12–25 shows protective sleeves manufactured by the Aeroquip Corporation. Fire sleeves are installed on hose in areas where high temperatures exist (e.g., in engine compartments). Abrasion sleeves are used where the hose may rub against parts of the aircraft.

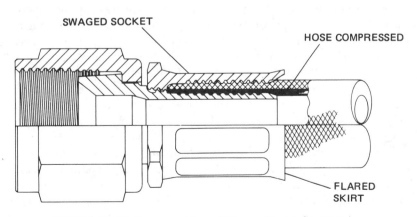

FIGURE 12–22 Permanent hose fitting. (Resistoflex Corp.)

STRAIGHT TO STRAIGHT

000AE 90-SIZE- *L*

STRAIGHT TO 45°

000AE 95-SIZE- *L*

STRAIGHT TO 90°

000AE 96-SIZE- *L*

45° TO 45°

180AE 91-SIZE- *L*

90° TO 45°

180AE 92-SIZE- *L*

90° TO 90°

180AE 93-SIZE- *L*

L = Length from seat to seat

FIGURE 12–24 Typical hose assemblies to mate with flareless fittings.   (Resistoflex Corp.)

FABRICATION, REPAIR, AND INSTALLATION OF FLUID LINES

Most of the plumbing jobs an aircraft technician will be called upon to do will be field and emergency repairs or replacements where specialized fabrication equipment may not be available. It is important that the technician knows basic techniques that can be used without specialized tools or equipment. The techniques covered in this text are basic operations that can be done with tools found at most aviation maintenance facilities.

Preparation of Tubing

When a section of tubing must be replaced, it should be replaced with a tube of the identical material, diameter, and wall thickness. The replacement section should be straight and round.

Silicone Coated Asbestos Fire Sleeve
−65°F. to +450°F.

Heat Shrinkable Polyolefin Abrasion Sleeve
−65°F. to +275°F.

Nylon Spiral Wrap Abrasion Sleeve
−65°F. to +200°F.

FEP100 Teflon Abrasion Sleeve
−65°F. to +400°F.

Neoprene Tubing Abrasion Sleeve
−65°F. to +250°F.

FIGURE 12–25 Protective sleeves for hose.   (Aeroquip Corp.)

The ends of the tube are cut to the correct dimension. It is important to make clean, square cuts at 90° to the center line of the tubing. When the technician wants to cut aluminum tubing or tubing of any comparatively soft metal, a tube cutter similar to that shown in Figure 12–26 should be used. The tube cutter will make a clean, right-angle cut without leaving burrs or crushing the tube. A hardened reamer is often included as part of a cutter. The reamer is used to smooth the inner edge of the cut where the metal has been pressed inward a small amount. If the tube ends are not properly cleaned and smoothed, the flares will not be satisfactory because any nick, cut, or scratch will be enlarged in the flaring operation.

When a section of tubing is to be replaced in an aircraft system, the section being replaced can be used as a pattern. If this is not possible, a piece of welding rod or stiff wire can be used. The wire is bent as required to conform to the

FIGURE 12–26 Hand-operated tube cutter.

TABLE 12–7 Bend Radii and Torque for Tube Installations

Tube OD, in	Torque Range for Tube Nuts, lb/in*		Minimum Bend Radii, in†	
	Aluminum Alloy	Stainless Steel	Aluminum Alloy	Stainless Steel
$\frac{1}{8}$	—	—	$\frac{3}{8}$	
$\frac{3}{16}$	—	30–70	$\frac{7}{16}$	$\frac{21}{32}$
$\frac{1}{4}$	40–65	50–90	$\frac{9}{16}$	$\frac{7}{8}$
$\frac{5}{16}$	60–80	70–120	$\frac{3}{4}$	$1\frac{1}{8}$
$\frac{3}{8}$	75–125	90–150	$\frac{15}{16}$	$1\frac{5}{16}$
$\frac{1}{2}$	150–250	155–250	$1\frac{1}{4}$	$1\frac{3}{4}$
$\frac{5}{8}$	200–350	300–400	$1\frac{1}{2}$	$2\frac{3}{16}$
$\frac{3}{4}$	300–500	430–575	$1\frac{3}{4}$	$2\frac{5}{8}$
1	500–700	550–750	3	$3\frac{1}{4}$
$1\frac{1}{4}$	600–900	—	$3\frac{3}{4}$	$4\frac{3}{8}$

*Pound-inches can be converted into newton-meters by multiplying by a factor of 0.1129.
†Inches can be converted into centimeters by multiplying by a factor of 2.54.

marked every few inches before it is bent, it can also be used to determine the length of the tubing needed for the replacement. While the pattern is being formed, special attention must be paid to clearance around obstructions and to the alignment of the ends at the point where they connect to the rest of the installation.

Short, straight sections of tubing between fixed parts of an aircraft should be avoided because of the danger of excessive stress when the tube expands or contracts with temperature changes. It is general practice to make installations with bends in the tubing to absorb any changes in length.

Hand Bending

The wall thickness and the outside diameter govern the minimum permissible bend radius for tubing, but it is advisable to make the bends as large as the installation will permit. It is also desirable to make all bends of the same radius in any one line. Minimum bend radii for aluminum-alloy and stainless-steel tubing installations for use on aircraft are provided in Table 12–7.

The method for determining the radius of a bend is shown in Figure 12–27. The radius of the bend is measured from the inner surface of the tubing. Tubing correctly bent will maintain a circular shape and present a smooth, uniform appearance without kinks or distortion. Figure 12–28

shows two bends in the upper part of the photo that are acceptable. The two lower bends exhibit excessive flattening and kinking. A small amount of flattening in bends is acceptable, but should not exceed an amount such that the small diameter of the flattened portion is less than 75% of the original outside diameter.

Use of Bending Tools. Soft tubing with less than a $\frac{1}{4}$-in diameter can be bent by hand without a bender. For larger sizes, specialized tools and equipment are needed. Bending tools are divided into two types: **hand benders**, which require a different tool for each tube OD, and **production benders**, which can be used for different tube sizes by changing the attachments. Production benders may be either manually or power operated.

The choice of the particular bender to be used depends upon the size and the material of the tubing to be bent, the kind of benders available, and the number of bends to be made. If only one or two bends are to be made, it is often more economical to use a hand bender rather than take the time to set up a production bender. The use of a typical hand bender is shown in Figure 12–29.

The bending mechanism of a typical production bender is shown in Figure 12–30 on page 284. The bending mechanism consists of a radius block, clamp block, and slide block.

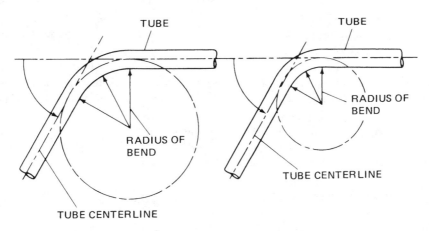

FIGURE 12–27 Radius of tubing bend.

FIGURE 12–28 Good and bad bends.

The clamp block secures the tubing at the point of the bend to the radius block. The clamp block and the slide block are adjusted for position by means of vise mechanisms.

When a section of tubing is to be bent, it is placed in the bender as shown in Figure 12–30. The radius block is selected for the radius desired and the diameter of the tubing to be bent. The clamp block is grooved to fit the tubing and, when the clamp block vise is tightened, holds the tube firmly against the radius block. The slide block vise is also tightened to bear against the tubing. When the handle is turned, the radius block and clamp block rotate, drawing the tubing around the radius block to form the bend. The slide block prevents distortion of the tubing as the bend is formed.

The production bender has a scale to indicate the degree of bend completed. This makes it possible to produce an accurate bend as required for the installation. In the operation of the tube bender, it is important to see that all parts are

clean and not scratched or nicked; that the radius block, the clamp block, and the slide block are of the correct sizes; and that the vises are correctly adjusted. Proper lubrication of tubing and all sliding parts is required.

Flaring

The purpose of a flare on the end of a tube is twofold. First, it provides a flange that is gripped between the sleeve or flare nut and the body of a fitting. This prevents the end of the tube from slipping out of the fitting. Second, the flare acts as a gasket between the sleeve and the cone of the fitting, thus providing a fluid-tight seal. The flare must be nearly perfect because minute cracks or irregularities would permit leakage. The flare must be neither too long nor too short. A flare that is too long will bear against the threads of the fitting and may cause damage to both the flare and the threads. A flare that is too short will not have enough material for a good metal-to-metal contact. A fluid-tight seal is marginal under either condition. A definition of maximum and minimum flare lengths is graphically presented in Figure 12–31.

Before beginning the flare, the sleeve and the nut should be slipped on the tubing, since it may be impossible to install them after the flare is formed.

It is not possible to make a satisfactory flare without the aid of a good tool. Several types of flaring tools are available at reasonable cost, but the technician must make sure that the tool selected will produce a suitable flare without damaging the tubing.

A practical hand flaring tool, shown in Figure 12–32, consists of parallel bars between which are split blocks with holes of various sizes. The blocks are split so that they can be separated for the insertion and removal of tubing. The holes are slightly less than the outside diameter of the tubing so that they will grip the tubing firmly when the clamping screw is tightened. A yoke, which carries the flaring cone, slides over the entire assembly.

To produce a flare with this tool, the clamping screw at the end of the tool is loosened so that tubing can be inserted through the correct-sized hole. About $\frac{1}{4}$ in [0.635 cm] of the tubing is extended above the clamping blocks. The clamping screw is then tightened to hold the tubing in place. Next, the yoke with the 37° flaring cone is slid over the tool and positioned so the cone is directly over the end of the tubing. When the flaring-cone screw is turned, the cone is forced into the end of the tubing until the desired amount of flare is formed.

Aluminum tubing with outside diameters less than $\frac{3}{8}$ in [0.952 cm] must have a double flare. Double flares can usually be made with a slight addition to or alteration of the flaring tools. The use of an adapter, shown in Figure 12–33, will make most of the flaring tools capable of forming double flares. A comparison of single and double flares is shown in Figure 12–5.

Tube Beading

In many low-pressure applications, it is permissible to make a hose-to-tube connection by sliding the hose over the tube

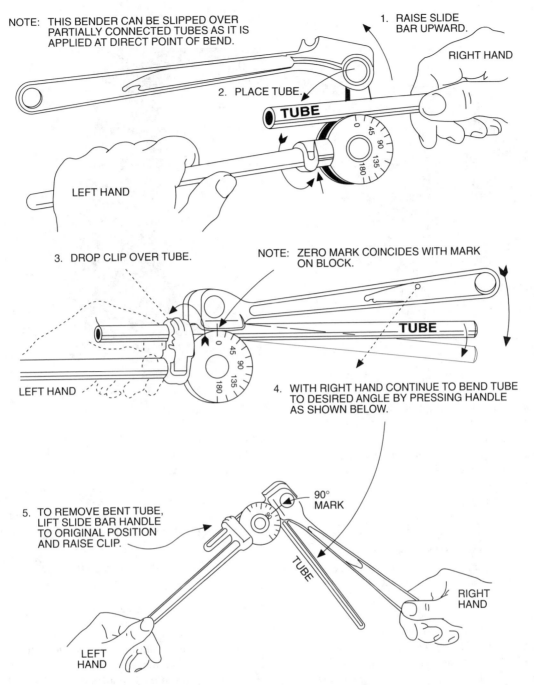

NOTE: THIS BENDER CAN BE SLIPPED OVER PARTIALLY CONNECTED TUBES AS IT IS APPLIED AT DIRECT POINT OF BEND.

1. RAISE SLIDE BAR UPWARD.

RIGHT HAND

2. PLACE TUBE.

TUBE

LEFT HAND

3. DROP CLIP OVER TUBE.

NOTE: ZERO MARK COINCIDES WITH MARK ON BLOCK.

TUBE

LEFT HAND

4. WITH RIGHT HAND CONTINUE TO BEND TUBE TO DESIRED ANGLE BY PRESSING HANDLE AS SHOWN BELOW.

5. TO REMOVE BENT TUBE, LIFT SLIDE BAR HANDLE TO ORIGINAL POSITION AND RAISE CLIP.

90° MARK

TUBE

RIGHT HAND

LEFT HAND

FIGURE 12–29 How to use a hand bender.

end and securing it with a hose clamp. Figure 12–34 illustrates this method of connection. If the tubing diameter is more than $\frac{3}{8}$ in [0.952 cm], a bead must be formed on the tubing.

A manufactured hand beading tool consists of a body with inner and outer beading rollers. An adjusting screw in the body adjusts the position of the two rollers to form the bead. Stop nuts are placed on this screw for limiting the depth of the bead. To make a bead with the tool, the tubing is placed over the inner roller of the correct size for the tube being beaded. The tube end is placed against the face of the beading tool to locate the bead in the correct position. The tool is adjusted so that the outer rollers make contact against the tube. A clamp assembly is attached to the tube to hold it stationary. The clamp assembly is held in the left hand while the beading tool is turned with the right hand. This operation is shown in Figure 12–35. The rollers will be tightened slightly every revolution, forming the bead.

Installation of Flareless Fittings

A flareless fitting consists of a fitting, a sleeve, and a nut, as illustrated in Figure 12–36. When installed, the cutting edge of the sleeve is embedded in the tubing to which it is attached. The recommended method for installing a flareless fitting is to use a **presetting tool** to make the installation of the sleeve on the tube. The presetting tool is held in a vise and duplicates the fitting to which the tubing will be attached. A cross section of such a tool is shown in Figure 12–37. The bottom of the counterbore correctly positions

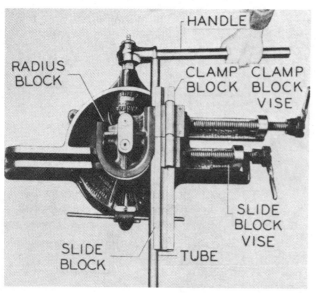

FIGURE 12–30 Tubing in the production bender ready for bending.

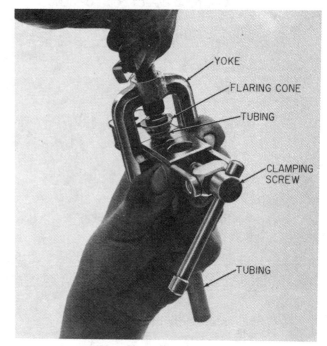

FIGURE 12–32 Hand flaring tool.

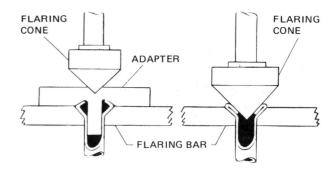

FIGURE 12–33 Making a double flare.

the tubing so the sleeve will be in the right position. The taper in the counterbore engages the pilot lip on the end of the sleeve to force the cutting edge of the sleeve into the tubing. The coupling nut engages the shoulder on the sleeve to hold the tube firmly in place.

The following is a typical procedure for installing or presetting a flareless fitting on a tube:

1. See that the end of the tube is properly cut, deburred, and dressed.

2. Select a presetting tool of the correct size for the tube being used. Mount the presetting tool in a vise.

3. Select the correct size of sleeve and nut. Slide them onto the end of the tube, the nut first with the threads out toward the end of the tube, then the sleeve with the pilot and the cutting edge toward the end of the tube.

4. Select the correct lubricant for the type of system in which the tubing will be installed. For example, if the tube is being put in a hydraulic system, the lubricant should be the hydraulic fluid used for the system. A petroleum-based oil should be used for fuel-system fittings. Lubricate the fitting threads, tool seat, and shoulder sleeve.

5. Insert the tube end into the presetting tool until it is firmly against the bottom of the counterbore. Slowly screw the nut on the tool threads until the tube cannot be turned with the thumb and fingers. At this point, the cutting edge of the sleeve is gripping the tube sufficiently to prevent tube

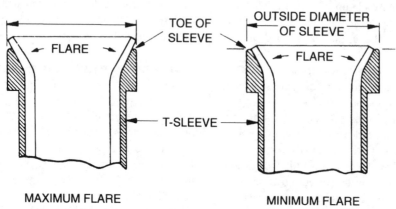

FIGURE 12–31 Maximum and minimum flare lengths.

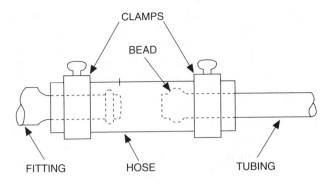

FIGURE 12–34 Low-pressure hose-to-tube connection.

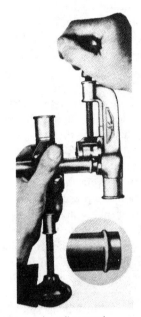

FIGURE 12–35 Using a beading tool.

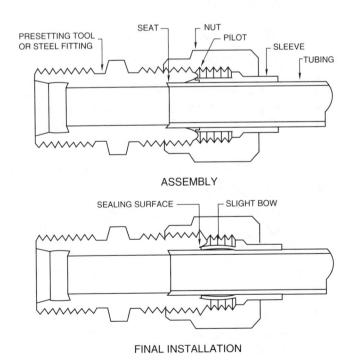

PRESETTING TOOL
OR STEEL FITTING   SEAT   NUT
                            PILOT
                                    SLEEVE
                                    TUBING

ASSEMBLY

SEALING SURFACE   SLIGHT BOW

FINAL INSTALLATION

FIGURE 12–36 Flareless-fitting installation.

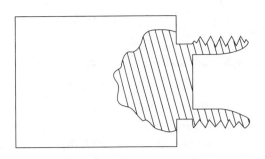

FIGURE 12–37 Cross section of a presetting tool.

rotation, and the fitting is ready for the final tightening necessary to set the sleeve on the tube.

6. Tighten the nut the number of turns specified for the size and material of tubing involved. The number of turns will usually be $\frac{5}{6}$, 1, or $1\frac{5}{6}$. The sleeve is now permanently set, with the cutting edge seated into the outer surface of the tube. Sleeves should not be removed from tubing and reused under any circumstances.

After the sleeve for a flareless fitting has been seated on the tubing, the nut is loosened and the tube removed from the presetting tool. If the section of tubing is not to be installed at once, the fitting should be capped to prevent the entrance of any foreign matter.

All installations should be carefully inspected and tested before being installed. The interior of the tubing should be checked for metal chips, dirt, or other foreign materials.

The inspection procedures for flareless fittings after they have been preset is generally as follows:

1. The cutting edge and the pilot of the sleeve should be checked. The cutting edge should be embedded into the tube's outside surface approximately 0.003 to 0.008 in [0.076 to 0.203 mm], depending upon the size and the material of the tubing. A lip of material will be raised under the pilot. The pilot of the sleeve should be in contact with or very close to the outside surface of the tube. The tube projection from the pilot of the sleeve to the end of the tube should conform to the appropriate specifications.

2. The sleeve should be bowed slightly.

3. The sleeve may rotate on the tube but with a longitudinal movement of not more than $\frac{1}{64}$ in [0.396 mm].

4. The sealing surface of the sleeve which contacts the 24° angle of the fitting seat should be smooth, free from scores, and showing no longitudinal or circumferential cracks.

5. The minimum internal diameter of the tube at the point where the sleeve cut is made should be checked against the specification for the size of tubing used.

The tube assembly should be tested at a pressure equal to twice the intended working pressure.

Plumbing Installation

The proper functioning of the many fluid systems in aircraft is assured by the original design and manufacture of the systems; continued satisfactory operation depends on the proper maintenance, service, and installation of replacement parts.

Installation of Tubing. An important step in the installation of tubing is the proper lubrication of the fittings. While not essential to all fittings, lubrication must be applied to some and is a good practice for others. In the application of a lubricant, it is important that none of the lubricant enter the tubing unless the lubricant is the same material that will be used in the system. Figure 12–38 shows the points of lubrication for typical fittings. The following general rules apply:

- Lubricate nuts and fittings on the outside of the sleeve and on the male threads of the fittings, except for the starting threads.
- Lubricate coupling nuts and fittings on the outside of the flare, and lubricate the female threads, except for the starting threads.

Petroleum-based lubricant may not be used for the fittings of oxygen systems. A special lubricant conforming to AN-C-86 or MIL-T-5542-B may be used.

Several lubricants may be used on hydraulic fittings, including the fluid to be used in the system. Straight threads of brass or steel may be left dry or may be lubricated with the system fluid. If the threads are an aluminum alloy, petrolatum (petroleum jelly) may be used.

For pipe threads, the lubricant must be of a type that is not soluble in the fluid being carried in the system. The lubricant used with a pipe fitting also serves as a seal and fills the space at the roots of the threads. If a petroleum-based lubricant is used on a fuel system carrying gasoline or jet fuel, the lubricant will be dissolved and a leak will develop.

Before tubing assemblies are installed, a final inspection should be made. Flares and sleeves must be concentric and free of cracks. The tubing must not be appreciably dented or scratched. Each assembly must be in initial alignment with the fitting to which it is to be attached. A fitting or an assembly must never be forced into position. A section that must be forced to line up is under initial stress and may fail in operation.

The tubing should be pushed against the fitting snugly and squarely before starting to turn the coupling nut. The tubing should not be drawn up to the fitting by tightening the nut because the flare may be easily sheared off. To make sure that a snug fit is effected, all nuts should be started by hand.

Tubing installed in an airplane must not be used as a footrest or as a ladder, and lamp cords and other weights should not be suspended from it.

The most important of all operations for tubing installation is that of tightening or torquing the nuts. The most common mistake is to overtighten the nuts in order to ensure a leak-free union in a pressure system. Overtightening causes damage to the tubing and fittings and may cause a failure in flight. Correct torque values are given in Table 12–7. The effects of an overtightened connection are illustrated in Figure 12–39.

To obtain correct torque values when tubing sections are installed, it is desirable to use a torque wrench. Special wrenches for fittings are available, and some of these are constructed with snap releases so that they automatically release when the proper torque is attained.

Installation of Hose Fittings

There are a large number of different types and designs of reusable hose fittings for various levels of fluid pressure. The basic principle of the reusable fitting is to clamp the hose between the socket and the nipple with sufficient force to prevent a separation at the maximum pressure for which

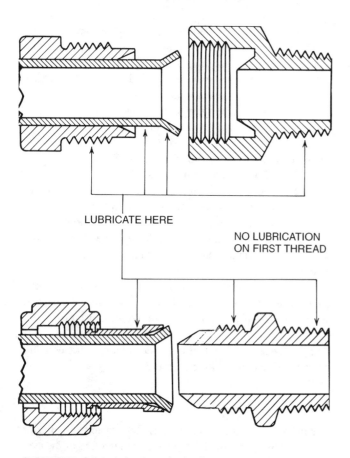

LUBRICATE HERE

NO LUBRICATION ON FIRST THREAD

FIGURE 12–38 Lubrication of tube fittings.

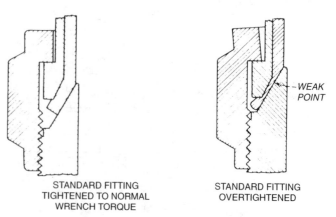

STANDARD FITTING TIGHTENED TO NORMAL WRENCH TORQUE

WEAK POINT

STANDARD FITTING OVERTIGHTENED

FIGURE 12–39 Fittings properly and improperly tightened.

the hose is designed. In this section, we shall describe the installation procedures for a medium-pressure and a high-pressure hose end. The manufacturer's instructions or FAA *Advisory Circular 43.13-1A* should be consulted for an actual installation.

A reusable fitting for a medium-pressure hose, MIL-H-8794, is shown in Figure 12–21. This fitting consists of a coupling nut, a nipple, and a matching socket. The fitting may be made of cadmium-plated steel or anodized aluminum alloy. The fitting is installed on a hose as shown in Figure 12–40. The procedure is as follows:

1. Place the hose in a suitable holding fixture and cut squarely with a fine-toothed hacksaw or a hose cutter.
2. Place the socket in a vise and screw the hose into it by turning the hose counterclockwise. Turn the hose until it bottoms in the socket.
3. Use an installation mandrel or a tube fitting to temporarily "lock" the coupling nut and the nipple together.
4. Lubricate the nipple threads and the inside of the hose with an appropriate lubricant.
5. Screw the nipple into the socket using a wrench on the nut to turn the nipple. Leave a clearance of 0.005 to 0.031 in [0.127 to 0.79 mm] between the coupling nut and the socket. This clearance is to allow freedom of movement for the coupling nut.
6. Remove the mandrel or the fitting from the coupling nut.

An end fitting for a high-pressure rubber hose, MIL-H-8788, is shown in Figure 12–41. Several differences can be seen between this fitting and the one in Figure 12–21. To withstand the higher pressures, this fitting has a longer hex-shaped socket with notches on the edges of the flats. The nipple has a permanently attached hex-shaped surface for a wrench. The coupling nut cannot be removed from the nipple. Figure 12–42 is from FAA *Advisory Circular 43.13-1A* and shows instructions for installing an end on a high-pressure hose.

Note that for a high-pressure hose, the cover must be removed from a length of the hose equal to the distance from the end of the socket to the notches.

The preceding instructions apply to only two of the many hose and end-fitting combinations. While some end fittings will have more complex looking components, the installation procedures are very similar to those already discussed.

After completing the installation of the end fittings, the hose assembly should be tested to the required pressure. If the assembly is not being immediately installed on an aircraft, the ends should be capped to prevent contamination.

Maintenance Practices for Aircraft Hose

Hose for aircraft fluid systems requires reasonable care and an understanding of the conditions that can cause damage, deterioration, or malfunction. The primary purpose of the

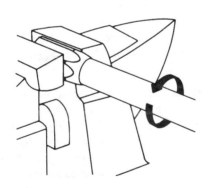

INSERTING HOSE
IN SOCKET

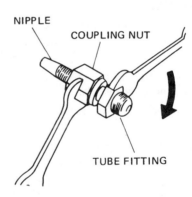

ASSEMBLING NIPPLE
AND NUT WITH TUBE FITTING

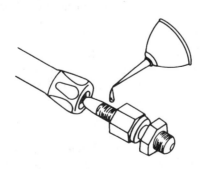

LUBRICATING NIPPLE
THREADS

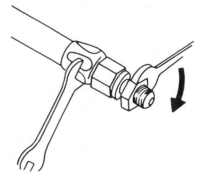

SCREWING THE NIPPLE
INTO THE HOSE AND SOCKET

FIGURE 12–40 Installation of a reusable fitting on a hose. *(Aeroquip Corp.)*

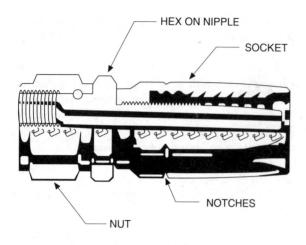

FIGURE 12–41 High-pressure reusable hose-end fitting.

hose is to carry a fluid at a required pressure and flow rate to serve the functions of the system involved. The following practices are recommended for the care of hose:

- Do not use hose assemblies as footholds or handholds.
- Do not lay hose where it may be stepped upon or run over by a vehicle.
- Do not lay objects on top of hose assemblies.
- When loosening or tightening hose fittings, turn the swivel nut only. Do not turn the hexes that form part of the socket or nipple assembly. Hold the socket with a wrench to prevent it from turning.
- Hold the fitting to which a hose assembly is to be connected to prevent it from turning. Use an end wrench of the correct size.
- Cover open ends of hose assemblies with caps or plugs until the assemblies are to be installed.
- Check the hose and the fittings for cleanliness, inside and out, before installation.

When inspecting hose in aircraft systems, the principal conditions to check for are leaks, wear or damage to the outer surface, broken wire strands in the metal braid, corrosion of the metal braid, evidence of overheating, bulges, twists in the hose alignment, damage or wear of the chafe guards, damage or wear of the fire sleeves, damage to the end fittings, separation of the plies, blisters, cracks in the outer cover, and any other indication of damage or deterioration. Any appreciable defect in the condition of the hose or the fittings is usually reason for replacement. A leak may be caused by a loose fitting. This may be corrected by loosening and inspecting the fitting; if there is no sign of damage to the fitting, tighten it to the proper torque. A fitting must not be overtorqued to stop a leak. Leaks or seepage from the hose surface requires replacement of the hose assembly.

If there is more than one broken wire per plait in the covering braid or if there are more than six broken wires per lineal foot, the hose should be replaced.

Hose that is reinforced with carbon-steel wire braid is subject to corrosion. This is easily detected by a rust color on the surface. If the corrosion is appreciable, the hose should be replaced. Stainless-steel wire braid often turns a golden yellow to brown color when subjected to heat. This condition should not be confused with corrosion. If the coloring is extreme, it is possible that the hose has been overheated and may require replacement.

The hose mountings in the aircraft should be inspected for the condition of the clamps, any bulging of the hose or other damage to the hose at the clamps, the condition of the cushioning in the clamps, the position of the hose and the cushion in the clamp, and the security of the clamp screws. The positioning of the cushion material in the clamp must be such that the material does not lodge between the end tabs of the clamp when the clamp is closed.

Hose that is twisted, as indicated by the lay line along the hose, can be corrected by loosening one of the fittings, straightening the hose, and retorquing the fitting.

Fire sleeves are mounted on hose to protect the hose from excessive heat and flame. If the fire sleeve is worn through, torn, cut, or oil soaked, the hose assembly should be replaced. The removed hose assembly may be inspected and tested and, if found to be serviceable, may have a new fire sleeve installed and then be returned to service.

End fittings are checked for corrosion, cleanliness, nicks, scratches, cracks, damage to threaded areas, damage to cone-seat sealing surfaces, damage to flanges, and backed-out retaining wires on swivel nuts. The hose assembly should be replaced if any condition found could cause malfunction or deterioration.

In all inspections of hose installations, the technician should consult the applicable manufacturer's manual to assure that specified conditions are met. There are many different types and designs of fittings, and it is essential that the instructions and specifications for the particular type of fitting being inspected are understood.

Installation Practices for Aircraft Hose. Before installation of a section of aircraft hose, the hose should be thoroughly inspected as previously explained. If the hose is straight, the inside can be examined by looking through it toward a light source. If there is an elbow on one end, a flashlight or other light source can be used to illuminate the inside of the elbow, and the interior of the tube can be examined by looking in the opposite end. If it is not possible to look inside the hose, a steel ball slightly smaller than the ID of the hose should be passed through the tube. The ball should roll freely through the tube from one end to the other.

Hose that is preformed to fit certain installations should not be straightened out. Straightening causes undue stresses, wrinkling inside the hose, and other possible defects. To prevent the straightening of preformed hose, a wire or cord can be attached to each end and pulled taut.

Some installation practices for hose assemblies are illustrated by the drawings of Figure 12–43. The installation of flexible hose assemblies requires that the hose be of a length that will not be subjected to tension. The hose section should be of sufficient length to provide about 5 to 8% slack. The hose should be installed without twisting by keeping the lay line on the hose straight. Bends in the hose should not have a radius less than 12 times the ID of the

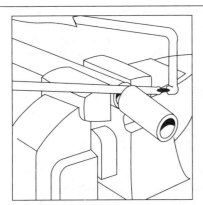

1. PLACE HOSE IN VICE AND CUT TO DESIRED LENGTH USING FINE TOOTH HACKSAW OR CUT OFF WHEEL.

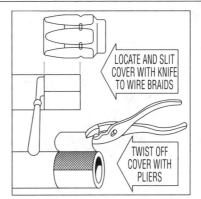

LOCATE AND SLIT COVER WITH KNIFE TO WIRE BRAIDS

TWIST OFF COVER WITH PLIERS

2. LOCATE LENGTH OF HOSE TO BE CUT OFF AND SLIT COVER WITH KNIFE TO WIRE BRAID. AFTER SLITTING COVER, TWIST OFF WITH PAIR OF PLIERS. (SEE NOTE BELOW)

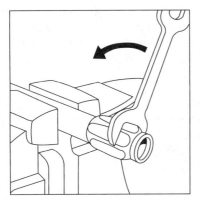

3. PLACE HOSE IN VISE AND SCREW SOCKET ON HOSE COUNTERCLOCKWISE.

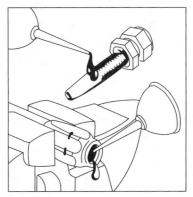

4. * LUBRICATE INSIDE OF HOSE AND NIPPLE THREADS LIBERALLY.

NOTE:
HOSE ASSEMBLIES FABRICATED PER MIL-H-8790 MUST HAVE EXPOSED WIRE BRAID COATED WITH A SPECIAL SEALANT.

NOTE:
STEP 2 APPLIES TO HIGH PRESSURE HOSE ONLY.

* **CAUTION:**
DO NOT USE ANY PETROLEUM PRODUCT WITH HOSE DESIGNED FOR SYNTHETIC FLUIDS "SKYDROL." FOR A LUBRICANT DURING ASSEMBLY, USE A VEGETABLE SOAP LIQUID.

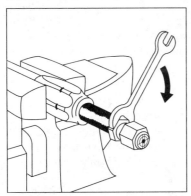

5. SCREW NIPPLE INTO SOCKET USING WRENCH ON HEX OF NIPPLE AND LEAVE .005 INCHES TO .031 INCHES CLEARANCE BETWEEN NIPPLE HEX AND SOCKET.

DISASSEMBLE IN REVERSE ORDER

FIGURE 12–42 Installation of an end on a high-pressure hose.

hose for normal installations. The coupling nuts for flexible hose assemblies should be torqued to the correct value as specified by the manufacturer.

When a plain hose is used to provide a flexible joint between two sections of tubing, the ends of the tubing should be beaded. Clamps should not be overtightened because of the danger of damaging the hose. A good practice is to tighten the clamp fingertight plus a one-quarter turn. It must be emphasized that plain hose and clamps should not be used where the fluid in the system is under pressure.

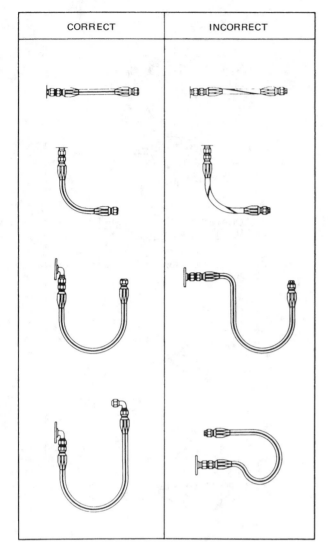

CORRECT	INCORRECT

FIGURE 12–43 Correct and incorrect hose installations. *(Stratoflex Inc.)*

ciled on the hose, and storage life for hose assemblies does not exceed four years. Storage life may vary, and it is necessary to consult the manufacturer's information to determine the exact condition for any particular type of hose.

Inspection of Fluid-Line Systems

Lines and fittings should be inspected carefully at regular intervals for leaks, damage, loose mountings, cracks, scratches, dents, and other damage. Flexible lines (hose) should be checked for cracks, cuts, abrasions, soft spots, and any other indication of deterioration. Parts with defects should be either replaced or repaired. A damaged metal line should be replaced in its entirety if the damage is extensive. If the damage is localized, it is permissible to cut out the damaged section and insert a new section with approved fittings. Care must be taken that no foreign material enters the line during the repair operation. When soft aluminum tubing using flared fittings is replaced, a double flare should be used on all tubing with a $\frac{3}{8}$-in [0.952-cm] OD or smaller.

The following defects are not acceptable for metal lines:

- Cracked flare
- Scratches or nicks greater in depth than 10% of the tube wall thickness or in the heel of a bend
- Severe die marks, seams, or splits
- A dent of more than 20% of the tube diameter or in the heel of a bend

Color Codes for Plumbing Lines

Color codes for aircraft and missile plumbing lines are established by AND10375. These codings replace the plain-color system used before August 1949. Because of fading of the colors and the fact that the color perception of some persons is not sufficiently acute, the plain colors were found to be subject to misinterpretation under adverse conditions. For this reason, new color coding was established with black-and-white symbols.

Figure 12–44 shows codes used to identify various systems in an aircraft. The code bands are colored as indicated in the illustration. The symbols are black against a white background.

Storage of Aircraft Hose. Synthetic rubber hose and hose assemblies should be stored in a dark, cool, dry area and be protected from circulating air, sunlight, fuel, oil, water, dust, and ozone. Storage life of synthetic rubber hose normally does not exceed five years from the cure date sten-

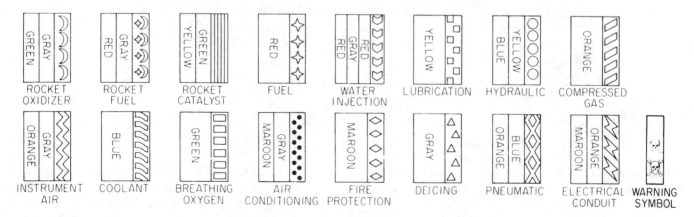

FIGURE 12–44 Color-coding bands for plumbing lines.

1. What metals and alloys are commonly used for tubing in plumbing systems?
2. Describe a pipe thread.
3. How are pipe-thread dimensions designated?
4. Explain how a flareless fitting provides a seal against fluid leakage.
5. How may a flareless fitting be identified by inspection?
6. Describe a swaged tube fitting.
7. What are the advantages of swaged fittings?
8. At what pressure is a hose assembly pressure tested?
9. Describe the differences in the construction of low-pressure, medium-pressure, and high-pressure hose.
10. How may hose be identified?
11. Why is it poor practice to install short, straight sections of tubing between fixed parts of an aircraft?
12. Explain the importance of the flare in a flared tube fitting.

13. On what tubing is double flaring required?
14. Why are beads used on some tubing?
15. What lubricant may be used when presetting a flareless-fitting sleeve for use in a hydraulic system?
16. What is the purpose of lay line on aircraft hose?
17. How may a twist in a section of hose be corrected?
18. Describe the installation of reusable hose fittings.
19. When installing a reusable fitting on high-pressure hose with a synthetic rubber outer ply, how can the amount of the outer ply to be removed be determined?
20. How can the interior of a preformed hose be checked for blisters or collapse of the inner tubing?
21. Why should tube fittings not be forced into position?
22. What is the effect of overtightening flared fittings?
23. How much slack should be allowed in the installation of aircraft hose assemblies?
24. What defects are not acceptable in metal tubing?
25. What are the tubing color codes for hydraulic, fuel, and oxygen lines?

13 Federal Aviation Regulations

INTRODUCTION

When the Federal Aviation Administration (FAA) is mentioned, most people think in terms of its responsibilities in the operation and maintenance of the world's largest and most advanced air-traffic-control system. Almost half of the agency's work force is engaged in some phase of air-traffic control. No air-traffic-control system, no matter how automated, can function safely and efficiently unless the people and the machines within the system measure up to certain prescribed standards. The FAA has been charged with establishing and enforcing standards relevant to the training and testing of aviation personnel and the manufacture and continued airworthiness of aircraft.

There are more than 200 000 civil aircraft in the United States, and the FAA requires that each be certificated as airworthy. The original design and each subsequent aircraft constructed from that design must be approved by FAA inspectors. Even home-built aircraft require FAA certification. All civil aircraft certificated for operation by the Federal Aviation Administration must be maintained in accordance with the requirements of the Federal Aviation Regulations (FAR) issued by the FAA. Federal Aviation Regulations affect aircraft design, operation, maintenance, repair, and alteration. In this chapter, regulations of particular interest to aviation maintenance technicians will be discussed.

HISTORY OF THE FEDERAL AVIATION ADMINISTRATION

Twenty-three years after the Wright brothers' flight at Kitty Hawk on December 1, 1903, Congress recognized the potential of a new industry—air transportation—and enacted legislation to bring it within federal control. Now an operating arm of the **Department of Transportation**, the **Federal Aviation Administration** traces its ancestry back to the Air Commerce Act of 1926, which led to the establishment of the Aeronautics Branch (later reorganized as the Bureau of Air Commerce) in the Department of Commerce, with the authority to certificate pilots and aircraft, develop air-navigation facilities, promote flying safety, and issue flight information.

Since that date, pilots and aircraft have had to be certificated by the federal government as qualified to fly. In a few short years, radio aids, communications, and ground lights were improved and expanded; airport construction was encouraged; and the first Civil Air Regulations were written.

The government acted just in time. In May 1927, Charles Lindbergh bridged the northern Atlantic in $33\frac{1}{2}$ hours, generating new interest and enthusiasm for aviation in both Europe and America.

Aviation continued to grow and expand at a very rapid rate in the decade following Lindbergh's flight, creating a need for new machinery to regulate civil flying. The result was the Civil Aeronautics Act of 1938, which established the independent Civil Aeronautics Authority with responsibilities in both safety and economic areas. In 1940, the powers previously vested in the Civil Aeronautics Authority were assigned to a new **Civil Aeronautics Administration (CAA)**, which was placed under an assistant secretary in the Department of Commerce. In addition, a semi-independent **Civil Aeronautics Board (CAB)**, which had administrative ties with the Department of Commerce but reported directly to the Congress, was created to regulate airfares and routes.

The CAA performed very well during World War II, but it proved unequal to the task of managing the airways in the years after the war because of the tremendous surge of civil air traffic and the introduction of new high-performance aircraft. In 1958, the same year American jets entered commercial service, Congress passed the **Federal Aviation Act**, which created an independent Federal Aviation Agency with broad new authority to regulate civil aviation and provide for the safe, efficient use of the nation's airspace.

In April 1967, with the passage of the **Department of Transportation Act**, the Federal Aviation Agency became the Federal Aviation Administration and was incorporated into the newly created Department of Transportation (DOT). The DOT was established to give unity and direction to a coordinated national transportation system. The FAA's basic responsibility, to provide for safety in flight, was unchanged. While working with other administrations in the Department of Transportation for long-range transportation planning, the FAA today continues to concern itself primarily with the promotion and regulation of civil aviation to insure safe and orderly growth.

AVIATION SAFETY REGULATION

The **Federal Aviation Act of 1958** forms the legal basis for the present system of Federal Aviation Regulations. The provisions of this act are contained in 15 sections referred to as *titles*.

The **Title VI, Safety Regulation of Civil Aeronautics**, is the section that is of most interest to aviation maintenance technicians from an operational point of view. The first section of Title VI details the duties of the secretary of transportation and thereby the FAA administrator in promoting safety:

> Sec. 601 (a) The Secretary of Transportation is empowered and it shall be his duty to promote safety of flight of civil aircraft in air commerce by prescribing and revising from time to time:
>
> 1. Such minimum standards governing the design, materials, workmanship, construction, and performance of aircraft, engines, and propellers as may be required in the interest of safety.
> 2. Such minimum standards governing appliances as may be required in the interest of safety.
> 3. Reasonable rules and regulations and minimum standards governing, in the interest of safety: (A) the inspection, servicing, and overhaul of aircraft, aircraft engines, propellers, and appliances; (B) the equipment and facilities for such inspection, servicing, and overhaul; and (C) in the discretion of the Secretary of Transportation, the periods for, and the manner in which such inspection, servicing, and overhaul shall be made, including provision for examinations and reports by properly qualified private persons whose examinations or reports the Secretary of Transportation may accept in lieu of those made by its officers and employees.
> 4. Reasonable rules and regulations governing the reserve supply of aircraft, aircraft engines, propellers, appliances, and aircraft fuel and oil required in the interest of safety, including the reserve supply of aircraft fuel and oil which shall be carried in flight.
> 5. Such reasonable rules and regulations, or minimum standards, governing other practices, methods, and procedure as the Secretary of Transportation may find necessary to provide adequately for national security and safety in air commerce.

Most safety standards and regulations are carried out by setting **minimum standards** and issuing **certification** that the standards have been met. Six basic certificates have been created in order to enforce and maintain these standards:

1. Airman certificates (pilots, mechanics, air-traffic-control personnel, repairmen, and similar certificates)
2. Aircraft certificates (type, production, airworthiness)
3. Air-carrier operating certificates
4. Air-navigation facility ratings
5. Air-agency ratings (schools, repair stations, and similar facilities)
6. Airport operating certificates

ORGANIZATION OF THE FAA

The Federal Aviation Act provides for an administrator, a deputy administrator, and the necessary personnel to carry out the duties and responsibilities as stated in the law.

A large number of personnel and a complex organizational structure are presently involved in regulating aviation. An aircraft technician would be most directly involved with the **Office of Airworthiness (AWS)**. The Office of Airworthiness includes the Aircraft Engineering Division, the Aircraft Manufacturing Division, and the Aircraft Maintenance Division. The Aircraft Maintenance Division has a General-Aviation Branch and an Air-Carrier Branch. The Office of Airworthiness develops rules, standards, and policies to be carried out in the various geographic regions.

Regional Offices

The **FAA regional offices** serve as an extension of the national headquarters and handle the day-to-day problems that arise in the various geographic regions. The regional offices plan the functions that will occur in the region, such as compiling statistics, providing navigational aids and regional personnel, and administering examinations and inspections. They are responsible for the standardization of the maintenance and engineering practices of the airlines and manufacturers located within the region. There are ten regional offices, as shown in Figure 13–1 (a list of FAA regional offices is provided in the Appendix).

District Offices

The technician's primary contact with the FAA will be with airworthiness inspectors assigned to the local district offices. There are two types of district offices, the **General Aviation District Office (GADO)** and the **Flight Stan-**

FIGURE 13–1 FAA regional offices.

dards District Office (FSDO). The FSDO has both an air-carrier and a general-aviation section, while the GADO deals only with general-aviation activities. (See the Appendix for a list of district offices.)

Aeronautical Center

The FAA Mike Monroney Aeronautical Center in Oklahoma City consists of a number of FAA departments. The Airmen Certification Branch has the records of all individuals that have received airman certificates. One of the largest centers for aviation training is located there, as is the department that develops the various FAA tests. FAA aircraft are also maintained and repaired at this installation.

Accident Investigation

The FAA participates with the National Transportation Safety Board (NTSB) in the investigation of major aircraft accidents to determine if any immediate action is needed to correct deficiencies and prevent a recurrence. In addition, the agency investigates most nonfatal and many fatal general-aviation accidents on behalf of the NTSB, although the responsibility of determining probable cause remains with the NTSB. The FAA also investigates accidents to see if any Federal Aviation Regulations have been violated.

FEDERAL AVIATION REGULATIONS

The Federal Aviation Regulations are published as Chapter 14 of the United States Code of Federal Regulations. The various regulations are the minimum standards which have been set to ensure, as much as possible, aviation safety. The regulations are divided into parts relating to a specialized area, such as FAR Part 25, Airworthiness Standards: Transport-Category Aircraft; or FAR Part 147, Aviation Maintenance Technician Schools.

The FAA is continually researching ways of improving and updating a regulation. Once a proposal has been developed for a new regulation or a revision, the public is notified by a notice of proposed rule making (NPRM). The NPRM states the proposed changes and the logic behind them. The public is invited to comment on the proposal. After public opinion has been gathered, the proposal is either passed, modified, or dropped. FAR Part 11 contains the procedures for regulatory changes, including how a citizen may petition the FAA for a rule change.

Of the many parts of the FARs, certain ones are of particular interest to aviation maintenance technicians, repair stations, and others involved in airframe and powerplant maintenance and certification. A person seeking certification is expected to know the appropriate sections of three parts:

Part 43 Maintenance, Preventive Maintenance, Rebuilding, and Alteration
Part 65 Certification: Airmen Other Than Flight Crew Members
Part 91 General Operating and Flight Rules

A number of the FARs are concerned with the certification processes for aircraft and parts. In the process of certifying an aircraft as airworthy, a technician may have to refer to the following:

Part 21 Certification Procedures for Products and Parts
Part 23 Airworthiness Standards: Normal-, Utility-, Acrobatic-, and Commuter-Category Airplanes
Part 25 Airworthiness Standards: Transport-Category Airplanes
Part 27 Airworthiness Standards: Normal-Category Rotorcraft
Part 29 Airworthiness Standards: Transport-Category Rotorcraft
Part 33 Airworthiness Standards: Aircraft Engines
Part 35 Airworthiness Standards: Propellers
Part 36 Noise Standards: Aircraft-Type Certification
Part 39 Airworthiness Directives

Other parts of the FARs that the technician may be involved with would include the following:

Part 1 Definitions and Abbreviations
Part 45 Identification and Registration Marking
Part 145 Repair Stations

A complete list of all FAR parts may be found in the Appendix.

General Operating and Flight Rules (Part 91)

Part 91 establishes rules and procedures for the operation of all aircraft within the United States. As such it is of primary importance to the aircraft owner and operator to know and comply with these rules. Included in this part are the requirements for inspecting the aircraft and maintaining it in an airworthy condition.

Subpart C of Part 91, **Equipment, Instrument, and Certificate Requirements**, requires that an aircraft must have an appropriate and current airworthiness certificate displayed in the aircraft before it can be operated.

Subpart C also specifies instrument and equipment requirements for the various types (VFR day, VFR night, and IFR) of operations. These requirements are for operation and not the same as those for airworthiness certification. It is the responsibility of the operator to know and comply with these requirements. However, the technician will often be called upon to advise the operator or will be expected to see that the proper equipment is installed. It is not within the scope of this text to present the equipment required for various types of flight operations, nor is it required knowledge for the technician. It would be helpful for the technician to know that the requirement for instruments may be found in Section 91.205. In addition, Section 91.207 sets forth requirements for emergency locator transmitters, and the rules for operation with inoperative instruments and equipment may be found in Section 91.213.

Subpart E of Part 91, **Maintenance, Preventive Maintenance, and Alterations**, prescribes rules governing the maintenance, preventive maintenance, and alteration of U.S. registered civil aircraft operating within or outside the United States.

The owner or operator of an aircraft is primarily responsible for maintaining that aircraft in an airworthy condition, including compliance with airworthiness directives (FAR 39). All work done to the aircraft must be done as prescribed in the various regulations.

The owner must have the aircraft inspected as prescribed in this part, and between such inspections she or he must see that the aircraft will not be operated unless any defects have been repaired in accordance with FAR 43.

No person may operate a rotorcraft for which a rotorcraft maintenance manual containing an airworthiness limitations section has been issued, unless the replacement times, inspection intervals, and related procedures specified in that section of the manual are followed.

The owner or operator is also responsible for seeing that maintenance personnel have made appropriate entries in the maintenance records releasing the aircraft for service. The owner will also ensure that the maintenance records are maintained as set forth in Sections 91.417 and 419.

Required inspections for the aircraft are covered in Section 91.409. Special inspection requirements for the altimeter system may be found in Section 91.411 and for the ATC transponders in Section 91.413. Detailed information on inspection requirements and maintenance records (Sections 91.415, 417, 419, and 421) will be found in other chapters of this text.

Maintenance, Preventive Maintenance, Rebuilding, and Alteration (Part 43)

FAR Part 43 prescribes standards and procedures for the maintenance, preventive maintenance, rebuilding, and alteration of all aircraft with a U.S. airworthiness certificate, with the exception of aircraft issued an experimental airworthiness certificate. This exception does not apply if the aircraft has previously had a different type of airworthiness certificate issued.

Part 43 establishes who is authorized to perform maintenance, what performance rules are to be observed, and what procedures are to be followed in approving the aircraft for return to service. It also provides for work on U.S.-registered aircraft by Canadian personnel.

The term *maintenance* is commonly used in referring to all types of repairs and service work performed on an aircraft. However, to understand and comply with the provisions of Part 43, it is essential to know the technical definitions of the terms used to describe the various types of work performed. These terms will usually be found in FAR Part 1, Definitions and Abbreviations, or in Appendix A of Part 43. FAR Part 1 provides the following definitions:

"**Maintenance**" means inspection, overhaul, repair, preservation, and the replacement of parts, but excludes preventive maintenance.

"**Preventive Maintenance**" means simple or minor preservation operations and the replacement of small or standard parts not involving complex assembly op-

erations. [Examples of preventive maintenance may be found in Part 43, Appendix A. An advisory circular, 43.12, is also available on this subject. For the technician, there is no difference between performing maintenance and preventive maintenance in either terms of procedure or record entry.]

"**Alteration**" refers to changing the design of the aircraft from that originally certificated. Alterations range from major changes in the structure or engine to the addition or removal of relatively minor equipment. Alterations are divided into two categories: major and minor.

"**Major alteration**" means an alteration not listed in the aircraft, aircraft engine, or propeller specification:

1. that might appreciably affect weight, balance, structural strength, performance, powerplant operation, flight characteristics, or other qualities affecting airworthiness; or

2. that is not done according to accepted practices or cannot be done by elementary operations.

"**Minor alteration**" means an alteration other than a major alteration.

Appendix A of Part 43 provides additional criteria that can be used in determining whether or not an alteration would be classified as major or minor.

Although not specifically defined, a repair can be considered as the correction of a defective condition. Repairs are also classified as major or minor, as defined in FAR Part 1:

"**Major repair**" means a repair

1. that, if improperly done, might appreciably affect weight, balance, structural strength, performance, powerplant operation, flight characteristics, or other qualities affecting airworthiness; or

2. that is not done according to accepted practices or cannot be done by elementary operations.

"**Minor repair**" means a repair other than a major repair.

If this definition is taken literally, virtually all repairs could be considered major. Appendix A again provides further criteria for determining the proper classification. The technician must be able to apply these criteria to the operations he or she is performing.

Accepted Practices and Elementary Operations. The terms *accepted practices* and *elementary operations* appear in the definitions of both major repairs and major alterations. Although formal definitions are not given, an understanding of these terms will be helpful for the technician to interpret the regulations and to classify repairs and alterations.

As the term implies, **accepted practices** are those procedures which are widespread in the industry. They are often found in manufacturer's service manuals and in *Advisory Circular 43.13-1A*. An example of an acceptable practice would be the riveting of a sheet of aluminum to an aircraft structure during a repair. Welding the aluminum to the structure would usually not be considered as an accepted practice and could be done only under special approval.

The term **elementary operations** is a little harder to define. An analysis of Appendix A leads to the conclusion that an elementary operation is one requiring a low skill level to perform and is easily inspected to verify that it has been properly done. A "nut and bolt" type of operation would therefore be considered an elementary operation. A process involving riveting and welding would probably not be considered an elementary operation.

Even with the guidelines listed in Appendix A, the technician will often have to make an interpretation of whether a repair is major or minor. If in doubt, he or she should contact the manufacturer or the local FAA district office for advice.

Approved Data and Acceptable Data. The terms *approved data* and *acceptable data* are frequently encountered when working with the regulations. **Acceptable data** refers to information that may be used as a basis for FAA approval for major repairs and alterations. Examples would include *Advisory Circulars 43.13-1A* and *43.13-2A,* manufacturer's technical publications, or military technical orders. **Approved data** refers to data which have previously been found to be acceptable and thus have been approved by the FAA. If an alteration or repair duplicates an operation previously approved, approved data are being used. In such cases, it is usually only necessary to have the workmanship approved and not the data. An example of approved data would be a supplemental type certificate, which will be discussed in a later chapter of this text.

Authorization to Perform Maintenance. A number of individuals or agencies are authorized by Part 43 to perform maintenance operations on aircraft. As this textbook was being written, the term **mechanic** is the legal term used in all parts of the regulation. The majority of those working in aviation maintenance prefer the term **aviation maintenance technician**. The FAA has acknowledged that such a change should be considered and is in the preliminary stages of producing an NPRM that would include this change. In the belief that such a change is imminent, the term *aviation maintenance technician* is used in this text, in place of *mechanic.*

The holder of an **aviation maintenance technician (mechanic) certificate** may perform maintenance, preventive maintenance, and alterations in accordance with the privileges and limitations set forth in Part 65.

Part 43 also allows persons working under the supervision of a certificated technician to perform the maintenance, preventive maintenance, and alterations that his or her supervisor is authorized to perform. The supervisor must personally observe the work being done to the extent necessary to ensure that it is being done properly, and he or she must be readily available, in person, for consultation. The performance of a 100-hour or annual inspection or the inspection performed after completion of a major repair or alteration cannot be supervised. They must be done personally by the properly certificated personnel.

A **repairman** can perform maintenance or preventive maintenance consistent with the privileges and limitations of Part 65. He or she may also supervise others doing the work with the same conditions as those for the mechanic.

The holder of a **repair-station certificate** may perform maintenance, preventive maintenance, and alterations as allowed by the ratings that the repair station holds. The various repair-station ratings and their privileges are covered in Part 145.

The holder of a **pilot certificate** issued under Part 61 may perform preventive maintenance on any aircraft owned or operated by that pilot which is not used in air carrier service. There are two limitations in this authorization. The first is that the aircraft be owned or operated by the pilot. The second is that the aircraft is not used in air-carrier service; this limitation includes aircraft used in air-taxi operations, except for limited exceptions.

The **manufacturer** may rebuild or alter any aircraft or component that it has manufactured under a Type Certificate, Production Certificate, Technical Standard Order (TSO) Authorization, FAA Parts Manufacturer Approval (PMA), or Product and Process Specification issued by the FAA Administrator. The manufacturer may also perform any inspections required by Part 91 or Part 125 while currently operating a production certificate or under a currently approved production inspection system for such aircraft.

The holder of an **air-carrier operating certificate** may perform maintenance, preventive maintenance, and alterations as provided in Parts 121, 127, and 135.

Performance Rules. The section of regulations on performance rules is short and concise, but effectively sets forth the standards to follow in performing maintenance, preventive maintenance, and alterations. The technician must ensure that his or her work meets these criteria.

Paragraph 43.13(a) sets forth the criteria for work methods and techniques:

Each person maintaining or altering, or performing preventive maintenance shall use methods, techniques, and practices acceptable to the Administrator. He/she shall use the tools, equipment, and test apparatus necessary to assure completion of the work in accordance with accepted industry practices. If special equipment or test apparatus is recommended by the manufacturer involved, he/she must use that equipment or apparatus or its equivalent acceptable to the Administrator.

Paragraph 43.13(b) sets forth criteria for determining the quality of the work performed:

Each person maintaining or altering, or performing preventive maintenance, shall do that work in such manner and use materials of such quality that the condition of the aircraft, airframe, aircraft engine, propeller, or appliance worked on will be at least equal to its original or properly altered condition (with regard to aerodynamic function, structural strength, resistance to vibration, and deterioration, and other qualities affecting airworthiness).

Advisory Circulars 43.13-1A and *2A* and the various manufacturer's maintenance manuals provide more details on how to meet the criteria listed in the preceding paragraphs.

Paragraph 43.13(c) states that the methods, techniques, and practices contained in the maintenance manual of a certificated air carrier, required by its operating specifications to provide a continuous airworthiness maintenance and inspection program, are considered to be acceptable means for compliance with the requirements of FAR 43.13.

Sections 43.15 and 43.16 set forth the standards to be observed when performing inspections. The details of these sections will be discussed in Chapter 16 of this text.

Return to Service. Section 43.5 states that an aircraft or a component may not be returned to service after it has undergone maintenance, rebuilding, or alteration until it has been approved for return to service by an authorized person and the appropriate paperwork has been completed. The paperwork consists of the required maintenance-record entries, necessary major repair or alteration forms, and appropriate changes in the aircraft flight manual or operating limitations.

A **flight test** is not required in Part 43; however, Section 91.407(b) states that no person may carry any persons other than crew members in an aircraft that has been repaired or altered in a manner that may have appreciably changed its flight characteristics until it has been approved for return to service in accordance with Part 43 and an appropriately rated pilot, with at least a private pilot's certificate, flies the aircraft, makes an operational check of the repaired or altered part, and logs the flight in the aircraft records.

The test flight will not be necessary if ground tests or inspections show conclusively that the operation of the aircraft has not been adversely affected.

Those authorized by Section 43.7 to approve the aircraft or component for return to service include the following:

1. The holder of an aviation maintenance technician certificate or inspection authorization as authorized in Part 65.
2. The holder of a repair-station ratings consistent with the privileges of his or her ratings.
3. The manufacturer of any aircraft or component on which it is approved to perform maintenance.
4. An air carrier or commercial operator under the provisions of Parts 121, 127, and 135.
5. The holder of a pilot certificate for preventive maintenance he or she has performed.

Section 43.9 provides the requirements for the content, form, and disposition of records of maintenance, preventive maintenance, and alterations. Section 43.11 does the same for inspection records. This information is covered in Chapter 16 of this text.

Section 43.2 defines the use of the terms *overhaul* and *rebuilding* in maintenance records.

Section 43.12 states that no person may make (or cause to be made) fraudulent record entries, reproduce records for fraudulent purposes, or alter records for fraudulent purposes. The commission of an act prohibited under this section is cause for suspension or for revoking the applicable certificate or authorization.

Section 43.17 deals with the approval of work performed by Canadian personnel on U.S.-registered aircraft while they are in Canada. In general, a person holding a valid Canadian Department of Transport license (an Aircraft Maintenance Engineer) with appropriate ratings has privileges similar to those of the U.S.-certificated maintenance technician.

Appendices. As discussed previously, Appendix A provides criteria for the classification of repairs and alterations as major or minor. It also lists the items considered preventive maintenance.

Appendix B provides instructions for the recording of major repairs and alterations using Form 337 or a major-repair-station work order.

Appendix D is a section with which all technicians must be familiar. It provides the items that must be included during a 100-hour or annual inspection.

Appendix E provides the procedures and standards for altimeter tests and inspections required by Section 91.411. Appendix F does likewise for ATC transponder tests required by Section 91.413.

Certification: Airmen Other Than Flight Crew Members (Part 65)

FAR Part 65 covers the application, eligibility, and privileges of air-traffic-control-tower operators, aircraft dispatchers, maintenance technicians, repairmen, and parachute riggers. This text will cover only the maintenance technician and the repairman certificates. The FAA issues an aviation maintenance technician certificate with two possible ratings: airframe (A) and powerplant (P).

Eligibility for Aviation Maintenance Technician Certificates. To be eligible for an aviation maintenance technician certificate, a person must be at least 18 years of age and be able to read, write, speak, and understand the English language. A person employed outside of the United States by a U.S. air carrier that does not meet the language requirement may be certificated but will have his or her certificate endorsed, "Valid only outside the United States."

Before the testing procedure can begin, the applicant must either show proof of having completed a program at an approved aviation maintenance technician school or show proof of appropriate experience for the rating or ratings he or she seeks. To qualify on the basis of experience, the applicant must have 18 months of experience doing duties appropriate to the rating sought. If he or she wishes to receive both ratings (A and P), the applicant must show documentary evidence of 30 months of experience doing appropriate work.

An aviation maintenance technician holding one rating may not apply for an additional rating during any period that his or her certificate is suspended. In the event that a certificate is revoked, that person may not reapply for one year after the date of revocation. A person that has been convicted of any federal or state crimes involving drug violations is not eligible to apply for certification until one year after the date of the final conviction.

Testing for Aviation Maintenance Technician Certificates. Once the appropriate eligibility requirements have been met, the applicant is ready to start the testing to demonstrate his or her knowledge and skill. The applicant must pass all the required tests appropriate to the rating he or she seeks within a period of 24 months.

The **written examinations** cover the construction and maintenance of aircraft appropriate to the rating sought and the applicable provisions of FAR Parts 43, 65, and 91. There are three written tests. A general test is required for the airframe rating and the powerplant rating, containing material common to both areas. The applicant for an airframe rating will take a test (100 questions) over airframe structures and systems. The Powerplant applicant will take a 100-question test over Powerplant theory, systems, and components. The minimum passing grade for each test is 70%.

In the event that the applicant fails one or more sections of a test, he or she may retake that section after waiting 30 days from the date of the test or by presenting a statement from a certificated aviation maintenance technician that he or she has had additional instruction in the areas failed. As stated previously, all tests must be passed in a 24-month period.

Although the test questions have been published, the integrity of the written exam is closely guarded. Any person found attempting to cheat or otherwise compromise an examination is not eligible for an airman certificate for a period of one year from the date of the violation. In addition, the commission of that act is a basis for suspending or revoking any airman certificate or rating held by that person.

After successfully completing the applicable written tests, the applicant is eligible to demonstrate his or her skills through **oral and practical examinations**. These are given by FAA-designated mechanic examiners. A student in an aviation maintenance technician school may be allowed to take the oral and practical tests as part of the final subjects of his or her training.

Certification of Aviation Maintenance Technicians.

When all examinations have been passed and an application for the certificate and ratings completed, a temporary airman certificate is issued. This certificate is good for 120 days. After processing the application, the FAA will mail a permanent certificate to the applicant.

An aviation maintenance technician certificate is effective until it is surrendered, suspended, or revoked. It has no expiration date. In the event a certificate is suspended or revoked, it must be returned to the administrator.

A name change or the replacement of a lost certificate can be taken care of by contacting the FAA in Oklahoma City. Details of this procedure are given in Section 65.16. The FAA must be notified, in writing, within 30 days of any change of address. The details for this notification are provided in Section 65.21.

Each person who holds an aviation maintenance technician certificate must keep it within the immediate area where he or she normally exercises the privileges of the certificate. In addition, he or she must present it for inspection upon the request of any member of the FAA, NTSB, or any federal, state, or local law enforcement officer.

Privileges and Limitations of Aviation Maintenance Technician Certificates.

A certificated aviation maintenance technician is not allowed to do major repairs to, or major alterations of, propellers, or any repair or alteration of instruments. Beyond this, his privileges are limited chiefly by his ratings, experience, and knowledge.

A certificated aviation maintenance technician may *perform* maintenance or alteration of an aircraft, appliance, or part for which he or she is rated. The technician may also perform the 100-hour inspection on the portion of the aircraft for which he or she is rated (airframe or powerplant).

A certificated aviation maintenance technician may *supervise* the maintenance or alteration of an aircraft, appliance, or part for which he or she is rated, provided that he or she has satisfactorily performed the work at an earlier date.

A certificated aviation maintenance technician may *approve for return to service* an airframe or powerplant for which he or she is rated after performing or supervising its maintenance or alterations, provided that he or she has satisfactorily performed the work at an earlier date. This privilege does not include major repairs and alterations. The technician may also approve for return to service an airframe or powerplant for which he or she is rated after performing the 100-hour inspection.

If an aviation maintenance technician has not previously performed a task, the technician may show his or her ability to do it by performing it to the satisfaction of FAA personnel or by doing it under the direct supervision of an appropriately rated aviation maintenance technician or repairman with previous experience in that task. It should be noted that this portion of the regulation requires no documentation or other record of the performance of these tasks. Whether or not an individual has previously performed a task depends upon proper interpretation and good judgment on the part of that individual.

An aviation maintenance technician may not exercise the privileges of his or her certificate and ratings unless he or she understands the instructions of the manufacturer and in the maintenance manuals for the specific operation concerned.

An aviation maintenance technician may not exercise the privileges of his or her certificate unless during the past 24 months he or she has demonstrated his or her ability to the FAA or has spent at least 6 months working as an aircraft aviation maintenance technician, technically supervising other aviation maintenance technicians, or supervising in an executive capacity the maintenance or alteration of aircraft.

Inspection Authorizations. Inspection authorization is not a rating but an authorization to perform certain inspection functions in the area of aircraft maintenance.

Eligibility for Inspection Authorizations. To be eligible for inspection authorization, a person must hold a currently effective airframe and powerplant aviation maintenance technician certificate, both of which must have been continuously in effect for the past three years. He or she must also have been actively engaged, for at least the two-year period prior to the date he or she applies, in maintaining aircraft certified and maintained under the Federal Air Regulations.

The applicant must have a fixed base of operations at which he or she may be located in person or by telephone during the normal working week. This does not have to be the place where the applicant will exercise his or her inspection authority. The applicant must also have available equipment, facilities, and inspection data necessary to inspect aircraft properly.

The applicant must also pass a written test on his or her ability to inspect according to safety standards for approving aircraft for return to service after major repairs, major alterations, and annual and progressive inspections.

Duration and Renewal of Inspection Authorizations.
Each inspection authorization expires on March 31 of each year. The authorization is not effective if the holder's Airframe and powerplant aviation maintenance technician certificate is suspended or revoked. It also becomes ineffective if the authorization itself is suspended or revoked, or if the holder no longer has a fixed base of operations or fails to meet any of the other requirements for eligibility.

An inspection authorization can be renewed each year if the holder shows that he or she still meets the eligibility requirements and has either performed four annual inspections, performed inspections of eight major repairs or alterations, or performed or supervised at least one progressive inspection.

When exercising the privileges of an inspection authorization, the holder must keep the authorization available for inspection by the aircraft owner or the aviation maintenance technician submitting the aircraft repair or alteration for approval, and he or she must present it upon the request of any representative of the FAA, NTSB, or any federal, state, or local law enforcement officer.

Privileges and Limitations of Inspection Authorizations. The holder of an inspection authorization may inspect and approve for return to service any aircraft or related part or appliance after a major repair or alteration to it in accordance with FAR 43, if the work was done in accordance with technical data approved by the FAA. This privilege does not apply to aircraft maintained in accordance with a continuous airworthiness program under FAR 121 or 127. He or she may also perform an annual inspection or perform or supervise a progressive inspection.

If the holder of an inspection authorization changes his or her fixed base of operation, he or she may not exercise the privileges of the authorization until he or she has given written notification of the change to the FAA district office for the area in which the new base is located.

Repairman Certificates. A repairman certificate is a specialized certificate dependent upon the person's employment in that area of specialty.

Eligibility for Repairman Certificates. To be eligible for a repairman certificate, a person must be 18 years old, be specifically qualified to perform the tasks necessary for the certificate desired, and have 18 months of practical experience in the duties of his or her job. The applicant must also be employed for a specific job by a certificated repair station, commercial operator, or air carrier, and he or she must be recommended for certification by his or her employer.

Privileges and Limitations of Repairman Certificates. A certificated repairman may perform or supervise the maintenance of aircraft or related components appropriate to the job for which he or she is employed and certificated, but only in connection with duties for the repair station, commercial operator, or air carrier by whom he or she was employed and recommended.

A repairman may not perform or supervise duties under his or her certificate unless he or she understands the current instructions of both the operator and the manufacturer relating to the specific operations concerned.

The FAR parts that have been covered to this point (Parts 43, 65, and 91) include regulations that the aviation maintenance technician must know. In the next few sections of this text the emphasis will be placed on the type of material in other parts and how to use such information. Much of the material in these parts is of little interest to the aviation maintenance technician and will be covered superficially, if at all.

CERTIFICATION OF PRODUCTS AND PARTS

Title VI of the Federal Aviation Act of 1958 mandates the FAA to prescribe such minimum standards governing the design, materials, workmanship, construction, and performance of any aircraft, aircraft engines, and propellers as may be required in the interest of safety. The FAA has prescribed such minimum standards by establishing airworthiness standards for a number of categories. When an aircraft, engine, or propeller meets or exceeds these requirements, it is issued a certificate. The FAA issues type certificates, production certificates, and airworthiness certificates for aircraft. It also issues Parts Manufacturer Approvals (PMAs) and Technical Standard Order (TSO) Authorizations for the manufacture of parts.

Type Certificates

The aircraft type certificate is issued for a certain design, or type, of aircraft, aircraft engine, or propeller. A person seeking a type certificate for a product must make formal application in the manner prescribed in Part 21, Certification Procedures for Products and Parts. The FAA then designates the standards and any special conditions to be met by that product. The applicant does all necessary tests and calculations and provides all documentation to show that the product does meet or exceed all requirements. If the product passes these tests, the FAA will issue a type certificate.

The type certificate consists of the type design, the operating limitations, the type certificate data sheet, the applicable regulations with which the FAA requires compliance, and any other conditions or limitations prescribed for the product. The type design consists of the following:

1. The drawings and specifications necessary to show the configuration of the product and the design features specified in the applicable regulatory requirements.

2. Information on dimensions, materials, and processes needed to define the structural strength of the product.

3. Any other data necessary to allow the determination of the airworthiness and noise characteristics of the product.

4. The airworthiness limitations section of the instructions for continued airworthiness as required in the appropriate airworthiness standards for that product.

Although any interested person may apply for a type certificate, the FAA does not normally issue one for a product with manufacturing facilities outside of the United States. An exception can be made if the administrator finds that the location places no undue burden on the FAA in administering applicable airworthiness requirements.

The type certificate may be transferred or made available to a third person by licensing agreement. The FAA must be notified of the execution or termination of any licensing agreements involving type certificates.

The holder of design approval, including either the type certificate or a supplemental type certificate for a product for which application was made after January 28, 1981, shall furnish at least one set of complete **Instructions for Continued Airworthiness** to the owner of each type product upon its delivery. The instructions will be made available to any other person required by the regulations to comply with any part of them. Changes to the instructions for continued airworthiness will be made available to any person required to comply with them.

A type certificate is effective until surrendered, suspended, or revoked, unless a termination date is otherwise established by the administrator.

Privileges of Type Certificates. The holder of a type certificate for a product may do the following:

1. In the case of an aircraft, obtain an airworthiness certificate.

2. In the case of an aircraft engine or propeller, obtain approval for installation on certificated aircraft.

3. In the case of any product, apply for a production certificate for that product.

4. Obtain approval of replacement parts for that product.

Aircraft that are produced under a type certificate are subject to individual inspection by the FAA for conformity with type design. The manufacturer must also maintain adequate records, provide an appropriate production inspection system, and perform adequate tests to ensure that the aircraft conforms to the original type design.

Changes to Type Design. Changes to a type design are classified as major or minor. A minor change is one which has no appreciable effect upon the weight, balance, structural strength, reliability, operational characteristics, or other characteristics affecting the airworthiness of the prod-

uct. Minor changes to the type design may be approved under a method acceptable to the administrator before submitting any substantiating or descriptive data.

A major change in type design requires the submission of substantiating data and the necessary descriptive data for inclusion in the type design. The applicable regulations for the design change will be those incorporated in the original type certificate or those applicable and in effect on the date of application for the change, plus any special provisions that the administrator deems necessary for the proposed change.

If an airworthiness directive (FAR 39) is issued for a product and the administrator finds that design changes are necessary to correct the unsafe condition, the holder of the type certificate will submit appropriate design changes for approval. Upon approval of these changes, the holder will make available the descriptive data covering the changes to all operators of the products previously certificated under the type certificate. Any person who proposes to change a product must make an application for a new type certificate if the administrator finds that the proposed change is so extensive that a complete investigation of compliance with the applicable regulations is required. A new type certificate will be required for normal-, utility-, acrobatic-, commuter-, and transport-category aircraft if the proposed change is

1. in the number of engines or rotors; or

2. to engines or rotors using different principles of propulsion or to rotors using different principles of operation.

In the case of an aircraft engine, a new type certificate will be required if the proposed change is in the principle of operation. A new type certificate will be required for a propeller if the proposed change is in the number of blades or the principle-of-pitch change mechanism.

Issue of Type Certificates. Upon proof of compliance with the required regulations, type certificates are issued to normal-, utility-, acrobatic-, commuter-, and transport-category aircraft, aircraft engines, and propellers.

A **normal-category aircraft** is an airplane with a seating configuration, excluding pilot seats, of nine or less and a maximum certificated takeoff weight of 12 500 pounds or less. It is intended for nonacrobatic operation. Nonacrobatic operation includes any maneuver incident to normal flying: stalls (except whip stalls), lazy eights, chandelles, and steep turns when the angle of bank is not more than 60°.

A **utility-category aircraft** is an airplane with a seating configuration, excluding pilot seats, of nine or less and a maximum certificated takeoff weight of 12 500 pounds or less. It is intended for limited acrobatic operations, including spins (if approved for the particular type of airplane), lazy eights, chandelles, and steep turns in which the angle of bank is more than 60°.

An **acrobatic-category aircraft** is an airplane with a seating configuration, excluding pilot seats, of nine or less and a maximum certificated takeoff weight of 12 500 pounds or less. It is intended for use without restrictions other than those shown to be necessary as the result of flight tests.

A **commuter-category aircraft** is a propeller-driven multiengine airplane with a seating configuration, excluding pilot seats, of 19 or less and a maximum certificated takeoff weight of 19 000 pounds or less. It is intended for nonacrobatic operations.

The term **transport-category aircraft** applies to airplanes with more than 12 500 pounds maximum certificated takeoff weight, except for those in the commuter category.

The term **aircraft engine** means an engine that is used or intended to be used for propelling aircraft. It includes turbosuperchargers, appurtenances, and accessories necessary for its functioning, but it does not include propellers.

The term **propeller** means a device for propelling an aircraft that has blades on an engine-driven shaft and that, when rotated, produces by its action on the air a thrust approximately perpendicular to its plane of rotation. It includes control components normally supplied by the manufacturer but does not include the main and auxiliary rotors or the rotating airfoils of the engines.

Aircraft may be certificated in more than one category if they meet all applicable requirements.

Restricted-Category Aircraft. A restricted type certificate will be issued for an aircraft that has been shown to have no feature or characteristic which makes it unsafe when operated under the limitations prescribed for its intended use. The aircraft must also meet airworthiness requirements of an aircraft category except for those requirements that the administrator finds inappropriate for the special purpose for which the aircraft is to be used. An aircraft that has been accepted for use by an armed force of the United States and has later been modified for a special purpose may also be given a restricted type certificate.

Special-purpose operations include the following:

1. Agricultural uses (spraying, dusting, and seeding as well as livestock and predatory-animal control)
2. Aerial surveying (topography, mapping, and oil and mineral exploration)
3. Forest and wildlife conservation
4. Patrolling (pipelines, power lines, and canals)
5. Weather control (cloud seeding)
6. Aerial advertising (skywriting, banner towing, airborne signs, and public address systems)
7. Any other operation specified by the administrator

Military-Surplus Aircraft. An applicant for a type certificate for a surplus military aircraft will be entitled to a certificate if the aircraft is a counterpart of a previously type-certificated aircraft and shows compliance with the applicable regulations for that type certificate. If the aircraft does not have a previously certificated counterpart, it must be shown that the aircraft complies with the applicable regulations specified in Section 21.27(f).

Import Products. A type certificate may be issued for a product that is manufactured in a foreign country with which the United States has a trade agreement for these products and that is to be imported into the United States if the following conditions are met:

1. The country in which the product was manufactured certifies that the product has been examined, tested, and found to meet the applicable airworthiness, noise, and special provisions designated by the administrator.
2. The applicant has submitted the technical data required by the administrator.
3. The manuals, placards, listings and instrument markings required by the applicable requirements are presented in the English language.

Each holder or licensee of a U.S. type certificate for an aircraft engine or propeller manufactured in a foreign country with which the United States has a trade agreement for these products must furnish with each engine or propeller a certificate of airworthiness for export issued by the country of manufacture. This certificate must certify that the individual engine or propeller conforms to its U.S. type certificate, is in condition for safe operation, and has been subjected by the manufacturer to a final operational check.

Supplemental Type Certificates. Any person who alters a product by introducing a major change in type design, not great enough to require a new type certificate, should apply to the administrator for a supplemental type certificate (STC). The holder of the type certificate for that product can alternatively apply for an amendment of the original type certificate. Supplemental type certificates are covered in Chapter 14.

Production Certificates

A person may apply for a production certificate if he or she holds a current type certificate, a license agreement, or an STC for the product concerned.

An applicant must show that he or she has established and can maintain a quality-control system for the product so that each article will meet the design provisions of the pertinent type certificate. The details of this quality-control program are outlined in Section 21.143.

The holder of a production certificate may obtain an airworthiness certificate without further showing, except that the administrator may inspect the aircraft for conformity with the type design; or in the case of other products, obtain approval for installation on certificated aircraft.

Airworthiness Certificates

Section 91.203 requires that no person may operate an aircraft unless it has an appropriate and current airworthiness certificate. Any registered owner of a U.S.-registered aircraft may apply for an airworthiness certificate for that aircraft. Airworthiness certificates are issued in two classes, standard and special.

The first class of certificates, **standard airworthiness certificates**, are issued for aircraft type certificated in the normal, utility, acrobatic, commuter, or transport category; for manned free balloons; and for aircraft designated by the administrator as being in a special class of aircraft. The second class, **special airworthiness certificates**, are restricted, limited, and provisional airworthiness certificates; special flight permits; and experimental certificates. Airworthiness certificates are transferred with the aircraft.

Standard Airworthiness Certificates. Standard airworthiness certificates are effective as long as the maintenance, preventive maintenance, and alterations are performed in accordance with Parts 43 and 91 of the regulations and the aircraft is registered in the United States. New aircraft are eligible for an airworthiness certificate once they have met all the necessary provisions of their production certificates, their import certifications, or their inspections if they are produced only under a type certificate.

The owner of an other-than-new aircraft must present evidence that the aircraft conforms to a type design under a type certificate or a supplemental type certificate and to applicable airworthiness directives. The aircraft also must have had a 100-hour inspection by a person authorized to perform such an inspection. An airworthiness certificate will be issued if the administrator finds, after inspection, that the aircraft conforms to the type design and is in condition for safe operation.

Restricted Airworthiness Certificates. The issue and duration of restricted airworthiness certificates are similar to those for standard airworthiness certificates.

Limited Airworthiness Certificates. An airworthiness certificate is issued in the limited category when it is shown that the aircraft has previously been issued and conforms to a limited-category type certificate. The aircraft must also be in a good state of preservation and repair and in a condition for safe operation.

Section 91.315 states that "No person may operate a limited-category civil aircraft carrying persons or property for compensation or hire."

Special Flight Permits. A special flight permit may be issued for an aircraft that may not currently meet applicable airworthiness requirements but is capable of safe flight for the purpose of

1. flying the aircraft to a base where repairs, alterations, or maintenance are to be performed or to a point of storage;
2. delivering or exporting the aircraft;
3. production flight testing new production aircraft; and
4. evacuating the aircraft from areas of impending danger.

A special flight permit may also be issued to authorize the operation of the aircraft at a weight in excess of its maximum certificated takeoff weight for flight beyond the normal range over water or over land areas where adequate landing facilities or appropriate fuel is not available. The excess weight that may be authorized under this rule is limited to the additional fuel, fuel-carrying facilities, and navigational equipment necessary for the flight.

An applicant for a special flight permit must submit a statement indicating

1. the purpose of the flight;
2. the proposed itinerary;
3. the crew required to operate the aircraft and its equipment;
4. the ways, if any, in which the aircraft does not comply with applicable airworthiness requirements;
5. any restriction the applicant considers necessary for the safe operation of the aircraft; and
6. any other information considered necessary for the administrator to know for the purpose of prescribing operating limitations.

The administrator may make or require the applicant to make appropriate inspections and tests for safety.

A special flight permit is effective for the period of time specified in the permit.

Experimental Certificates. Experimental certificates are issued for research and development, for showing compliance with regulations, for crew training, for exhibitions, for air racing, for market surveys, and for operating amateur-built aircraft. An applicant for an experimental certificate must submit the following information:

1. A statement setting forth the purpose for which the aircraft is to be used.
2. Enough data to identify the aircraft.
3. Upon inspection of the aircraft, any pertinent information found necessary by the administrator to safeguard the general public.
4. In the case of an aircraft to be used for experimental purposes, the purpose of the experiment; the estimated time or number of flights required for the experiment; the areas over which the experiment will be conducted; and, except for aircraft converted from a previously certificated type without appreciable change in the external configuration, three-view drawings or three-view dimensioned photographs of the aircraft.

An experimental certificate for research and development, for showing compliance with regulations, for crew training, or for market surveys is effective for one year after the date of issue or renewal, unless a shorter period is prescribed by the administrator. The duration of an experimental certificate for amateur-built airplanes, for exhibitions, and for air racing are unlimited unless the administrator finds for good cause that a specific period should be established.

Approval of Materials, Parts, Processes, and Appliances

Replacement or modification parts for installation on type-certificated aircraft must be one of the following five types:

1. Parts produced under a type certificate or production certificate.
2. Parts produced by an owner or operator for maintaining or altering his or her own product.

3. Standard parts (such as bolts or nuts) conforming to established industry or United States specifications. Standard hardware is covered in Chapter 10 of this text.

4. Parts produced under an FAA technical standard order (TSO) authorization.

5. Parts produced under a parts manufacturer approval (PMA).

Technical Standard Orders. Subpart O of Part 21 prescribes requirements for the issue of **technical standard order (TSO) authorizations** and describes how to obtain authorization to manufacture a part under a TSO. Technical standard orders contain minimum performance and quality-control standards for specified materials, parts, or appliances used on civil aircraft. An article produced under a TSO authorization is considered to be an approved part.

The holder of a TSO authorization, in addition to following the standards for each product, must establish and maintain a quality-control program to ensure that the article meets the standards and is safe for operation. A complete file of technical data and records for each product model must be maintained. Each part is to be marked permanently and legibly with the following:

1. The name and address of the manufacturer.

2. The name, type, part number, or model designation of the article; the nominal; and the weight of the article.

3. The serial number or date of manufacture of the article, or both.

4. The applicable TSO number.

A TSO authorization is not transferable and is effective until surrendered, withdrawn, or otherwise terminated by the manufacturer.

Parts Manufacturer Approval. The procedure to receive **parts manufacturer approval (PMA)** is in Part 21, Subpart K. An applicant must submit the following to the FAA regional office in his or her area:

1. The identity of the product on which the part is to be installed.

2. The name and address of the manufacturing facilities at which the part is to be manufactured.

3. The design of the part, which consists of drawings and specifications necessary to show the configuration of the part and information on the dimensions, material, and processes necessary to define the structural strength of the part.

4. Test reports and computations necessary to show that the design of the part meets the airworthiness requirements of the Federal Aviation Regulations applicable to the product on which the part is to be installed, unless the applicant shows that the design of the part is identical to the design of a part that is covered under a type certificate.

The applicant must also have a fabrication inspection system that ensures that each completed part conforms to its design data and is safe for installation on applicable type-certificated products.

Parts manufacturer approval is not transferable and is effective until surrendered, withdrawn, or otherwise terminated by the administrator.

Reporting of Failures, Malfunctions, and Defects. Section 21.3 requires that the holder of certificates, approvals, or authorizations for any part must report any failure, malfunction, or defect that has resulted in any of 13 listed occurrences (fire, brake-system failure, and so on). The holder must also report if he or she determines that a part that has left the quality-control system could result in one or more of these occurrences.

AIRWORTHINESS STANDARDS

The FAA publishes the minimum standards for aircraft certification in several different parts of the Federal Aviation Regulations depending upon the use of the aircraft. In this section we will look at the format and organization of these parts as well as the type of information they contain. You are not expected to know details of the various standards, but you should be capable of locating specific information in the appropriate part.

The administrator will determine which part or parts of the regulation are effective when an application for a type certificate is filed. The proposed aircraft must meet those standards or a type certificate will not be issued. Since the airworthiness standards are constantly being revised, it is important to be able to determine exactly which amendments apply to a particular aircraft. This information is found in the type certificate data sheets.

Applicability

The various parts which prescribe airworthiness standards for the issuance of type certificates and changes to those certificates are as follows:

Part 23, Airworthiness Standards: Normal, Utility, Acrobatic, and Commuter Aircraft

Part 25, Airworthiness Standards: Transport-Category Aircraft

Part 27, Airworthiness Standards: Normal-Category Rotorcraft

Part 29, Airworthiness Standards: Transport-Category Rotorcraft

Part 31, Airworthiness Standards: Manned Free Balloons

Part 33, Airworthiness Standards: Aircraft Engines

Part 35, Airworthiness Standards: Propellers

Part 36, Noise Standards: Aircraft Type Certification

The FAA recodified the Civil Air Regulations into the Federal Aviation Regulations in the early 1960s. Aircraft that had applications for type certificates prior to this change are certified under the Civil Air Regulation airworthiness standards. The relationship between the parts of the Federal Aviation Regulations and the Civil Air Regulations is as follows:

FAR	CAR
23	3
25	4b
27	6
29	7
33	13
35	14

The aviation maintenance technician working on an aircraft certificated under the Civil Air Regulations may have to refer to parts of the Federal Aviation Regulations to find the information on the standards that the aircraft must meet.

Airworthiness Standards, Aircraft (Parts 23, 25, 27, 29)

The four parts pertaining to aircraft (Parts 23, 25, 27, and 29) are arranged in such a manner that the section number for a specific topic will be the same in all four parts. For example, Sections 23.1529, 25.1529, 27.1529, and 29.1529 all require the applicant for a type certificate to prepare instructions for continued airworthiness. Some material may not be applicable in one or more of the parts, and thus there are gaps in the numbering.

The airworthiness standards are also organized into a number of subparts:

General. This subpart lists the applicability for the part and other information necessary to clarify that applicability.

Flight. This subpart details the flight requirements that the aircraft must be able to perform and maintain, such as stability, stalls, spins, and performance under various power conditions. The requirements for weight, center of gravity, and ground handling are also listed in this subpart.

Structure. This subpart sets the minimum requirements for the various loads that the structure must be able to withstand in flight, during takeoff and landing, and while on the ground.

Design and construction. In this subpart are factors and considerations to be used in designing various parts of the aircraft. Included in this section are the design characteristics of the landing-gear system, personnel and cargo accommodations, and fire protection. Specific examples are the size of the control cables used, the need for a flap-position indicator, and the motion and effect of the powerplant throttle.

Powerplant. The powerplant installation is covered in this subpart; the engine itself is type certificated under Part 33. One of the requirements of this subpart is that the engine and propeller be type certificated under their respective parts. Also included are fuel systems and components, oil systems, cooling systems, induction systems, exhaust systems, powerplant controls and accessories, and powerplant fire protection. Specific examples are the size of fuel strainers, the pressure requirements of oil tanks, the propeller ground clearance, and the use of induction-system screens.

Equipment. This subpart covers the remainder of the equipment installed in the aircraft other than that covered under the powerplant section. Major coverage is given to instruments, electrical systems, lights, and safety equipment. A section entitled Miscellaneous Equipment sets the standards for such items as electronics, hydraulics, oxygen, and pressurization equipment. Specific examples of standards in this section are the required flight instruments, the number of spare fuses, the color of position lights, and the use of chemical oxygen generators.

Operating limitations and information. As the name suggests, the standards in this subpart establish operating limitations for the equipment and the personnel operating the aircraft. Included are the requirements for markings and placards, an airplane flight manual, and instructions for continued airworthiness. Specific examples of this subpart are airspeed operating limitations, required material in the maintenance manuals, airspeed-indicator markings, and minimum flight crew.

Each part has **appendices** which aid in either explaining or determining compliance with the various sections. For example, Part 23, Appendix B, Control Surface Loadings, provides data that may be used in determining control surface loads. Appendix G, Instructions for Continued Airworthiness, provides detailed information for the instructions, including format, content, and the requirement for an airworthiness limitations section.

Various **special Federal Aviation Regulations (SFAR)** may also be attached to a part. For example, SFAR 23 contains special standards to be used for an applicant seeking a normal-category type certificate for a reciprocating or a turbopropeller small airplane that would carry more than 10 occupants and is intended for use under Part 135.

Airworthiness Standards, Engines (Part 33)

Part 33, Aircraft Engines, consists of the following subparts:

Subpart A, General
Subpart B, Design and Construction: General
Subpart C, Design and Construction: Reciprocating Aircraft Engines
Subpart D, Block Tests: Reciprocating Aircraft Engines
Subpart E, Design and Construction: Turbine Aircraft Engines
Subpart F, Block Tests: Turbine Aircraft Engines
Appendix A, Instructions for Continued Airworthiness

The design and construction subparts contain requirements for items such as vibration, systems for fuel, induction, ignition, lubrication, and so on. Block-test requirements include vibration, calibration, endurance, operations, and teardown inspection requirements.

Airworthiness Standards, Propellers (Part 35)

Part 35 consists of the following subparts:

Subpart A, General
Subpart B, Design and Construction

Subpart C, Tests and Inspections
Appendix A, Instructions for Continued Airworthiness

Technical Standard Orders

A technical standard order (TSO) provides standards for manufacturing a certain type of component. The standard for a component will be assigned a TSO number, such as TSO-C22f (safety belts), TSO-C52a (flight directors), or TSO-C62c (aircraft tires). The complete list of items that have a TSO can be found in *Advisory Circular 20-110*.

Airworthiness Directives (Part 39)

Part 39 prescribes requirements for airworthiness directives that apply to aircraft, aircraft engines, propellers, or appliances (hereafter referred to as *products*) when an unsafe condition exists in a product and that condition is likely to exist or develop in other products of the same type design. No person may operate a product to which an airworthiness directive applies except in accordance with the requirements of the directive. Detailed information on airworthiness directives is given in a separate chapter of this text.

Identification and Registration Marking (Part 45)

Part 45 of the Federal Aviation Regulations prescribes the requirements for

1. the identification of aircraft and the identification of aircraft engines and propellers that are manufactured under the terms of a type or a production certificate;
2. the identification of certain replacement and modified parts produced for installation on type-certificated products; and
3. the nationality and registration marking of United States–registered aircraft.

Identification of Type-certificated Products. Aircraft must be identified by an identification plate secured in such manner that it will not likely be defaced or removed during normal service or lost or destroyed in an accident. It must be secured to the aircraft at an accessible location near an entrance, or, if it must be legible to a person on the ground, it may be located externally on the fuselage near the tail surfaces.

The engine must be identified by a fireproof plate and fireproof markings. The identification plate must be affixed to the engine at an accessible location and in such a manner that it will not likely be defaced or removed during normal service or lost in an accident.

Each person who manufactures a propeller, propeller blade, or propeller hub under the terms of a type or production certificate must identify that product by means of a plate, stamping, etching, or other approved method of fireproof identification. The identification must be placed on a noncritical surface and must not be likely to be defaced or removed during normal service or lost or destroyed in an accident.

Required Data for Type-certificated Products. The identification on type-certificated products must include the following information:

1. The builder's name
2. The model designation
3. The builder's serial number
4. The type-certificate number, if any
5. The production-certificate number, if any
6. For aircraft engines, the established rating
7. Any other information the administrator finds appropriate

No person may remove, replace, or change the above identification data on a certificated product without the approval of the administrator. If it is necessary to remove an identification plate while doing maintenance in accordance with Part 43, the exact plate and/or information must be used for replacement.

Markings for Parts and Components. Each person who produces a part for which a replacement time, inspection interval, or related procedure is specified in the airworthiness limitations section of a manufacturer's maintenance manual or in the instructions for continued airworthiness must permanently and legibly mark that component with a part number (or equivalent) and a serial number (or equivalent).

Each person who produces a replacement or a modification part under a parts manufacturer approval must permanently and legibly mark the part with

1. the letters FAA-PMA;
2. the name, trademark, or symbol of the holder of the parts manufacturer approval;
3. the part number; and
4. the name and model designation of each type-certificated product on which the part is eligible for installation.

If the administrator finds that a part is too small or that it is otherwise impractical to mark a part with any of the preceding information, a tag must be attached to the part or its container must include the information that could not be marked on the part.

Nationality and Registration Marks. A United States–registered aircraft cannot be operated unless the nationality and registration marks are displayed in accordance with the provisions of Subpart C of FAR 45. The marks must be painted on or affixed by other means insuring a similar degree of permanence, have no ornamentation, contrast in color with the background, and be legible. Unless otherwise authorized by the administrator, no person may place on any aircraft a design, mark, or symbol that modifies or confuses the nationality and registration marks.

Temporary markings may be affixed to an aircraft with a readily removable material if the aircraft is intended for immediate delivery to a foreign purchaser or if it is bearing a temporary registration number.

Exhibit and Antique Aircraft, Special Rules. When the display of the registration markings would be inconsistent with the exhibition of that aircraft, a United States–registered aircraft may be operated without those markings if it is operated for the purpose of exhibition, including in motion pictures, television productions, and airshows. The aircraft must be operated only at the exhibition sites, between such locations, for practice and test flights necessary for exhibition purposes, and between the aircraft's home base and the exhibition sites.

If the aircraft is at least 30 years old or has the same external configuration of an aircraft built at least 30 years ago, it may be operated with markings in accordance with the provisions set forth in Section 45.22(b) rather than the current requirements.

Display of Marks. Each operator of a U.S.-registered aircraft must display on the aircraft a series of marks consisting of the Roman capital letter *N* followed by the registration number of the aircraft. Each suffix letter used in the marks must also be a capital letter.

When the marks that include only the capital letter *N* and the registration number are displayed on limited- or restricted-category aircraft or experimental or provisionally certificated aircraft, the owner must also display on that aircraft, near the entrance to the cockpit and in letters not less than 2 inches nor more than 6 inches in height, the words *limited, restricted, experimental,* or *provisional airworthiness,* as the case may be.

The location and the size of markings for various types of aircraft is beyond the scope of this text. You should be aware that this information can be located in Section 45.25, 45.27, and 45.29 of the Federal Aviation Regulations.

The use of the capital letter *N* denotes United States registration. A list of the letters used by other countries is in the Appendix of this text.

Markings for Export Aircraft. A person who manufactures an aircraft in the United States for delivery outside the country may display on the aircraft any marks required by the state or registry of the aircraft. However, no person may operate an aircraft so marked within the United States, except for test and demonstration flights for a limited period of time or while in transit to the purchaser.

When an aircraft that is registered in the United States is sold to a person who is not a citizen of the United States, the seller must, before delivery to the purchaser, remove all United States registration marks from that aircraft.

Repair Stations (Part 145)

Part 145 prescribes the requirements for the issuance of repair-station certificates and associated ratings and the general operating rules for the holders of such certificates and ratings.

A certificated repair station located in the United States is called a *domestic repair station,* and one located outside the United States is known as a *foreign repair station.*

In general, a repair-station certificate gives an agency, rather than an individual, the authorization to perform maintenance, preventive maintenance, and alterations. A repair station may also authorize the work to be approved for return to service. The repair station is limited to doing only those operations for which it is rated. These ratings may be very broad or may be quite limited.

Repair-Station Ratings. Ratings are issued in a number of categories and further broken down into various classes. Airframe ratings are as follows:

Class 1, Composite Construction of Small Aircraft
Class 2, Composite Construction of Large Aircraft
Class 3, All-metal Construction of Small Aircraft
Class 4, All-metal Construction of Large Aircraft

Powerplant ratings consist of the following:

Class 1, Reciprocating Engines of 400 Horsepower or Less
Class 2, Reciprocating Engines of More Than 400 Horsepower
Class 3, Turbine Engines

Propeller ratings are as follows:

Class 1, All Fixed-Pitch and Ground-Adjustable Propellers of Wood, Metal, or Composite Construction
Class 2, All Other Propellers, by Make

Radio ratings are as follows:

Class 1, Communication Equipment
Class 2, Navigational Equipment
Class 3, Radar (Pulsed Frequency) Equipment

Instrument ratings consist of the following:

Class 1, Mechanical
Class 2, Electrical
Class 3, Gyroscopic
Class 4, Electronic

Accessory ratings include the following:

Class 1, Mechanical Accessories That Depend on Friction, Hydraulics, Mechanical Linkage, or Pneumatic Pressure for Operation
Class 2, Electrical Accessories That Depend on Electrical Energy for Their Operation, and Generators
Class 3, Electronic Accessories That Depend on the Use of an Electron Tube, Transistor, or Similar Device, Including Electronic Controls.

Repair-Station Limited Ratings. A limited rating may be issued for a repair station. This rating may allow the station to perform work on a particular make of aircraft, a particular model of that aircraft, or even a specific component, by manufacturer or model, of an aircraft. A limited rating may also be issued for special services such as welding or nondestructive testing and inspection.

Repair-Station Facilities and Equipment. The applicant for a repair-station rating must have sufficient equipment to provide the services for the rating or ratings sought.

Appendix A of Part 145 lists the operations that the holder of the various ratings must be able to perform.

The facilities must include adequate housing to ensure proper separation of activities such as cleaning, inspection, and assembly. There must be adequate space and storage of work in progress, new parts and materials, and equipment in such a manner as to not compromise the work being done. An applicant for an airframe rating must provide suitable permanent housing for at least one of the heaviest aircraft within the weight class of the rating she or he seeks. If the location of the station is such that climatic conditions allow the work to be done outside, permanent work docks may be used if they meet the other applicable requirements for facilities.

Repair-Station Personnel. A repair station must provide adequate personnel to perform, supervise, and inspect the work for which the station is rated. The officials of the repair station must carefully consider the qualifications and the abilities of their employees and must determine the abilities of their noncertificated employees on the basis of the practical tests or their employment records. The repair-station officials are primarily responsible for the satisfactory work of their employees.

The number of employees may vary, but the number must be sufficient to perform the volume of work in process and efficiently produce airworthy work.

The repair station officials must determine the abilities of their supervisors and must provide enough of them for all phases of their activities. The ability of each supervisor is subject to evaluation, if requested by the FAA. When apprentices or students are involved in the operation, the repair station must have at least one supervisor for each ten apprentices or students.

Each person who is directly in charge of the maintenance functions of a repair station must be appropriately certificated as a repairman or a mechanic under Part 65. In addition, at least one of the persons in charge of maintenance functions for a station with an airframe rating must have had experience in the methods and procedures prescribed by the administrator for returning aircraft to service after 100-hour, annual, and progressive inspections.

Each certificated repair station must maintain a roster of the following:

1. Its supervisory personnel, including the names of the officials of the station that are responsible for its management and the names of its technical supervisors, such as foreman and crew chiefs.

2. Its inspection personnel, including the name of the chief inspector and those inspectors who make the final airworthiness determinations before releasing an article for service.

The station must also provide and keep current a summary of the employment of each person whose name is on the roster. The summary must include the following:

1. His or her present title

2. His or her total years of experience in the type of work he or she is doing

3. His or her past employment record with the names of places and terms of employment by month and year

4. The scope of his or her present employment

5. The type and number of the mechanic or repairman certificate that he or she holds

Repair-Station Inspection Procedures. At the time application is made for a repair-station certificate, the applicant must provide a manual containing the inspection procedures and thereafter maintain it in current condition at all times. The manual must explain the internal inspection system of the repair station in a manner easily understood by any employee of the station. It must state in detail the inspection requirements of 145.45 (a) through (e), and the repair station's inspection system. The manual must refer whenever necessary to the manufacturer's inspection standards for the maintenance of the particular article.

The repair station must give a copy of the manual to each of its supervisory and inspection personnel and make it available to other personnel. The repair station is responsible for seeing that all supervisory and inspection personnel thoroughly understand the manual.

Privileges and Limitations of Repair-Station Certificates. A certificated repair station may maintain or alter any airframe, powerplant, propeller, instrument, radio, accessory, or related part for which it is rated. It may also approve for return to service any article for which it is rated after maintenance or alteration. If the station has an airframe rating, it may perform the 100-hour, annual, or progressive inspection and approve the aircraft for return to service.

A repair station may not approve for return to service a major repair or alteration unless it has been performed in accordance with data approved by the administrator.

A repair station that performs work for an operator certificated under Part 121 must perform that work under the applicable standards of Part 121 and the air carrier's commercial operator's manual.

Repair-Station Performance Standards. Each certificated repair station must perform its maintenance and alteration operations in accordance with the standards in Part 43. It also must maintain in current condition all manufacturers' service manuals, instructions, and service bulletins that relate to the articles that it maintains or alters.

Repair-Station Approval for Return to Service. Each repair station, before approving an item for return to service after maintenance or alteration, must have that article inspected by a qualified inspector. The station must certify on the maintenance or alteration record of that item that it is airworthy.

A qualified inspector is a person employed by the station who has shown by experience that she or he understands the inspection methods, techniques, and equipment used in determining the airworthiness of the article concerned. She or he must also be proficient in using various types of mechanical and visual inspection aids appropriate for the article being inspected.

Repair-Station Records and Reports. Each certificated repair station must maintain adequate records of all work that it does, naming the certificated mechanic or repairman who performed or supervised the work and the inspector of that work. The station must keep these records for at least two years after the work is done.

The preceding paragraph refers to the records for the repair station. The aircraft records must have entries as described in Parts 43 and 91.

Each domestic repair station must report to the administrator within 72 hours after it discovers any serious defect or other recurring nonairworthy condition in an aircraft, powerplant, propeller, or related component.

FAR 145 also prescribes rules for foreign repair stations and limited ratings for manufacturers. That information is not covered in this text.

Operations Certification

The FAA requires those performing certain types of operations to comply with rules over and above those found in Part 91. The parts covering these operations include the following:

Part 121, Certification and Operations: Domestic, Flag, and Supplemental Air Carriers and Commercial Operation of Large Aircraft

Part 125, Certification of Operations: Airplanes Having a Seating Capacity of 20 or More Passengers or a Maximum Payload of 6000 Pounds or More

Part 127, Certification and Operations of Scheduled Air Carriers with Helicopters

Part 135, Air-Taxi and Commercial Operators

In general, these operators are required to develop and have approved by the FAA a set of operating specifications. These specifications become their operations manual and set forth procedures for virtually all operations, including maintenance. The various regulatory parts provide guidelines for the development of those specifications.

A detailed study of the maintenance requirements of these parts is beyond the scope of this text. Some portions of Part 135 are covered in the next section to provide an overview of the type of requirements in these parts. Operators under these parts are often exempted from requirements in Parts 43 and 91. This is only done if there are more restrictive requirements in the specific operations part.

Air-Taxi and Commercial Operators (Part 135)

Part 135 prescribes rules governing air-taxi and commercial operators (ATCO) operations conducted under authority of Part 298: the transportation of mail by aircraft under a postal-service contract; the carriage in air commerce of persons or property for compensation or hire as a commercial operator (not an air carrier) in aircraft having a maximum passenger seating capacity of less than 20 passengers or a maximum payload capacity of less than 6000 pounds; or the carriage in air commerce of persons or property in common carriage operations solely between points entirely within

any state of the United States and in aircraft having a maximum passenger seating capacity of 30 seats or less or a maximum payload capacity of less than 7500 pounds.

A number of commercial operations are exempted from the requirements of this part. They include instruction and training, nonstop sight-seeing trips within 25 statute miles of the airport, and other similar operations. The complete list of exceptions may be found in Section 135.1(b).

Provisions exist for air-taxi operation with aircraft of more than 30 seats. In general, the requirements are the same as the operating and maintenance regulations required for supplemental air carriers under Part 121.

ATCO Operations Certification. A person desiring to operate aircraft under this part must make application and receive approval for an ATCO operating certificate. The certificate holder must prepare and keep current a manual setting forth the procedures and policies approved by the administrator. This manual must be used by the certificate holder's flight, ground, and maintenance personnel in conducting its operations. If the administrator finds that because of the size of the operation all or part of the manual is not necessary, she or he may authorize a deviation from the manual requirements. The manual must include the following maintenance-related items:

1. The name of each management person required under Section 135.37(a), including a director of maintenance.

2. Procedures for ensuring that the pilot in command knows that required airworthiness inspections have been made and that the aircraft has been approved for return to service in compliance with the applicable maintenance requirements.

3. Procedures to be followed by the pilot in command to obtain maintenance, preventive maintenance, and servicing of the aircraft at a place where previous arrangements have not been made by the operator, when the pilot is authorized to act for the operator.

4. Procedures for reporting and recording mechanical irregularities that come to the attention of the pilot in command before, during, and after the completion of a flight.

5. Procedures to be followed by the pilot in command for determining that mechanical irregularities or defects reported for previous flights have been corrected or that correction has been deferred.

6. The approved airworthiness program, when applicable.

Maintenance Required for ATCO Operations. Subpart J of Part 135 sets rules for maintenance, preventive maintenance, and alterations. Each certificate holder is primarily responsible for the airworthiness of his or her aircraft, must have it maintained as required, and must have defects repaired between required maintenance under Part 43.

The requirements for maintenance are divided into two categories, those for aircraft with nine passenger seats or less and those with ten or more passenger seats.

Both of these categories are subject to mechanical reliability reports (135.415) and a mechanical interruption summary (135.417).

Maintenance Required for ATCO Operations, Nine or Less Passenger Seats. Aircraft with a passenger seating capacity of nine seats or less must be maintained in accordance with the provisions of Parts 43 and 91, plus additional provisions required in Part 135. The operator must comply with the manufacturer's recommended maintenance programs or a program approved by the administrator for each item of equipment required. A manufacturer's recommended maintenance program is considered to be material required in maintenance manuals or instructions required by FAR Sections 23.1529, 33.4, and 35.4. A person operating the smaller aircraft may have the aircraft maintained under the provisions of Section 135.419, an approved aircraft inspection program, rather than Parts 43 and 91.

An approved aircraft inspection program must include the following:

1. Instructions and procedures for the conduct of the aircraft inspections, setting forth in detail the parts and areas that must be inspected.
2. A schedule for the performance of the aircraft inspections, expressed in terms of time in service, calendar time, number of systems operations, or any combination of these.
3. Instructions and procedures for recording discrepancies found during inspections and correction or deferral of discrepancies, including form or disposition of records.

Whenever the administrator finds that the aircraft inspection requirements of Part 91 are not adequate for this part, or upon application of the certificate holder, the administrator may amend the certificate holder's operations specifications to require or allow an approved aircraft inspection program for any make or model aircraft of which the certificate holder has exclusive use of at least one aircraft.

The registration number of each aircraft subject to this inspection must be listed in the operating specifications.

Maintenance Required for ATCO Operations, Ten or More Passenger Seats. Aircraft with ten or more passenger seats, excluding any pilot seat, are maintained under a program required by Part 135, Subpart J. An operator of this category aircraft must develop and have approved a maintenance program to provide for required inspections, maintenance, preventive maintenance, and alterations. The program must provide adequate organization, procedure, and instruction to ensure that the aircraft are being maintained as provided in the certificate holder's manual. This manual must include a chart or description of the organization and a list of all persons with whom the organization has arranged for the performance of the required inspections placed in the operator's manual.

It should be noted that what has been developed in this case is a continuous program to ensure airworthiness. The standards and procedures are the ones approved and in the manual. This is different from a continuous inspection program that involves only the inspection of the aircraft.

In the event that a certificate holder contracts maintenance under a program required by Part 135, she or he must ensure that the person performing the maintenance, preventive maintenance, or alteration is doing so under the provisions of the regulations and her or his (the certificate holder's) manual. This manual must include the following maintenance-related items:

1. The method of performing routine and nonroutine maintenance, preventive maintenance, and alterations.
2. A designation of the items of maintenance and alteration that must be inspected.
3. The method of performing required inspections and a designation by occupational title of personnel authorized to perform each required inspection.
4. Procedures for the reinspection of work performed under previous required-inspection findings.
5. Procedures standards and limits necessary for required inspections and the acceptance or rejection of the items required to be inspected.
6. Procedures for periodic inspection and calibration of precision tools, measuring devices, and test equipment.
7. Procedures to ensure that all required inspections are performed.
8. Instructions to prevent any person who performs any item of work from performing any required inspection on that work.
9. Instructions and procedures to prevent any decision of an inspector regarding any required inspection from being countermanded by persons other than supervision personnel of the inspection unit or a person at the level of administrative control that has overall responsibility for both maintenance and required inspection.
10. Procedures to ensure that required inspections, other maintenance, preventive maintenance, and alterations that are not completed as a result of work interruptions are properly finished before the aircraft is released to service.

Maintenance Records for ATCO Operations. Each certificate holder must have in his or her manual a suitable system that provides for the retention of the following information:

1. A description of the work performed.
2. The name of the person performing the work if the work is performed by a person outside the organization of the certificate holder.
3. The name or other positive identification of the individual approving the work.

The requirements of Section 43.9, the content, form, and disposition of maintenance records, apply to this type of operation. The holder of the certificate must also keep the records specified in Section 135.439(a) for the length of time specified in Section 135.439(b). These requirements are similar to the requirements of Section 91.417, which are covered in another chapter of this text.

Airworthiness Release for ATCO Operations. No certificate holder may operate an aircraft after maintenance, preventive maintenance, or alterations are performed unless an airworthiness release is prepared or an appropriate entry is made in the aircraft or logbook.

The airworthiness log or logbook entry must

1. be prepared in accordance with the certificate holder's manual;

2. include a certification that the work was required in accordance with the certificate holder's manual, that all items required to be inspected were inspected by an authorized person who determined that the work was satisfactorily completed, that no known condition exists that would make the aircraft unairworthy, and that so far as the work performed is concerned the aircraft is in condition for safe operation; and

3. be signed by an authorized certificated mechanic or repairman, as long as the repairman signs the release or entry only for the work for which she or he is employed and for which she or he is certificated.

Instead of stating each of the conditions of the certification as described in the preceding list, the certificate holder may state in his or her manual that the signature of an authorized certificated mechanic or repairman constitutes that certification.

Continuing Analysis and Surveillance for ATCO Operations. Each certificate holder must establish and maintain a system for the continuing analysis and surveillance of (1) the performance and effectiveness of his or her inspection program; (2) the program covering other maintenance, preventive maintenance, and alterations; and (3) the correction of any deficiencies in these programs, regardless of whether they are carried out by the certificate holder or another person.

REVIEW QUESTIONS

1. When was the FAA created?
2. In what FAA region do you live?
3. Why does the FAA investigate accidents?
4. What does Part 25 of the Federal Aviation Regulations deal with?
5. Who has the primary responsibility of seeing that airworthiness directives are complied with?
6. Who can supervise a 100-hour inspection?
7. Which Appendix of Part 43 provides information for filling out Form 337?
8. How long does an applicant have to complete the examinations for a technician's (mechanic's) rating?
9. What types of repairs may a certified technician perform on instruments?
10. What privileges does inspection authorization offer?
11. When does a manufacturer have to provide instructions for continued airworthiness?
12. What is a utility-category aircraft?
13. What would be the advantage of obtaining a production certificate?
14. How long is an experimental airworthiness certificate good for?
15. What does PMA stand for?
16. Which part of the Federal Aviation Regulations deals with airworthiness standards for aircraft engines?
17. What does Appendix G of Part 23 cover?
18. Which FAR part would you use to determine the size of registration markings to paint on an aircraft?
19. What classes of powerplant ratings might you find in a repair station?
20. What requirements must a repair station meet if it uses apprentices?
21. What FAR part does a domestic air carrier operate under?
22. What size of aircraft can be operated under Part 135?
23. What type of maintenance is required for an air-taxi aircraft with six seats?
24. What must be included in an approved aircraft inspection program under Part 135?
25. What information must be in the airworthiness release for a Part 135 operator?

Technical Publications 14

INTRODUCTION

Over an aircraft's service life virtually all of its components are involved in some form of inspection, preventive maintenance, overhaul, repair, or replacement. Even the best designed aircraft and components will occasionally develop defects. A system must be in place that will allow for the dissemination of technical material from both the FAA and aircraft manufacturers to correct these defects. The technical information needed to correct defects must be available in clear, concise language and organized in such a way as to be easy to use by the technician. This information may be in the form of a technical manual, or it may appear in notices of potential problems. The development and distribution of this needed technical information is a responsibility shared by the FAA and the manufacturers. The proper use of these publications by the technician is a must in maintaining today's complex aircraft.

ADVISORY CIRCULARS

The FAA issues **Advisory Circulars (AC)** (see Figure 14–1) to inform the aviation community, in a systematic way, about nonregulatory information of interest. Unless the advisory circular is incorporated into an FAR, its contents are not binding on the public. An advisory circular is issued to provide guidance and information in its designated subject area or to show a method acceptable to the FAA for complying with a related FAR.

Advisory Circulars are issued in a numbered-subject system corresponding to the subject areas of the FARs. The advisory circular numbers relate to the FAR parts and, when appropriate, to the specific sections of the Federal Aviation Regulations. For example, *AC43.13-1A, Acceptable Methods, Techniques, and Practices—Aircraft Inspection and Repair;* and *AC43.13-2A, Acceptable Methods, Techniques, and Practices—Aircraft Alterations*, correspond with FAR 43.13, which lists the performance rules for the repair and alteration of aircraft.

An Advisory Circular Checklist (AC00-2) is issued that lists all of the Advisory Circulars published and gives directions for obtaining them. Most Advisory Circulars are free and may be obtained from the superintendent of documents.

SERVICE-DIFFICULTY REPORTING PROGRAM

The primary function of the FAA is to ensure that aviation is a safe and dependable form of transportation. One of the means for promoting safety is the **service-difficulty reporting (SDR) program**. The SDR program is an information system designed to provide assistance to aircraft owners, operators, maintenance organizations, and manufacturers in identifying aircraft problems encountered during service. The program provides for the collection, organization, analysis, and dissemination of aircraft service information so as to improve the service reliability of aviation products.

Malfunction or Defect Report

One of the components of the SDR program is the collection of information that relates to safety problems that have been identified in the field. General-aviation service difficulties are generally reported to the FAA by maintenance personnel through the use of a **Malfunction or Defect Report (M or D)**. M or D reports provide the FAA with an essential service record of mechanical difficulties encountered in aircraft operations. Such reports contribute to the correction of conditions or situations which otherwise would continue to prove costly or could cause a serious accident or incident.

The FAA requests the cooperation of maintenance technicians in reporting discrepancies located during inspections. The reporting of malfunctions and defects found in aircraft and engines is an important step in detecting and eliminating unsafe conditions. All persons concerned with the maintenance, repair, inspection, or operation of aircraft are encouraged to make such reports on the Malfunction or Defect Report form and forward the form to the FAA. A copy of the malfunction or defect report form is shown in Figure 14–2.

Submission of service-difficulty information by the aviation public is voluntary. Certificated repair stations, air-taxi operators, and air carriers are required by the FARs to submit reports of malfunctions and defects to the FAA. Reports of defects are mailed to the local FAA district office, where they are reviewed and then forwarded to the Safety Data Branch in Oklahoma City for processing.

AC NO: 00-34A

DATE: 7/29/74

ADVISORY CIRCULAR

DEPARTMENT OF TRANSPORTATION
FEDERAL AVIATION ADMINISTRATION

SUBJECT: AIRCRAFT GROUND HANDLING AND SERVICING

FIGURE 14–1 Advisory circular.

Airworthiness Alerts

Malfunction or defect reports are retained in a computer bank for a period of five years, providing a base for the detection of trends and failure rates. Analysis of service-difficulty information is done primarily by the Safety Data Branch. When trends are detected, they are made available to maintenance personnel through **Advisory Circular 43-16, General Aviation Airworthiness Alerts** (see Figure 14–3). The intent of the alerts is to familiarize technicians with problems encountered in the field so that they may pay special attention to these areas when inspecting aircraft of a similar type. Compliance with the recommendations made in airworthiness alerts is voluntary.

Airworthiness alerts are published monthly and are distributed free of charge to all authorized inspectors, repair stations, air-taxi operators, and FAA-certified technician schools. Alerts may also be purchased as Advisory Circular 43-16 from the superintendent of documents.

AIRWORTHINESS DIRECTIVES

When an aircraft or engine is newly developed, engineers have to rely on the past history of similar products or analyses to predict when defects or unsafe conditions will arise. As service time is accumulated, various problems can arise that create potentially unsafe conditions. This happens even though the manufacturers have conducted extensive testing of the product. When an unsafe condition is discovered, a system must be in place that can evaluate the seriousness of the problem and, if necessary, notify the appropriate people of the corrective action that must be taken.

The FAA uses **Airworthiness Directives (ADs)** to notify aircraft owners of these unsafe conditions and to require their correction. ADs prescribe the conditions and the limitations, including inspections, repairs, or alterations, under which the product may continue to be operated.

The guidelines for when an airworthiness directive is issued is covered in **FAR Part 39.1, Applicability**, which says,

This part prescribes airworthiness directives that apply to aircraft, aircraft engines, propellers, or appliances (hereinafter referred to in this part as products) when,

(a) an unsafe condition exists in a product; and

(b) that condition is likely to exist or develop in other products of the same type design.

ADs are Federal Aviation Regulations and are published in the Federal Register as **amendments to FAR Part 39**. Depending on the urgency, ADs are published in the following categories:

Notice of Proposed Rule-Making (NPRM). An NPRM is issued and published in the Federal Register when an un-

1. REGISTRATION NO. N–	DEPARTMENT OF TRANSPORTATION FEDERAL AVIATION ADMINISTRATION **MALFUNCTION OR DEFECT REPORT**			Form Approved Budget Bureau No. 04–R0003	8. DATE SUB.	FOR FAA USE ONLY CONTROL NO.					
	A. MAKE	B. MODEL	C. SERIAL NO.	7A. COMMENTS *(Describe the malfunction or defect and the circumstances under which it occurred. State probable cause and recommendations to* **prevent** *recurrence.)*							
2. AIRCRAFT											
3. POWERPLANT											
4. PROPELLER											
5. APPLIANCE/COMPONENT *(assy. that includes part)*											
A. NAME	B. MAKE	C MODEL	D. SERIAL NO.								
6. SPECIFIC PART *(of component)* CAUSING TROUBLE											
A. NAME	B. NUMBER	C. PART/DEFECT LOCATION					Continue on reverse				
				SUBMITTED BY							
FAA USE	E. PART TT	F. PART TSO	G. PART CONDITION	B.	C.	D.	E.	F.	G.	H.	I.
D. ATA CODE				REP. STA.	OPER.	MECH.	AIR TAXI	MFG.	FAA	OTHER	

FAA FORM 8330-2 (9–70) SUPERSEDES PREVIOUS EDITIONS

FIGURE 14–2 Malfunction or defect report form.

**General Aviation
Airworthiness
Alerts**

U.S. Department
of Transportation
Federal Aviation
Administration

AC No. 43–16

ALERT NO. 167
JUNE 1992

Improve Reliability-
Interchange Service
Experience

FIGURE 14–3 Airworthiness alerts.

safe condition is discovered in a product. Interested persons are invited to comment on the NPRM by submitting such written data, views, or arguments as they may desire. The comment period is usually 60 days. Proposals contained in the notice may be changed or withdrawn in light of comments received. When an NPRM is adopted as a final rule, it is published in the Federal Register, printed, and distributed by first-class mail to the registered owners of the product affected.

Immediate Adopted Rule. ADs of an urgent nature are adopted without prior notice (without first being an NPRM) as *immediately adopted rules*. These ADs usually become effective less than 30 days after publication in the Federal Register and are distributed by first-class mail to the registered owners of the product affected. The majority of ADs are issued as immediate adopted rules.

Emergency AD. Emergency ADs are issued when immediate corrective action is required. Emergency ADs are distributed to the registered owners of the affected product by telegram or priority mail and are effective upon receipt. Emergency ADs are published in the Federal Register as soon as possible after initial distribution.

ADs issued to other than aircraft. ADs may be issued which apply to engines, propellers, or appliances installed on multiple makes or models of aircraft. When the product can be identified as being installed on a specific make or model aircraft, AD distribution is made to the registered owners of those aircraft. However, there are times when such a determination cannot be made, and direct distribution to the registered owners is impossible.

AD Distribution and Publication

As described in the preceding paragraphs, ADs are mailed to registered owners and published in the **Federal Register**. Most of the individuals involved in aircraft maintenance receive their notice of airworthiness directives through a subscription service to the FAA or through a private vendor.

The Airworthiness Directives (AD) summaries are published in four separate books. The Small Aircraft and Rotorcraft books relate to aircraft of 12 500 lb or less maximum certificated takeoff weight and all rotorcraft, regardless of weight. The Large Aircraft books relate to aircraft of more than 12 500 lb maximum certificated takeoff weight, except for rotorcraft. Each book is divided into four major subjects (aircraft, engines, propellers, and appliances). Within each subject, ADs are listed alphabetically by the manufacturers' or type certificate (TC) holders' names. The ADs applicable to engines and propellers are contained in the respective small or large books. Appliance ADs are duplicated in both the small and large books.

Each Book 1 contains ADs issued from 1941 through 1979 which are still in effect. These ADs will remain in effect until otherwise suspended, revoked, or amended. If an AD from Book 1 is amended, it will appear as a supplement to Book 2. Once purchased, Book 1 should be retained indefinitely.

Each Book 2 contains ADs beginning with 1980. Biweekly supplements of the AD activity are provided for a 2-year period to update Book 2.

Each of the four books may be purchased separately. The titles and contents are:

1. *Small Aircraft and Rotorcraft—Book 1, Summary of Airworthiness Directives*. (ADs issued between 1943 and 1979 and still in effect.)

2. *Small Aircraft and Rotorcraft—Book 2, Summary of Airworthiness Directives*. (ADs beginning with 1980 and continuing to the present time.) Ordering this book includes biweekly supplements for a 2-year period.

3. *Large Aircraft—Book 1, Summary of Airworthiness Directives*. (ADs issued between 1941 and 1979 and still in effect.) Reminder: All rotorcraft are contained in Small Aircraft and Rotorcraft books.

4. *Large Aircraft—Book 2, Summary of Airworthiness Directives*. (ADs beginning with 1980 and continuing to the present time.) Ordering this book includes biweekly supplements for a 2-year period. Reminder: All rotorcraft are contained in Small Aircraft and Rotorcraft books.

The total collection of the ADs is referred to as the *Summary of Airworthiness Directives* and is published every two years.

AD Indexes

The *Summary of Airworthiness Directives* contains two indexes. The first is alphabetical by the manufacturer of the product, and the second is a numerical listing of all AD notes from the oldest to the most current.

The alphabetical index divides the ADs into the given subject areas (aircraft, engines, propellers, and appliances). Each model of the manufacturer's product is listed with the applicable AD notes for that model.

A sample of the alphabetical index is shown in Figure 14–4. The numerical index, organized sequentially by AD

Aircraft Make and Model	TC Number	AD Number		Subject	Page Number	Book Number
		48-05-04		Operator-limitations placard . . .	8	1
		48-07-01		Stabilizer attaching bolts	8	1
		48-25-02		Welded exhaust muffler	8	1
		48-25-03		Wing-drag wire system	8	1
		50-31-01		Fin-spar reinforcement	9	1
		51-21-01		Rudder rib flanges	11	1
		61-25-01		Met-Co-Aire landing gear	14	1
		62-24-03		Cabin heat system	15	1
		79-08-03		Electrical system	71	2
140A	5A2	61-25-01		Met-Co-Aire landing gear	14	1
		62-24-03		Cabin heat system	15	1
		79-10-14		Fuel-tank venting	73	2
150 Series	3A19	62-22-01		Vacuum-pump modification . .	15	1
		67-03-01		Exhaust-gas heat exchangers . . .	17	1
		68-17-04		Stall warning system	19	1
		71-22-02		Cracks in nose-gear fork	3	2
		79-08-03		Electrical system . :	71	2
		79-10-14		Fuel-tank venting	73	2
150	3A19	73-23-07		Defective spar attach fittings . .	16	2
		74-06-02		AVCON mufflers	18	2
		75-15-08		Engine lubrication	23	2
		77-02-09		Wing-flap system	49	2
150A	3A19	75-15-08		Engine lubrication	23	2
150B	3A19	75-15-08		Engine lubrication	23	2
150C	3A19	75-15-08		Engine lubrication	23	2
150D	3A19	75-15-08		Engine lubrication	23	2
		83-17-06		Aileron balance weights	106	2
150E	3A19	75-15-08		Engine lubrication	23	2
		83-17-06		Aileron balance weights	106	2
150F	3A19	72-03-03	R3	Wing-flap jack screw	6	2
		75-15-08		Engine lubrication	23	2
		80-11-04		Cracked nutplates	90	2
		83-17-06		Aileron balance weights	106	2
150G	3A19	67-31-04		Removal of glove compartment	18	1
		72-03-03	R3	Wing-flap jack screw	6	2
		75-15-08		Engine lubrication	23	2
		80-11-04		Cracked nutplates	90	2
		83-17-06		Aileron balance weights	106	2
150H	3A19	67-31-04		Removal of glove compartment	18	1
		72-03-03	R3	Wing-flap jack screw	6	2
		75-15-08		Engine lubrication	23	2
		80-11-04		Cracked nutplates	90	2
		83-17-06		Aileron balance weights	106	2
150J	3A19	72-03-03	R3	Wing-flap jack screw	6	2
		75-15-08		Engine lubrication	23	2
		80-11-04		Cracked nutplates	90	2
		83-17-06		Aileron balance weights	106	2

FIGURE 14–4 Airworthiness directive alphabetical index.

AD Number	Manufacturer		Page Number
82-08-06	Piper	Superseded by 82-27-13	
82-09-01	Teledyne		31
82-09-03	Beech		107
82-09-52	Bell	Superseded by 82-09-53	
82-09-53	Bell		58
82-10-01	Aerospatiale		23
82-10-05	AiResearch, Garrett	Superseded by 82-27-07	
82-10-06	Hiller	Superseded by 82-15-02	
82-11-01	Robinson		4
82-11-04	Piper		134
82-11-05	Bendix, Equipment		13
82-12-04	Detroit Diesel Allison, Equipment		1
82-12-06	Government Aircraft Factories		7
82-12-07	Aerospatiale		23
82-13-01	Bendix, Equipment		14
82-13-02	Piccard	Superseded by 83-15-03	
82-13-03	Allison		8
82-13-04	Hiller		18
82-13-05	Aerospatiale		24
82-13-06	Schweizer		8
82-14-01	Hughes		33
82-15-01	Hughes		33
82-15-02	Hiller		19
82-15-03	Hughes		34
82-15-06	Empresa Brasileira	Superseded by 82-20-02	
82-15-07	Robinson		5
82-16-05	Piper		BW
82-16-06	Bell		58
82-16-07	Hiller		20
82-16-08	Helic		1
82-16-09	Enstrom		11
82-16-11	Government Aircraft Factories		8
82-16-12	Bell		59
82-17-01	Hughes		34
82-17-02	Rolladen Schneider		1
82-17-03	Sikorsky		17

FIGURE 14–5 Airworthiness directive numerical index.

number, provides the name of the manufacturer and, in the event that the AD has been superseded, the new AD to which reference should be made (see Figure 14–5).

The indexes are updated every six months during the two-year period that the summary is effective. The FAA intends these indexes to be used as a guide only. They should not be relied upon as conclusive evidence of AD applicability.

New and revised ADs are compiled by the FAA every two weeks and mailed out in what is called the **Biweekly Listing**. The purchase of the AD summary includes a subscription to the biweekly revisions until the ADs are republished. Because of their importance, airworthiness directives may be purchased directly from the FAA in Oklahoma City.

AD Numbers

The AD summary notes are amendments of FAR Part 39 and therefore carry an amendment number, such as 39-3184. This number changes when an AD note is revised, at which time a new amendment number is issued. The AD is normally identified in maintenance records and for other purposes by a six-digit number, such as 92-08-03. The *92* identifies the year of issue (1992), the *08* identifies the period as being in the eighth biweekly period, which is the fifteenth or sixteenth week of the year. The *03* is a sequential

number and simply means that it was the third AD adopted in that period.

An AD note is often **revised**, meaning that something of substance within the AD note has been changed. A revised AD note will usually retain its six-digit number; however, an *R* will be added after the six-digit number. Number 92-08-03R$_2$ indicates that this AD has been revised twice.

AD Content

The content of AD notes can be divided into several categories, such as applicability, compliance time and date, effective dates, and required actions. (Figure 14–6 is a sample AD.)

Applicability of ADs. Each AD contains an applicability statement specifying the product (aircraft, aircraft engine, propeller, or appliance) to which it applies. The applicability statement is usually identified in the first part of the AD. Applicability is given by product model and, as applicable, serial number and category of certification. In some ADs with multiple parts, the applicability statements may be made at different places, depending upon the required action. It is very important for the technician to research carefully the applicability statements to see if part or all of the AD applies.

81-04-07 R1 *PIPER:* Amendment 39-4044 as amended by Amendment 39-4272. Applies to Model PA-38-112, Serial Nos. 38-78A0001 thru 38-80A0198, certificated in all categories.

To avoid possible hazards in flight associated with cracks in the fin forward spar, P/N 77601-03, and the fuselage bulkhead assembly, P/N 77553-02, accomplish the following:

a. Within the next 25 hours in service from the effective date of this AD or upon the attainment of 300 hours in service, whichever is later, unless accomplished within the previous 75 hours in service and thereafter, at intervals not to exceed 100 hours in service from the last inspection, accomplish the following:

(1) Inspect the forward surface of the fin forward spar web in the area of the vertical edges of the fin forward spar attachment fitting P/N 77553-05 for cracks using a dye penetrant method or equivalent. Remove two forward fin attachment bolts and displace fin spar ⅛″ laterally in each direction to increase visibility of spar area adjacent to edge of attachment fitting. Remove any scuff marks on spar by sanding prior to applying dye penetrant.

(2) Inspect the fuslage bulkhead assembly P/N 77553-02, at fuselage station 221.42, in the area of the fin forward spar attachment fitting, P/N 77553-05, for cracks using a dye penetrant method or equivalent. Access to aft side of bulkhead may be obtained by removing rudder and adjacent access door and to front side by removing the luggage compartment rear partition. When using luggage compartment, provide a stand to support the aft fuselage and, in order to assure that no associated damage will occur during the inspection, provide a support board for the mechanic.

b. Replace fin spars with cracks exceeding one-half inch and bulkheads with cracks exceeding three-quarter inch prior to further flight with undamaged parts of the same part number, an equivalent part, or accomplish a repair which must be approved by the Chief, Engineering and Manufacturing Branch, FAA, Eastern Region. Parts which have cracks less than these values must be replaced or repaired within 25 hours in service. A ferry flight may be authorized under FAR 21.197 to a place of repair with prior approval of the Chief, Engineering and Manufacturing Branch, FAA, Eastern Region.

c. Inspect replacement and repaired parts in accordance with the AD unless otherwise authorized by the Chief, Engineering and Manufacturing Branch, FAA, Eastern Region.

d. Equivalent inspections and parts must be approved by the Chief, Engineering and Manufacturing Branch, FAA, Eastern Region.

e. Upon submission of substantiating data by an owner or operator through an FAA Maintenance Inspector, the Chief, Engineering and Manufacturing Branch, FAA, Eastern Region may adjust the compliance times specified in this AD.

Amendment 39-4044 was effective February 19, 1981.

This amendment 39-4272 is effective December 9, 1981.

FIGURE 14–6 Sample airworthiness directive.

Some aircraft owners and operators are of the opinion that ADs are not applicable to aircraft certificated in certain categories, such as experimental or restricted. This is not true; if an AD does not specifically exempt a category, then the AD applies.

Compliance Time or Date. Compliance requirements specified in ADs are established for safety reasons and may be stated in numerous ways. Some ADs are of such a serious nature that they require compliance before further flight. In some instances these ADs authorize flight, providing a ferry permit is obtained, but without such authorization in the ADs, further flight is prohibited. Other ADs express compliance time in terms of a specific number of hours of operation, for example, "Compliance required within the next 50 hours time in service after the effective date of this AD." Compliance times may also be expressed in operational terms, such as "within the next 10 landings after the effective date of this AD."

For turbine engines, compliance times are often expressed in terms of cycles. A cycle normally consists of an engine start, a takeoff operation, a landing, and an engine shutdown. When a direct relationship between airworthiness and calendar time is identified, compliance time may be expressed as a calendar date. Another aspect of compliance times to be emphasized is that not all ADs have a one-time compliance. **Repetitive inspections** at specified intervals after initial compliance may be required. Repetitive inspections are used in lieu of a permanent fix because of costs or until a permanent fix is developed.

Effective Dates. The effective date of the AD is usually published toward the end of the note. In the case of a revision, the effective date of the revision and the date of the original amendment are given. Care should be taken to ensure that when a revision becomes effective, it does not require additional action on a previously received AD.

Required Action. The required action may take the form of inspection, replacement of parts, modification of design, or changes in operating procedures. The required action could be totally described in the AD note, or reference may be made to a manufacturer's service bulletin which will give the details of the required action. Notification to the FAA upon compliance or other special record entries in the aircraft maintenance records may be required. The action may call for recurring inspections or other action until certain parts are replaced or modified.

Compliance Responsibility

Responsibility for AD compliance lies with the registered owner or operator of the aircraft. This responsibility is usually met by having appropriately rated maintenance technicians accomplish the maintenance required by the AD.

Maintenance technicians may also have direct responsibility for AD compliance, aside from the times when AD compliance is the specific work contracted for by the owner or operator. When a 100-hour, annual, or progressive inspection or an inspection required under Part 123 or 125 is accomplished, FAR Section 43.15(a) requires the person performing the inspection to perform it so as to determine that all applicable airworthiness requirements are met, which includes compliance with ADs.

Maintenance technicians should be aware that even though an inspection of the complete aircraft is not made, if the inspection conducted is a progressive inspection, determination of AD compliance for those portions of the aircraft inspected is required.

In conducting an AD search, the technician must first obtain the make, model, and serial number of the aircraft, engine, propeller, and appliances. The four subject indexes may be used as an aid in performing the AD search. In referring to Figure 14–4, it should be noted that there are 11 possible ADs that could apply to a Cessna 150G Airframe. The 11 ADs include 6 listed under the 150 Series and 5 listed under 150G. After noting the possible ADs from the index, it is then necessary to refer to the specific AD to check model and serial number applicability.

When performing an AD search, it is also necessary to check the biweekly listing, since ADs there are not included in the index.

AD Record Entries

FAR Part 91.417 requires that the aircraft records contain the current status of applicable airworthiness directives including, for each, the method of compliance, the AD number, and the revision date. If the AD involves recurring action, the time and date when the next action is due is required. An Airworthiness Directive Compliance Record, such as is shown in Figure 14–7, is recommended as a record of AD compliance.

TYPE CERTIFICATE DATA SHEETS

The FAA prescribes minimum standards governing the design, materials, workmanship, construction, and performance of aircraft, aircraft engines, and propellers. When an aircraft engine or propeller meets or exceeds these requirements, it is issued a Type Certificate. The Type Certificate consists of the type design, the operating limitations, the

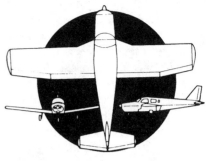

VOLUME I

SINGLE-ENGINE AIRPLANES

TYPE CERTIFICATE DATA SHEETS
AND SPECIFICATIONS

DEPARTMENT OF TRANSPORTATION
FEDERAL AVIATION ADMINISTRATION

FIGURE 14–8 Type certificate data sheets.

Type Certificate Data Sheet, and the applicable regulations with which the FAA requires compliance.

Type Certificate Data Sheets (TCDSs) (see Figure 14–8) were originated and first published in January 1958. A Type Certificate Data Sheet is evidence that the product has been type certificated. Data sheets contain information on the design specifications of the aircraft. In addition, they show the applicable regulations under which the aircraft was certificated.

The technician makes extensive use of the TCDS during an inspection to see that the aircraft does conform to its certificated design.

Aircraft that are type certificated under the old civil air regulations have **Aircraft Specifications** rather than Type Certificate Data Sheets. Specifications originated during the implementation of the Air Commerce Act of 1926.

AIRWORTHINESS DIRECTIVE COMPLIANCE RECORD
AIRCRAFT, POWER PLANT, PROPELLER, OR APPLIANCE/COMPONENT

Make _____ Model _____ S/N _____ N _____

Date of Manufacture _____ Tach Time _____ Total Time _____

Date _____ ADs OK'd thru _____ Name and A&P No. _____

| AD number | AD rev. date | Subject | Compliance | | Method of compliance | One-time | Reoccuring | Next compliance due date/time | Authorized signature and number |
			Date	Total time					
74-20-05		EXHAUST MUFFLER	3-2-82	1471.0	BY INSP PER PAR. (A)(1)		x	1571.0	A&P 2033266
79-12-02		CONTROL CABLES	3-2-82	1471.0	BY VISUAL INSP PER PAR (B)		x	1671.0	A&P 2033266
81-14-07		LIFT STRUT FORKS	4-6-81	1380.2	BY REPLACEMENT PER S.B. 121	x			A&P 2033266
82-05-01R₁	08-02-81	INSPECT MAIN SPAR	—	—	N/A BY SERIAL NUMBER			—	—

FIGURE 14–7 Airworthiness directive compliance record.

Specifications are FAA record-keeping documents issued for both type-certificated and non-type-certificated products which have been found eligible for U.S. airworthiness certificates. Specifications are basically the same as TCDSs with one major difference. Aircraft Specifications have the required and optional equipment listed for each model of aircraft. To find this information on aircraft with a Type Certificate Data Sheet would require using the equipment list for that particular aircraft. Although they are no longer issued, specifications remain in effect and are further amended. Specifications may be converted to Type Certificate Data Sheets at the option of the type certificate holder. However, to do so, the type certificate holder must provide an equipment list, usually in the form of a service letter or bulletin.

The Aircraft, Engine, and Propeller Type Certificate Data Sheets and Specifications are issued in six volumes, each of which may be purchased separately in paper copy form or as a complete library on microfiche. Information concerning the current prices and how to order may be obtained from the U.S. Department of Transportation, Publications Section, Washington, D.C. The volumes are as follows:

Volume I, Single-Engine Airplanes. Contains documents for all single-engine fixed-wing airplanes, regardless of takeoff weight.

Volume II, Small Multiengine Airplanes. Contains data sheets for multiengine fixed-wing airplanes of 12 500 lb or less maximum takeoff weight.

Volume III, Large Multiengine Airplanes. Contains data sheets for multiengine fixed-wing airplanes of more than 12 500 lb maximum takeoff weight.

Volume IV, Rotorcraft, Gliders, and Balloons. Contains data sheets for all rotorcraft, gliders, and piloted balloons.

Volume V, Aircraft Engines and Propellers. Contains data sheets for engines and propellers of all types and models.

Volume VI, Aircraft Listing and Aircraft Engine and Propellers Listing. Contains information pertaining to older aircraft models of which 50 or less are shown on the FAA Aircraft Registry. It also includes engine and propeller information for which approvals have expired or for which the manufacturer no longer holds a production certificate.

The Type Certificate Data Sheets and Specifications are listed in the master alphabetical index under the name of the current type certificate holder. The index is cross-referenced to assist in locating a specific data sheet or specification that has been transferred to a person other than the original manufacturer (see Figure 14–9). A revised master index is issued annually in January and incorporates all changes made during the preceding calendar year.

Supplements containing newly issued data sheets and revisions to existing ones are issued monthly. For a revised data sheet specification, the revision number appears directly below the data sheet or specification number. The revised material is indicated by either a bracket or a solid vertical bar along the left-hand margin.

Content of Data Sheets

The content of the various Type Certificate Data Sheets and Specifications varies, but for the purposes of this book it will be broken down into the following subparts (see Figure 14–10):

1. Identification
2. Data for a specific model
3. Data pertinent to all models
4. Notes

Type Certificate Data Sheets are issued for aircraft, aircraft engines, and propellers. Since the information for each of these three products is different, they will be covered separately.

Aircraft Data Sheets

Identification. The upper right-hand corner of the Type Certificate Data Sheet or Specification contains a box with the data sheet number, the latest revision number and date, and the various aircraft models covered under that type certificate. The Type Certificate Data Sheet number is the same as the type certificate number. The name and address of the current type certificate holder is also given.

Data for a Specific Model. A Type Certificate may have more than one model of an aircraft approved under it due to changes in the type design. In most cases, the manufacturer will designate the changes with a new model number, such as Cessna 150E. In a few cases, the manufacturer may stay with the basic model designation and indicate the changes with serial numbers. All of the information which pertains specifically to that model is presented first, and then the next change or model of aircraft is covered. The aircraft is identified by model, type certificate category or categories, and date of approval.

Many Aircraft Specifications and some Type Certificate Data Sheets carry coded information to describe the general characteristics of the product. This information may be found in the model caption line or a separate line entry titled "Type" or "Designation." Aircraft codes (designations) are as follows for the example 2 PO-CLM:

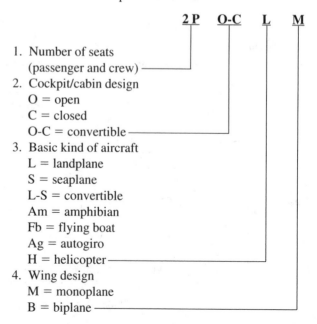

                                    2 P   O-C   L   M

1. Number of seats
   (passenger and crew)
2. Cockpit/cabin design
   O = open
   C = closed
   O-C = convertible
3. Basic kind of aircraft
   L = landplane
   S = seaplane
   L-S = convertible
   Am = amphibian
   Fb = flying boat
   Ag = autogiro
   H = helicopter
4. Wing design
   M = monoplane
   B = biplane

The other information contained for a particular model will include

- Engine(s)
- Fuel grade and capacity
- Engine operational limits
- Propeller and propeller limits
- Airspeed limits
- Number of seats and location
- Baggage capacity
- Oil capacity
- Control-surface movement
- Center-of-gravity weight information

This section also contains a list of serial numbers produced for each model, which is very important when determining conformity with Type Certificate Data Sheets.

Data Pertinent to All Models. This section contains the location of the datum and the leveling means. It also includes the certification basis, production basis, and required equipment. The certification basis lists the exact regulations and the special conditions, if any, that apply to each of the models. The production basis lists whether or not a production certificate was used and other pertinent information.

On aircraft that are still listed on an Aircraft Specification, the equipment will be listed and numbered to correspond to the required equipment listed for specific models. The list also shows the weight of the equipment and the arm for weight-and-balance purposes. Optional equipment is also included in this list.

A Type Certificate Data Sheet contains the statement, "The basic required equipment as prescribed in the applicable airworthiness regulations (see Certification Basis) must be installed in the aircraft for certification." Any special equipment not covered by this statement will be listed. The required equipment is listed in the aircraft's equipment list.

Notes. This section contains information that may apply to all models or certain aircraft within a given model. The information listed is very important and should not be ignored.

In all cases, **Note 1** reads, "Current weight-and-balance report together with the list of equipment included in certificated empty weight and loading instructions when necessary must be provided for each aircraft at the time of original certification." This note on an Aircraft Specification is slightly

Type Certificate Holder Name, Location, Aircraft Model No.	Controlling Region*	Approval No.	Data Sheet/ Spec. Listing Page No.	Certification Basis	Volume No.
LOCKHEED-GEORGIA COMPANY A Div. of Lockheed Aircraft Corp. Marietta, GA	CE-A				
Jetstar 1329-23A, -23D, -23E, 1329-25		2A15	2A15-10	CAR 4b	III
382, 382B, 382E, 382F, 382G		A1SO	A1SO-11	CAR 1 (CAR 9a)	III
300-50A-01 (USAF C-141A)		A2SO	A2SO-3	CAR 1 (CAR 4b)	III
282-44A-05 (C-130B)		A5SO	A5SO	FAR 21.25 (a) (2)	III
LONGREN AIRCRAFT COMPANY (See Mitchell)					
LUSCOMBE AIRCRAFT CORPORATION Atlanta, GA (See Temco)	CE-A				
(Larsen Luscombe) 8, 8A, 8B, 8C, 8D, 8E, 8F, T-8F		A-694	A-694-22	Part 4a	I
LUSCOMBE AIRPLANE CORPORATION Dallas, TX (See Luscombe Aircraft Corp., Temco)	SW				
Phamtom 1, 1S		ATC 552	Page 266 (124)	—	VI
4		TC 687	Page 267 (124)	—	VI
MACCHI (See Aeronautica Macchi & Aerfer)					
MAEL AIRCRAFT CORPORATION Portage, WI	CE-C				
(Burrs) BA-42		A6SO	A6SO-2	FAR 21 (FAR 23)	II

The page number in parenthesis in the "Data Sheet/Spec/ Listing Page No." column identifies the page number of the original publication. This page number is sometimes referenced in the TCDS and on the application for airworthiness certification.

*Region in which the technical data are filed. Addresses on last page of index.

FIGURE 14-9 Type certificate master index.

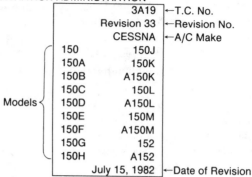

DEPARTMENT OF TRANSPORTATION
FEDERAL AVIATION ADMINISTRATION

		3A19	←T.C. No.
		Revision 33	←Revision No.
		CESSNA	←A/C Make
Models {	150	150J	
	150A	150K	
	150B	A150K	
	150C	150L	
	150D	A150L	
	150E	150M	
	150F	A150M	
	150G	152	
	150H	A152	
		July 15, 1982	←Date of Revision

TYPE CERTIFICATE DATA SHEET NO. 3A19

This data sheet which is part of Type Certificate No. 3A19 prescribes conditions and limitations under which the product for which the type certificate was issued meets the airworthiness requirements of the Federal Aviation Regulations.

Type Certificate Holder   Cessna Aircraft Company
Pawnee Division   (Does not have to be the manufacturer)
Wichita, Kansas 67201
Class

First Models
1 { -Model 150, 2 PCLM (Utility Category), Approved July 10, 1958
  Model 150A, 2 PCLM (Utility Category), Approved June 14, 1960
  Model 150B, 2 PCLM (Utility Category), Approved June 20, 1961 ←Date Approved
  Model 150C, 2 PCLM (Utility Category), Approved June 15, 1962

Engine	Continental 0-200-A
*Fuel	80/87 Min. grade aviation gasoline
*Engine Limits	For all operations, 2750 rpm (100 hp.)
Propeller and Propeller Limits	1. Sensenich M69CK 24 lb. (-32) Diameter: not over 69 in., not under 67.5 in. Static rpm at maximum permissible throttle setting: not over 2470, not under 2320 No additional tolerance permitted 2. McCauley 1A100/MCM 21 lb. (-32) Diameter: not over 69 in., not under 67.5 in. Static rpm at maximum permissible throttle setting: not over 2475, not under 2375 No additional tolerance permitted

*Airspeed limits (CAS)		
Never exceed	157 mph	(136 knots)
Maximum structural cruising	120 mph	(104 knots)
Maneuvering	106 mph	(92 knots)
Flaps extended	85 mph	(74 knots)

C.G. Range	($+33.4$) to ($+36.0$) at 1500 lb. ($+32.2$) to ($+36.0$) at 1250 lb. or less Straight line variation between points given
Empty Wt. C.G. Range	None
Leveling Means	Top Edge of fuselage splice plate
*Maximum Weight	1500 lb.
No. of Seats	2 at ($+39$); (for child's optional jump seat refer to Equipment List)
Maximum Baggage	80 lb. at ($+65$)

FIGURE 14–10 Sample type certificate data sheet.

| Fuel Capacity | 26 gal. (22.5 gal. usable, two 13 gal. tanks in wings at +42) |
| | See NOTE 1 for data on system fuel |

| Oil Capacity | 6 qt. (−13.5; unusable 2 qt.) |
| | See NOTE 1 for data on system oil |

Control Surface MOVEMENTS	Wing Flaps	Retracted	0°
		1st Notch	10°
		2nd Notch	20°
		3rd Notch	30°
		4th Notch	40°
	Ailerons	Up 20°	Down 15°
	Elevator	Up 25°	Down 15°
	Elevator tab	Up 10°	Down 20°
	Rudder	Right 16°	Left 16°

Serial Nos. Eligible	Model 150:	617, 17001 thru 17999, 59001 thru 59018
	Model 150A:	628, 15059019 thru 15059350
	Model 150B:	15059351 thru 15059700
	Model 150C:	15059701 thru 15060087

Data Pertinent to All Models

Datum Fuselage station 0.0 front face of firewall

Certification Basis Part 3 of the Civil Air Regulations dated May 15, 1956, as amended by 3-4. In addition, effective S/N 15282032 and on for 152 and S/N 681, A1520809 and on for A152, FAR 23.1559 effective March 1, 1978. FAR 36 dated December 1, 1969, plus Amendments 36-1 thru 36-5 for 152 and A152 only.

Application for Type Certificate dated August 13, 1956.

Type Certificate No. 3A19 issued July 10, 1958, obtained by the manufacturer under delegation option procedures.

Equivalent Safety Items	S/N 15077006 thru 15079405
	S/N 15279406 and on
	S/N A1500610 thru A1500734
	S/N 681, A1500433, A1520735 and on

| Airspeed Indicator | CAR 3.757 (See NOTE 4) |
| Operating Limitations | CAR 3.778(a) |

Production Basis Production Certificate No. 4. Delegation Option Manufacturer No. CE-1 authorized to issue airworthiness certificates under delegation option provisions of Part 21 of the Federal Aviation Regulations.

Equipment: The basic required equipment as prescribed in the applicable airworthiness regulations (see Certification Basis) must be installed in the aircraft for certification. This equipment must include a current Airplane Flight Manual effective S/N 15282032 and on, S/N 681, and S/N A1520809 and on. In addition, the following item of equipment is required:

1. Stall warning indicator, audible, Cessna Dwg. 0511062 (Model 150 thru 150E)

2. Stall warning indicator, audible, Cessna Dwg. 0413029 (Model 150F thru 150 M, 1977 Model) (A150K thru A150M, 1977 Model) (152 and on, A152 and on)

NOTE 1.   Current weight and balance report together with list of equipment included in certificated empty weight and loading instructions when necessary must be provided for each aircraft at time of original certification.

Serial Nos. 17001 thru 17999, 59001 thru 59018, 15059019 thru 15077005 and A1500001 thru A1500609
The certificated empty weight and corresponding center of gravity location must include unusable fuel of 21 lb. at (+40) for landplanes or 27 lb. at (+40) for seaplanes and an undrainable oil of (0) lb. at (−13.5) for both landplane and seaplane.

FIGURE 14–10 Continued.

NOTE 2. The following information must be displayed in the form of composite or individual placards.

    A. In full view of the pilot:

       (1) "This airplane must be operated in compliance with the operating limitations stated in the form of placards, markings and manuals."

       (2) (a) <u>Model 150A, 150B and 150C</u>
          "Acrobatic maneuvers are limited to the following":

Maneuver	Entry Speed
Chandelle	106 m.p.h. (92 knots)
Steep turns	106 m.p.h. (92 knots)
Lazy eights	106 m.p.h. (92 knots)
Stalls (except whip)	Use slow deceleration
Spins	Use slow deceleration

FIGURE 14–10 Continued.

different in wording but not in meaning or intent. Unique weight-and-balance information for this aircraft is also located in this note.

Another note, **Note 2**, pertains to the required placards for the aircraft and includes the following or a similar statement: "The following placards must be displayed in front of and in clear view of the pilot." Placards, if any, are listed and must be present if the aircraft is to conform to its type certificate.

Other notes contain information about equipment installed in specific models, special operating procedures for various models, modifications, conversion from one model to another, and similar information.

Aircraft Engine Data Sheets

Identification. This section is similar to that found on the Aircraft Type Certificate Data Sheets. Engine TCDSs also use coded information in identifying engines. Engine Codes (Types) are as follows for the example 4LIA (sometimes 4LAI):

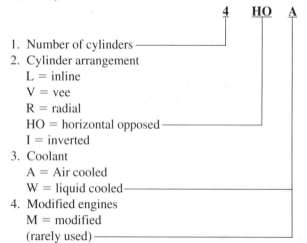

                4   HO   A

1. Number of cylinders
2. Cylinder arrangement
   L = inline
   V = vee
   R = radial
   HO = horizontal opposed
   I = inverted
3. Coolant
   A = Air cooled
   W = liquid cooled
4. Modified engines
   M = modified
   (rarely used)

Data Pertinent to Specific Models. Data in this section includes the following information for the various models of engines:

- Type
- Power rating (maximum continuous and takeoff)
- Displacement
- Timing
- Compression ratio

- Shaft (type)
- Fuel grade
- Oil grade versus air temperature (type)
- Bore and stroke
- Oil-sump capacity
- Weight
- Propeller
- Carburetion ignition
- Magnetos ignition

Data Pertinent to All Models. This section contains information on the certification and production basis of the engine.

Notes. The notes contain information on engine temperatures, fuel- and oil-pressure limits, spark plugs approved for use, and accessory drives. There is also information on the difference between the various models, approved optional equipment, procedures for conversion to other models, and special operating procedures, as applicable.

Propeller Data Sheets

The propeller Type Certificate Data Sheets contain identification information that is similar to the aircraft and engine sheets. They also include the type of propeller, the engine shaft it fits, the hub and blade material, the number of blades, and the hub models eligible for this certificate. A listing of the eligible blades, by model number, follows, with the power ratings with which they can be used, the diameter limits, and the weights (approximate) of the complete propellers with the various blades installed. The certification and production basis are also listed. The notes explain the hub and blade model designations and special provisions for pitch control, operation, and interchangeability of parts.

SUPPLEMENTAL TYPE CERTIFICATES

Part 21 of the Federal Aviation Regulations provides for the issuance of Supplemental Type Certificates (STCs) to persons who make major changes on a previously approved type design when the change is not so extensive as to require a new type certificate. An applicant for an STC must show that the altered product meets all applicable airworthiness requirements.

The holder of an STC may do the following:

1. In the case of aircraft, obtain an airworthiness certificate.
2. In the case of other products, obtain approval for installation on certificated aircraft.
3. Obtain a production certificate for the change in the type design that was approved by the supplemental type certificate.

An STC is considered an amendment to the original Type Certificate. After an STC is issued for a particular change, others may make the same modification, provided that it is done in accordance with the data that has been FAA approved.

A summary is published that lists those STCs in which the holders have indicated that the design, parts, or kits will be made available to the public. Figure 14–11 shows an example from the summary of STCs. The publication contains three sections identified as (1) aircraft, (2) engines, and (3) propellers. These sections, arranged alphabetically by the aircraft manufacturer's name, contain the following information:

Column 1 identifies the aircraft's model and the type certificate number for which the STC was issued.
Column 2 lists the assigned STC number.
Column 3 provides a brief description of the modification.
Column 4 identifies the FAA certificating region.
Column 5 identifies the individual or organization to whom the STC was issued.

Some of the STCs are listed as applying to a series of aircraft, engines, or propellers. STCs are applicable to a given aircraft model and may not be eligible to be used in conjunction with another STC or with other approved aircraft modifications. Therefore, prior to undertaking multiple STC installations, it should be determined that the proposed changes will introduce no adverse effects upon the airworthiness of the aircraft.

INSTRUCTIONS FOR CONTINUED AIRWORTHINESS

The manufacturer of an aircraft that applied for the issuance of a type certificate after January 28, 1981, must develop and furnish at least one set of complete **Instructions for Continued Airworthiness** to the owner of each aircraft upon its delivery.

The Instructions for Continued Airworthiness must contain the following information:

A. Airplane Maintenance Manual.
  1. Introduction information that includes an explanation of the airplane's features and data to the extent necessary for maintenance or preventive maintenance.
  2. A description of the airplane and its systems and installations, including its engines, propellers, and appliances.
  3. Basic control and operation information describing how the airplane components and systems are controlled and how they operate, including any special procedures and limitations that apply.
  4. Servicing information that covers details regarding servicing points, tank capacities, reservoirs, types of fluids to be used, pressures applicable to the various systems, location of access panels for inspection and servicing, locations of lubrication points, lubricants to be used, equipment required for servicing, tow instructions and limitations, mooring information, jacking instructions, and leveling information.
B. Maintenance instructions.
  1. Scheduling information for each part of the airplane and its engines, auxiliary power units, propellers, accessories, instruments, and equipment that provides the recommended periods at which they should be cleaned, inspected, adjusted, tested, and lubricated as well as the degree of inspection, the applicable wear tolerances, and the work recommended at these periods. The recommended overhaul periods and necessary cross-reference to the airworthiness limitations section of the manual must also be included. In addition, an inspection program that includes the frequency and extent of the inspections necessary to provide for the continued airworthiness of the airplane must be included.
  2. Troubleshooting information describing probable malfunctions, how to recognize those malfunctions, and the remedial action for those malfunctions.
  3. Information describing the order and method of removing and replacing products and parts, with any necessary precautions to be taken.
  4. Other general procedural instructions, including procedures for system testing during ground running, symmetry checks, weighing and determining the center of gravity, lifting and shoring, and storage limitations.
C. Diagrams of structural access plates and the information needed to gain access for inspections when access plates are not provided.
D. Details for the application of special inspection techniques, including radiographic and ultrasonic testing where such processes are specified.
E. Information needed to apply protective treatments to the structure after inspection.
F. All data relative to structural fasteners, such as identification, discard recommendations, and torque values.
G. A list of special tools needed.
H. In addition, for commuter-category airplanes, the following information must be furnished:
  1. Electrical loads applicable to the various systems.
  2. Methods of balancing control surfaces.
  3. Identification of primary and secondary structures.
  4. Special repair methods applicable to the airplane.

The Instructions for Continued Airworthiness must contain a section of **airworthiness limitations** that is segregated and clearly distinguishable from the rest of the document. This section must set forth each mandatory replacement

Aircraft Make, Model and TC Number	STC Number	Description	RG	STC Code	STC Holder
PA-28-140, -150, -151, -160, -161; TC 2A13	SA793CE	Installation of Lycoming O-360-A1A engine and Hartzell HC-C2YK-1B/7666A-2 or HC-C2YK-1BF/F7666A-2 propeller. Amended 10/20/83.	CE	86195	Robert L. and Barbara V. Williams Box 654 Udall, KS 67146
PA-28-140, -150, -151; TC 2A13	SA802GL	Modify airplane to fly on unleaded automotive gasoline, 87 minumum antiknock index. Issued 7/30/84.	CE-W	64274	Petersen Aviation, Inc. Route 1, Box 18 Minden, NE 68959
PA-28-140, -150, -160, -180, -235, 28-S-160, S-180, 28R-180, -200; TC 2A13	SA819EA	Alteration to replace standard anti-collision light with Grimes Series 550 single anti-collision red or white strobe kit. P/N 34-0001-1 or 34-0001-2. Amended 3/30/78.	GL	39500	Grimes Manufacturing Company 515 North Russell Street Urbana, OH 43078
PA-28-140, -150, -160, -180, -235, -S-160, -S-180, -R-180; TC 2A13	SA822SW	Mitchell automatic flight system AK217 consisting of Century III autopilot with optional automatic pitch trim, aileron stabilizer, and radio coupler.	SW	60000	Mitchell Industries, Inc. P.O. Box 610 Municipal Airport Mineral Wells, TX 76067
PA-28-140, -150, -160, -180, -28R-180, 28R-200; TC 2A13	SA880WE	Installation of new design fiberglass wing tip. Amended 1/4/83.	NM	58500	Met-Co Aire P.O. Box 2216 Fullerton, CA
PA-28-140, -150, -160, -180, PA-28R-180, -200, PA-34-200; TC 2A13, A7SO	SA979EA	Installation of L/R wing tip transparent fairings and sensor mounting bracketry. Issued 4/5/74.	EA	68700	Rock Avionic Systems, Inc. 412 Avenue of the Americas New York, NY 10011
PA-28-140, -28-150, -28-160, -28-180, 28R-180, -28R-200; TC 2A13	SA1072CE	Installation of wing leading edge cuffs and droop tips, dorsal fin and vertical stabilizer vortex generator. Amended 11/6/84.	CE-W	42560	Horton Stol-Craft, Inc. Wellington Municipal Airport Wellington, KS 67152

FIGURE 14–11 Summary of STCs index.

time, structural inspection interval, and related structural inspection procedures required for type certification.

If a set of Instructions for Continued Airworthiness has been developed for an aircraft, then it must be followed whenever an inspection or some maintenance is performed on the aircraft.

MANUFACTURERS' PUBLICATIONS

Service Bulletins

The medium used by manufacturers to communicate with owners and operators is the **service bulletin**. Service bulletins are issued to notify others of design defects, possible modifications, or a change in approved maintenance practices. Service bulletins may contain information on special inspections or checks that are needed to maintain the aircraft, engine, or accessory in safe operating condition.

In addition, **alert service bulletins** may be issued by the manufacturer on matters requiring the urgent attention of the operator and are generally limited to items affecting safety. Compliance with service bulletins is generally considered voluntary by the general-aviation operator. A service bulletin can be made mandatory by incorporating it into an airworthiness directive. Service-bulletin compliance may also be made mandatory by requiring compliance in the inspection program that is being used, or, for commercial operators, in the operating guidelines that they have had approved. In addition, most engine or component overhaul procedures require compliance with service bulletins when the unit is overhauled.

Service bulletins are usually mailed to registered owners and obtained by maintenance personnel on a subscription basis.

TECHNICAL MANUALS

Maintenance Manual

The technical publication that is most frequently used by the technician is the **aircraft maintenance manual**. The maintenance manual contains the information necessary to enable the technician to check, service, troubleshoot, and repair the airplane and its associated systems. It also includes information necessary for the technician to perform maintenance or make minor repairs to components while they are installed on the airplane. The maintenance manual does not contain information relative to work normally performed on components when they are removed from the aircraft. Maintenance manuals are prepared for the technician who normally performs work on units, components, and systems while they are installed on the aircraft. For instance, the airplane maintenance manual provides the instructions for installing a magneto and timing it to the engine, but information on overhauling the magneto is located in the overhaul manual. Most maintenance manuals

will make frequent use of illustrations and schematic diagrams that make the text easier to understand. Troubleshooting information describing probable malfunctions, how to recognize those malfunctions, and the corrective action required are included. Detailed information on more complex specialized subjects, such as wiring diagrams or structural repair, may be included in the maintenance manual for smaller aircraft but is generally separated into separate manuals for larger aircraft.

Overhaul Manual

The manufacturer's **overhaul manual** contains detailed step-by-step instructions covering work normally performed on a unit away from the aircraft. The overhaul manual is prepared for the technician who normally performs shop work, not for the aircraft service technician. The manual, which includes the approved and recommended overhaul procedures, is written to provide a technician with the information necessary to overhaul and test a part, thus making it available for reinstallation on an aircraft.

Structural Repair Manual

The **structural repair manual** contains descriptive information for the identification and repair of the aircraft's primary and secondary structure. The manual assists in defining damage that has a significant effect on the strength or life of the aircraft and that which does not. For significant damage, the structural repair manual provides the technical data for repairing the aircraft. Procedures, processes, and limitations pertaining to structural repair are covered. Also included are such items as typical damage repairs, critical-area limitations, and structural alignment information. On some smaller aircraft, structural repair information is included in the aircraft's maintenance manual.

Illustrated Parts Catalog

An **illustrated parts catalog** is designed to assist maintenance personnel in the ordering, storing, and issuing of aircraft parts. The parts manual is a companion document to the maintenance manual and contains all of the individual parts, such as seals, screens, filters, pulleys, fittings, and so on that are required to perform the procedures detailed in the maintenance manual.

In addition, a parts catalog is often helpful in locating and identifying specific aircraft components with which the technician is unfamiliar. A typical parts illustration is shown in Figure 14–12. When ordering parts, it is important that the technician pay careful attention to the applicable serial numbers and codes.

Wiring Diagram Manual

Because of the increased use of electronics systems on aircraft, it has become necessary to produce a separate **wiring diagram manual** that provides information on all of the electrical and electronic circuits. Schematic diagrams are

GAMA/ATA CODE & INDEX NO.	PART NO.	DESCRIPTION	UNITS PER ASSY
46-20	4500S-A1	VALVE ASSY,PARKING BRAKE	2
	4504	VALVE STEM ASSY	1
- 1	4505	STEM	1
- 2	4507	SPRING.	1
- 3	4506-1	HEAD,VALVE	1
- 4	21S062-0375	PIN	1
- 5	4511	SEAT,VALVE	1
- 6	AN6227B5	O RING	1
- 7	AN6227B11	O RING	1
- 8	4503	SPACER	1
- 9	4208	ARM ASSY,BRAKE VALVE	1
- 10	3/32 X 5/8	RIVET	2
- 11	4222	WASHER	4

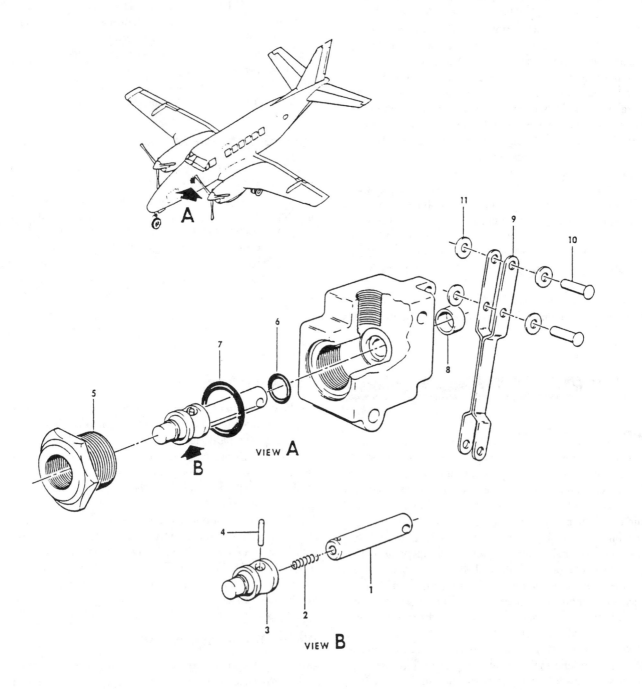

FIGURE 14–12 Typical parts catalog illustration.   *(Beech Aircraft Co.)*

used to illustrate complex electrical systems, such as power generation, autopilot, lighting, warning, and so forth. Basic logic symbols are used to depict the internal circuitry of components and to illustrate the basic signal flow in a system.

Weight-and-Balance Manual

The **weight-and-balance-manual** is used to transmit weight-and-balance data to the operator. The manual contains the information required to analyze and establish weight-and-balance procedures. The manual also includes the procedures used to weigh the aircraft.

Nondestructive Testing Manual

Nondestructive testing plays an important role in the determination of an aircraft's airworthiness. The airframe manufacturer's **nondestructive testing manual** contains information and specific instructions and data pertaining to the nondestructive testing of the aircraft structure.

Tool and Equipment List

Maintaining an aircraft in airworthy condition not only requires detailed technical information, but it also frequently requires specialized tools and equipment. On smaller aircraft, special tools and equipment may be listed in the aircraft maintenance manual, but most transport-category aircraft will have a separate **tool and equipment list** for this purpose. This list provides a listing of all the specialized tools and equipment required for the maintenance, service, and overhaul of the aircraft. The list also includes those pieces of test equipment required to properly test the aircraft or its components after performing maintenance.

Vendor Data

Due to frequent changes and varying combinations of some types of aircraft equipment and accessories, such as avionics, filters, and instrumentation, it is impossible to include information on these types of items in the aircraft maintenance manual. The manufacturers of these components produce technical information related to the installation, maintenance, and testing of their products. This information is commonly referred to as **vendor data**.

Technical-Data Presentation Systems

Over the years, various systems have been developed for organizing and presenting technical material. For many years, while airplanes and their systems remained relatively simple, the only manual that was developed, if any, was a maintenance manual. The length of such a manual might have totaled only fifty pages. As aircraft became more complex, the need for more detailed technical information became apparent. Also, a system to standardize the organization of this material was needed to assist the technician in readily locating the required information, even when working on several different models of aircraft. Thus, several **technical-data**

presentation systems have been developed to meet this need.

Air Transport Association Specification 100. The Air Transport Association of America (ATA) has developed a specification to establish a standard for the presentation of technical data by aircraft, engine, or component manufacturers that is required for their respective products. This specification is identified as **ATA Specification 100**.

It is the primary intent of ATA Specification 100 to clarify the general requirements of the airline industry with reference to the coverage, preparation, and organization of technical data. ATA Specification 100 details the names and contents of the various manuals that must be prepared by the manufacturer. The standards which provide for a uniform method of arranging technical materials is an effort to simplify the technician's problem in locating instructions and parts. The ATA Standard Numbering System, which uses a three-part number, is designed to provide numbers for the identification of all systems, subsystems, and subjects. A standard number is composed of three elements that consist of two digits each, as shown in Figure 14–13. Whether a technician is working on a Boeing-747 or a McDonnell Douglas MD-11, information on similar components will be found in the same chapter and section. A complete table of the ATA numbering system, subsystem, and titles is included in the Appendix. Not only are all of the technical manuals arranged according to the Specification 100 numbering system, but items such as service bulletins are also referenced by the appropriate specification number. Many airlines also organize parts rooms according to the ATA numbering system.

The ATA has also developed other specifications which are related to aircraft maintenance. The specifications include the following:

Specification 101, technical data on ground support equipment
Specification 102, a computer software manual, which provides a standard for the presentation of digital computer software documentation
Specification 103, a specification for airport fuel inspection and testing
Specification 104, guidelines for aircraft maintenance training
Specification 105, guidelines for training and qualifying personnel in nondestructive testing

GAMA Specification 2. Another system, **GAMA Specification 2**, was developed by the General Aviation Manufacturers Association (GAMA) for use in preparing manufacturer's maintenance data in a standardized format. General aviation manufacturers may prepare maintenance data in conformance with either Air Transport Association of America Specification 100 or with GAMA Specification 2, which closely follows and is compatible with ATA Specification 100. GAMA Specification 2 provides for the preparation and issuance of less complex manuals for less complex airplanes than those operated by the large scheduled airlines, but it still achieves industrywide standardization of nomenclature and format.

First Element, Chapter (system)		Second Element, Section (subsystem)		Third Element, Subject (unit)	Coverage
26 (System) Fire protection	—	00	—	00	Material which is applicable to the system as a whole.
26	—	20 (subsystem) Extinguishing	—	00	Material which is applicable to the subsystem as a whole.
26	—	22 (subsubsystem) Engine fire extinguishing	—	00	Material which is applicable to the subsubsystem as a whole. This number (digit) is assigned by the manufacturer.
26	—	22	—	03 (Unit) Bottles	Material which is applicable to a specific unit of the subsubsystem. Both digits are assigned by the manufacturer.

FIGURE 14–13 ATA Standard Numbering System.

Because this specification can apply to a wide range of aircraft, from small single-engine aircraft to complex business jets, GAMA Specification 2 has the provision for simple handbooks for simple airplanes and much more comprehensive handbooks for complex airplanes. Although the flexibility built into Specification 2 allows it to be adopted to a variety of aircraft, its use results in a high degree of standardization by providing for uniformity of the arrangement of systems and components.

Microfiche. Microfiche is fast becoming the industry standard for producing and storing technical information. **Microfiche**, often referred to as *aerofiche* by manufacturers, is a piece of photographic film, measuring 4 by 6 inches, which is capable of storing up to 288 pages of information. This tremendous reduction in size makes it possible to store a roomful of information on a table top within one's reach. Microfiche is less expensive to produce than traditional hard-copy manuals and is also much easier to revise as technical information is updated. A disadvantage of microfiche is that reading it requires the use of a microfiche reader (see Figure 14–14). Machines called *reader printers* are capable of printing out pages from the microfiche. Technical information is available on microfiche and may be purchased directly from the government printing office, from manufacturers, or from independent publishers who reproduce technical information and add features which make it easier to use.

MANUFACTURERS' OPERATING PUBLICATIONS

Manufacturers produce technical publications that are required for the operation of an aircraft. While such publica-

tions are primarily intended for flight crew members, the technician must be familiar with the aspects of these publications that are related to the maintenance and inspection of an airplane.

Approved Flight Manual

An **FAA Approved Flight Manual** contains all the pertinent information essential to proper operation of an airplane. An approved flight manual listed on an aircraft's Type Certificate Data Sheet is considered a required piece of equipment that must be kept on board the airplane. The manual contains an airplane's standard operating procedures and techniques, operating limitations, emergency procedures, and weight-and-balance data.

Minimum Equipment List

An airplane is certified with all required equipment in place and in operating condition. It follows that, should any item

FIGURE 14–14 Microfiche reader.

on that aircraft become inoperative, the aircraft no longer meets its type certification standards and is, therefore, not airworthy.

If some deviations from the type-certificated configuration and equipment required by the FAR operating rules were not permitted, an aircraft could not be flown unless all such equipment were operating. Experience has proven, however, that the operation of every system or component installed on an aircraft is not necessary when the remaining operating instruments and equipment provide for continued safe operation. The confusion about just which items can be inoperative, and under what conditions, is clarified by the use of a **Minimum Equipment List (MEL)**. The Federal Aviation Regulations permit the publication of an MEL which provides for the operation of an aircraft with certain items or components inoperative, provided the FAA finds an acceptable level of safety is maintained by a transfer of the function to another operating component or by reference to other instruments or components providing the required information.

The MEL may not include obviously required items such as wings, rudders, flaps, engines, and landing gear. Also, the list may not include items which do not affect the airworthiness of the aircraft, such as galley equipment, entertainment systems, and passenger convenience items.

Unless otherwise specified in the "Remarks" column, the FAA does not define where or when an inoperative item is to be repaired or replaced. The failure of instruments or items of equipment in addition to those allowed to be inoperative by the MEL causes the aircraft to no longer be airworthy. The MEL is not intended to provide for continued operation of the aircraft for an indefinite period with inoperative items. The basic purpose of the MEL is to permit the operation of an aircraft with inoperative equipment within the framework of a controlled and sound program of repairs and parts replacement. It is important that the owner or operator make repairs at the first airport, since additional malfunctions may require the aircraft to be taken out of service. Operators are responsible for exercising the necessary control to assure that an aircraft is not flown with multiple MEL items inoperative without first determining that any interrelationship between inoperative systems or components will not result in a potential safety problem.

An MEL can only be obtained for an aircraft for which the manufacturer has published a **Master MEL**. A Master MEL, such as the one shown in Figure 14–15, is adopted by the aircraft operator to fit the equipment requirements of his or her specific aircraft. After an MEL is prepared from the Master MEL, it is submitted to the local FAA Flight Standards District Office for approval. An approved MEL constitutes a supplemental type certificate for an aircraft. For that reason, an MEL must be approved for each specific aircraft.

U.S. Department of Transportation Federal Aviation Administration		MASTER MINIMUM EQUIPMENT LIST	

AIRCRAFT Cessna CE-650	REVISION NO 1 DATE April 10, 1987		PAGE 33-1

SYSTEM & SEQUENCE NUMBERS ITEM	1.	2. NUMBER INSTALLED	
		3. NUMBER REQUIRED FOR DISPATCH	
			4. REMARKS OR EXCEPTIONS
33 LIGHTS			
−1 Anticollision Light System (strobe)	1	0	* May be inoperative provided a. aircraft is not operated at night.
−2 Position Light System	−	0	* May be inoperative provided a. aircraft is not operated at night. OR
	2	1	* One may be inoperative.
−3 Wing Inspection Light	2	1	* One may be inoperative. OR
	2	0	* Both may be inoperative provided a. aircraft is not operated in known or forecast icing conditions at night.
−4 Landing Lights	2	1	* One may be inoperative provided a. both Taxi Lights (33–11) are operative. OR
	2	0	* Both may be inoperative provided a. aircraft is not operated at night.

FIGURE 14–15 Master minimum equipment list.   *(Cessna Aircraft Co.)*

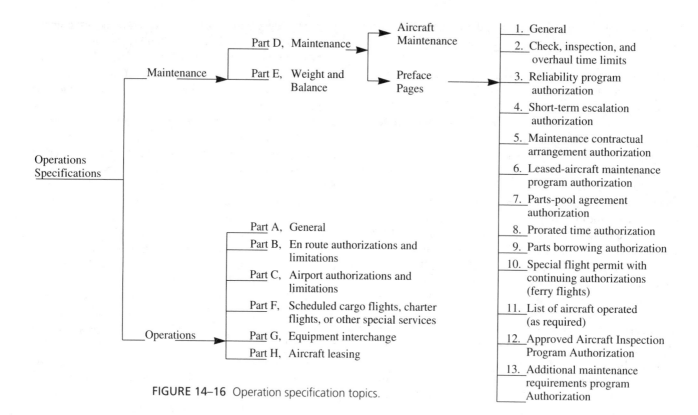

FIGURE 14–16 Operation specification topics.

Operation Specifications

Operation Specifications are issued to supplement air-carrier and air-taxi rules by listing additional privileges and limitations that are not specifically covered by the regulations. Operation specifications are divided into different subparts such as the kinds of operations authorized, airport limitations, weight-and-balance data, and maintenance procedures (see Figure 14–16).

Operation Specifications for maintenance cover items such as the aircraft's inspection program, overhaul time limits, component replacement times, and contract maintenance.

Operation specifications are prepared by the applicant and submitted to the local FAA Flight Standards District Office for approval. When approved, the provisions of the operation specifications are as legally binding as those of the FAR.

REVIEW QUESTIONS

1. What is the purpose of an Advisory Circular?
2. How may an Advisory Circular be made mandatory?
3. How are general-aviation service difficulties generally reported to the FAA?
4. What is the purpose of Airworthiness Alerts?
5. Alerts are published as what Advisory Circular?
6. When does the FAA issue an Airworthiness Directive?
7. Airworthiness Directives are published as amendments to which FAR?

8. How are ADs indexed?
9. How often is the AD Summary published?
10. What are the subject areas into which ADs are divided?
11. How often are newly issued and revised AD Notes mailed out?
12. What do the digits *06* signify in the AD number 92-06-07?
13. When is a technician responsible for performing an AD search?
14. What items need to be included in an AD record entry?
15. What is the difference in content between a TCDS and Aircraft Specifications?
16. TCDSs are published in how many volumes?
17. Technical information about older aircraft models of which no more than 50 remain in service would be found in which volume?
18. What does the designation code 4 P-CSM mean?
19. To what will Note 2 on a TCDS always refer?
20. What is the purpose of an STC?
21. What publication is issued by manufacturers to notify owners of design defects?
22. To which manufacturer's publication would you refer in locating information on troubleshooting an aircraft system?
23. What is the purpose of the ATA Specification 100 code?
24. What publication provides a list of the aircraft equipment that must be operable in order to initiate a flight?
25. What is the purpose of *Operation Specifications*?

Ground Handling and Safety 15

INTRODUCTION

Most aviation technicians devote a portion of their time to ground-handling and taxiing aircraft. The complexity of today's aircraft and accompanying ground-support equipment creates an expensive and dangerous work environment. The purpose of this chapter is to give you information on the procedures, techniques, and safety precautions involved in handling and servicing aircraft on the ground. Because of the many different types of aircraft and aircraft operations, it is possible to provide only general information and standard practices. For the specific operations applying to a particular aircraft, the technician should consult the aircraft's service manual.

GENERAL SAFETY PRECAUTIONS

By their very nature and design, airplanes make for a dangerous work environment. The danger of this environment is further increased by the wide variety of machines, tools, and materials required to support and maintain aircraft. Developing and following a set of personal work safety habits is in a technician's own self-interest. Personal safety begins with the proper use of eye and ear protection as well as proper dress for the job being performed. Technicians should only operate equipment with which they are familiar and can operate safely. Hand tools should be kept in proper working order. Technicians should know the location of first-aid and emergency equipment.

Good housekeeping in hangars, shops, and on the flight line is essential to safe and efficient maintenance. The highest standards of orderly work arrangements and cleanliness should be observed while maintaining an aircraft. When a maintenance task is complete, the technician should remove and properly store maintenance stands, hoses, electrical cords, hoists, crates, boxes, and anything else used to perform the work.

Pedestrian lanes and fire lanes should be marked and used as a safety measure to prevent accidents and to keep pedestrian traffic out of work areas.

Power cords and air hoses should be straightened, coiled, and properly stored when not in use. Oil, grease, and other substances spilled on hangar or shop floors should be immediately cleaned or covered with an absorbent material to prevent fire or personal injury. Drip pans should be placed beneath engines and engine parts wherever dripping exists. *Under no circumstances should oil or cleaning fluid be emptied into floor drains.* Fumes from this type of disposal may be ignited and cause severe property damage. Gasoline spills on the hangar floor should be flushed away with water. Sweeping these fuel spills with a dry broom could cause static electricity that might ignite the fuel.

Aircraft finishes should be applied in a controlled environment (paint room) whenever possible. A technician should never do this type of work near an open flame or in the presence of lights that are not explosion proof. No other work should be done on an aircraft while it is being painted.

Welding should only be performed in designated areas. Any part to be welded should be removed from the aircraft, if possible. The repair can then be done in a welding shop under a controlled environment. A welding shop should be equipped with proper tables, ventilation, tool storage, and fire-prevention and -extinguishing equipment.

COMPRESSED-GAS SAFETY

Compressed gases are frequently used in the maintenance and servicing of aircraft. The use of compressed gases requires a special set of safety measures. The following rules apply for the use of compressed gases:

1. Handle cylinders of compressed gases as you would high-energy sources and therefore potential explosives.

2. Always use safety glasses when handling and using compressed gases.

3. Never use a cylinder that cannot be positively identified.

4. When storing or moving a cylinder, have the cap securely in place to protect the valve stem.

5. When large cylinders are moved, strap them to a properly designed wheeled cart to insure stability.

6. Use the appropriate regulator on each gas cylinder. Adapters or homemade modifications can be dangerous.

7. Never direct high-pressure gases at a person.

8. Do not use compressed gas or compressed air to blow away dust or dirt, since the resultant flying particles are dangerous.

9. Release compressed gas slowly; the rapid release of a compressed gas will cause an unsecured gas hose to whip

dangerously and also may build up a static charge which could ignite a combustible gas.

10. Clean gas cylinders; oil or grease on an oxygen cylinder can cause an explosion.

FIRE SAFETY

Fire is one of humanity's greatest discoveries. For all its many advantages, however, fire is capable of producing disaster in a matter of seconds. Fires continue to take their toll even though the technological knowledge and capability exists to prevent and retard fires.

Nature and Classification of Fires

Fire results from the chemical reaction that occurs when oxygen combines rapidly with fuel to produce heat and light (see Figure 15–1). The essentials of this process are (1) fuel—a combustible gas, liquid, or solid; (2) oxygen—sufficient in volume to support the process of combustion and usually supplied from the surrounding air; and (3) heat—sufficient in volume and intensity to raise the temperature of fuel to its ignition or kindling point.

There are four classes of fires, each determined by what is burning. The most common is that which occurs in ordinary combustible materials: paper, wood, textiles, and rubbish. It is designated as a Class A fire. Fires in combustible liquids, such as gasoline, alcohol, oil, grease, and oil-based paint, form a second category known as Class B fires. The third, Class C fires, are those occurring in live electrical equipment, such as fuse boxes, switches, appliances, motors, or generators. Class D fires are those of high intensity that may occur in certain metals such as magnesium, sodium, potassium, titanium, and zirconium. The greatest hazard exists when these metals are in a molten state or in finely divided forms of dust, chips, turnings, or shavings.

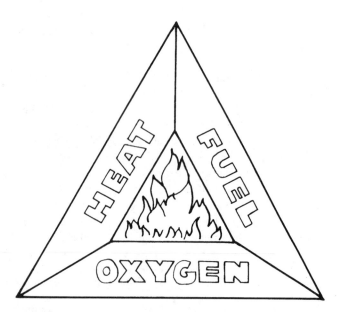

FIGURE 15–1 The fire triangle consists of heat, fuel, and oxygen.

Spontaneous Ignition

Aviation technicians need to be particularly aware of **spontaneous ignition** caused by the lubricants and solvents that are used in maintaining aircraft. Certain materials, such as rags soaked with oil or solvents, are capable of generating sufficient heat to cause combustion. These rags should be disposed of in airtight cans.

Principles of Extinguishing Fires

Based on the principle of the fire triangle (see Figure 15–1), there are three ways to extinguish fires: *(1) cooling the fuel below its kindling point, (2) excluding the oxygen supply, and (3) separating the fuel from the oxygen.* These methods have led to the development of different types of extinguishers for different types of fires.

Extinguishing Agents

Class A fires respond best to water or water-type extinguishers which cool the fuel below combustion temperatures. Class B and C extinguishers are effective but not equal to the wetting/cooling action of the Class A extinguisher.

Class B fires respond to carbon dioxide (CO_2), halogenated hydrocarbons (halons), and dry chemicals, all of which displace the oxygen in the air, thereby making combustion impossible. Foam is effective, especially when used in large quantities. Water is ineffective on Class B fires and in fact will cause the fire to spread.

Class C fires involving electrical wiring, equipment, or current respond best to carbon dioxide (CO_2), which displaces the oxygen in the atmosphere, making combustion impossible. The CO_2 extinguisher must be equipped with a nonmetallic horn to be approved for use on electrical fires. Two reasons for this requirement must be considered:

1. The discharge of CO_2 through a metal horn can generate static electricity. The static discharge could reignite the fire.

2. The metal horn, if in contact with the electric current, would transmit that current to the extinguisher's operator.

Halogenated hydrocarbons are very effective on Class C fires. The vapor reacts chemically with the flame to extinguish the fire. Dry chemicals are effective but have the disadvantage of contaminating the local area with powder. Also, if used on wet and energized electrical equipment, they may aggravate current leakage.

Water or foam are not acceptable agents for use on electrical equipment, as they also may aggravate current leakage.

Class D fires respond to the application of dry powder, which prevents oxidation and the resulting flame. The application may be from an extinguisher, a scoop, or a shovel. Special techniques are needed in combating fires involving metal. Manufacturers' recommendations should be followed at all times. Areas which could be subjected to metal fires should have the proper protective equipment installed. Under no conditions should a person use water on a metal fire. It will cause the fire to burn more violently and can cause explosions.

Identification of Fire Extinguishers

Applying the incorrect material on a specific fire can do more harm than good and may actually be dangerous. It is important that fire extinguishers be well marked for quick identification under emergency conditions. In the excitement of a fire, it is very easy to grab the wrong type of extinguisher. Before using an extinguisher, read the labels. Usually, fire extinguishers are marked with decals, paint codes, or similar labels. The National Fire Protection Association recommends the following markings for various extinguishers:

1. Extinguishers suitable for use on Class A fires should be identified by a triangle containing the letter *A*. If the triangle is colored, it should be colored green.
2. Extinguishers suitable for use on Class B fires should be identified by a square containing the letter *B*. If the square is colored, it should be colored red.
3. Class C fire extinguishers should be identified by a circle containing the letter *C*. If colored, the circle should be colored blue.
4. Extinguishers suitable for fires involving metal should be identified by a five-point star that contains the letter *D*. If the star is colored, it should be yellow.

Extinguishers suitable for more than one class of fire may be identified by multiple symbols. Figure 15–2 shows typical fire extinguisher markings in use today. The Underwriters Laboratories (UL) is a nationally recognized testing agency established and supported by manufacturers and insurance companies. Throughout the years, the UL label has become a symbol of quality and efficiency in fire-extinguishing equipment. One of the functions of the UL is that of testing and listing all types of electrical and fire-fighting equipment. Therefore, all such equipment should have the UL label.

All fire extinguishers, whether in hangars, airplanes, or elsewhere, should be checked frequently to be sure they are in working order. This check should determine if the extinguishers are full and whether they have been tampered with. Proper maintenance, including checking for needed repairs, recharging, or replacement, should be carried out annually.

ORDINARY
COMBUSTIBLES
1. WATER

FLAMMABLE
LIQUIDS

ELECTRICAL
EQUIPMENT

2. CARBON DIOXIDE, DRY CHEMICAL BROMOCHLORODIFLUOROMETHANE AND BROMOTRIFLUOROMETHANE

ORDINARY
COMBUSTIBLES

FLAMMABLE
LIQUIDS

ELECTRICAL
EQUIPMENT

3. MULTIPURPOSE DRY CHEMICAL

FLAMMABLE
LIQUIDS

ELECTRICAL
EQUIPMENT

A
CAPABILITY

4. MULTIPURPOSE DRY CHEMICAL (INSUFFICIENT AGENT FOR CLASS A FIRE)

COMBUSTIBLE
METALS

5. DRY POWDER

FIGURE 15–2 Typical extinguisher markings.

FLIGHT-LINE SAFETY

The source of most accidents on the flight line is that of propellers or rotors. A propeller or rotor is difficult to see when in operation. Even personnel familiar with the danger of a turning propeller or rotor sometimes forget about its presence. Propeller-and-rotor-to-person accidents differ from other aircraft accidents in that they usually result in fatal or serious injury. This is because a propeller or rotor rotating under power, even at slow, idling speed, has sufficient force to inflict serious injury. It should be remembered that a rotating propeller or rotor is extremely dangerous and should be treated with extreme caution. Some manufacturers of propeller and rotor blades use paint schemes to increase the visibility of the blades. Technicians should give strong consideration to maintaining the visibility paint scheme of the original manufacturer.

Persons directly involved with aircraft service are most vulnerable to injuries by propellers or rotors. Working around aircraft places these individuals in the most likely position for possible propeller or rotor accidents. Aircraft service personnel should develop the following safety habits:

1. Treat all propellers as though the ignition switches are on.
2. Chock airplane wheels before working around aircraft.
3. Use wheel chocks and parking brakes before starting engines.
4. Attach pull ropes to pull chocks from wheels close to rotating propellers or rotor blades.

5. After an engine run and before the engine is shut down, perform an ignition-switch test to detect a faulty ignition switch.

6. Before moving a propeller or connecting an external power source to an aircraft, be sure that the aircraft is chocked, ignition switches are in the OFF position, throttle is closed, mixture is in IDLE CUT-OFF position, and all equipment and personnel are clear of the propeller or rotor. Faulty diodes in aircraft electrical systems have caused starters to engage when external power was applied, regardless of the ignition-switch position.

7. Remember, when removing an external power source from an aircraft, keep the equipment and yourself clear of the propeller or rotor.

8. Always stand clear of rotor and propeller blade paths, especially when moving the propeller. Particular caution should be practiced around warm engines.

Ground-support personnel in the vicinity of aircraft that are being run up need to wear proper eye and ear protection. Ground personnel must also exercise extreme caution in their movements about the ramp; a great number of serious accidents have occurred involving personnel in the area of turbojet engine air inlets. The turbojet engine intake and exhaust hazard areas are illustrated in Figure 15–3. Care should also be taken to ensure that the run-up area is clear of all items such as nuts, bolts, rocks, rags, or other loose debris.

TOWING AIRCRAFT

Particular care must be exercised when pulling or pushing an aircraft. Persons performing towing operations should be thoroughly familiar with the procedures that apply to the type of aircraft to be moved. When towing aircraft, the proper tow bar must be used. The wrong type of tow bar or other makeshift equipment can cause damage to the aircraft.

As illustrated in Figure 15–4, the tow bar is usually made with fittings for attachment to the ends of the axles or to the tow fittings on the landing-gear strut or axle. Large aircraft are towed by means of specially designed towing vehicles. Figure 15–5 shows a Boeing 747 airplane being towed by such a vehicle. The aircraft is attached to the towing vehicle

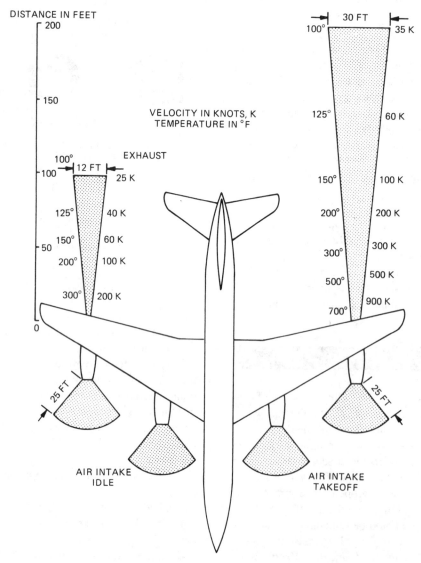

FIGURE 15–3 Engine intake and exhaust hazard areas.

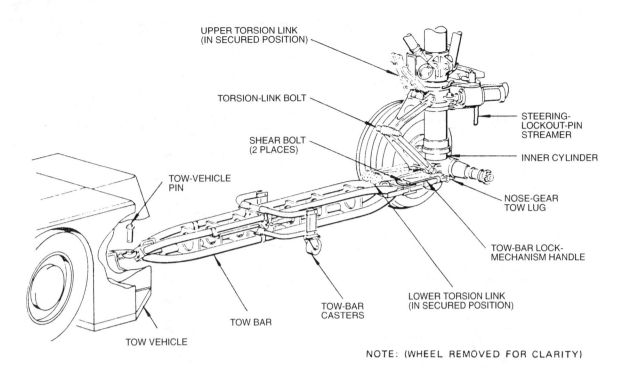

FIGURE 15–4 Tow-bar attachment.

by a tow bar designed for the purpose. In this type of towing, the operator of the towing vehicle must be in constant communication with the operator of the aircraft in the cockpit. The operator of the aircraft is also in radio contact with ground-traffic control in the tower.

During the towing of aircraft, care must be taken to make sure that the following precautions are followed, otherwise damage to the aircraft may result:

1. The tow-vehicle operator should never tow an aircraft in congested areas without someone to assist in determining that there is adequate clearance. When moving a large aircraft, a ground-handling chart, such as the one shown in Figure 15–6, should be consulted to verify that adequate turning space exists.

2. When a small aircraft is being moved manually, care must be taken to apply hand pressure to the aircraft at only those areas that have structural strength sufficient to withstand the pressure without damage.

3. When towing, the nose gear should not be turned beyond its steering radius in either direction, as this will result in damage to the nose gear and steering mechanism. The turning limits are often marked on the nose gear, as shown in Figure 15–7.

4. Transport-category aircraft equipped with nose-wheel steering are often equipped with a steering lockout pin. When inserted, the pin will lock out the hydraulic steering and allow a sharper turning radius. A typical lockout pin installation is shown in Figure 15–8 on page 338.

5. On light aircraft equipped with mechanical nose-wheel steering, the airplane control locks must be removed before towing the airplane. Serious damage to the steering linkage can result if the airplane is towed while the control locks are installed.

6. Prior to movement of any aircraft, all landing-gear struts and tires should be properly inflated and brake pressure built up, when applicable.

7. The aircraft's center of gravity must be within towing limits.

8. The tow-vehicle operator should avoid sudden starts and stops. Although the tug will control the steering of the airplane, someone should be positioned in the pilot's seat to operate the brakes in case of an emergency. Once an airplane starts to roll, a steady, slow movement should be maintained.

9. Any movement of aircraft in a hangar should be backed up with wheel chocks to prevent roll-back.

10. Clearance must be obtained from the airport control tower, either by appropriate radio frequency or by prior arrangement through other means, before moving aircraft across runways or taxiways.

FIGURE 15–5 Towing a large airplane.

| X TURN RADIUS | CLEARANCE RADIUS | | | | | Z MINIMUM WIDTH FOR 180° TURN |
| | A WING TIP | B NOSE GEAR | C WING GEAR | D TAIL TIP | E NOSE | |
FEET / METERS	FEET / METERS	FEET / METERS	FEET / METERS	FEET / METERS	FEET / METERS	FEET / METERS
0 / 0	113 / 34.44	85 / 25.91	23 / 7.01	122 / 37.18	110 / 33.53	108 / 32.92
20 / 6.10	131 / 39.93	89 / 27.13	42 / 12.80	132 / 40.23	111 / 33.83	131 / 39.93
40 / 12.19	149 / 45.41	96 / 29.26	62 / 18.90	142 / 43.28	116 / 35.36	158 / 48.16
60 / 18.29	168 / 51.21	106 / 32.31	82 / 24.99	153 / 46.63	125 / 38.10	188 / 57.30
80 / 24.38	186 / 56.69	118 / 35.97	102 / 31.09	167 / 50.90	136 / 41.45	220 / 67.06
100 / 30.48	205 / 62.48	133 / 40.54	121 / 36.88	181 / 55.17	149 / 45.41	254 / 77.42
120 / 36.58	225 / 68.58	157 / 47.85	141 / 42.98	197 / 60.04	163 / 49.68	298 / 90.83
140 / 42.67	244 / 74.37	166 / 50.60	161 / 49.07	213 / 64.92	178 / 54.25	327 / 99.67
160 / 48.77	264 / 80.47	183 / 55.78	181 / 55.17	230 / 70.10	195 / 59.44	364 / 110.95

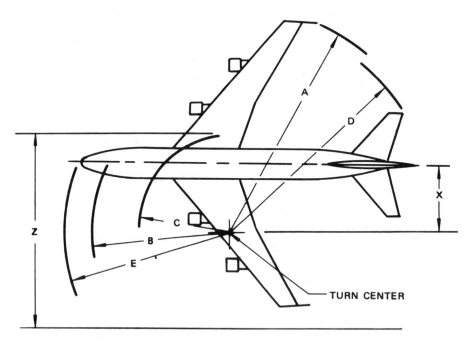

NOTE: RADII *B* AND *C* ARE TO THE OUTSIDE EDGE OF OUTBOARD TIRES

FIGURE 15–6 Boeing 747 turning clearances. *(Boeing Commercial Aircraft Co.)*

TAXIING AND STARTING

The taxiing of an airplane can be a relatively simple matter, or it can be a very complex and critical operation, depending upon the size and the type of aircraft being taxied. Before any airplane can be taxied, it is necessary to start the engine or engines. Again, this operation can be simple or complex, depending upon the size and the type of engine.

Starting Small Reciprocating Engines

The following procedures are typical of those used to start reciprocating engines. There are, however, wide variations in the procedures for the many different types of reciprocating engines. *No attempt should be made to use the methods presented here for actually starting an engine.* Instead, always refer to the procedures contained in the applicable manufacturer's instructions.

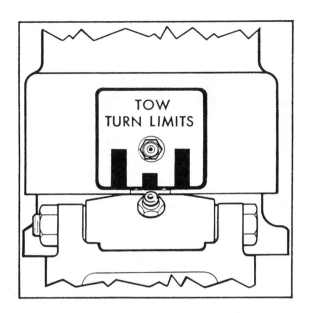

FIGURE 15–7 Tow turn-limits indicator.

Three conditions are necessary for the starting of any internal-combustion engine: (1) the presence of fuel in the combustion chamber, (2) a source of ignition for the fuel, and (3) a method for rotating the engine to start the sequence of operating events. The methods by which these conditions are satisfied depend to some extent upon the type of engine and the design of the fuel, ignition, and starting systems.

For a light-aircraft engine equipped with a float-type carburetor, the starting procedure may be as follows:

1. Be sure that the engine is supplied with oil and an adequate supply of fuel is available in the fuel tank or tanks.
2. With the ignition switch OFF, rotate the engine at least three complete revolutions by hand or with the starter to see that the engine is clear of obstructions. When the engine is rotated by hand, the person turning the propeller must remember to stand well out of the range of the propeller in case the engine should fire. In some cases, the engine can be "hot" even though the ignition switch is in the OFF position.
3. Prime the engine with the priming pump, using approximately three strokes, provided that the engine is cold. The fuel valve must be ON.
4. Place the mixture control in the FULL RICH position.
5. Open the throttle to about one-eighth of the FULL OPEN position.
6. Turn the ignition switch on and press the starter button, or turn the starter switch key to the start position. *If prolonged cranking is necessary, allow the starter motor to cool at frequent intervals*, since excessive heat may damage the armature.
7. After the engine starts, adjust the throttle for approximately 800 to 1000 rpm [84 to 105 rad/s] and check the oil-pressure gauge for pressure. If oil pressure does not show within about 30 seconds, shut the engine down and check for the trouble. With some small engines, after a long period of storage, the oil pump will lose its prime, and it is neces-

sary to reprime the pump by removing the oil screen and pouring oil into the opening.

When starting a light engine with a pressure-type carburetor or a fuel-injection system, it is not usually necessary to use a primer. In these cases, fuel will be supplied as soon as the mixture control is moved from the IDLE CUTOFF position toward the FULL RICH position. The fuel booster pump must be turned on to provide fuel pressure. The mixture control should be moved to the FULL RICH position after the engine is started. Another procedure recommended for starting certain engines with fuel injection is to start the engine on the priming fuel only and then move the mixture control to the FULL RICH position as soon as the engine starts.

Starting Large Reciprocating Engines

Large radial engines installed on the DC-3, DC-6, Constellation, and other large aircraft should be started according to the manufacturer's instructions. The steps in starting are similar to those used for light-aircraft engines, but additional precautions are necessary. First, a fire guard should be placed to the rear and outboard of the engine being started, in case the engine backfires and fire burns in the induction system of the engine. The fire guard should have an adequate supply of carbon-dioxide gas in suitable fire-extinguisher bottles in order to immediately direct the gas into the induction system of the engine. The engine should be kept turning so that the fire will be drawn into the cylinders. Very often, it is not even necessary to use the extinguisher because the air rushing into the engine carries the fire with it and, as the engine starts, the fire cannot continue to burn in the induction system.

Before attempting to start a radial engine, the engine should be rotated several complete revolutions to eliminate the possibility of **liquid lock** caused by oil in the lower cylinders. If the engine stops suddenly while being rotated by hand or with the starter, oil has collected in a lower cylinder, and before the engine can be started, the oil must be removed. This is best accomplished by removing a spark plug from the cylinder. It is not recommended that the engine rotation be reversed to clear the oil. After the oil is drained from the cylinder, the spark plug can be replaced and the engine can be started.

For large reciprocating engines, priming is usually accomplished by means of a fuel-pressure pump and an electrically operated priming valve. Priming is accomplished by pressing the priming switch while the fuel booster pump is turned on.

Large reciprocating engines may have direct-cranking starters similar to those used on light-aircraft engines but much more powerful, or they may have inertia starters in which the cranking energy is stored in a rapidly rotating flywheel. With the inertia starter, the flywheel must be energized by means of an electric motor or by hand crank until enough energy is stored to turn the engine for several revolutions. The ENGAGE switch is then turned on to connect the flywheel-reduction gearing to the crankshaft through the

FIGURE 15–8 Steering-lockout-pin installation. *(Boeing Commercial Aircraft Co.)*

starter jaws. A plate clutch, located between the flywheel and the starter jaws, allows slippage to avoid damage due to inertial shock when the starter is first engaged.

With the engine properly primed, the throttle set, and the ignition switch ON, the engine should start very soon after it is rotated by the starter. The throttle is then adjusted for proper warmup speed.

Hand Propping

Hand propping a starter-equipped engine with a low battery or a defective starter, although convenient, can expose per-

sonnel to a possible accident. For safety reasons, the replacement of the faulty starter and the use of a ground power source should be considered rather than hand cranking. *Only experienced persons should do the hand propping, and a reliable person should be in the cockpit.* Hand propping with the cockpit unoccupied has resulted in many serious accidents.

If the aircraft has no self-starter, the engine must be started by swinging the propeller. The person who is turning the propeller calls, "Fuel on, switch off, throttle closed, brakes on." The person operating the engine will check these items and repeat the phrase. The switch and the throttle must not be touched again until the person swinging the prop calls, "Contact." The operator will repeat "Contact" and then turn on the switch. *Never turn on the switch and then call, "Contact."*

When swinging the prop, a few simple precautions will help to avoid accidents. When touching a propeller, always assume that the ignition is ON. The switches which control the magnetos operate on the principle of short-circuiting the current to turn the ignition off. If the switch is faulty, it can be in the OFF position and still permit current to flow in the magneto primary circuit.

Be sure the ground is firm. Slippery grass, mud, grease, or loose gravel can lead to a fall into or under the propeller. Never allow any portion of your body to get in the way of the propeller. This applies even though the engine is not being cranked.

Stand close enough to the propeller to be able to step away as it is pulled down. Stepping away after cranking is a safeguard in case the brakes fail. Do not stand in a position that requires leaning toward the propeller to reach it. This throws the body off balance and could cause you to fall into the blades when the engine starts. In swinging the prop, always move the blade downward by pushing with the palms of the hand. Do not grip the blade with the fingers curled over the edge, since "kickback" may break them or draw your body into the blade path.

Starting Turbine Engines

Before starting a gas-turbine engine, it is important to see that the areas in front of and to the rear of the engines are clear. The airplane should be resting on clean concrete or asphalt so that no loose material can be drawn into the inlet of the engines. The area to the rear of a gas-turbine engine is exposed to intense heat and high-velocity gases when the engine is running, as is illustrated in Figure 15–3. Materials or objects that can be damaged by heat should be cleared from the exhaust area.

With a gas-turbine engine, care must be taken to follow the manufacturer's instructions for starting. Since these engines have several different types of fuel-control units, the operator must know the procedures to be followed for the particular engine. Basically, the engine must be rotated to approximately 10% of full speed, ignition must be provided, and the proper amount of fuel must be delivered to the fuel nozzles. A manual start for some types of jet engines can be accomplished as follows:

1. Turn on the master switch.
2. Turn on the fuel booster pump.
3. Turn on the starter switch and hold it until the engine speed as shown on the percent-of-power gauge is approximately 10%.
4. Turn on the ignition switch.
5. Move the power-control lever (throttle) slowly forward while watching the exhaust-gas temperature (EGT) gauge. As soon as the fuel ignites, the EGT gauge will show a rapid rise in temperature.
6. Stop moving the power-control lever until the temperature stabilizes, and then slowly move it forward, being careful to see that the EGT does not exceed the proper limit (approximately 600°C). The engine should accelerate smoothly as the power-control lever is moved forward.
7. If the engine does not accelerate properly or if the EGT exceeds the safe limit, the power-control lever should be retarded to cut off the fuel and stop the engine.

When, during the starting of a gas-turbine engine, the EGT exceeds the prescribed safe limit, the engine is said to have a **hot start**. As just explained, when this occurs, the engine should be immediately shut down. Hot starts are usually caused by an excess of fuel entering the combustion chamber. With an automatic fuel-control unit, a hot start will not occur unless the unit is malfunctioning.

If the engine fails to accelerate properly, it is said to have a **false start**, also called a **hung start**. This type of start is caused by attempting to ignite the fuel before the engine has been accelerated sufficiently by the starter. The automatic systems used on modern engines prevent this type of problem unless the fuel-control unit is malfunctioning.

With an automatic starting system such as is currently found on most turbine aircraft, the starting procedure is comparatively simple. With the electric power turned on in the aircraft and the ground-starter unit connected to the aircraft, the pilot or the flight engineer needs merely to initiate the starting sequence with the START switch and the power-control lever. The starting sequence will then take place under the control of the fuel-control unit and associated equipment. After one engine is started on a multiengine aircraft by means of the ground-starter unit, the other engines can be started by using bleed air from the running engine.

Taxiing Aircraft

The taxiing of small aircraft is not difficult, but it does require watchfulness and care. Before starting to taxi, the pilot or the technician observes carefully to see that no other aircraft, persons, vehicles, or obstructions are likely to be in or near the taxi route. A small aircraft's direction is controlled during taxiing with the use of rudder-pedal steering, brakes, and, in the case of twin-engine aircraft, differential engine power. A large aircraft is controlled during taxiing by the use of selective engine thrust, the nose-wheel steering system, and the brakes. The geometric arrangement of the main landing gear can result in different ground-maneuvering characteristics for different aircraft. Basic factors that influence the geometry of a turn are as follows:

1. Degree of the nose-wheel steering angle
2. Engine power settings
3. Center-of-gravity location
4. Gross weight
5. Pavement surface conditions
6. Ground speed
7. Amount of differential braking

Manufacturers often supply detailed instructions for the taxiing, towing, and handling of aircraft. Typical of such instructions are the following for the Cessna Model 421 airplane: Before attempting to taxi the aircraft, ground personnel should be checked out by qualified persons. When it is determined that the propeller blast area is clear, apply power, start the taxi roll, and perform the following checks:

1. Taxi forward a few feet and apply the brakes to determine their effectiveness.

2. While taxiing, make slight turns to determine the effectiveness of the nose-gear steering.

3. Check the operation of the turn-and-bank indicator and the directional gyro.

4. Check for sluggish instruments during taxiing.

5. In cold weather, make sure all instruments have warmed sufficiently for normal operation.

6. Minimum turning distance must be strictly observed when the aircraft is being taxied close to buildings or other stationary objects (see Figure 15–9).

7. Do not operate the engine at high rpm when taxiing over ground containing stones, gravel, or any loose material that may cause damage to the propeller blades.

When a large aircraft is being taxied in the vicinity of other aircraft or near terminal buildings, one or more people to watch for proper clearance should be standing on the ground forward and to the right or left of the aircraft so as to be clearly visible to the aircraft operator. The lineman should use standard FAA hand signals to indicate to the operator what action is to be taken. Some of the standard signals are shown in Figure 15–10. Note that these signals are easily understood because the motions represent the action to be taken by the operator of the aircraft. On a field with an operating control tower, taxiing of the aircraft must be accomplished in accordance with directions from the ground-control operator in the airport tower. The operator of the aircraft must keep the radio tuned to the ground-control frequency at all times while taxiing. In the case where the aircraft is not radio equipped or in the event of a radio failure, lights are used to regulate taxiing. The following signal lights from the tower are used when necessary:

Steady red	Stop
Flashing red	The aircraft should immediately taxi clear of the runway it is on
Steady green	OK to taxi
Alternating red and green	OK to taxi but with extreme caution
Flashing white	Return to starting point

TYING DOWN OR MOORING AIRCRAFT

Small aircraft should be tied down after each flight to preclude damage from sudden wind gusts. Aircraft should be headed, as nearly as possible, into the wind, depending on the locations of the fixed, parking-area tie-down points.

When parking an airplane, it should be sufficiently protected from adverse weather conditions and checked to see that it presents no danger to other aircraft. When parking the

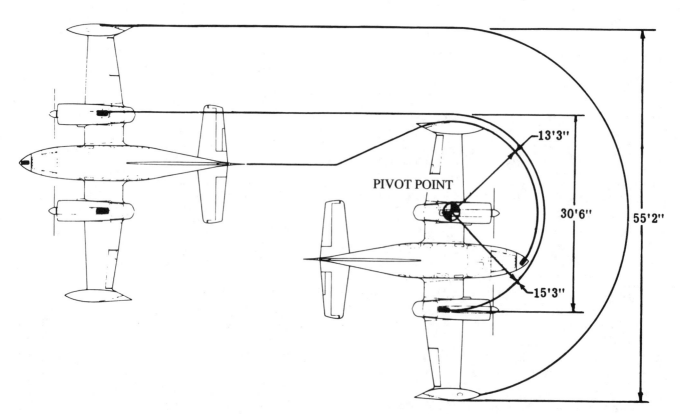

FIGURE 15–9 Minimum turning distance for a light twin-engine airplane. (Cessna Aircraft Co.)

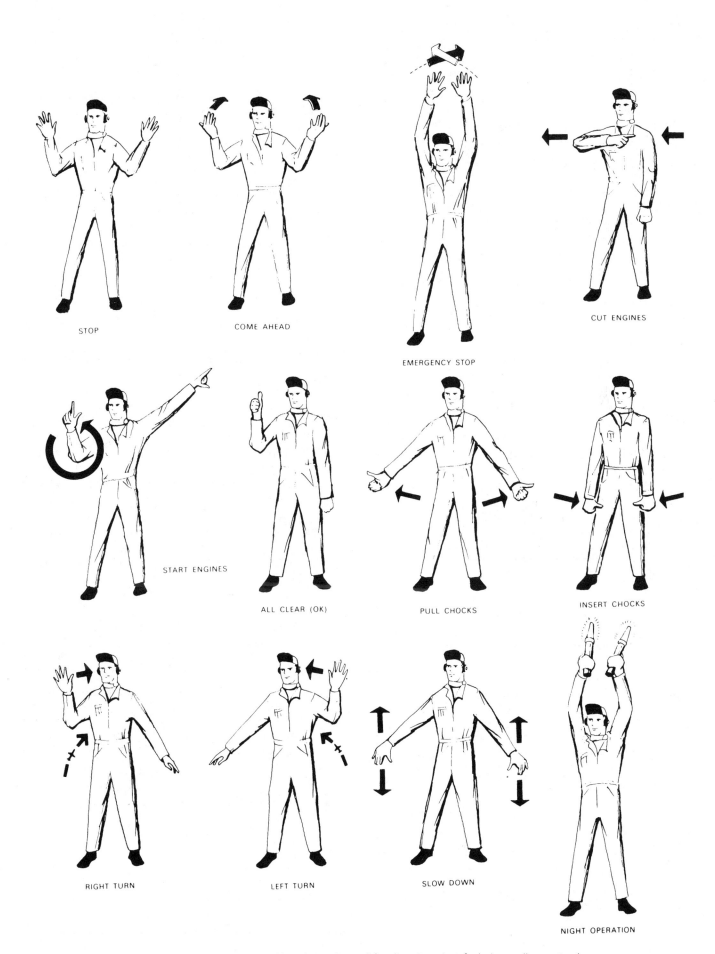

STOP

COME AHEAD

EMERGENCY STOP

CUT ENGINES

START ENGINES

ALL CLEAR (OK)

PULL CHOCKS

INSERT CHOCKS

RIGHT TURN

LEFT TURN

SLOW DOWN

NIGHT OPERATION

FIGURE 15–10 Some standard hand signals used for directing aircraft during taxiing or towing.

airplane for any length of time or overnight, it should be moored securely.

Airport tie-down areas are usually equipped with three-point rings, hooks, or other devices embedded in concrete for the purpose of attaching tie-down ropes, chains, or cables. The location of tie downs are usually indicated by some means, such as white or yellow paint markings.

Tie-down devices are often left attached to the fitting in the concrete or other pavement, and it is only necessary to attach the rope, chain, or cable to the aircraft after parking. If a rope is used, it should be attached to the aircraft with a nonslip knot, such as a bowline or a square knot. Examples of these knots are shown in Figure 15–11. It is recommended that the tie-down rope or cable be attached in such a manner that it is at an angle of approximately 45° with the ground, if possible.

Tie-down ropes may be made of Manila hemp, cotton, nylon, Dacron, or some other synthetic material. The synthetic ropes are less likely to deteriorate than hemp or cotton. Hemp has the disadvantage of shrinking when it becomes damp, and, for this reason, care must be taken to see that the rope is not pulled tight when the aircraft is tied down. A slack of 2 or 3% of the rope length normally should be sufficient to prevent aircraft damage due to

	Minimum Tensile Strength, lb					
			Dacron		Yellow Polypropylene	
Size, in	Manila	Nylon	Twist	Braid	Twist	Braid
$\frac{3}{16}$	—	960	850	730	800	600
$\frac{1}{4}$	600	1 500	1 440	980	1 300	1 100
$\frac{5}{16}$	1 000	2 400	2 200	1 650	1 900	1 375
$\frac{3}{8}$	1 350	3 400	3 120	2 300	2 750	2 025
$\frac{7}{16}$	1 750	4 800	4 500	2 900	—	—
$\frac{1}{2}$	2 650	6 200	5 500	3 800	4 200	3 800
$\frac{5}{8}$	4 400	10 000	—	—	—	—
$\frac{3}{4}$	5 400	—	—	—	—	—
1	9 000	—	—	—	—	—

FIGURE 15–12 Comparison of common tie-down ropes.

shrinking of the rope. The strength of various types of commonly used tie-down ropes are compared in Figure 15–12.

When an aircraft is being secured in a tie-down area, **control locks** and **gust locks** should be installed to prevent the controls and surfaces from being damaged by wind. Wind can cause control surfaces to move violently from one extreme to the other, and this often results in damage. A twin-engine airplane with gust locks installed is shown in Figure 15–13. On some smaller aircraft, the aileron and elevator controls may be secured by using the front seat belts. Care must be taken to see that all such locks are removed before attempting to taxi or fly the airplane.

When an aircraft is operated from an airport where tie-down facilities are not installed, steel tie-down stakes should be carried in the aircraft. One of the most effective types of tie-down stakes has the appearance of a large corkscrew with a loop at the top to which the tie-down rope is attached. The stakes are screwed into the ground and are normally adequate to hold the aircraft secure under all conditions of wind except hurricanes or tornados. If the ground is saturated with water, the ability of the stakes to hold will be impaired. Tie-down kits containing stakes and ropes or chains are available from aircraft-supply companies.

Airplanes should be headed into the wind when parked; however, since tie-down spaces are usually laid out by the airport management, there may be no choice as to the direction in which the aircraft will be headed when it is tied down. All parked aircraft, whether tied down or not, should have chocks placed against the front and rear of the main wheels to prevent the aircraft from rolling.

Parking brakes should be set except when the brakes are overheated or during cold weather when accumulated moisture may freeze a brake. When parking a helicopter, the ground-handling wheels should be retracted, thus allowing the helicopter to rest on the skid-type landing gear. The

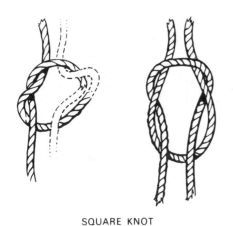

SQUARE KNOT

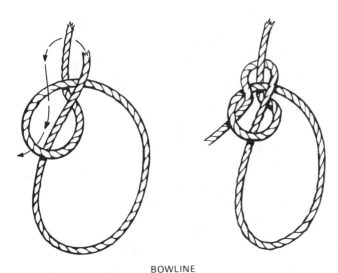

BOWLINE

FIGURE 15–11 Knots that may be used to tie down aircraft.

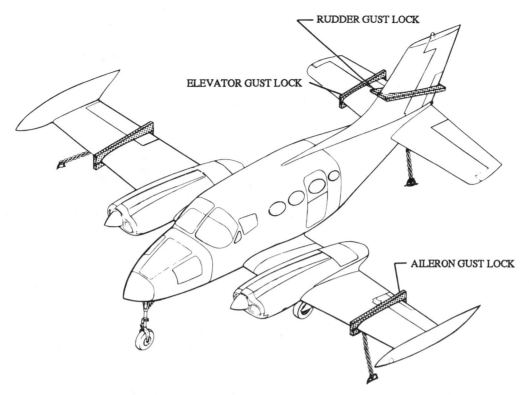

FIGURE 15–13 Gust locks on an airplane. *(Cessna Aircraft Co.)*

main and tail rotor blades should be secured if a helicopter is parked in an area subjected to turbulence created by jet, prop, or rotor blast from other aircraft.

JACKING AND HOISTING AIRCRAFT

From time to time in performing maintenance and inspection procedures, it becomes necessary to jack or hoist the aircraft. In the jacking or hoisting of aircraft, it is essential that the operation be performed according to the manufacturer's instructions. The points where hoisting or jacking fittings are attached to the aircraft are designed to withstand the stresses imposed when the aircraft is hoisted or jacked. If the aircraft is lifted or jacked at a point other than those designed for the operation, severe damage will often be caused. The locations of jacking points for a light twin-engine aircraft are shown in Figure 15–14.

Jacking Aircraft

Since jacking procedures and safety precautions vary for different types of aircraft, only general jacking procedures and precautions are discussed here. Consult the applicable aircraft manufacturer's maintenance instructions for specific jacking procedures. In jacking one wheel only, a small jack can be used at the landing-gear jack pad. This procedure is employed when tires are being changed and when brakes are being repaired or serviced. When only one set of wheels has to be raised, a low single-base jack is used, as shown in Figure 15–15. Before the wheel is raised, the remaining wheels must be chocked fore and aft to prevent

movement of the aircraft. The wheel should be raised only high enough to clear the floor.

For jacking the complete aircraft, wide-base jacks, such as the large tripod types shown in Figure 15–16, should be used because of the greater stability they afford. The size and configuration of the aircraft will dictate the type and number of jacks needed to raise it. Many small aircraft are raised by using a jack under each wing spar and a weighted tail stand, as shown in Figure 15–17 on page 346. If this method is used, be sure to consult the manufacturer's recommendations on the amount of weight needed. Transport-category aircraft may use several jacks, with three or four jacks being used to raise the aircraft and additional jacks being inserted to stabilize it after it has been jacked up. Figure 15–18 on page 346 illustrates the jack points on a Boeing 747. The airplane is provided with three main jacking points and five stabilizing jacking points. The primary jacking points are at the wing-body junction and the tail. The five stabilizing points are located with one at the nose and two under each wing.

Many aircraft have jack pads located at the jack points. Others have removable jack pads or jacking adapters that are inserted into receptacles prior to jacking (see Figure 15–19 on page 347). The correct jack pad should be used in all cases. The function of the jack pad is to ensure that the aircraft load is properly distributed at the jack point and to provide a convex bearing surface to mate with the concave jack stem.

Typical jacking procedures and precautions for raising an aircraft are as follows:

1. Place the aircraft in a hangar, if possible, to avoid the effects of wind. (Head the airplane into wind if it is to be jacked outside.)

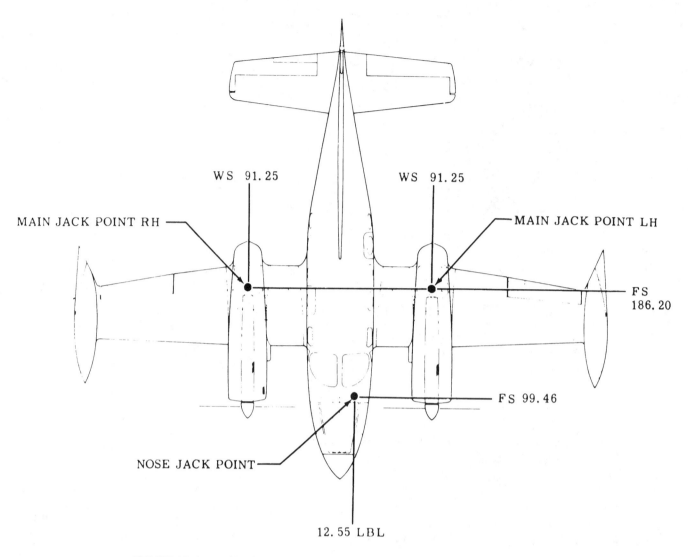

WS 91.25

WS 91.25

MAIN JACK POINT RH

MAIN JACK POINT LH

FS 186.20

FS 99.46

NOSE JACK POINT

12.55 LBL

FIGURE 15–14 Jacking points for a light twin-engine aircraft. (*Cessna Aircraft Co.*)

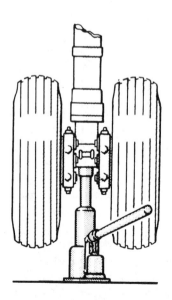

FIGURE 15–15 Jacking a single set of wheels.

2. See that the areas under and around the airplane are clear of obstructions (such as ladders, work platforms, or entry stands).

3. If any other work is in progress on the aircraft, ascertain if any critical panels have been removed. On some aircraft the stress panels or plates must be in place when the aircraft is jacked to avoid structural damage.

4. Install the landing-gear ground locks if applicable.

5. Place the jacks directly under the center of each jack point and extend the jacks until they touch the jack points (most accidents during jacking are the result of misaligned jacks).

6. Check the legs of the jacks to see that they will not interfere with the operations to be performed after the aircraft is jacked, such as retracting the landing gear.

7. Station one person at each jack, if possible. Operate all jacks evenly so that the airplane remains as nearly level as possible.

8. Keep the amount of lift to an absolute minimum and always within the safe limits of the jack.

9. Set the locking devices on the jacks to prevent accidental lowering of the airplane.

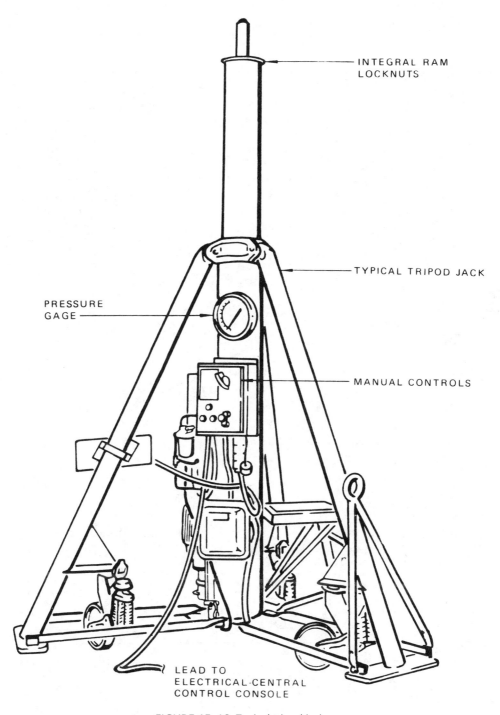

INTEGRAL RAM
LOCKNUTS

TYPICAL TRIPOD JACK

PRESSURE
GAGE

MANUAL CONTROLS

LEAD TO
ELECTRICAL-CENTRAL
CONTROL CONSOLE

FIGURE 15–16 Typical tripod jack.

10. Hold any climbing on the aircraft to an absolute minimum, and avoid violent movements by persons who are required to go aboard.

To lower an aircraft, the following jacking instructions apply:

1. Before lowering the aircraft, make certain that the landing gear is down and locked and that all ground-locking devices are properly installed.

2. Check the area under and around the aircraft to insure that it is free from all equipment and workstands.

3. Release the mechanical locks, then slowly and carefully release the pressure, lowering the jacks evenly. *Cau-*

tion: As the aircraft is lowered, watch that the oleo struts do not bind up.

4. As soon as possible, remove the jacks out from under the aircraft.

Hoisting

It is often necessary to hoist airplanes and helicopters in order to perform certain service and maintenance operations. When hoisting the entire airplane or any of the airplane components, it is recommended that hoisting slings, manufactured specifically for the airplane, be used. These slings are designed to lift the airplane or components from the approximate center of gravity. Most fuselage hoist

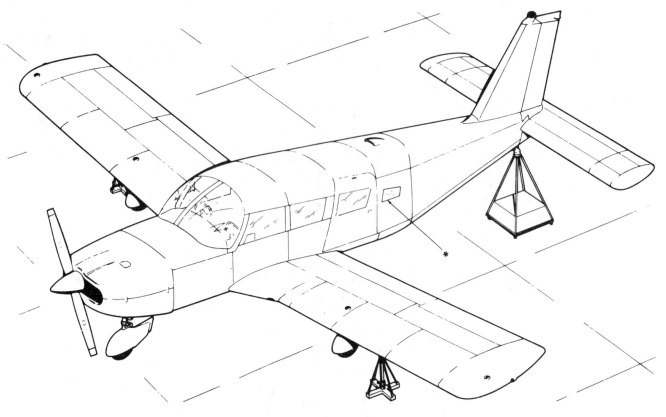

FIGURE 15–17 Jacking a small airplane. *(Piper Aircraft Co.)*

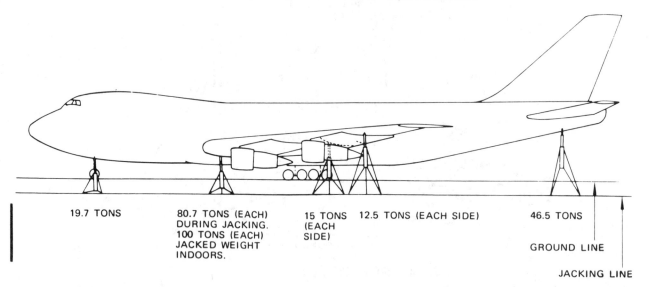

| 19.7 TONS | 80.7 TONS (EACH) DURING JACKING. 100 TONS (EACH) JACKED WEIGHT INDOORS. | 15 TONS (EACH SIDE) | 12.5 TONS (EACH SIDE) | 46.5 TONS |

GROUND LINE

JACKING LINE

FIGURE 15–18 Jacking arrangement for a Boeing 747 aircraft. *(Boeing Commercial Aircraft Co.)*

slings are adjustable to allow for different weight and center-of-gravity variations. Figure 15–20 illustrates an aircraft and its components being hoisted with a sling.

GROUND-SUPPORT EQUIPMENT

Ground-support equipment is needed primarily for the operation of aircraft on the ground when the aircraft engines are not operating. In some cases, small ground-support units, such as battery carts, preoil units, and test units, are used

with light aircraft, but this is not usually a regular and ongoing procedure, as it is with large aircraft. Ground-support units for large aircraft include electrical power supplies, air-conditioning and/or air-supply units, hydraulic test units, and various service units. The units used on a regular basis are those supplying electric power and air for starting engines, ventilating, heating, and cooling.

Electrical Power Supplies

Electrical power for light aircraft is usually supplied by means of a **battery cart**. Such carts contain one or more

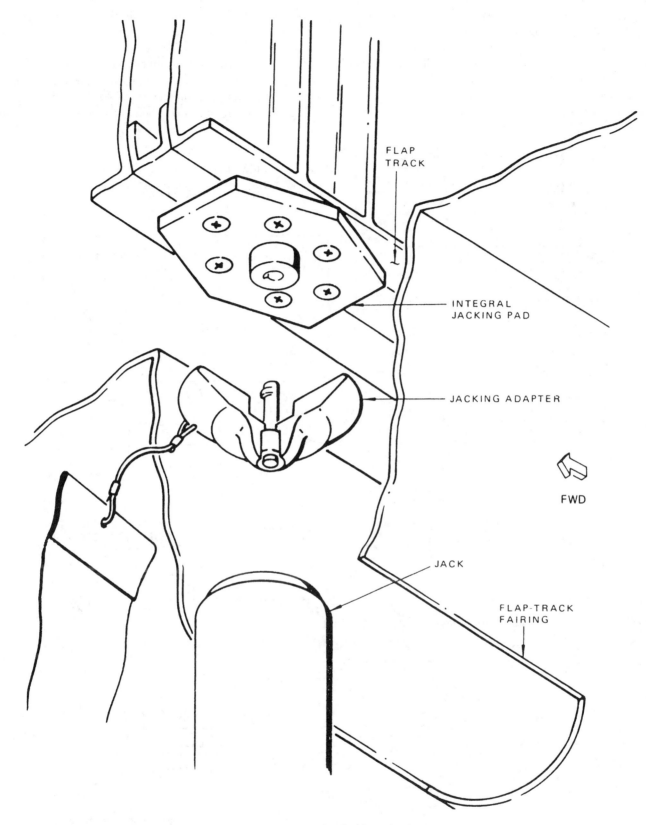

FLAP
TRACK

INTEGRAL
JACKING PAD

JACKING ADAPTER

FWD

JACK

FLAP-TRACK
FAIRING

FIGURE 15–19 Wing jacking adapter.

storage batteries connected to supply 12, 24, or 28.5 volts (V) dc and having sufficient capacity to provide several engine starts before recharging (see Figure 15–21). Battery carts should be kept clean and fully charged. Since they are not used on a regular basis, they are often overlooked and not properly cared for. It is good practice to schedule a regular time for checking and servicing ground-power batteries.

When a battery cart is to be used, the technician must determine that the voltage and capacity of the unit is compatible with the system voltage of the aircraft being serviced. A 12-V power supply cannot service a 24-V system, and a 24-V power supply will cause severe damage if connected to a 12-V system.

A specially designed receptacle is usually installed in the

aircraft system to make it impossible to connect the power supply with the wrong polarity (+/−). The plug on the power-supply cable cannot be inserted into such an aircraft receptacle unless the polarity is correct. In all cases, the master switch of the aircraft should be off when the battery cart is being connected or disconnected to the aircraft.

A **mobile electrical power unit** consists of engine-driven generators mounted on a trailer or a truck, such as the power supply pictured in Figure 15–22. It is designed to supply the correct type of power (ac or dc) and the correct voltage and capacity for the aircraft that it is to service.

All large aircraft, such as commercial airliners, require electrical power when in flight and when on the ground preparing for flight. It is therefore necessary to have

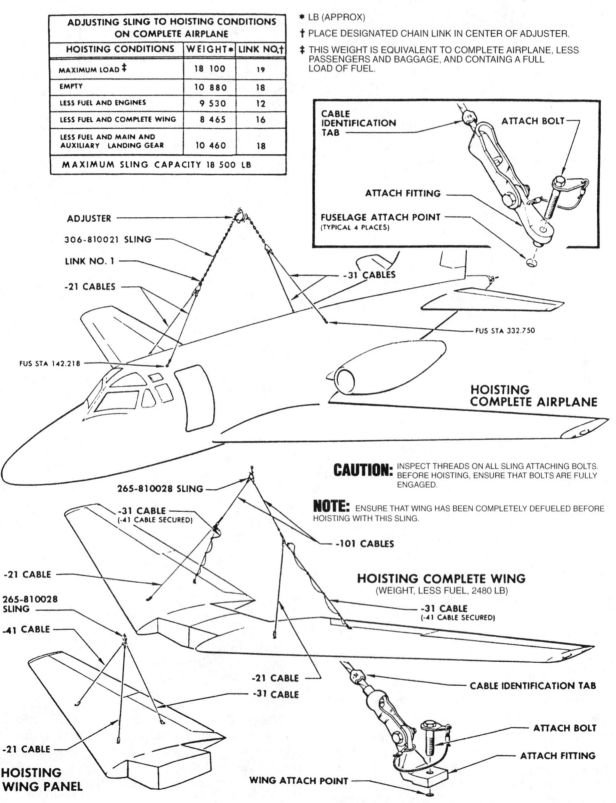

ADJUSTING SLING TO HOISTING CONDITIONS ON COMPLETE AIRPLANE		
HOISTING CONDITIONS	WEIGHT✱	LINK NO.†
MAXIMUM LOAD ‡	18 100	19
EMPTY	10 880	18
LESS FUEL AND ENGINES	9 530	12
LESS FUEL AND COMPLETE WING	8 465	16
LESS FUEL AND MAIN AND AUXILIARY LANDING GEAR	10 460	18
MAXIMUM SLING CAPACITY 18 500 LB		

✱ LB (APPROX)

† PLACE DESIGNATED CHAIN LINK IN CENTER OF ADJUSTER.

‡ THIS WEIGHT IS EQUIVALENT TO COMPLETE AIRPLANE, LESS PASSENGERS AND BAGGAGE, AND CONTAING A FULL LOAD OF FUEL.

CABLE IDENTIFICATION TAB

ATTACH BOLT

ATTACH FITTING

FUSELAGE ATTACH POINT
(TYPICAL 4 PLACES)

ADJUSTER

306-810021 SLING

LINK NO. 1

-21 CABLES

-31 CABLES

FUS STA 332.750

FUS STA 142.218

HOISTING COMPLETE AIRPLANE

CAUTION: INSPECT THREADS ON ALL SLING ATTACHING BOLTS. BEFORE HOISTING, ENSURE THAT BOLTS ARE FULLY ENGAGED.

NOTE: ENSURE THAT WING HAS BEEN COMPLETELY DEFUELED BEFORE HOISTING WITH THIS SLING.

265-810028 SLING

-31 CABLE
(-41 CABLE SECURED)

-101 CABLES

-21 CABLE

265-810028 SLING

HOISTING COMPLETE WING
(WEIGHT, LESS FUEL, 2480 LB)

-41 CABLE

-31 CABLE
(-41 CABLE SECURED)

-21 CABLE

-31 CABLE

CABLE IDENTIFICATION TAB

ATTACH BOLT

-21 CABLE

HOISTING WING PANEL

WING ATTACH POINT

ATTACH FITTING

FIGURE 15–20 Aircraft sling. *(Rockwell International)*

FIGURE 15–21 Battery cart.

FIGURE 15–22 Ground-power unit. *(Hobart Brothers Co.)*

electrical power available at the gate where an airliner loads and unloads passengers. Most large aircraft are equipped with auxiliary power units (APUs), which are small gas-turbine engines that drive generators (alternators) and also supply bleed air for engine starting and air conditioning. These units are expensive to operate, and so ground-power supplies are used as much as possible.

A typical electrical power supply for ground service of aircraft consists of a diesel-engine-driven or turbine-engine-driven alternator that produces 400 Hz 115/200 V ac. The power unit is plugged into the aircraft as soon as it is parked at the gate and assumes the electrical load of the aircraft when the engines are shut down.

Mobile ground-power units, even though widely used for many years, are now being replaced by **fixed power supplies** where possible. The fixed power supply is less costly, eliminates the air pollution and noise caused by mobile units, and reduces the clutter of equipment around the aircraft. Until recently, a separate fixed power supply was used at each airline terminal; however, most large airports are now equipped with a single, central power station, and electrical power is delivered through cables to the various terminals on the airport. This has been made possible by transmitting the power from the central station at a high voltage (4160 V) to the terminal, then reducing it by means of transformers to the correct voltage for the aircraft (115/200 V). If 400 Hz of power is transmitted any appreciable distance at 115/200 V, the line losses are excessive.

At 4160 V, 400 Hz of power can be transmitted 2 or 3 mi [3 to 5 km] without serious power loss.

Fixed electrical power is delivered to the aircraft by means of a multiple cable. This cable is either mounted on the side of the passenger bridge or is stored in a pit adjacent to the aircraft parking space. When an aircraft arrives at the gate, the ground-service personnel can quickly and easily connect the power cable and supply the aircraft with electrical power.

It must be noted that many smaller airports do not yet have fixed electrical power available, and it is therefore necessary to use mobile units or APUs. Regardless of what type of power supply is employed, the technician must carefully follow the procedures specified for the airplane that is being serviced.

Ground Air Supply

In addition to electrical power, a transport-category passenger aircraft requires an adequate air supply. Low-pressure (35 to 45 psi) [241 to 310 kPa], high-volume air is required for starting engines, ventilating, heating, and cooling.

Air-supply units are made in a variety of different types. Typical of these are turbine-engine compressors, diesel-engine-driven compressors, and electric motor-driven compressors. The compressor may be axial flow, centrifugal, or screw type. The compressor and the driving unit may be mounted on a trailer that requires towing, or they may be on a self-propelled vehicle, such as a small truck. A unit of this type is shown in Figure 15–23.

The air needed for servicing an aircraft must be warm and oil free. Some types of compressors require that oil separators be installed in the outlet line to eliminate oil from the air. The screw-type compressor does not require an oil separator because no oil is used to lubricate the screw elements. The screws are precision machined and timed so there is no contact between them as they rotate.

A cutaway view of a screw compressor is shown in Figure 15–24. It consists of two screw helical rotors that mate with a very small clearance that does not allow an appreciable amount of air leakage. The male rotor has four lobes and turns 50% faster than the female rotor. The effect is to continuously compress the air as it moves from the inlet to the

FIGURE 15–23 Mobile air-supply unit. *(Atlas Copco Inc.)*

FIGURE 15–24 Cutaway view of a screw compressor. *(Atlas Copco Inc.)*

outlet. This is because the air space between the rotor lobes decreases continuously from the inlet to the outlet.

Air is drawn into the compressor through the inlet port, and it then moves into the space between the lobes. As the rotors revolve further, the air inlet is sealed and the compression of the air begins. The rotary motion produces a smooth compression that continues until each groove in turn reaches the beginning of the outlet port. The air is then forced smoothly out of the compressor, and the outlet end is sealed again, ready for the next cycle. No special inlet and outlet valves are needed because both the inlet and the outlet ports are automatically covered and uncovered by the ends of the rotors.

The most effective and economical air supply for large aircraft is the stationary type. A system of this kind is limited to new airports or those that are being extensively remodeled. This is because the air piping must be installed under the runways and aprons. The initial cost of this installation is higher than for the use of mobile units; however, the savings in operating costs quickly make up for the difference.

With a fixed air supply, there is one location for all the compressors, and the compressors are driven by efficient electric motors. Since electric-power costs are substantially lower than fuel costs, the economic advantages are considerable. The compressors are large screw-type units capable of starting several turbine engines at once. Maintenance costs are very low because of the trouble-free electric motors and the screw-type compressors.

The fixed air system eliminates the need for ground units adjacent to the airplanes. This eliminates air and noise pollution and reduces traffic clutter around the airplanes.

FUELING

Perhaps the most common service operation on aircraft is the filling of fuel tanks. Improper fueling procedures have caused many aircraft accidents and in-flight incidents. Fueling personnel should be familiar with the fuel requirements for the models and types of aircraft they are servicing. The

following paragraphs contain a description of the problems that may be encountered in fueling aircraft and the recommended procedures for combating these problems.

Fuel Contamination

Fuel is contaminated when it contains any material that was not provided under the fuel specification. This material generally consists of water, rust, sand, dust, microbial growth, and certain additives that are not compatible with the fuel, fuel-system materials, and engines.

Water Contamination. All aviation fuels absorb moisture from the air and contain water in both suspended-particle and liquid form. The amount of suspended particles varies with the temperature of the fuel. Whenever the temperature of the fuel is decreased, some of the suspended particles are drawn out of the solution and slowly fall to the bottom of the tank. Whenever the temperature of the fuel increases, water is drawn from the atmosphere to maintain a saturated solution. Changes in fuel temperature, therefore, result in a continuous accumulation of water. During freezing temperatures, this water may turn to ice, restricting or stopping the fuel flow.

There is no way of preventing the accumulation of water formed through condensation in fuel tanks. The accumulation is certain, and the rate of accumulation will vary; therefore, storage tanks, fuel-truck tanks, and aircraft fuel tanks should be checked daily for the presence of water. Any water discovered should be removed immediately. Adequate settling time is necessary for accurate testing. The minimum settling time for aviation gasoline is 15 min per foot-depth of fuel and 60 min per foot-depth of turbine fuel. Testing storage tanks and fuel trucks may be done by attaching water-detecting paste or litmus paper to the bottom of the tank dipstick. The procedure is to push the dipstick to the bottom of the tank and hold it there for 30 s. When the dipstick is removed, the detecting paste or litmus paper will have changed color if water is present.

Microorganism Contamination. Many types of microbes have been found in unleaded fuels, particularly in the turbine-engine fuels. The microbes, which may come from the atmosphere or the storage tanks, live at the interface between the fuel and the liquid water in the tank. These microorganisms of bacterial and fungi rapidly multiply and cause serious corrosion in tanks and may clog filters, screens, and fuel-metering equipment. The growth and corrosion are particularly serious in the presence of other forms of contamination.

Other Contaminations. Pipelines, storage tanks, fuel trucks, and drum containers tend to produce rust that can be carried in the fuel in small-sized particles. Turbine fuels tend to dislodge rust and scale and carry them in suspension. Fuel may also be contaminated with dust and sand, which enters through openings in the tanks and through the use of fuel-handling equipment that is not clean.

Contamination Control

The presence of any contamination in fuel systems is dangerous. Laboratory and field tests have demonstrated that when water is introduced into a gasoline tank, it immediately settles to the bottom. Fuel tanks are constructed with sumps to trap this water. Most fuel-dispensing equipment includes filters to remove the liquid, dust, and rust particles from the fuel. It is advisable when fueling from drums to use a 5-micron filter or, as a last resort, a chamois-skin filter and filter funnel.

It is impossible to keep water from forming in airplane tanks; therefore, it is necessary to regularly drain the airplane's fuel sumps in order to remove all water from the system. It may be necessary to gently rock the wings of some aircraft while draining sumps to completely drain all the water. On certain tailwheel-type aircraft, raising the tail to a level-flight attitude may result in additional flow of water to the gascolator or main fuel strainer. If left undrained, the water accumulates and will pass through the fuel line to the engine where it may cause the engine to stop operating. The elimination of contaminants from aviation fuel may not be entirely possible, but these contaminants can be controlled by the application of good "housekeeping" habits.

Aviation Fuel Grades

Normal engine operation depends on the use of the proper fuel grade. If the proper grade of fuel is not available, it is sometimes permissible to use the next higher grade of fuel (for example, 100LL rather than 80.) Never allow a grade lower than that specified for normal operation to be used. To do so may cause extensive damage to the engine. Aviation fuels are color coded to help prevent the fueling of aircraft with the wrong fuel. Fuels are colored as follows:

Octane	Color
80	Red
100LL	Blue
100	Green
115 (military use only)	Purple
Jet A	Clear or straw colored

In recent years, the availability of 80-octane (red) fuel has diminished. Instead, fuel suppliers have been providing 100LL (100 octane, low-lead) for use in all light, piston-engine airplanes. Some engines designed for 80-octane fuel require modifications for the continuous use of 100LL. Additionally, maintenance problems in some engines, primarily exhaust-valve erosion and spark-plug fouling, have been attributed to the higher lead fuel (despite its name, 100LL contains four times the lead of 80 octane). Engine manufacturers advise strict adherence to their latest recommended operating procedures to help assure the proper operation of the affected engines.

Autogas

Many owners of airplanes designed to use 80-octane aviation gas (avgas) have switched to automobile gas (autogas) fuel instead of using 100LL. Aside from the economic advantages, many 80-octane engines seem to run better on autogas than on either 80- or 100-octane avgas. Moreover, in an FAA study, autogas was found to have no detrimental effects on low-compression engines, and although some questions have been raised about the ability of autogas to resist vapor lock, the FAA found no conclusive evidence that quality autogas would cause vapor-lock problems in aircraft. The key is in knowing the quality of the fuel. Alcohols are often added to auto fuel between the refinery and the local merchant. Alcohol in autogas causes several problems in airplanes. It causes the fuel to be less resistant to vapor lock, to retain water in suspension, and to deteriorate rubber seals in the fuel system.

Automobile fuel has been approved for use in the engines of numerous aircraft. These aircraft may be operated on automobile fuel under the provisions of supplemental type certificates.

Turbine Fuels

There are two types of turbine fuel in common use today: (1) kerosene-grade turbine fuel, now named Jet A, and (2) a blend of gasoline and kerosene fractions, designated Jet B. There is a third type, called Jet A-1, which is made for operation at extremely low temperatures.

There is very little physical difference between Jet A (JP-5) fuel and commercial kerosene. Jet A was developed as a heavy kerosene, having a higher flash point and lower freezing point than most kerosenes. It has a very low vapor pressure, so there is little loss of fuel from evaporation or boil-off at higher altitudes. It contains more heat energy per gallon than does Jet B (JP-4).

Nevertheless, Jet B is similar to Jet A. It is a blend of gasoline and kerosene fractions. Most commercial turbine engines will operate on either Jet A or Jet B fuel. However, the difference in the specific gravity of the fuels may require fuel-control adjustments. Therefore, the fuels cannot always be considered interchangeable. Both Jet A and Jet B fuels are blends of heavy distillates and tend to absorb water. The specific gravity of jet fuels, especially kerosene, is closer to water than that of aviation gasoline; thus, any water introduced into the fuel, either through refueling or condensation, will take an appreciable time to settle out. At high altitudes, where low temperatures are encountered, water droplets combine with the fuel to form a frozen substance referred to as **gel**. The mass of gel, or "icing," that may be generated from the moisture held in suspension in jet fuel can be much greater than that in gasoline.

Turbine-Fuel Additives

Certain turbine-engine-powered aircraft require the use of fuel containing **anti-icing additives**. Therefore, fuel personnel must know whether or not the fuels they dispense contain additives. When anti-icing additives are to be added to the fuel, the manufacturer's instructions (usually printed on the container) should be followed to assure the proper mixture. Anti-icing-additive content in excess of 0.15% by volume of fuel is not recommended, as higher concentrations can cause the aircraft fuel-capacitance system to give erroneous indications.

Certain oil companies, in developing products to cope with aircraft fuel icing problems, found that their products also checked "bug" growth. These products, known as **biocides**, are usually referred to as additives. However, some additives may not be compatible with the fuel or the materials in the fuel system and may be harmful to other parts of the engine with which they come in contact. Additives that have not been approved by the manufacturer and the FAA should not be used.

Fuel-Tank Markings

Federal Aviation Regulations Part 23, Section 23.1557(c)(1), requires that aircraft fuel filler openings be marked with the word *FUEL*, and the minimum fuel grade or designation for the engines. Typical fuel-tank markings are illustrated in Figure 15–25. It is equally important that tank vehicles be conspicuously marked to show the type of fuel carried. The marking should be of a color in sharp contrast to that of the vehicle and in lettering at least 12 in tall. This marking should be on each side and on the rear of the tank vehicles. Additionally, the tank vehicles' hose lines should be marked by labels next to the nozzle and every 6 ft.

Fueling Precautions and Procedures

Fuel-dispensing vehicles and stationary facilities should be equipped with appropriate fire extinguishers, fire blankets, static-grounding cables, explosion-proof flashlights, and ladders.

Fueling vehicles should be positioned as far from the aircraft as permitted by the length of the fuel-dispensing hose. Mobile units should be parked parallel to or heading away from the aircraft wing leading edge so they may be moved away quickly in the event of an emergency.

Fueling personnel should first check with the flight crew to determine the type, grade, and quantity of fuel required, including additives, for the aircraft. In the absence of the flight crew, fueling personnel should check the placard located near the aircraft fuel-tank filler port or the aircraft flight manual to determine the type and grade of fuel required. They should also check to make sure that the following procedures are maintained:

1. No electrical or radio equipment in the aircraft should be energized or be maintained while fuel is being dispensed into the aircraft, except those switches that may require energizing to operate fuel-selector valves and quantity-gauge systems.

2. Qualified personnel should be stationed at the aircraft fuel-control panel during pressure-fueling operations.

3. Fueling personnel should not carry objects in the breast pockets of their clothing when servicing aircraft or filling fuel-service vehicles because loose objects may fall into fuel tanks.

4. Every piece of equipment used in doing the work should be absolutely clean. The refueling-hose nozzle should be wiped clean with a cloth free of lint and dirt. The hose nozzle should never be dragged over any part of the airplane or the ground.

5. Matches or lighters should never be carried during fueling operations.

6. Because of the high lead content, direct fuel contact with the skin or the wearing of fuel-saturated clothing should be avoided. Skin irritation or blisters may result from direct contact with fuel.

7. Immediate medical attention should be sought if fuel enters the eyes.

8. Care must be exercised to avoid damaging the filler neck of the airplane fuel tank when the tank is being filled with gasoline. The fuel-hose nozzle should be carefully supported and the fuel introduced slowly so that there is absolutely no danger of splashing or overfilling the fuel tank.

9. In the event of fuel spillage, discontinue fueling operations until the spill can be removed using proper safety precautions.

Gasoline flowing through a hose may build up a charge of static electricity. When the refueling-hose nozzle is withdrawn from the tank, the charge of static could ignite the fuel. For this reason, before starting the refueling operation, the nozzle of the hose must be grounded.

Usually, refueling hoses have ground wires attached, in which case all that needs to be done is to connect the wire (by means of the clip provided) to some metal part of the airplane. The airplane grounding wire should be attached to some metal part permanently embedded in the earth.

The following is a typical sequence that should be followed by a fueling crew in grounding the aircraft:

1. Connect a grounding cable from the fueling vehicle to a satisfactory ground. Grounding posts usually consist of pipes or rods driven far enough into the ground to result in a zero potential.

2. Connect a grounding cable from the ground to the aircraft (on a landing-gear axle or other unpainted surface). Do not attach ground cables to the propeller or the radio antenna.

3. Connect a grounding cable from the fueling vehicle to the aircraft. The fueling vehicle may be equipped with a T- or Y-cable permitting ground attachment first and then the grounding of the aircraft with the other end.

4. Connect a grounding cable from the fuel nozzle to the aircraft before removing the aircraft tank cap. This bond is

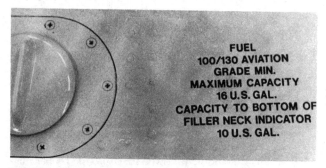

FUEL
100/130 AVIATION
GRADE MIN.
MAXIMUM CAPACITY
16 U.S. GAL.
CAPACITY TO BOTTOM OF
FILLER NECK INDICATOR
10 U.S. GAL.

FIGURE 15–25 Typical fuel-tank markings.

most essential and needs to be maintained throughout the fueling operation and until the fuel cap is replaced.

Overwing Fueling

When using the **overwing method**, the fuel-filler hose should be draped over the wing's leading edge (as is demonstrated in Figure 15–26). Never lay the fuel-filler hose over the wing's trailing edge because aircraft structural damage may result. A simple rubber shower mat may be used to provide protection for the wing's leading edge during fuel operation. Step ladders or padded upright ladders may be used to provide easy access to high-wing and large aircraft. Avoid standing on wing surfaces, and never stand on wing struts. Hold the fuel nozzle firmly while it is inserted in the fuel-tank filler neck, and never block the nozzle lever in the open position. Be sure that the fuel filler caps are replaced and securely latched when fueling is completed.

Pressure Fueling

Most large transport-category aircraft are fueled from a single point under the wing or fuselage, as demonstrated in Figure 15–27. This means that a fuel hose can be connected at one point under the wing or fuselage of the aircraft to refuel the entire system or that a fueling station is provided under each wing so that all the tanks in one wing can be fueled from one station.

The fueling operation of such a system consists essentially of coupling the fuel-hose nozzle to the receptacle, opening the fueling shutoff valves, and pumping fuel into the tanks. The fuel tanks may be filled completely by opening the fueling shutoff valves and allowing the fueling level-

FIGURE 15–27 Pressure fueling.

control valves to shut off the fuel flow. If the operator wants to fill the tanks partially, the preset indicators on the fueling control panel, such as the one illustrated in Figure 15–28, may be employed to show when the desired quantity of fuel has entered a particular tank and to automatically shut off the flow of fuel.

This type of **pressure fueling** greatly reduces the time required to fuel large aircraft. Other advantages of the pressure-fueling process are that it eliminates aircraft skin damage and hazards to personnel as well as reduces the chances for fuel contamination. Pressure fueling also reduces the chance of static electricity igniting fuel vapors; thus the need for the static ground wire is eliminated.

Defueling Aircraft

The **defueling** of aircraft is critical in that the danger of fire is increased by empty fuel tanks filled with flammable vapors and by the vapors produced by spilled fuel, particularly gasoline. The technician involved with defueling should take every precaution to reduce the possibility of fire and should observe every rule and regulation established to prevent and combat fires. The safety of the procedure is the prime consideration.

Defueling procedures are usually given in manufacturer's aircraft maintenance manuals; however, these cannot always give consideration to the conditions that may exist at the time and place of defueling. The technician involved must, therefore, exercise great care in her or his preparations and performance to assure that all precautions have been observed.

FIGURE 15–26 Overwing fueling.

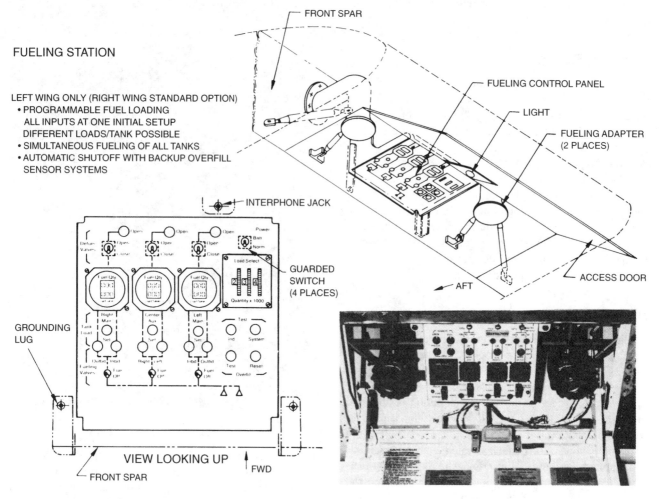

FUELING STATION

LEFT WING ONLY (RIGHT WING STANDARD OPTION)
- PROGRAMMABLE FUEL LOADING
  ALL INPUTS AT ONE INITIAL SETUP
  DIFFERENT LOADS/TANK POSSIBLE
- SIMULTANEOUS FUELING OF ALL TANKS
- AUTOMATIC SHUTOFF WITH BACKUP OVERFILL
  SENSOR SYSTEMS

FIGURE 15–28 Underwing fueling station on a Boeing 767.   *(Boeing Commercial Aircraft Co.)*

When an airplane is to be defueled, the following general rules should be followed unless other procedures which are just as effective have been established by the aircraft operator or operational organization:

1. Move the aircraft to be defueled to an open area at a distance from any other aircraft, structure, or vehicle such that a fire could not jump from the defueling aircraft to any other unit or structure.

2. Provide fire extinguishers of a type approved for fuel fires in sizes and numbers adequate for the size of the aircraft and the volume of fuel involved.

3. If possible, provide a supply of inert gas (nitrogen or CO_2) for purging the tank or tanks as they are being drained.

4. Ground the aircraft.

5. Personnel in the area should wear shoes with rubber or plastic soles which cannot cause sparks. Cigarette lighters, matches, and other sources of ignition should be removed from the area.

6. Disconnect or remove the aircraft battery.

7. Use no electrical equipment in the area except that which has been certified flameproof.

8. Provide for fuel disposal in a fuel truck or in clean, closed containers. Metal fuel containers which could cause sparks in handling should be avoided.

9. Do not allow fuel to be spilled on the surface beneath the aircraft.

10. If tools are needed to open fuel-drain valves, use tools which cannot cause sparks.

11. As soon as the fuel containers have been filled, install the container covers or caps. If open containers are used, they should be emptied immediately into containers or tanks which can be closed.

REVIEW QUESTIONS

1. What method should be used to remove spilled gasoline from a hangar floor?

2. What are some safety precautions that should be followed when moving compressed-gas cylinders?

3. What three ingredients are essential for a fire?

4. List the different classifications of fire.

5. Why are oil-soaked rags a fire hazard?

6. What type of fire-extinguisher agent is recommended for a Class B fire?

7. What three conditions are necessary for the starting of an internal-combustion engine?

8. Why should a reciprocating engine be rotated by hand or with the starter before it is started?

9. What should be checked with regards to a radial engine before starting?

10. Describe the precautions that should be taken with respect to the surrounding area before starting a gas-turbine engine.

11. What is a hot start and what action should be taken when such a start occurs?

12. What is a hung start?

13. What factors influence the geometry of a turn when taxiing?

14. By what means are aircraft tied down when parked?

15. What precaution must be observed in using hemp rope to tie down an airplane?

16. What is the purpose of gust locks?

17. Under what conditions is it advisable to leave the parking brakes off on a parked aircraft?

18. Why is it important to use specially designed jacking points when jacking aircraft?

19. What are the advantages of fixed electrical power supplies over mobile units?

20. For what purposes is a ground air supply required on large aircraft?

21. List some examples of fuel contamination.

22. What color is 100LL avgas?

23. What markings should be found adjacent to a fuel filler opening on an aircraft?

24. What are two methods of fueling aircraft?

25. What are the advantages of pressure fueling?

16 Aircraft Inspection and Servicing

INTRODUCTION

Airplanes are designed and built to provide many years of service. For an airplane to remain airworthy and safe to operate, it should be operated in accordance with the recommendations of the manufacturer and cared for with sound inspection and maintenance practices. The Federal Aviation Regulations (FARs) require the inspection of all civil aircraft at specific intervals to make sure that the aircraft's condition is equal to its original or properly altered condition with regard to aerodynamic function, structural strength, and resistance to vibration.

Aircraft inspection may range from a casual "walk around" to a detailed inspection involving complete disassembly and the use of complex inspection aids. This chapter will discuss aircraft inspection requirements and practices as well as review activities, such as servicing and lubrication, that generally accompany inspections.

REQUIRED AIRCRAFT INSPECTIONS

In establishing an aircraft's inspection requirements, it is necessary to consider the aircraft's size and type as well as the purpose for which it is used, and its operating environment.

Some aircraft must be inspected each 100 hours of time in service, while others must be inspected only once every 12 calendar months.

The inspection requirements for aircraft in various types of operation are stated in FAR 91.409. Small aircraft usually fall under the requirements of annual, 100-h, and progressive inspections. Large aircraft (over 12 500 lb) and turbine-powered multiengine airplanes (turbojet and turboprop) fall under the jurisdiction of a different set of inspection programs.

Annual and 100-Hour Inspections

The **annual** and **100-h inspections** are designed to provide a complete inspection of aircraft at specified intervals. These inspections determine the condition of an aircraft and the maintenance required to return the aircraft to an acceptable condition of airworthiness.

For aircraft operating under FAR Part 91, the maximum interval between annual inspections is **12 calendar months**,

meaning the aircraft will again become due for inspection on the last day of the same month, 12 months later. In addition to an annual inspection, aircraft operated commercially are also required to have a 100-hour (h) inspection. The procedures and scope of these inspections are set forth in Appendix D of FAR Part 43 and should be followed in detail (see Figure 16–1). The regulations speak of 100-h and annual inspections as being of identical scope; the only difference between the two is the persons authorized to perform them. Certificated airframe and powerplant maintenance technicians are authorized to perform a 100-h inspection. A certificated airframe and powerplant maintenance technician holding an **inspection authorization (IA)** issued by the FAA may perform the annual inspection.

An annual inspection may be substituted for a 100-h inspection. The 100-h time limitation may be exceeded by not more than 10 h, if necessary, to reach a place where the inspection can be performed. The excess time, however, is included in computing the next 100 h of time in service. As an example, an aircraft that flew 105 h between inspections would have only 95 h until the next inspection is due. However, the reverse does not apply. For example, an aircraft that has been inspected after only 90 h does not have 110 h before the next inspection. There is no provision for exceeding an annual inspection. To move an aircraft that is "out of annual" requires a special flight permit from the local FAA flight standards district office (FSDO).

FAR 43.15 provides a list of rules for persons performing inspections. One of the rules is that a **checklist** must be used by a person performing an inspection. The technician may use the checklist in FAR 43 (see Figure 16–1), Appendix D, the manufacturer's inspection checklist, or a checklist designed by the technician that includes the items listed in Appendix D to check the condition of the entire aircraft. In most instances, it is preferable to use the manufacturer's checklist since it was written specifically to include the procedures and details necessary to adequately inspect that particular make and model of aircraft (see Figure 16–2). The checklist will usually be found in the aircraft maintenance manual.

Progressive Inspections

The **progressive inspection system** has been designed to schedule inspections of aircraft on a predetermined basis. The purpose of the program is to allow maximum use of the

(a) Each person performing an annual or 100-hour inspection shall, before that inspection, remove or open all necessary inspection plates, access doors, fairing, and cowling. He shall thoroughly clean the aircraft and aircraft engine.

(b) Each person performing an annual or 100-hour inspection shall inspect (where applicable) the following components of the fuselage and hull group:

(1) Fabric and skin—for deterioration, distortion, other evidence of failure, and defective or insecure attachment of fittings.

(2) Systems and components—for improper installation, apparent defects, and unsatisfactory operation.

(3) Envelope, gas bags, ballast tanks, and related parts—for poor condition.

(c) Each person performing an annual or 100-hour inspection shall inspect (where applicable) the following components of the cabin and cockpit group:

(1) Generally—for uncleanliness and loose equipment that might foul the controls.

(2) Seats and safety belts—for poor condition and apparent defects.

(3) Windows and windshields—for deterioration and breakage.

(4) Instruments—for poor condition, mounting, marking, and (where practicable) for improper operation.

(5) Flight and engine controls—for improper installation and improper operation.

(6) Batteries—for improper installation and improper charge.

(7) All systems—for improper installation, poor general condition, apparent and obvious defects, and insecurity of attachment.

(d) Each person performing an annual or 100-hour inspection shall inspect (where applicable) components of the engine and nacelle group as follows:

(1) Engine section—for visual evidence of excessive oil, fuel, or hydraulic leaks, and sources of such leaks.

(2) Studs and nuts—for improper torquing and obvious defects.

(3) Internal engine—for cylinder compression and for metal particles or foreign matter on screens and sump drain plugs. If there is weak cylinder compression, for improper internal condition and improper internal tolerances.

(4) Engine mount—for cracks, looseness of mounting, and looseness of engine to mount.

(5) Flexible vibration dampeners—for poor condition and deterioration.

(6) Engine controls—for defects, improper travel, and improper safetying.

(7) Lines, hoses, and clamps—for leaks, improper condition, and looseness.

(8) Exhaust stacks—for cracks, defects, and improper attachment.

(9) Accessories—for apparent defects in security of mounting.

(10) All systems—for improper installation, poor general condition, defects, and insecure attachment.

(11) Cowling—for cracks, and defects.

(e) Each person performing an annual or 100-hour inspection shall inspect (where applicable) the following components of the landing gear group:

(1) All units—for poor condition and insecurity of attachment.

(2) Shock absorbing devices—for improper oleo fluid level.

(3) Linkage, trusses, and members—for undue or excessive wear, fatigue, and distortion.

(4) Retracting and locking mechanism—for improper operation.

(5) Hydraulic lines—for leakage.

(6) Electrical system—for chafing and improper operation of switches.

(7) Wheels—for cracks, defects, and condition of bearings.

(8) Tires—for wear and cuts.

(9) Brakes—for improper adjustment.

(10) Floats and skis—for insecure attachment and obvious or apparent defects.

(f) Each person performing an annual or 100-hour inspection shall inspect (where applicable) all components of the wing and center section assembly for poor general condition, fabric or skin deterioration, distortion, evidence of failure, and insecurity of attachment.

(g) Each person performing an annual or 100-hour inspection shall inspect (where applicable) all components and systems that make up the complete empennage assembly for poor general condition, fabric or skin deterioration, distortion, evidence of failure, insecure attachment, improper component installation, and improper component operation.

(h) Each person performing an annual or 100-hour inspection shall inspect (where applicable) the following components of the propeller group:

(1) Propeller assembly—for cracks, nicks, binds, and oil leakage.

(2) Bolts—for improper torquing and lack of safetying.

(3) Anti-icing devices—for improper operations and obvious defects.

(4) Control mechanisms—for improper operation, insecure mounting, and restricted travel.

(i) Each person performing an annual or 100-hour inspection shall inspect (where applicable) the following components of the radio group:

(1) Radio and electronic equipment—for improper installation and insecure mounting.

(2) Wiring and conduits—for improper routing, insecure mounting, and obvious defects.

(3) Bonding and shielding—for improper installation and poor condition.

(4) Antenna including trailing antenna—for poor condition, insecure mounting, and improper operation.

(j) Each person performing an annual or 100-hour inspection shall inspect (where applicable) each installed miscellaneous item that is not otherwise covered by this listing for improper installation and improper operation.

FIGURE 16–1 FAR 43, Appendix D.

aircraft, to reduce inspection costs, and to maintain a maximum standard of continuous airworthiness. This system is particularly adaptable to larger multiengine aircraft and aircraft operated by companies and corporations where high use is demanded. A progressive inspection satisfies the complete airplane inspection requirements of both the 100-h

and annual inspections. The instructions and schedule for a progressive inspection must be approved by a representative of the local district office of the FAA having jurisdiction over the area in which the applicant for the progressive inspection is located. Approval for such an inspection system requires that a person holding an inspection authorization

Perform inspection or operation at each of the inspection intervals as indicated by a circle (O).

A. PROPELLER GROUP

Nature of Inspection	50	100	500	1000
1. Inspect spinner and back plate for cracks	O	O	O	O
2. Inspect blades for nicks and cracks	O	O	O	O
3. Check for grease and oil leaks	O	O	O	O
4. Lubricate propeller per Lubrication Chart		O	O	O
5. Check spinner mounting brackets for cracks		O	O	O
6. Check propeller mounting bolts and safety (Check torque if safety is broken)		O	O	O
7. Inspect hub parts for cracks and corrosion		O	O	O
8. Rotate blades of constant speed propeller and check for tightness in hub pilot tube			O	O
9. Remove constant speed propeller; remove sludge from propeller and crankshaft				O
10. Inspect complete propeller and spinner assembly for security, chafing, cracks, deterioration, wear and correct installation		O	O	O
11. Check propeller air pressure (at least once a month)		O	O	O
12. Overhaul propeller				O

B. ENGINE GROUP

CAUTION: Ground Magneto Primary Circuit before working on engine.

Nature of Inspection	50	100	500	1000
1. Remove engine cowl	O	O	O	O
2. Clean and check cowling for cracks, distortion and loose or missing fasteners		O	O	O
3. Drain oil sump (See Note 2)	O	O	O	O
4. Clean suction oil strainer at oil change (Check strainer for foreign particles)	O	O	O	O
5. Clean pressure oil strainer or change full flow (cartridge type) oil filter element (Check strainer or element for foreign particles)		O	O	O
6. Check oil temperature sender unit for leaks and security		O	O	O
7. Check oil lines and fitting for leaks, security, chafing, dents and cracks (See Note 4)		O	O	O
8. Clean and check oil radiator cooling fins		O	O	O
9. Remove and flush oil radiators			O	O
10. Fill engine with oil per information on cowl or Lubrication Chart	O	O	O	O
11. Clean engine				O

CAUTION: Use caution not to contaminate vacuum pump with cleaning fluid. Refer to Lycoming Service Letter 1221A.

Nature of Inspection	50	100	500	1000
12. Check condition of spark plugs (Clean and adjust gap as required: adjust per Lycoming Service Instruction No. 1042)		O	O	O
13. Check cylinder compression (Refer to AC 43.13-1A)		O	O	O
14. Check ignition harness and insulators (High tension leakage and continuity)			O	O

Nature of Inspection	50	100	500	1000
15. Check magneto points for proper clearance (Maintain clearance at .018 ± .006)			O	O
16. Check magneto for oil seal leakage				O
17. Check breaker felts for proper lubrication			O	O
18. Check distributor block for cracks, burned areas or corrosion and height of contact springs			O	O
19. Check magnetos to engine timing		O	O	O
20. Overhaul or replace magnetos (See Note 3)	O			O
21. Remove air filters and tap gently to remove dirt particles (Replace as required)		O	O	O
22. Clean fuel injector inlet line strainer (Clean injector nozzles as required) (Clean with acetone only)	O		O	O
23. Check condition of injector alternate air doors and boxes		O	O	O
24. Remove induction air box valve and inspect for evidence of excessive wear or cracks. Replace defective parts (See Note 7)			O	O
25. Inspect fuel injector attachments for tightness (See Note 8)			O	O
26. Check intake seals for leaks and clamps for tightness		O	O	O
27. Inspect all air inlet duct hoses (Replace as required)		O	O	O
28. Inspect condition of flexible fuel lines		O	O	O
29. Replace flexible fuel lines (See Note 3)				O
30. Check fuel system for leaks		O	O	O
31. Check fuel pumps for operation (Engine driven and electric)		O	O	O
32. Overhaul or replace fuel pumps (Engine driven and electric) (See Note 3)				O
33. Check vacuum pumps and lines		O	O	O
34. Overhaul or replace vacuum pumps (See Note 3)				O
35. Check throttle, alternate air, mixture and propeller governor controls for travel and operating condition	O		O	O
NOTE: Visually inspect the exhaust system per Piper Service Bulletin No. 373A at each 25 hours of operation. (See Note 10.)				
36. Inspect exhaust stacks, connections and gaskets for cracks and loose mounting (Replace gaskets as required)	O	O	O	O
37. Inspect muffler, heat exchange, baffles and "augmentor" tube (See Note 6)	O	O	O	O
38. Check breather tubes for obstructions and security		O	O	O
39. Check crankcase for cracks, leaks and security of seam bolts		O	O	O
40. Check engine mounts for cracks and loose mountings		O	O	O
41. Check engine baffles for cracks and loose mounting		O	O	O
42. Check rubber engine mount bushings for deterioration (Replace as required)			O	O
43. Check fire wall seals		O	O	O
44. Check condition and tension of alternator drive belt	O	O	O	O
45. Check condition of alternator and starter	O	O	O	O
46. Check fluid in brake reservoir (Fill as required)		O	O	O
47. Inspect all lines, air ducts, electrical leads and engine attachments for security, proper routing, chafing, cracks, deterioration and correct installation		O	O	O
48. Lubricate all controls		O	O	O
49. Overhaul or replace propeller governor (See Note 3)				O
50. Complete overhaul of engine or replace with factory rebuilt (See Note 3)	O			O
51. Reinstall engine cowl		O	O	O

NOTES:

1. Both the annual and 100 hour inspections are complete inspections of the airplane, identical in scope, while both the **500 and 1000 hour** inspections are extensions of the annual or 100 hour inspection, which require a more detailed examination of the airplane, and overhaul or replacement of some major components. Inspection must be accomplished by persons authorized by the FAA.
2. Intervals between oil changes can be increased as much as 100% on engines equipped with full flow (cartridge type) oil filters - provided the element is replaced each 50 hours of operation.
3. Replace or overhaul as required or at engine overhaul. (For engine overhaul, refer to Lycoming Service Instructions No. 1009.)
4. Replace flexible oil lines as required, but no later than 1000 hours of service.
5. Refer to Piper Service Letter No. 597 for flap control cable attachment bolt use.
6. Refer to Piper Service Bulletin No. 373A for exhaust system inspection.
7. Refer to Piper Service Bulletin No. 358.
8. Torque all attachment nuts to 135 to 150 inch-pounds. Seat "Pal" nuts finger tight against plain nuts and then tighten an additional 1/3 to 1/2 turn.
9. Inspect rudder trim tab for "free play": must not exceed .125 inches. Refer to Service Manual for procedure, Section V. Refer to Piper Service Bulletin No. 390A.
10. Compliance with Piper Service Letter No. 673 eliminates repetitive inspection requirements of Piper Service Bulletin No. 373A and FAA Airworthiness Directive No. 73-14-2.
11. Piper Service Letter No. 704 should be complied with.

FIGURE 16–2 Inspection checklist for a light airplane. *(Piper Aircraft Corp.)*

C. CABIN GROUP (continued)

#	Nature of Inspection	50	100	500	1000
19.	Check stabilator attachments		O	O	O
20.	Check stabilator and tab hinge bolts and bearings for excess wear (Replace as required)		O	O	O
21.	Check stabilator trim mechanism (Check tab free play) (Refer to Section V)		O	O	O
22.	Check aileron, rudder, stabilator, stabilator trim cables, turnbuckles, guides and pulleys for safety, damage and operation		O	O	O
23.	Inspect all control cables, air ducts, electrical leads and attaching parts for security, routing, chafing, deterioration, wear and correct installation		O	O	O
24.	Clean and lubricate stabilator trim drum screw	O			
25.	Clean and lubricate all exterior needle bearings	O			
26.	Lubricate per Lubrication Chart		O	O	O
27.	Check rotating beacon for security and operation		O	O	O
28.	Check condition and security of Autopilot bridle cable and clamps		O	O	O
29.	Reinstall inspection plates and panels		O	O	O

E. WING GROUP

#	Nature of Inspection	50	100	500	1000
1.	Remove inspection plates and fairings		O	O	O
2.	Check surfaces and tips for damage, loose rivets, and condition of walk-way		O	O	O
3.	Check aileron hinges and attachments		O	O	O
4.	Check aileron cables, pulleys and bellcranks for damage and operation		O	O	O
5.	Check flaps and attachments for damage and operation		O	O	O
6.	Check condition of bolts used with hinges (Replace as required)		O	O	O
7.	Lubricate per Lubrication Chart	O			
8.	Check wing attachment bolts and brackets		O	O	O
9.	Inspect all control cables, air ducts, electrical leads, lines and attaching parts for security, routing, chafing, deterioration, wear and correct installation		O	O	O
10.	Check fuel tanks and lines for leaks and water		O	O	O
11.	Remove, drain and clean fuel gascolator bowls (Drain and clean at least every 90 days)		O	O	O
12.	Fuel tanks marked for capacity		O	O	O
13.	Fuel tanks marked for minimum octane rating		O	O	O
14.	Check fuel tank vents (Refer to Piper Service Bulletin No. 382)		O	O	O
15.	Reinstall inspection plates and fairings		O	O	O

C. CABIN GROUP

#	Nature of Inspection	50	100	500	1000
1.	Inspect cabin entrance, doors and windows for damage and operation		O	O	O
2.	Check upholstery for tears		O	O	O
3.	Check seats, seat belts, security brackets and bolts		O	O	O
4.	Check trim operation		O	O	O
5.	Check operation and condition of rudder pedals		O	O	O
6.	Check parking brake handle and toe brakes for operation and cylinder leaks		O	O	O
7.	Check control wheels, column, pulleys and cables		O	O	O
8.	Check flap control cable attachment bolt per Note 5		O	O	O
9.	Check landing, navigation, cabin and instrument lights		O	O	O
10.	Check instruments, lines and attachments		O	O	O
11.	Check manifold pressure gauge filters (2); replace as required		O	O	O
12.	Check gyro operated instruments and electric turn and bank (Overhaul or replace as required)		O	O	O
13.	Replace filters on gyro horizon and directional gyro or replace central air filter	O			
14.	Clean or replace vacuum regulator filter		O	O	O
15.	Check altimeter (Calibrate altimeter system in accordance with FAR 91.170, if appropriate)		O	O	O
16.	Perform pitot-static tests, if appropriate (Refer to FAR 91.170)	O			
17.	Check operation of fuel selector valves		O	O	O
18.	Check operation of fuel drains (See Note 11)		O	O	O
19.	Check condition of heater controls and ducts		O	O	O
20.	Check condition and operation of air vents		O	O	O

D. FUSELAGE AND EMPENNAGE GROUP

#	Nature of Inspection	50	100	500	1000
1.	Remove inspection plates and panels		O	O	O
2.	Check baggage doors, latches and hinges		O	O	O
3.	Check battery, box and cables (Check at least every 30 days. Flush box as required and fill battery per instructions on box)	O			
4.	Check electronic installation		O	O	O
5.	Check bulkheads and stringers for damage		O	O	O
6.	Check antenna mounts and electric wiring	O			
7.	Check hydraulic pump fluid level (Fill as required)		O	O	O
8.	Check hydraulic pump lines for damage and leaks		O	O	O
9.	Check fuel lines, valves and gauges for damage and operation		O	O	O
10.	Check security of all lines		O	O	O
11.	Check vertical fin and rudder surfaces for damage		O	O	O
12.	Check rudder hinges, horn and attachments for damage and operation		O	O	O
13.	Check vertical fin attachments		O	O	O
14.	Check ELT installation and condition of battery and antenna		O	O	O
15.	Check rudder, tab hinge bolts for excess wear (Replace as required)		O	O	O
16.	Check rudder trim mechanism (See Note 9)		O	O	O
17.	Check stabilator surfaces for damage		O	O	O
18.	Check stabilator, tab hinges, horn and attachments for damage and operation		O	O	O

NOTES:

1. Both the annual and 100 hour inspections are complete inspections of the airplane, identical in scope, while both the **500 and 1000** hour inspections are extensions of the annual or 100 hour inspection, which require a more detailed examination of the airplane, and overhaul or replacement of some major components. Inspection must be accomplished by persons authorized by the FAA.

2. Intervals between oil changes can be increased as much as 100% on engines equipped with full flow (cartridge type) oil filters - provided the element is replaced each 50 hours of operation.

3. Replace or overhaul as required or at engine overhaul. (For engine overhaul, refer to Lycoming Service Instructions No. 1009.)

4. Replace flexible oil lines as required, but no later than 1000 hours of service.

5. Refer to Piper Service Letter No. 597 for flap control cable attachment; bolt use.

6. Refer to Piper Service Bulletin No. 373A for exhaust system inspection.

7. Refer to Piper Service Bulletin No. 358.

8. Torque all attachment nuts to 135 to 150 inch-pounds. Seat "Pal" nuts finger tight against plain nuts and then tighten an additional 1/3 to 1/2 turn.

9. Inspect rudder trim tab for "free play;" must not exceed .125 inches. Refer to Service Manual for procedure, Section V. Refer to Piper Service Bulletin No. 390A.

10. Compliance with Piper Service Letter No. 673 eliminates repetitive inspection requirements of Piper Service Bulletin No. 373A and FAA Airworthiness Directive No. 73-14-2.

11. Piper Service Letter No. 704 should be complied with.

FIGURE 16–2 Continued.

F. LANDING GEAR GROUP

	Nature of Inspection	50	100	500	1000
1.	Check oleo struts for proper extension (Check for proper fluid level as required)	○	○	○	○
2.	Check nose gear steering control and travel		○	○	○
3.	Check wheel alignment		○	○	○
4.	Put airplane on jacks (Refer to Section II)	○	○	○	○
5.	Check tires for cuts, uneven or excessive wear and slippage		○	○	○
6.	Remove wheels, clean, check and repack bearings		○	○	○
7.	Check wheels for cracks, corrosion and broken bolts		○	○	○
8.	Check tire pressure (N-31 psi/M-50 psi)		○	○	○
9.	Check brake lining and disc		○	○	○
10.	Check brake backing plates		○	○	○
11.	Check brake lines and retaining clamps		○	○	○
12.	Check condition of center spring		○	○	○
13.	Check gear forks for damage		○	○	○
14.	Check oleo struts for fluid leaks and scoring		○	○	○
15.	Check gear struts, attachments, torque links, retraction links and bolts for condition and security		○	○	○
16.	Check downlocks for operation and adjustment		○	○	○
17.	Check torque link bolts and bushings (Rebush as required)			○	○
18.	Check drag end side brace link bolts (Replace as required)			○	○
19.	Check gear doors and attachments		○	○	○
20.	Check gear warning horn and light for operation		○	○	○
21.	Check hydraulic fluid level in pump reservoir		○	○	○
22.	Retract gear - check operation		○	○	○
23.	Retract gear - check doors for clearance and operation	○	○	○	○
24.	Check operation of squat switch		○	○	○
25.	Check downlock switches, up-switches and electrical leads for security		○	○	○
26.	Lubricate per Lubrication Chart		○	○	○
27.	Remove airplane from jacks		○	○	○

G. OPERATIONAL INSPECTION

	Nature of Inspection	50	100	500	1000
1.	Check fuel pumps, fuel tank selector and crossfeed operation	○	○	○	○
2.	Check fuel quantity and pressure or flow gauges	○	○	○	○
3.	Check oil pressure and temperatures	○	○	○	○
4.	Check alternator output	○	○	○	○
5.	Check manifold pressure	○	○	○	○
6.	Check alternate air	○	○	○	○
7.	Check parking brake and toe brakes	○	○	○	○
8.	Check vacuum gauge	○	○	○	○
9.	Check gyros for noise and roughness	○	○	○	○
10.	Check cabin heater operation	○	○	○	○
11.	Check magneto switch operation	○	○	○	○
12.	Check magneto RPM variation	○	○	○	○
13.	Check throttle and mixture operation	○	○	○	○
14.	Check propeller smoothness	○	○	○	○
15.	Check constant speed propeller action	○	○	○	○
16.	Check engine idle	○	○	○	○
17.	Check electronic equipment operation	○	○	○	○
18.	Check operation of controls	○	○	○	○
19.	Check operation of flaps	○	○	○	○

H. GENERAL

	Nature of Inspection	50	100	500	1000
1.	Aircraft conforms to FAA Specifications	○	○	○	○
2.	All FAA Airworthiness Directives complied with	○	○	○	○
3.	All Manufacturers Service Letters and Bulletins complied with	○	○	○	○
4.	Check for proper Flight Manual	○	○	○	○
5.	Aircraft papers in proper order	○	○	○	○

5. Refer to Piper Service Letter No. 597 for flap control cable attachment bolt use.
6. Refer to Piper Service Bulletin No. 373A for exhaust system inspection.
7. Refer to Piper Service Bulletin No. 358.
8. Torque all attachment nuts to 135 to 150 inch-pounds. Seat "Pal" nuts finger tight against plain nuts and then tighten an additional 1/3 to 1/2 turn.
9. Inspect rudder trim tab for "free play;" must not exceed .125 inches. Refer to Service Manual for procedure, Section V. Refer to Piper Service Bulletin No. 390A.
10. Compliance with Piper Service Letter No. 673 eliminates repetitive inspection requirements of Piper Service Bulletin No. 373A and FAA Airworthiness Directive No. 73-14-2.
11. Piper Service Letter No. 704 should be complied with.

NOTES:

1. Both the annual and 100 hour inspections are complete inspections of the airplane, identical in scope, while both the **500 and 1000** hour inspections are extensions of the annual or 100 hour inspection, which require a more detailed examination of the airplane, and overhaul or replacement of some major components. Inspection must be accomplished by persons authorized by the FAA.
2. Intervals between oil changes can be increased as much as 100% on engines equipped with full flow (cartridge type) oil filters - provided the element is replaced each 50 hours of operation.
3. Replace or overhaul as required or at engine overhaul. (For engine overhaul, refer to Lycoming Service Instructions No. 1009.)
4. Replace flexible oil lines as required, but no later than 1000 hours of service.

FIGURE 16–2 Continued.

supervise the inspection program and that an inspection-procedures manual be available and readily understandable to the pilot and the maintenance personnel.

The frequency and detail of the progressive inspection must provide for the complete inspection of the aircraft within each 12 calendar months and be consistent with the manufacturer's recommendations, field-service experience, and the kind of operation in which the aircraft is engaged. The progressive inspection schedule must ensure that the aircraft, at all times, is airworthy and conforms to all applicable Aircraft Specifications, Type Certificate Data Sheets, Airworthiness Directives, and other approved data such as service bulletins and service letters issued by the manufacturer. If the progressive inspection is discontinued, the owner or operator must immediately notify the local FAA district office in writing. After the discontinuance, the first annual inspection is due within 12 calendar months after the last complete inspection under the progressive inspection schedule. The 100-h inspection required by FAR Part 91 is due within 100 h of operation after that same complete inspection.

A typical progressive inspection schedule is shown in Figure 16–3. Under this program, the airplane is inspected and maintained in four operations, called **events**, scheduled at 50-h intervals. The events are arranged so that a 200-h flying cycle results in a complete aircraft inspection.

An event inspection is a group of several predetermined location inspections, both **routine** and **detailed**, as shown in Figure 16–4. A routine inspection consists of a visual examination or check of the aircraft and its components and systems insofar as is practicable without disassembly. A detailed inspection consists of a thorough examination of the aircraft and its components and systems with such disassembly as is necessary. Figure 16–4 shows typical routine and detailed inspections for a propeller.

An inspection event may be conducted and recorded in the event record by an airframe and powerplant technician. When the four events are complete and recorded, an entry should be made in the cycle record by the person holding an IA who is supervising the inspection.

With a schedule such as is shown in Figure 16–3, unless an aircraft is operated a minimum of 200 h a year, it would not be beneficial to place it on a progressive inspection, since one of the requirements of a progressive schedule is that it be completed at least every 12 calendar months. The decreased maintenance costs and increased use stem from the fact that the inspection is divided up into smaller segments than a 100-h inspection and also that some parts of the airplane, such as the fuselage, wings, and empennage, receive a detailed inspection only every 200 h.

LARGE AND TURBINE-POWERED MULTIENGINE AIRPLANES

Inspection Programs

Large aircraft (over 12 500 lb) and turbine-powered multiengine aircraft are excluded from using 100-h or annual and progressive inspections. Due to their complexity, the inspection programs for these aircraft tend to be more specific

---

Event Inspection #1

To be performed at the 50–250–450–650–850 Flying Hour Intervals.
  Consist of:
  1. Engine Routine   4. Cabin Detail
  2. Propeller Routine   5. Operational Inspection
  3. Fuselage Detail

Event Inspection #2

To be performed at the 100–300–500–700–900 Flying Hour Intervals.
  Consist of:
  1. Engine Detail   4. Landing Gear Routine
  2. Empennage Detail   5. Operational Inspection
  3. Propeller—Routine

Event Inspection #3

To be performed at the 150–350–550–750–950 Flying Hour Intervals.
  Consist of:
  1. Engine Routine   4. Fuselage—Routine
  2. Propeller Detail   5. Landing Gear—Detailed
  3. Wings Detailed   6. Operational Inspection

Event Inspection #4

To be performed at the 200–400–600–800–1000 Flying Hour Cycles.
  Consist of:
  1. Engine Detail   3. Landing Gear—Routine
  2. Propeller Routine   4. Operational Inspection

FIGURE 16–3 Sample progressive inspection schedule.

Routine Propeller Inspection	Detailed Propeller Inspection
1. Check spinner attachments. 2. Inspect propeller for condition.	1. Remove and inspect spinner and back plate. 2. Inspect blades for nicks and cracks. 3. Check for grease and oil leaks. 4. Lubricate per lubrication chart. 5. Check spinner mounting brackets for cracks. 6. Check propeller mounting bolts for safety. 7. Inspect hub parts for cracks and corrosion. 8. Check blades for tightness in hub pilot tube. 9. Check propeller air pressure. Refer to Pressure Temperature Chart. 10. Check condition of propeller deicer system.

FIGURE 16–4 Sample routine and detailed propeller inspections.

than the 100-h or annual inspections. FAR Section 91.409 provides four options to the owner or operator in the selection of an inspection program:

Option 1. A continuous airworthiness inspection program that is a part of a continuous airworthiness maintenance program currently in use by a person holding a certificate issued under FAR Part 121.

Option 2. An approved aircraft inspection program currently in use by a person holding an air-taxi certificate under FAR Part 135.

Option 3. A current inspection program recommended by the manufacturer.

Option 4. Any other inspection program established by the registered owner or operator of the airplane and approved by the FAA administrator.

In the first three options listed, the provision *current* is mentioned. This requirement is intended to prevent the use of obsolete programs.

Continuous Airworthiness Inspection Programs

Air carriers operating under FAR Part 121 are required to have a **continuous airworthiness maintenance program**. A continuous airworthiness maintenance program is a compilation of the individual maintenance and inspection functions used by an operator to fulfill his or her total maintenance needs. Continuous airworthiness maintenance programs are included in the maintenance section of an air carrier's operations specifications approved by the Federal Aviation Administration. These specifications prescribe the scope of the program, including its limitations, and the reference manuals and other technical data to be used as supplements to the specifications. The following are the basic elements of continuous airworthiness maintenance programs:

1. Aircraft inspection This element deals with the routine inspections, servicing, and tests performed on the aircraft at prescribed intervals. It includes detailed instructions and standards (or related references) by work forms, job cards, and other records which also serve to control the activity and to record and account for the tasks that comprise this element. Each airline is free to develop its own terminology, which is assigned to the different parts that comprise the inspection program. The use of terms such as *A-Check* and *D-Check* as is illustrated in Figure 16–5 is common in a continuous inspection program. Figure 16–5 provides an example of the arrangement and terminology that may be used to comprise a continuous inspection program.

2. Scheduled maintenance This element concerns maintenance tasks performed at prescribed intervals. Some are accomplished concurrently with the inspection tasks that are part of the inspection element and may be included on the same form. Other tasks are accomplished independently. The scheduled tasks include the replacement of life-limited items and components requiring replacement for periodic overhaul; special inspections such as X-rays, checks, or tests for on-condition items; lubrications; and so on. Special work forms can be provided for accomplishing these tasks, or they can be specified by a work order or some other document. In any case, instructions and standards for accomplishing each task should be provided to ensure that it is properly accomplished and that it is recorded and signed for.

3. Unscheduled maintenance This element provides instructions and standards for the accomplishment of maintenance tasks generated by the inspection and scheduled maintenance elements, pilot reports, failure analyses, or other indications of a need for maintenance. Procedures for reporting, recording, and processing inspection findings, operational malfunctions, or abnormal operations, such as hard landings, are an essential part of this element. A continuous aircraft logbook can serve this purpose for occurrences and resultant corrective action between scheduled inspections. Inspection discrepancy forms are usually used for processing unscheduled maintenance tasks in conjunc-

Type	When	Who	What
No. 1 service "walk-around"	Before each flight	Mechanic and pilot	Exterior check of aircraft and engines for damage and leakage; includes specific checks such as brake and tire wear
No. 2 service	During overnight layovers at maintenance locations; at least every 45 hours of domestic flying or 65 hours of international flying	Mechanics	Same as No. 1 service plus specific checks including oils, hydraulics, oxygen, and unique needs by aircraft type
A-check	Approximately every 200 flying hours, or about every 15 to 20 days—depending on type of aircraft	3–5 Mechanics	More detailed check of aircraft and engine interior including specific checks, services, and lubrications of systems such as ■ Ignition ■ Hydraulics ■ Generators ■ Structure ■ Cabins ■ Landing gear ■ Air conditioning
B-/M-/L-checks	Heaviest level of routine line maintenance; approximately every 550 flying hours or every 40–50 days; work performed overnight	12–80 Mechanics	Similar to A-Check but in greater detail, with specific aircraft and engine needs such as torque tests, internal checks, and flight controls
C-check	Every 12–15 months, depending on aircraft type; airplane out of service for 3–5 days	From 150–200 mechanics and inspectors—depending on aircraft type	Detailed inspection and repair of aircraft, engines, components, systems and cabin, including operating mechanisms, flight controls, and structural tolerances
D-check	Most intensive inspection; every 4–5 years, depending on aircraft type; airplane out of service up to 30 days	From 150–300 mechanics and inspectors—depending on aircraft type	Major structural inspections for detailed needs which include attention to fatigue corrosion; aircraft is dismantled, repaired and rebuilt as required; systems and parts are tested, repaired or replaced

FIGURE 16–5 Typical inspection format for a continuous airworthiness inspection program.

tion with scheduled inspections. Instructions and standards for unscheduled maintenance are normally provided by the operator's technical manuals. The procedures to be followed in using these manuals and for recording and certifying unscheduled maintenance are included in the operator's procedural manual.

Approved Aircraft Inspection Programs

The **Approved Aircraft Inspection Program (AAIP)** concept was first developed for the benefit of FAR Part 135 airtaxi operators who requested the regulatory authority to develop and use inspection programs more suitable to aircraft in their operating environments than the environments suitable to conventional 100-h or annual inspections required by Part 91.

The AAIP allows each operator to develop a program which is tailored to meet his or her particular needs in satisfying aircraft inspection requirements. It provides for the operator to adjust the intervals between individual inspec-

tion tasks in accordance with the needs of the aircraft rather than repeat all tasks at each 100-h increment. It also allows the operator to develop procedures and standards for the accomplishment of those tasks. Along with these benefits is the responsibility to achieve an acceptable level of equivalent safety to the conventional Part 91 inspection requirements. The AAIP serves as the operator's specification for each segment of the program. This is in contrast to the 100-h or annual inspection, where the performing technician or repair station determines, in accordance with Appendix D of Part 43, what work is required. Under the AAIP, the operator is responsible for the program content and standards, and the performing mechanic or repair station is responsible for the accomplishment of the inspection as specified by work sheets and other criteria designated by the program.

An approved aircraft inspection program should encompass the total aircraft, including all installed equipment such as communications and navigational gear, cargo provisions, and so forth. It should include a schedule of the individual tasks or groups of tasks that comprise the program and their frequency of accomplishment. A well-developed AAIP should result in more effective maintenance at less cost than the more rigid 100-h or annual inspection systems.

An approved aircraft inspection program should include the following elements:

1. A schedule of the individual items or groups of items that comprise the inspection program and their frequency of accomplishment.

2. Work forms designating the inspection items with a sign-off provision for each item. These forms should also serve to coordinate and control the work in progress.

3. Instructions regarding methods, techniques, tools, and equipment needed for accomplishing each inspection item.

4. A system for recording discrepancies and the resulting corrective action taken.

5. A system of record keeping that provides information regarding the aircraft's current inspection status and records of past inspections.

Manufacturer's Recommended Inspection Programs

One of the more popular ways to satisfy the inspection requirements of FAR 91.409 is by the adoption of an **aircraft manufacturer's inspection program**. Under this arrangement, the aircraft manufacturer's program, including its methods, techniques, practices, standards of accomplishment, and inspection intervals, is adopted in its entirety.

The aircraft manufacturer's program in most cases contains the frequency and the extent of maintenance necessary for the aircraft, engine, propeller, and rotors. It may also include the frequency of overhauls and the life limit of the components requiring replacement.

A manufacturer's inspection program is a comprehensive program that will include all the necessary forms and manuals required to conduct the program. Manufacturers often assign names to these programs, such as The Continuous Airworthiness Program or The Continuous Inspection Program. These names should not be confused with the FAA use of the term *continuous inspection program*, which was previously described.

Operator-Developed Inspection Programs

An operator-developed inspection program is developed and published by the operator. It must include the methods, techniques, practices, and standards necessary for the proper accomplishment of the program. These programs are usually not developed from scratch but instead are manufacturer's programs that have been modified to suit the operator's particular needs. If the program is, in effect, a manufacturer's maintenance program with variations (such as a higher engine overhaul period), those variations categorize it as an operator-developed program, not an adoption of a manufacturer's program. An operator-developed program bears no prior FAA approval.

Significant variation from the manufacturer's recommendations have to be fully substantiated by the applicant, and

the program being approved must provide alternative actions which will ensure an equivalent level of safety. These actions include such things as manufacturer-recommended special inspections and special structural inspections.

Walk-Around Inspections

To keep an aircraft in proper operating condition and to locate defects that arise between major inspections, manufacturers recommend various types of **walk-around** and **preflight inspections**. Frequent minor inspections of airliners are conducted by flight engineers or maintenance personnel. The inspections are usually conducted at every stop at which time permits (see Figure 16–5). For example, the inspector might carry a flashlight and check the interior of the tailpipe for the condition of the turbines and thrust reversers. The inspector further might check the tires, look for fluid leaks, and examine the fuselage and control surfaces for wrinkles and any other condition that indicates deterioration or damage.

Daily and Preflight Inspections

Before the first flight of an airplane each day, a daily inspection should be performed. This inspection will usually involve the following instructions:

1. Check the fuel tanks for quantity by removing the fuel caps and observing the level of fuel in the tanks. A dipstick is sometimes necessary.

2. Check the quantity of oil in the oil tank by means of a dipstick or sight gauge.

3. Drain a small amount of fuel from each of the fuel drains and strainers. This is to ensure that sediment and water are removed from the tanks.

4. Check the inflation of all tires.

5. Check the extension of the shock struts to ensure that they are properly inflated.

6. Check the engine compartment for loose wires and fittings. Visually inspect spark-plug leads, nuts, air ducts, exhaust pipes, controls, and accessories. Check for oil leaks.

7. Examine the propeller blades for nicks, cuts, and evidence of any other damage.

8. With the ignition switch off, turn the propeller by hand at least two revolutions and note any unusual noises or other indications of malfunction. Stand clear of the propeller's plane of rotation.

9. Examine the cowling and the inspection plates for proper fastening.

10. Inspect the hinges and the control attachments for all control surfaces. Test each control for freedom of movement.

11. Visually inspect the exterior of the aircraft for damage, loose parts, or any other unsatisfactory condition. Check for fluid leaks.

12. Inspect the windshield, windows, and doors for damage. Check the door or doors for proper latching and locking.

13. Inspect the interior of the cabin and cockpit, including the seat belts, the seats, and any loose parts on the floor.

14. Test the operation of all controls from within the cockpit, including the operation of the brakes.

15. After starting the engine, check the operation of all instruments, the radio, and the propeller (if the propeller is a constant-speed type). See that the engine-oil pressure is indicated within 30 s after the engine starts.

16. Test the operation of the landing, navigation, cabin, and instrument lights.

If the aircraft is flown several times during the day, it is not necessary to make all the inspections just listed; however, a general check of the aircraft, including the quantity of fuel and oil, should be made before each flight.

Special Inspections

In addition to the regularly scheduled inspections, many manufacturers provide for special inspections that are to be performed in the event that the aircraft is subjected to stresses outside of its normal operating environment. These inspections may be for such events as lightning strikes, sudden engine stoppages, severe wind-gust loads, or extremely hard landings. Since these inspections tend to be very specific in nature, it is necessary to refer to the manufacturer's maintenance manual for performance details.

Altimeter and Transponder Inspections

In order to operate an airplane under instrument-flight rules (IFR) in controlled airspace, it must have its altimeters and the static-pressure system inspected in accordance with FAR Part 43, Appendix E, every 24 calendar months. This test is usually performed by an appropriately rated repair station; however, airframe technicians may perform the test and inspection on the static-pressure system.

ATC transponders are required to have an inspection every 24 calendar months in accordance with FAR 43, Appendix F, if they are to be operated. Although the altimeter and transponder inspections are primarily the pilot's responsibility, technicians can provide valuable assistance by pointing out overdue inspections when they are reviewing the maintenance records.

CONDUCTING A 100-HOUR OR ANNUAL INSPECTION

Although specific inspection practices and procedures will vary depending on the size of the aircraft and the type of inspection being conducted, the basic fundamentals followed in conducting an inspection do not change.

Inspection Preparation

The inspection process commences with the owner requesting that an inspection be performed on the airplane. At this point a work order should be filled out itemizing the work that is to be accomplished. A firm understanding should be reached about the cost of the inspection, what is included in this cost (e.g., servicing, lubrication, airworthiness directive compliance, and so on), and the approximate time period planned for completing the inspection. In order to gain a better understanding of the history and the present condition of the aircraft, it is necessary to obtain the airplane's maintenance records for thorough study and review.

An important part of the preinspection process is researching airworthiness directives (ADs) and service bulletins. The technician must determine whether all applicable airworthiness directives on the aircraft, powerplant, propeller, instruments, and appliances have actually been accomplished.

If the maintenance records indicate compliance with an AD, the technician should make a reasonable attempt to verify this. The reason for this verification is that it is not uncommon for a component that was brought into compliance with an AD and properly recorded to then be replaced by another component which had not yet been brought into compliance with the AD. Additional information on airworthiness directives is located in Chapter 14.

The FAA general aviation airworthiness alerts (AC43-15) are also an important source of service experience. These alerts are selected service difficulties reported to the FAA on malfunction or defect reports. It makes sense to use the experiences other persons have had on similar products. These publications help ensure that a problem area is not overlooked.

Prior to beginning the inspection, the checklist to be used should be located, along with discrepancy forms, the appropriate maintenance manual, and the Type Certificate Data Sheet.

The tools needed to perform the inspection should then be readied. Inspection tools can be many and varied, ranging from a pocket-sized magnifying glass to a complete X-ray machine. The principal tools for most inspectors are a flashlight and an inspection mirror; however, additional items such as a magneto-timing light, compression tester, and jacks are among the other tools usually required.

Opening and Cleaning

FAR 43, Appendix D, which lists the scope and detail of the items to be included in a 100-h or annual inspection, begins by saying:

> Each person performing an annual or 100-h inspection shall, before that inspection, remove or open all necessary inspection plates, access doors, fairing, and cowling. He shall thoroughly clean the aircraft and aircraft engine.

New technicians often have difficulty in knowing which inspection plates and panels must be removed. Many manufacturers provide assistance in the form of an inspection-panel diagram, such as the one illustrated in Figure 16–6. When opening inspection plates and cowlings, the technician should take note of any oil or other foreign-material accumulation which may offer evidence of fluid leakage or some other abnormal condition that should be corrected.

An engine and accessories wash-down should be accomplished prior to each inspection to remove oil, grease, salt

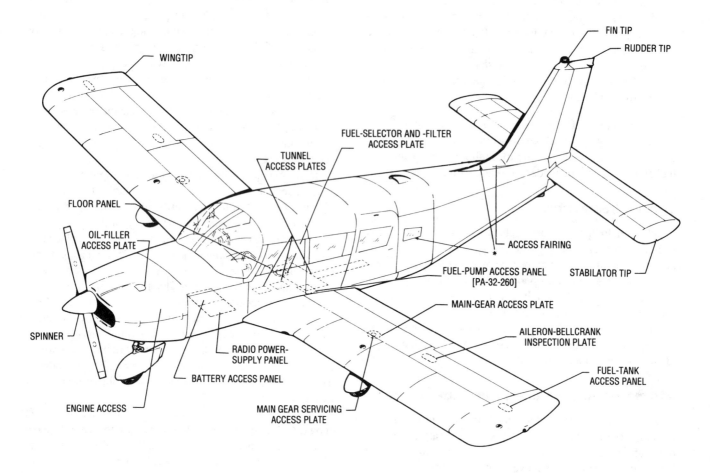

WINGTIP

FIN TIP

RUDDER TIP

FUEL-SELECTOR AND -FILTER
ACCESS PLATE

TUNNEL
ACCESS PLATES

FLOOR PANEL

OIL-FILLER
ACCESS PLATE

ACCESS FAIRING

*

FUEL-PUMP ACCESS PANEL
[PA-32-260]

STABILATOR TIP

SPINNER

MAIN-GEAR ACCESS PLATE

AILERON-BELLCRANK
INSPECTION PLATE

FUEL-TANK
ACCESS PANEL

RADIO POWER-
SUPPLY PANEL

BATTERY ACCESS PANEL

ENGINE ACCESS

MAIN GEAR SERVICING
ACCESS PLATE

*1974 MODELS AND UP

FIGURE 16–6 Access plates and panels.

corrosion, or other residue that might conceal component defects during inspection. Precautions, such as wearing rubber gloves, an apron or coveralls, and a face shield or goggles, should be taken when working with cleaning agents. Use the least toxic of available cleaning agents that will satisfactorily accomplish the work. These cleaning agents include (1) Stoddard solvent; (2) a water-based alkaline detergent cleaner mixed as 1 part cleaner, 2 to 3 parts water, and 8 to 12 parts Stoddard solvent; or (3) a solvent-based emulsion cleaner mixed as 1 part cleaner and 3 parts Stoddard solvent.

WARNING: Do not use gasoline or other highly flammable substances for a wash-down.

Perform all cleaning operations in well-ventilated work areas, and ensure that adequate fire-fighting and safety equipment is available. Compressed air, used for cleaning-agent application and drying, should be regulated to the lowest practical pressure. Use of a stiff-bristle brush is recommended if cleaning agents do not remove excess grease or grime during spraying.

Before cleaning the engine compartment, place a strip of tape on the magneto vents to prevent any solvent from entering these units.

Place a large pan under the engine to catch waste. With the engine cowling removed, spray or brush the engine with solvent or a mixture of solvent and degreaser. In order to remove especially heavy dirt and grease deposits, it may be necessary to brush areas that were sprayed.

WARNING: Do not spray solvent into the alternator, vacuum pump, or air intakes.

Allow the solvent to remain on the engine from 5 to 10 min, then rinse the engine clean with additional solvent and allow it to dry. Cleaning agents should never be left on engine components for an extended period of time. Failure to remove them may cause damage to components such as neoprene seals and silicone fire sleeves and could cause additional corrosion. Completely dry the engine and accessories using clean, dry, compressed air. If desired, the engine cowling may be washed with the same solvent.

Remove the protective tape from the magnetos and lubricate the controls, bearing surfaces, and other components in accordance with the lubrication chart.

Other parts of an airplane that often need cleaning prior to inspection are the landing gear and the underside of the aircraft. Most compounds used for removing oil, grease, and surface dirt from these areas are emulsifying agents. These compounds, when mixed with petroleum solvents, emulsify the oil, grease, and dirt. The emulsion is then removed by rinsing with water or by spraying with a petroleum solvent. Openings such as air scoops should be covered prior to cleaning.

Inspecting the Airplane

The most important part of any inspection is the visual examination given to the airplane to determine its airworthiness. All the related steps in the inspection process, such as service and repair, are dependent on a thorough visual exam. On a 100-h or annual inspection, this step cannot be supervised but instead must be performed by the appropriately certificated person who is returning the aircraft to service. As required by FAR 43.15, a checklist must be used and should be carefully followed. While performing the visual inspection, the inspector should avoid getting sidetracked on related service and repair problems but instead should make a written list of all discrepancies found on the aircraft. A **discrepancy report**, such as the one shown in Figure 16–7, should provide a location for the discrepancy write-up, the corrective action taken, the technician making the repair, and the inspector returning the aircraft to service.

The principal purpose in performing an inspection is to determine if the aircraft is airworthy. In order for an airplane to be declared airworthy, it must meet two criteria: (1) it must be in condition for safe operation, and (2) it must conform to its Type Certificate Data Sheet. *Safe operation* refers to the condition of the aircraft with relation to wear and deterioration. Some general inspection guidelines that may be followed in making this determination are inspection of

1. Metal parts for security of attachment, cracks, metal distortion, broken spotwelds, corrosion, condition of paint, and any other apparent damage

2. Movable parts for lubrication, servicing, security of attachment, binding, excessive wear, safetying, proper operation, proper adjustment, correct travel, cracked fittings, security of hinges, defective bearings, cleanliness, corrosion, deformation, sealing, and tension

3. Bolts in critical areas for proper installation, correct safetying, and correct torque values when visual inspection indicates the need for a torque check

4. Wiring for security, chafing, burning, defective insulation, loose or broken terminals, heat deterioration, and corroded terminals

5. Fluid lines and hoses for leaks, cracks, dents, kinks, chafing, proper radius, security, corrosion, deterioration, obstruction, and foreign matter

6. Filters, screens, and hoses for cleanliness, contamination, and replacement at specified intervals

The manufacturer's checklist and maintenance manual will explain what needs to be checked during the inspection.

An airplane's conformity to its Type Certificate Data Sheet is considered attained when the required and proper components are installed and they are consistent with the drawings, specifications, and other data that are a part of the type certificate. In addition to providing for required equipment, the data sheet provides such information as static rpm, control-surface travel limits, and whether or not a flight manual is considered required equipment. It also identifies limitations which must be displayed on the aircraft in the form of markings and placards. Any deviation from the aircraft's

PAGE _____		DISCREPANCY FORM				
A/C MAKE/MODEL _____		S/N _____		N- _____		
TYPE INSP. _____	TACH TIME _____		DATE _____			

ITEM NO.	DISCREPANCY	AUTH.	CORRECTIVE ACTION	MECH.	INSP.
1	Cabin door seal is loose	*Gℓ*	Reglued seal with #1300L	*K*	*MK*
2	Copilot's seat has rip in back	N/A	No action	—	—
3	Left magneto drop 250 rpm	*Gℓ*	Installed new points, performed internal and external timing	*K*	*MK*
4	Left flap chaffing inboard side	*Gℓ*	Adjusted flap clearance, checked travel	*K*	*MK*
5	Left aileron outboard hinge nutplate cracked	*Gℓ*	Replaced nutplate #PN13762-5	*K*	*MK*

FIGURE 16–7 Sample discrepancy report.

certificated type design is considered a major alteration and should be provided for in an FAA Form 337. *An aircraft that does not conform to its Type Certificate Data Sheet is considered to be unairworthy.*

An additional item included on most checklists is a review of the aircraft paperwork. The **aircraft registration** and **airworthiness certificate** should be located on board the aircraft. If these certificates are not available, the aircraft should not be reported as unairworthy; instead, the owner should be informed that the documents must be in the aircraft, with the airworthiness certificate displayed as required in FAR 91.203, when the aircraft is operated.

Other documents often needed but not a part of the airworthiness requirements might be the state registration, FCC radio-station licenses, and similar papers. The owner or operator is responsible for the proper display of these documents. However, the technician will be performing an appreciated service by informing the operator of any deficiencies in the display and carriage of these documents.

At the conclusion of the visual inspection, a meeting with the airplane's owner is usually in order to discuss the discrepancies located during the inspection. Discrepancies need to be divided into two categories: those that affect the airworthiness of the airplane and those that may be cosmetic in nature and do not affect the aircraft's airworthiness. Those items not affecting the airworthiness of the airplane may have corrective action delayed to a later date. For those discrepancies that affect airworthiness, the owner has the option of having them repaired by the maintenance technician performing the inspection or using the procedures specified in FAR 43.11 and having the aircraft declared unairworthy, as is shown in Figure 16–8. This will permit an owner to assume responsibility for having the discrepancies corrected prior to operating the aircraft. The discrepancies

can be cleared by an A & P technician unless they are major repairs or major alterations. If repairs are preventive maintenance, they can be cleared by the owner or pilot.

There is no stigma attached to the aircraft because it is reported to be unairworthy. In effect, the report says that the aircraft is airworthy with the exception of the items on the discrepancy list. When those listed items are corrected, the aircraft is eligible to be operated. The owner may want the aircraft flown to another location to have repairs completed, in which case the owner should be advised that a **Special Airworthiness Certificate**, FAA Form 8130-7 (formerly referred to as a ferry permit), is necessary. A special airworthiness certificate may be obtained at the local FAA district office.

LUBRICATION AND SERVICING

The visual inspection is required to determine the current condition of the aircraft and its components. The repair of discrepancies is required to bring the aircraft back up to airworthy standards. In an effort to keep the aircraft in airworthy condition, the manufacturer may recommend that certain services be performed at various operating intervals. While servicing and lubrication are often conveniently accomplished during an inspection, they should not be considered a part of the inspection itself.

Lubrication

As is true of any item of machinery where moving parts bear against one another, lubrication is required at many locations in an aircraft. The type of lubrication for each point is determined by the type of bearing, the bearing loads, the

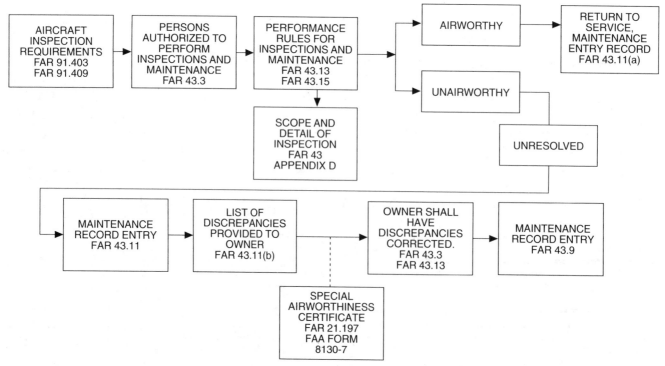

FIGURE 16–8 Annual-inspection flow chart.

frequency and speed of the movement, the temperatures at the bearing, and the materials that bear against one another. Lubricants used for aircraft may be ordinary lubricating oil such as that used in the engine, lightweight lubricating oil, various weights of greases, high-pressure (HP) grease, low-temperature grease, high-temperature grease, graphite, silicone, and other special lubricants.

The frequency of lubrication for each lubrication point in an aircraft is specified by the manufacturer together with the type of lubricant, the method of application, and any special instructions. Information regarding lubrication is provided in the manufacturer's maintenance manual for each model of aircraft.

Commercial airlines develop their own schedules and procedures for lubrication of their aircraft. Lubrication is usually detailed and accomplished as a part of the continuous airworthiness maintenance program.

Lubrication information for light airplanes is often presented in the form of charts and tables. Figure 16–9 is a chart for a light twin-engine airplane. Information and instructions for the use of the chart are shown in Figure 16–10. The parts nomenclature is given in Figure 16–11.

Some general guidelines that should be followed in the application of lubricants are as follows:

1. Cleanliness is essential to good lubrication. Lubricants and dispensing equipment must be kept clean. Use only one lubricant in a grease gun or oil can.
2. Store lubricants in a protected area. Containers should be closed at all times when not in use.
3. Wipe grease fittings, oil holes, and other lubrication points, with clean, dry cloths before lubricating.

4. When lubricating bearings which are vented, force grease into fittings until all old grease is extruded, unless otherwise noted.
5. When lubricating sealed bearings, use extreme care not to dislodge the seals.
6. Work moving parts, if practical, to assure thorough lubrication.
7. After any lubrication, clean surplus lubricant from all but the actual working surfaces.

SERVICING AIRCRAFT

Servicing aircraft requires great care and attention to detail, regardless of the type of aircraft being serviced or the particular service being performed. The servicing of particular aircraft and their components is usually described in great detail in the manufacturers' maintenance and service manuals. Instructions given in these manuals should be followed for the satisfactory and safe performance of aircraft.

Engine-Oil Service

The oil quantity for small aircraft should be checked daily or before each flight. The dipstick is marked to show the maximum quantity of oil permitted in the engine sump or tank. If the dipstick shows that the oil is near or below the required minimum level, oil must be added. In no case should the oil level be permitted to rise above the maximum line. The oil sump or tank must have sufficient **foaming space** to allow for the expansion of the oil and the development of foam.

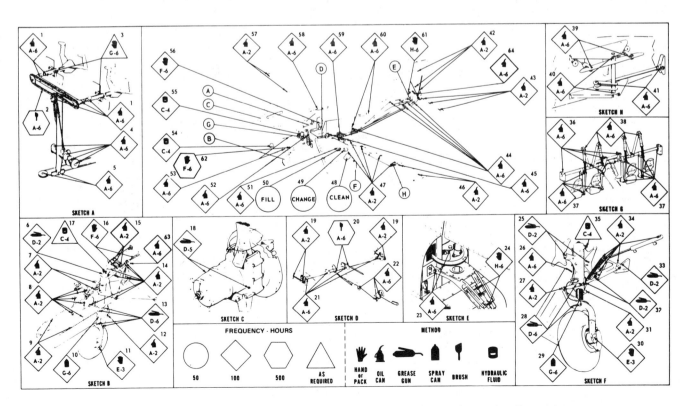

FIGURE 16–9 Lubrication chart for a light twin-engine airplane. *(Piper Aircraft Corp.)*

NOTES

1. PILOT AND PASSENGER SEATS—LUBRICATE TRACK ROLLERS AND STOP PINS AS REQUIRED (TYPE OF LUBRICANT: "A")
2. WHEEL BEARINGS REQUIRE CLEANING AND REPACKING AFTER EXPOSURE TO AN ABNORMAL QUANTITY OF WATER.
3. SEE LYCOMING SERVICE INSTRUCTIONS NO. 1014 FOR USE OF DETERGENT OIL.
4. FUEL SYSTEM—SERVICE REGULARLY—FUEL PUMP STRAINER INJECTOR SCREEN—FILTER BOWL—QUICK DRAIN UNIT.
5. BATTERY—FLUID LEVEL & CONDITION CHECK EVERY 25 HOURS
6. MIL-C-6529 TYPE 2/OIL IS THE OIL THE ENGINE IS SERVICED WITH AT INSTALLATION. THE ENGINE MUST OPERATE ON THIS OIL 25 HOURS MINIMUM, 50 HOURS MAXIMUM. (SEE LYCOMING SERVICE LETTER NO. L121A)
7. THIS CHART IS FOR INITIAL FACTORY LUBRICATION AND SERVICE.

CAUTIONS

1. DO NOT USE HYDRAULIC FLUID WITH A CASTOR OIL OR ESTER BASE.
2. DO NOT OVER-LUBRICATE COCKPIT CONTROLS.
3. DO NOT APPLY LUBRICANT TO RUBBER PARTS.
4. DO NOT LUBRICATE CABLES—THIS CAUSES SLIPPAGE.

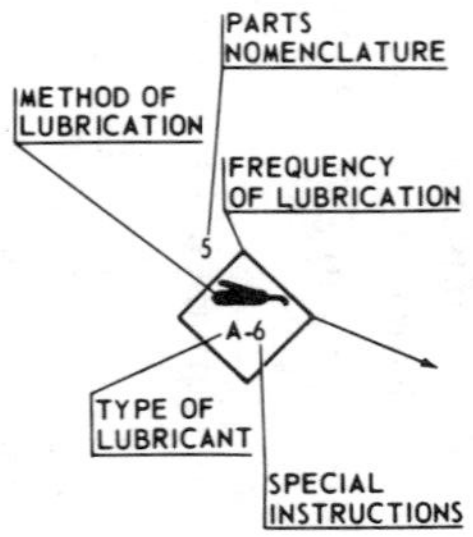

Diagram labels: PARTS NOMENCLATURE — METHOD OF LUBRICATION — FREQUENCY OF LUBRICATION — TYPE OF LUBRICANT — SPECIAL INSTRUCTIONS (5, A-6)

TYPE OF LUBRICANT

IDENTIFICATION LETTER	SPECIFICATION	LUBRICANT
A	MIL-L-7870	LUBRICATING OIL, GENERAL PURPOSE, LOW TEMPERATURE
B	MIL-L-6082 7	LUBRICATING OIL, AIRCRAFT RECIPROCATING ENGINE (PISTON) GRADE AS SPECIFIED SAE 50 ABOVE 60°F AIR TEMP. SAE 40 30° TO 90° AIR TEMP. SAE 30 0° TO 70°F AIR TEMP. SAE 20 BELOW 10°F AIR TEMP.
C	MIL-H-5606	HYDRAULIC FLUID, PETROLEUM BASE (OR UNIVIS—40 OR MOBIL AERO-HF)
D	(NONE)	GREASE, AIRCRAFT AND INSTRUMENT, GREASE AND ACTUATOR SCREW
E	MIL-G-23827	TEXACO MARFAX ALL PURPOSE GREASE OR MOBIL MOBIL GREASE 77 (OR MOBIL EP2 GREASE)
F	MIL-H-7711	GREASE—LUBRICATION, GENERAL PURPOSE, AIRCRAFT
G		FLUOROCARBON RELEASE AGENT DRY LUBRICANT #MS-122 (PURCH)
H		AERO LUBRIPLATE (PURCH) FISKE BROS. REFINING CO.

SPECIAL INSTRUCTIONS

1. AIR FILTER—TO CLEAN FILTER, TAP GENTLY TO REMOVE DIRT PARTICLES. DO NOT BLOW OUT WITH COMPRESSED AIR OR USE OIL. REPLACE FILTER IF PUNCTURED OR DAMAGED.
2. BEARING AND BUSHINGS—CLEAN EXTERIOR WITH A DRY TYPE SOLVENT BEFORE LUBRICATING.
3. WHEEL BEARINGS—DISASSEMBLE AND CLEAN WITH A DRY TYPE SOLVENT. ASCERTAIN THAT GREASE IS PACKED BETWEEN THE BEARING ROLLER AND CONE. DO NOT PACK GREASE IN WHEEL HOUSING.
4. OLEO STRUTS, HYDRAULIC PUMP RESERVOIR AND BRAKE RESERVOIR—FILL PER INSTRUCTIONS ON UNIT OR CONTAINER, OR REFER TO SERVICE MANUAL.
5. PROPELLER—REMOVE ONE OF THE TWO GREASE FITTINGS FOR EACH BLADE. APPLY GREASE THROUGH FITTING UNTIL FRESH GREASE APPEARS AT HOLE OF REMOVED FITTING.
6. LUBRICATION POINTS—WIPE ALL LUBRICATION POINTS CLEAN OF OLD GREASE, OIL, DIRT, ETC. BEFORE LUBRICATING.

FIGURE 16–10 Information and instructions for the lubrication chart in Figure 16–9.

1. CONTROL COLUMN FLEX JOINT, SPROCKET AND O-RING
2. AILERON AND STABILATOR CONTROL CHAIN
3. O-RING CONTROL SHAFT BUSHING
4. TEE BAR PIVOT POINTS, AILERON AND STABILATOR CONTROL PULLEYS
5. STABILATOR CONTROL ROD AND IDLER PULLEY
6. NOSE GEAR STRUT HOUSING
7. NOSE GEAR PIVOT POINT AND HYDRAULIC CYLINDER ROD END
8. NOSE GEAR DOOR RETRACTION MECHANISM
9. NOSE GEAR DOOR HINGES
10. EXPOSED OLEO STRUT (NOSE)
11. NOSE WHEEL BEARINGS
12. NOSE GEAR TORQUE LINK ASSEMBLY
13. NOSE GEAR TORQUE LINK ASSEMBLY AND STRUT HOUSING
14. NOSE GEAR PIVOT POINT, DRAG LINK ASSEMBLY, DOWNLOCK AND CYLINDER ASSEMBLY, STEERING ROLLER AND CENTERING SPRING PIVOT POINTS
15. STEERING BELLCRANK PIVOT POINTS AND ROD ENDS
16. NOSE GEAR ROLLER TRACK AND BUNGEE
17. NOSE GEAR OLEO STRUT FILLER POINT
18. PROPELLER ASSEMBLY
19. FLAP CONTROL ROD END BEARINGS
20. FLAP RETURN AND TENSION CHAIN
21. FLAP TORQUE TUBE BEARING BLOCK
22. FLAP HANDLE PIVOT POINT LOCK MECHANISM AND CABLE PULLEY
23. RUDDER SECTOR AND STABILATOR TRIM PIVOT POINTS
24. STABILATOR TRIM SCREW
25. MAIN GEAR PIVOT POINTS
26. MAIN GEAR DOOR HINGE
27. MAIN GEAR TORQUE LINKS
28. MAIN GEAR TORQUE LINKS
29. EXPOSED OLEO STRUT (MAIN)
30. MAIN GEAR WHEEL BEARINGS
31. MAIN GEAR DOOR CONTROL ROD ENDS
32. MAIN GEAR SIDE BRACE LINK ASSEMBLY
33. UPPER SIDE BRACE SWIVEL FITTING
34. MAIN GEAR DOWNLOCK ASSEMBLY RETRACTION FITTING AND CYLINDER ATTACHMENT POINTS
35. OLEO STRUT FILLER POINT (MAIN GEAR)
36. RUDDER TUBE CONNECTIONS, TUBE CABLE ENDS AND STEERING ROD ENDS
37. BRAKE ROD ENDS
38. TOE BRAKE ATTACHMENTS
39. AILERON BELLCRANK CABLE ENDS
40. AILERON BELLCRANK PIVOT POINTS
41. AILERON CONTROL ROD END BEARINGS
42. RUDDER HINGE AND TAB HINGE BEARINGS
43. STABILATOR TRIM TAB HINGE PINS
44. CONTROL CABLE PULLEYS
45. TRIM CONTROL WHEELS · STABILATOR AND RUDDER
46. AILERON HINGE PINS
47. FLAP HINGE BEARINGS AND ALTERNATE AIR DOORS
48. INDUCTION AIR FILTERS
49. CARTRIDGE TYPE OIL FILTERS
50. ENGINE OIL SUMP (8 QTS. CAPACITY)
51. GOVERNOR CONTROLS
52. CONTROL QUADRANT CONTROLS
53. FORWARD BAGGAGE DOOR HINGE AND LATCH PINS
54. HYDRAULIC PUMP RESERVOIR
55. BRAKE RESERVOIR
56. PILOT AND COPILOT SEAT ADJUSTMENT
57. AILERON HINGE PINS
58. MAIN DOOR HINGES AND LATCH MECHANISM
59. CONTROL CABLE PULLEYS
60. BAGGAGE AND REAR DOOR HINGES AND LATCH MECHANISM
61. RUDDER AND STABILATOR TRIM SCREWS
62. LATCH MECHANISM
63. LINK BUSHING
64. ARM BUSHING

FIGURE 16–11 Parts nomenclature for the lubrication chart in Figure 16–9.

Aircraft manufacturers usually give specific instructions regarding the engine-oil service. Typical of such instructions are those provided by the Cessna Aircraft Company for the Model 421 airplane:

Check oil level before each flight. Capacity for each engine-oil sump is 13 U.S. quarts, which includes 1 quart for oil filter. (Do not take off on less than 9 quarts.) When preflight shows less than 9 quarts, service with aviation-grade engine oil, SAE 30 below 40°F, SAE 50 above 40°F. The aircraft is delivered from the factory with straight mineral oil; therefore it will be necessary during the break-in period or first 50 hours of operation to add straight mineral oil. Multiviscosity oil with a range of SAE 10W30 is recommended for improved starting and turbocharger controller operation in cold weather. Detergent or dispersant oil conforming to Continental Motors Specification MHS-24A must be used after the first 50 hours of new or overhauled engine operation.

For many reciprocating engines that use petroleum oils, it is recommended that the oil be changed after every 50 h of operation. At the same time, the oil screens should be cleaned and the filters should be replaced. If an aircraft is not operated frequently, the oil should be changed at least every 4 months, even though the aircraft has not operated for 50 h. Some manufacturers recommend changing the oil at intervals of not more than 90 days. If the oil on the dipstick appears dirty when the oil quantity is being checked, the oil should be changed, the screens cleaned, and the filter replaced.

Gas-turbine engines generally use high-temperature synthetic lubricants conforming to MIL-L-7808 or MIL-L-23699. These lubricants are now commonly tested by chemical means to determine their quality. Oil changes are made only if the chemical test indicates that the oil is no longer satisfactory for lubrication.

When changing or adding oil to a gas-turbine-engine system, the technician must be certain that the correct type and grade of lubricant are used. In handling synthetic lubricants, care must be taken to prevent the lubricant from spilling or coming into contact with the skin. If the lubricant should come into contact with the hands or any other part of the body, it should be wiped off immediately and the affected area should be washed with soap and water.

Oil changes for turbine aircraft are governed by the approved service and maintenance procedures established by the airline operating the aircraft. These procedures are developed through cooperative planning with the manufacturers of the engines and the aircraft.

Servicing Brake and Hydraulic Systems

Brake and hydraulic systems must be serviced in accordance with the manufacturers' instructions. As long as these systems operate satisfactorily, service is not normally required at intervals of less than 100 h of operation. When it is necessary to add fluid, it is most important that the correct fluid be added. If the wrong fluid is used, it will be necessary to drain and flush the system and probably change all the seals in the components of the system. Petroleum solvents can be used to flush a system that uses petroleum fluids. Many large aircraft use a synthetic, fire-resistant hydraulic fluid. If these fluids or any of a similar type are put into a system designed for petroleum fluid, considerable damage can result.

Oxygen-System Service

Oxygen supplies should be replaced whenever the oxygen-pressure gauge (or gauges) show that the supply is low. Oxygen bottles (tanks) should be serviced only with breathing oxygen. In removing and replacing the units of an oxygen system, oils, greases, or other petroleum products must not be used for the lubrication of parts or fittings. Only an approved oxygen-system lubricant should be used. A suitable lubricant for oxygen-system fittings is military specification MIL-T-5542-B or an equivalent.

Oxygen systems that use compressed gaseous oxygen are replenished by means of an oxygen-service unit containing breathing oxygen. Service is accomplished by connecting the ground-service unit to the remote fill line of the system and opening the valves required to permit oxygen to flow into the system. Care must be taken to observe the oxygen pressure and to avoid overfilling the system.

WARNING: Avoid making sparks and keep all burning cigarettes or fire away from the vicinity of the airplane when working on the oxygen system. Inspect the filler connection for cleanliness before attaching it to the filler valve. Make sure that your hands, tools, and clothing are clean, particularly of grease and oil, because these contaminants will ignite upon contact with pure oxygen under pressure. As a further precaution against fire, open and close all oxygen valves slowly during filling.

Some airliners are equipped with chemical-oxygen generators instead of supplying the system with gaseous oxygen from a tank. These systems use a sodium chlorate core, which, when fired, generates pure oxygen that is fed through a filter to the oxygen outlets. Servicing these units involves replacing the oxygen generator.

Servicing Batteries

In servicing batteries, it is most important to know whether a battery is a lead-acid type or an alkaline type, such as a nickel-cadmium battery. Service procedures for the different types vary considerably. The service manual or the instructions for the battery should be consulted unless the technician is thoroughly familiar with the procedure.

Lead-Acid Batteries. Whenever checking a battery, determine that all the connections are clean and tight and the fluid level is above the baffle plates. If it is necessary to add fluid, use distilled water. Do not overfill the battery. When the cells are overfilled, water and acid may spill on the lower portions of the fuselage. A hydrometer check should be performed to determine the percentage of charge present in the battery.

Corrosion on the battery terminals and connections may be neutralized by applying a solution of baking soda and water mixed to the consistency of thin cream. Do not allow any of this soda solution to enter the battery. Repeat this application until all bubbling action has ceased before washing the battery and box with clean water.

Nickel-Cadmium Batteries. The following list of battery servicing precautions and checks is meant to be a general guide. For specific details and complete procedures, refer to the battery manufacturer's instructions.

1. After each 100 h of operation or every 30 days, whichever occurs first, check the electrolyte level and clean the battery and the filler-vent plugs.

2. Periodically check that the cell vents are clean and open. Plugged vents may cause excessive internal cell pressure and cause leaks.

3. Never remove a cell from the battery case unless a replacement is immediately available; otherwise, the remaining cells may swell, making replacement of the removed cells difficult. Loosen the vents before cell replacement to eliminate the possibility of cells swelling from internal gas pressure.

4. Check the torque of the terminal screws securing the cross-links that connect the cells together (and then adjust according to the manufacturer's specifications).

5. Check that no carbon deposits have built up on the cross-links or between them and the battery case. If such deposits are present, clean the affected areas.

6. When there is any indication of oil in the battery, remove all cells from the case and check all rubber parts for deterioration. Remove the oil and replace all damaged rubber parts. To remove the oil, use soap and water only.

7. Keep nickel-cadmium and lead-acid batteries stored separately to prevent mutual contamination. Unless kept in closed storage containers, nickel-cadmium electrolyte (potassium hydroxide) will absorb enough carbon dioxide from the air to render it ineffective.

Tires

Maintaining proper tire inflation will minimize tread wear and help prevent tire ruptures caused from running over sharp stones and ruts. When inflating tires, visually inspect them for cracks and breaks. Reverse the tires on the wheels, if necessary, to produce even wear. All tires and wheels are balanced before installation and the relationship of tire, tube, and wheel should be maintained upon reinstallation. Out-of-balance wheels can cause extreme vibration in the landing gear during takeoff and landing. When landing, tires grow slightly due to the shock load. Normally, this growth is balanced by tread wear so there is no increase in the tire diameter.

Other Service Items

In addition to the services described in the foregoing paragraphs, it is necessary to service other items and systems from time to time as specified in the maintenance manual. Among these are the oleo struts, shimmy dampers, bungee cylinders, instrument filters, fuel screens, heaters, and numerous other parts.

The service of these items is described in the maintenance manual for the aircraft and may or may not be established on a regular basis. For example, oleo struts need service only when inspection shows that they have lost air or fluid. Otherwise, they may need attention only at a time of major reconditioning or overhaul.

OPERATIONAL INSPECTION

Before an aircraft may be approved for return to service after an annual or 100-h inspection, FAR 43.15 requires that the engines be operated

To determine satisfactory performance, in accordance with the manufacturer's recommendations, of

(i) Power output (static and idle rpm);
(ii) Magnetos;
(iii) Fuel and oil pressure; and
(iv) Cylinder and oil temperature.

In addition to FAR 43.15, most manufacturer's inspection checklists include an operational inspection which pro-

vides for the operational testing of a wide variety of equipment, ranging from brakes to radios.

After an operational check is performed, the engine should be given a brief visual inspection for oil leaks resulting from the inspection and the servicing performed.

The conclusion of the operational inspection is also an excellent time to give the aircraft a final walk-around visual inspection, checking to see that all the inspection panels that were removed to perform the inspection have been correctly reinstalled.

MAINTENANCE RECORDS

In the process of returning an aircraft to service after completing an inspection or performing maintenance, technicians will encounter a number of official forms, including maintenance records relating to their work. A properly completed maintenance record provides the information needed by the owner or operator and the maintenance personnel to determine when scheduled maintenance is to be performed. Aircraft maintenance record keeping is a responsibility shared by the owner and the maintenance technician, with the ultimate responsibility assigned to the owner by FAR 91.417.

A properly executed set of maintenance records will save the owner money since maintenance personnel will spend less time in research to establish the status of the item to be worked on. Good records are also invaluable to maintenance personnel in troubleshooting aircraft malfunctions.

Maintenance Record Format

Maintenance records may be kept in any format which provides record continuity and includes the required information. There is no requirement that the records be bound; however, bound records or records which have some system of page control normally have greater credibility. Many owners and operators have found it advantageous to keep separate records for the airframe, engine, and propeller, particularly on multiengine aircraft. Engines, propellers, and appliances are often changed from one aircraft to another. The use of individual records facilitates the transfer of the record with the item when ownership changes. The important thing is to have a record system that will provide for storing and retrieving the necessary information.

FAR 91.417 sets forth the requirements on retaining aircraft records. As illustrated in Figure 16–12, records are divided into two basic groups: those items that must be retained until the work is superseded by other work or for a period of 1 yr after the work was performed, and those records which are considered permanent and are kept for the life of the aircraft. Records which are considered permanent are transferred with the aircraft if it is sold.

Inspection Record Entries

FAR 43.11 contains the requirements for inspection entries. When a technician approves or disapproves an aircraft for

Records that shall be retained until work is repeated or superseded by other work or 1 year after work is performed.	Records that shall be retained and transferred with the aircraft.
Records of: ____ Maintenance ____ Alterations ____ 100-hour ____ Annual ____ Progressive ____ Other approved inspection programs Which shall include: ____ Description of work performed. ____ Date. ____ Certificate number and signature of person *approving for return* to service.	____ T.T. in service: Airframe Engine Propeller ____ Current status of life-limited parts of airframe, engine, propeller, rotor, appliance. ____ TSOH of all items requiring OH. ____ Identification of current inspection program and times since last inspection. ____ Current status of AD notes and recurring action. ____ Copies of FAA Form 337 for each major alteration of the airframe, engine, propeller, appliance.

FIGURE 16–12 Record-keeping requirements.

return to service after an inspection, an entry must be made including the following items:

1. The type of inspection and a brief description of the extent of the inspection.

2. The date of the inspection and the aircraft's total time in service.

3. The signature, the certificate number, and the kind of certificate held by the person approving or disapproving the aircraft's return to service.

4. If the aircraft is found to be airworthy and approved for return to service, the following or a similarly worded statement: "I certify that this aircraft has been inspected in accordance with [insert type] inspection and was determined to be in airworthy condition."

If an aircraft is disapproved for return to service because during the course of the inspection it was found to be unairworthy, a signed and dated list of the discrepancies must be provided to the aircraft owner according to FAR 43.11, as is shown in Figure 16–8 on page 368.

In addition to the just-described information for progressive or continuous-type inspections where only part of the inspection is conducted at a time, the record entry must identify which part of the inspection was accomplished. If the owner maintains separate records for the airframe, powerplants, and propellers, the entry for the inspection must be entered in each record.

Maintenance Record Entries

Maintenance which is performed on aircraft must be properly recorded in the maintenance records. This includes any discrepancies corrected during the course of an inspection. FAR 43.9 governs the recording of aircraft maintenance and states that any maintenance technician who maintains, rebuilds, or alters an aircraft must make an entry containing the following information:

1. A description of the work or some reference to data acceptable to the FAA.

2. The date the work was completed.

3. The maintenance technician's name.

4. If the aircraft is approved for return to service, the signature and the certificate number of the approving maintenance technician.

Notice that one of the main differences between an inspection record entry and a maintenance record entry is that the maintenance entry is not required to include the total time of the aircraft. Although a time reference is not required, it is usually included to provide a future reference of how long the aircraft has operated since the work was accomplished.

Major Repair and Alteration Form (FAA 337)

Major repairs and alterations, in addition to being recorded in the maintenance records according to FAR 43.9, must also be recorded on FAA Form 337. FAA Form 337 serves two purposes; one is to provide owners with a record of major repairs and major alterations to their aircraft, and the other is to provide the FAA with a copy for its records. A copy of a completed FAA Form 337 is shown in Figure 16–13. The maintenance technician who performed or supervised the major repair or alteration prepares the original FAA Form 337 in duplicate.

An airframe and power-plant technician who holds an inspection authorization rating will then further process the form. The IA holder will inspect the alteration or repair to see that it conforms to FAA-approved data, but the IA holder cannot approve the data for major repairs or alterations. The IA holder can, however, inspect to see that the repair or alteration conforms to previously approved FAA data.

Approved data that can be used for major repairs and major alterations may come from one of the following sources:

1. Type Certificate Data Sheets
2. Aircraft Specifications
3. Supplemental Type Certificates (STCs)
4. Airworthiness Directives
5. FAA Field Approval
6. Manufacturer's FAA-Approved Data

DEPARTMENT OF TRANSPORTATION
FEDERAL AVIATION ADMINISTRATION

MAJOR REPAIR AND ALTERATION
(Airframe, Powerplant, Propeller, or Appliance)

Form Approved Budget Bureau No. 04-R060.1
FOR FAA USE ONLY
OFFICE IDENTIFICATION

INSTRUCTIONS: Print or type all entries. See FAR 43.9, FAR 43 Appendix B, and AC 43.9-1 (or subsequent revision thereof) for instructions and disposition of this form.

1. AIRCRAFT	MAKE Cessna	MODEL 182
	SERIAL NO. 15-10521	NATIONALITY AND REGISTRATION MARK N-3763
2. OWNER	NAME (As shown on registration certificate) William Taylor	ADDRESS (As shown on registration certificate) 36 Main Street Cambria, Pennsylvania 15946

3. FOR FAA USE ONLY

4. UNIT IDENTIFICATION

UNIT	MAKE	MODEL	SERIAL NO.	5. TYPE REPAIR	5. TYPE ALTERATION
AIRFRAME	~~~~~~~~~ (As described in item 1 above) ~~~~~~~~~			X	
POWERPLANT					
PROPELLER					
APPLIANCE	TYPE				
	MANUFACTURER				

6. CONFORMITY STATEMENT

A. AGENCY'S NAME AND ADDRESS	B. KIND OF AGENCY	C. CERTIFICATE NO.
George Morris High Street Johnstown, Pennsylvania 15236	X U.S. CERTIFICATED MECHANIC FOREIGN CERTIFICATED MECHANIC CERTIFICATED REPAIR STATION MANUFACTURER	1305888

D. I certify that the repair and/or alteration made to the unit(s) identified in item 4 above and described on the reverse or attachments hereto have been made in accordance with the requirements of Part 43 of the U.S. Federal Aviation Regulations and that the information furnished herein is true and correct to the best of my knowledge.

DATE March 19, 1992	SIGNATURE OF AUTHORIZED INDIVIDUAL George Morris

7. APPROVAL FOR RETURN TO SERVICE

Pursuant to the authority given persons specified below, the unit identified in item 4 was inspected in the manner prescribed by the Administrator of the Federal Aviation Administration and is ☐ APPROVED ☐ REJECTED

BY	FAA FLT. STANDARDS INSPECTOR	MANUFACTURER	X INSPECTION AUTHORIZATION	OTHER (Specify)
	FAA DESIGNEE	REPAIR STATION	CANADIAN DEPARTMENT OF TRANSPORT INSPECTOR OF AIRCRAFT	

DATE OF APPROVAL OR REJECTION April 9, 1992	CERTIFICATE OR DESIGNATION NO. 237412	SIGNATURE OF AUTHORIZED INDIVIDUAL Donald Pauley

FAA Form 337 (7-67)   (8320)

FIGURE 16-13 FAA Form 337.

NOTICE

Weight and balance or operating limitation changes shall be entered in the appropriate aircraft record. An alteration must be compatible with all previous alterations to assure continued conformity with the applicable airworthiness requirements.

8. DESCRIPTION OF WORK ACCOMPLISHED *(If more space is required, attach additional sheets. Identify with aircraft nationality and registration mark and date work completed.)*

1. Removed right wing from aircraft and removed skin from outer 6 feet. Repaired buckled spar 49 inches from tip in accordance with attached photographs and Figure 1 of drawing dated March 6, 1992.

   DATE: March 15, 1992, inspected splice in Item 1 and found it to be in accordance with data indicated. Splice is okay to cover. Inspected internal and external wing assembly for hidden damage and condition.

   Donald Pauley, A&P 237412 IA

2. Primed interior wing structure and replaced skin P/Ns 63-0085, 63-0086, and 63-00878 with same material, 2024-T3, .025 inches thick. Rivet size spacing all the same as original and using procedures in Chapter 2, Section 3, of AC 43.13-1A, dated 1972.

3. Replaced stringers as required and installed 6 splices as per attached drawing and photographs.

4. Installed wing, rigged aileron, and operationally checked in accordance with manufacturer's maintenance manual.

5. No change in weight or balance.

END

☐ ADDITIONAL SHEETS ARE ATTACHED

☆U.S. GOVERNMENT PRINTING OFFICE: 1977–771-021/344

FIGURE 16–13 Continued.

7. Designated Engineering Representative–Approved Data (DER-approved data)

8. Designated Alteration Station–Approved Data (DAS-approved data)

9. Appliance Manufacturer's Manuals

10. AC 43.13-1A may be used as approved data when it is appropriate to the product being repaired, is directly applicable to the repair being made, and is not contrary to the manufacturer's data

After Form 337 is completed, the original copy should be given to the owner, with the duplicate copy forwarded to the local FAA district office within 48 h. It is generally a good idea for the technician performing the work and the IA holder signing Form 337 to retain copies for their files.

MALFUNCTION OR DEFECT REPORT

As was discussed in Chapter 14, the FAA requests the co-operation of maintenance technicians in reporting discrepancies located during inspections. At the conclusion of an inspection, discrepancies identified that affect the airworthiness of an aircraft should be reported on a **Malfunction or Defect Report (M or D report)** and forwarded to the FAA. A copy of a malfunction or defect report is shown in Figure 14-2.

INSPECTION REMINDER

Whenever a maintenance technician completes an inspection of an aircraft, she or he should prepare an inspection reminder, FAA Form 8320-2, and affix it to the aircraft inside the cockpit in a location where it cannot be overlooked. This provides a constant reminder to owners and operators that an inspection will be due on a certain date or at a certain time. Inspection reminder forms are available at all FAA general aviation district offices. A copy of an inspection reminder form is shown in Figure 16–14.

DEPARTMENT OF TRANSPORTATION
FEDERAL AVIATION ADMINISTRATION

INSPECTION REMINDER

The next inspection of this aircraft is required by Federal Aviation Regulation

Section: _____

Hours in
Service: _____

OR

Date due: _____

FAA FORM 8600-1 (4-78)
FORMERLY FAA FORM 8320-2

FIGURE 16–14 Inspection reminder form.

CLEANING AIRCRAFT AND PARTS

A most important factor in the maintenance of an aircraft is cleanliness. Although cleaning is not actually a part of the inspection process, it is an important step in preparing an aircraft for redelivery to the owner as well as being a good preventative-maintenance practice. A film of dirt on the surface of a metal or a painted aircraft will attract moisture and chemical pollutants from the air and permit these materials to react chemically with the surface of the aircraft. The result is oxidation, corrosion, and general deterioration of the surface. Cleanliness is particularly important in areas near the ocean, where the air is salty, or in areas where air pollution is comparatively severe. Regular cleaning and waxing of the exterior surfaces of an aircraft will effectively reduce corrosion and other forms of deterioration. It is desirable to keep an airplane hangared when possible. Since this cannot always be done, the owner or operator of an aircraft should try to keep it as clean as possible.

This section is not intended to describe all the methods for cleaning aircraft and their parts; however, some general principles will be discussed, with particular emphasis on the dangers involved in the use of improper cleaning methods and unsuitable cleaning materials.

Exterior Painted Surfaces

When an airplane is covered with dust, sand, or other types of dirt, no attempt should be made to wipe the surface clean with a dry cloth, no matter how soft the material is. Wiping a dirty surface with a dry material will scratch the metal or paint on the surface because of the abrasiveness of the fine particles of sand or dust. Scratching the surface will destroy the smooth finish and will increase the rate of deterioration, particularly if the pure-aluminum coating on clad material is scratched through to the base metal.

Generally, the painted surfaces can be kept bright by washing them with water and mild soap, rinsing with water, and drying with cloths or a chamois. Harsh or abrasive soaps or detergents, which cause corrosion or scratches, should never be used. Remove stubborn oil and grease with a cloth moistened with Stoddard solvent. Many of the new-type paints being used on aircraft, such as urethane, do not require waxing to keep them bright. However, if desired, the airplane may be waxed with a good automotive wax. Soft cleaning cloths or a chamois should be used to prevent scratches when cleaning or polishing. A heavier coating of wax on the leading surfaces will reduce the abrasion problems in these areas.

Cleaning Plastic Windows

In the cleaning of plastic windshields, it is especially important to avoid wiping a dirty surface with a dry cloth. The first step in cleaning the aircraft and windshield is to flush the surface with clean water, using the bare hand to dislodge any dirt or abrasives. This will lessen the possibility of scratching the surface during the washing procedure. Wash the windshield thoroughly with a mild soap solution, taking care that the water is free from all possible abrasives. A soft cloth, sponge, or chamois may be used to apply the soap solution. Remove oil and grease with a cloth moistened with kerosene, naphtha, or methanol.

Never use gasoline, benzine, alcohol, acetone, anti-ice fluid, lacquer thinner, or glass cleaner to clean the plastic. These materials will attack the plastic and may cause it to craze.

Waxing with a good commercial wax will finish the cleaning job. A thin, even coat of wax, polished out by hand with clean, soft, flannel cloths, will fill in minor scratches and help prevent further scratching.

Cleaning Tires

Tires should be cleaned with an emulsion cleaner or with soap or detergent and water. Petroleum solvents cause the deterioration of natural rubber and therefore should not be applied to rubber tires. Small oil or grease spots can be cleaned off with a cloth dampened with a petroleum solvent, provided that the spot is wiped dry immediately.

Cleaning Deice Boots

To prolong the life of deice boots, they should be washed and serviced on a regular basis. The boots should be kept clean and free from oil, grease, and other solvents which cause rubber to swell and deteriorate. Clean the boots with mild soap and water, then rinse them thoroughly with clean water.

Isopropyl alcohol can be used to remove grime which cannot be removed using soap. When isopropyl alcohol is used for cleaning, wash the area with mild soap and water, then rinse thoroughly with clean water.

Interior Cleaning

To remove dust and loose dirt from the upholstery and carpet, clean the interior regularly with a vacuum cleaner.

Oily spots may be cleaned with a household spot remover, used sparingly. Before using any solvent, read the instructions on the container and test it on an obscure place on the fabric to be cleaned. Never saturate the fabric with a volatile solvent because it may damage the padding and backing materials.

Soiled upholstery and carpet may be cleaned with a foam-type detergent, used according to the manufacturer's instructions. To minimize wetting the fabric, keep the foam as dry as possible and remove it with a vacuum cleaner.

If the airplane is equipped with leather seating, clean the seats by using a soft cloth or a sponge dipped in mild soap suds. The soap suds, used sparingly, will remove traces of dirt and grease. The soap should be removed with a clean, damp cloth.

The plastic trim, headliner, instrument panel, and control knobs need only be wiped off with a damp cloth. Oil and grease on the control wheel and control knobs can be removed with a cloth moistened with Stoddard solvent. Volatile solvents, such as were mentioned in the paragraphs on the care of plastic windshields, must never be used since they soften and craze plastic.

REVIEW QUESTIONS

1. What is the difference between an annual inspection and a 100-h inspection? How are they alike?
2. What condition requires the performance of 100-h inspections on aircraft?
3. How often is an annual inspection required?
4. Why is a checklist required when making an annual or a 100-h inspection?
5. Who may perform a 100-h inspection?
6. Who is authorized to perform an annual inspection?
7. What inspection requirements does a progressive inspection satisfy?
8. Who must supervise a progressive inspection?
9. What inspection-program options are available to the operators of turbine-powered multiengine airplanes?
10. From what sources may a technician select an inspection checklist?
11. What condition requires the performance of an altimeter and static-pressure-system inspection?
12. How often must a transponder be inspected?
13. What are the principal tools used for inspecting an aircraft?
14. What is the purpose of a discrepancy list?
15. What two criteria must be met for an aircraft to be declared airworthy?
16. What should be done if the Airworthiness Certificate cannot be located during the inspection?
17. What precautions should be taken when servicing an oxygen system?
18. What items are required to be checked during the performance of an operational inspection?
19. What are the advantages of a well-organized set of maintenance records?
20. What happens to permanent aircraft records when an aircraft is sold?
21. How long must temporary aircraft records be kept?
22. What items must be included in an inspection record entry?
23. What items must be included in a maintenance record entry?
24. What is the purpose of FAA Form 337?
25. Give the number and distribution of copies required when Form 337 is prepared.

Appendix

27 FLIGHT CONTROLS

00 General
10 Aileron and Tab
20 Rudder/Ruddervator and Tab
30 Elevator and Tab
40 Horizontal Stabilizers/Stabilator
50 Flaps
60 Spoiler, Drag Devices, and Variable Aerodynamic Fairings
70 Gust Lock and Dampener
80 Lift Augmenting

28 FUEL

00 General
10 Storage
20 Distribution/Drain Valves
30 Dump
40 Indicating

29 HYDRAULIC POWER

00 General
10 Main
20 Auxiliary
30 Indicating

30 ICE AND RAIN PROTECTION

00 General
10 Airfoil
20 Air Intakes
30 Pilot and Static
40 Windows and Windshields
50 Antennas and Radomes
60 Propellers/Rotors
70 Water Lines
80 Detection

31 INDICATING/RECORDING SYSTEMS

00 General
10 Unassigned
20 Unassigned
30 Recorders
40 Central Computers
50 Central Warning System

32 LANDING GEAR

00 General
10 Main Gear
20 Nose Gear/Tail Gear
30 Extension and Retraction, Level Switch
40 Wheels and Brakes
50 Steering
60 Position, Warning, and Ground-Safety Switch
70 Supplementary Gear/Skis/Floats

33 LIGHTS

00 General
10 Flight Compartment and Annunciator Panels
20 Passenger Compartments
30 Cargo and Service Compartments
40 Exterior Lighting
50 Emergency Lighting

34 NAVIGATION

00 General
10 Flight-Environment Data
20 Attitude and Direction
30 Landing and Taxiing Aids
40 Independent Position Determining

50 Dependent Position Determining
60 Position Computing

35 OXYGEN

00 General
10 Crew
20 Passenger
30 Portable

36 PNEUMATIC

00 General
10 Distribution
20 Indicating

37 VACUUM/PRESSURE

00 General
10 Distribution
20 Indicating

38 WATER/WASTE

00 General
10 Potable
20 Wash
30 Waste Disposal
40 Air Supply

39 ELECTRICAL/ELECTRONIC PANELS AND MULTIPURPOSE COMPONENTS

00 General
10 Instrument and Control Panels
20 Electrical and Electronic Equipment Racks
30 Electrical and Electronic Junction Boxes
40 Multipurpose Electronic Components
50 Integrated Circuits
60 Printed Circuit-Card Assemblies

49 AIRBORNE AUXILIARY POWER

00 General
10 Power Plant
20 Engine
30 Engine Fuel and Control
40 Ignition/Starting
50 Air
60 Engine Controls
70 Indicating
80 Exhaust
90 Oil

51 STRUCTURES

00 General

52 DOORS

00 General
10 Passenger/Crew
20 Emergency Exit
30 Cargo
40 Service
50 Fixed Interior
60 Entrance Stairs
70 Door Warning
80 Landing Gear

53 FUSELAGE

00 General
10 Main Frame
20 Auxiliary Structure

30 Plates/Skin
40 Attach Fittings
50 Aerodynamic Fairings

54 NACELLES/PYLONS

00 General
10 Main Frame
20 Auxiliary Structure
30 Plates/Skin
40 Attach Fittings
50 Fillets/Fairings

55 STABILIZERS

00 General
10 Horizontal Stabilizers/Stabilator
20 Elevator/Elevon
30 Vertical Stabilizer
40 Rudder/Ruddervator
50 Attach Fittings

56 WINDOWS

00 General
10 Flight Compartment
20 Cabin
30 Door
40 Inspection and Observation

57 WINGS

00 General
10 Main Frame
20 Auxiliary Structure
30 Plates/Skin
40 Attach Fittings
50 Flight Surfaces

61 PROPELLERS

00 General
10 Propeller Assembly
20 Controlling
30 Braking
40 Indicating

65 ROTORS

00 General
10 Main Rotor
20 Antitorque Rotor Assembly
30 Accessory Driving
40 Controlling
50 Braking
60 Indicating

71 POWER PLANT

00 General
10 Cowling
20 Mounts
30 Fireseals and Shrouds
40 Attach Fittings
50 Electrical Harness
60 Engine Air Intakes
70 Engine Drains

72 (T) TURBINE/TURBOPROP

00 General
10 Reduction Gear and Shaft Section
20 Air-Inlet Section
30 Compressor Section
40 Combustion Section

50 Turbine Section
60 Accessory Drives
70 Bypass Section

73 ENGINE FUEL AND CONTROL

00 General
10 Distribution
20 Controlling/Governing
30 Indicating

74 IGNITION

00 General
10 Electrical Power Supply
20 Distribution
30 Switching

75 BLEED AIR

00 General
10 Engine Anti-Icing
20 Accessory Cooling
30 Compressor Control
40 Indicating

76 ENGINE CONTROLS

00 General
10 Power Control
20 Emergency Shutdown

77 ENGINE INDICATING

00 General
10 Power
20 Temperature
30 Analyzers

78 ENGINE EXHAUST

00 General
10 Collector/Nozzle
20 Noise Suppressor
30 Thrust Reverser
40 Supplementary Air

79 ENGINE OIL

00 General
10 Storage (Dry Sump)
20 Distribution
30 Indicating

80 STARTING

00 General
10 Cranking

81 TURBINES (RECIPROCATING ENG)

00 General
10 Power Recovery
20 Turbosupercharger

82 WATER INJECTION

00 General
10 Storage
20 Distribution
30 Dumping and Purging
40 Indicating

83 REMOTE GEAR BOXES (ENG DR)

00 General

SYSTÈME INTERNATIONAL D'UNITÉS (INTERNATIONAL-METRIC SYSTEM)

Numerical Prefixes

Exponential Expression	Decimal Equivalent	Prefix	Symbol
10^9	1 000 000 000	giga	G
10^6	1 000 000	mega	M
10^3	1 000	kilo	k
10^2	100	hecto	h
10	10	deka	da
10^{-1}	0.1	deci	d
10^{-2}	0.01	centi	c
10^{-3}	0.001	milli	m
10^{-6}	0.000 001	micro	μ

Common Units of the SI-Metric System

Parameter Measured	Unit	Description
Area	square meter (m^2)	10.76 ft^2
	square centimeter (cm^2)	0.1550 in^2
Density	kilograms per cubic meter (kg/m^3)	0.0624 lb/ft^3
Distance	meter (m)	39.37 in
	kilometer (km)	0.6214 mi
	centimeter (cm)	0.3937 in
Energy (work)	erg	1 dyn of force acting through 1 cm
	joule (J)	10^7 erg
		0.7377 ft-lb
Force	dyne (dyn)	1 dyn will accelerate 1 g 1 cm/s^2
	newton (N)	0.2248 lb
		100 000 dyn
Power	watt (W)	1 J/s
		0.001 34 hp
	kilowatt (kW)	1.340 48 hp
Pressure	pascal (Pa)	1 N/m^2
		0.000 145 psi
	kilopascal (kPa)	0.145 032 psi
	megapascal (MPa)	1000 kPa
	centimeter of mercury (cmHg)	0.1934 psi
Torque	newton-meter (N-m)	A force of 1 N with an arm of 1 m
		0.7375 lb-ft
		8.85 lb-in
Velocity	kilometers per hour (km/h)	0.6214 mph
		0.9113 ft/s
	meters per second (m/s)	3.281 ft/s
		2.237 mph
Volume	cubic centimeters (cm^3)	0.061 02 in^3
	liter (L)	1000 cm^3
	cubic meter (m^3)	1 000 000 cm^3
		35.31 ft^3
Weight (mass)	gram (g)	Weight (mass) of 1 cm^3 of water at its greatest density
		0.035 27 oz avdp
	kilogram (kg)	1000 g
		2.205 lb avdp

	English System	SI-Metric System
Gravity (g)	32.174 ft/s^2	9.806 65 m/s^2
Absolute zero	–459.7°F	–273.1°C
Horsepower	550 ft-lb/s	76.04 kg-m/s
		746W
π (pi)	3.141 59	
Density (ρ_0)	0.002 37 slug/ft^3	
	1.221 kg/m^3	
STANDARD ATMOSPHERE AT SEA LEVEL:		
Temperature	59°F	15°C
Absolute temperature	518.4°R	288.2°K
Specific weight (gρ_0)	0.076 51 lb/ft^3	1.2255 kg/m^3
Pressure (P_0)	29.92 inHg	760 mmHg
	2 116 lb/ft^3	10 332 kg/m^2
		101.33 kPa
STANDARD VALUES AT ALTITUDE:		
Isothermal level (Z_i)	35 332 ft	10 769 m
Isothermal temperature (t_i)	–69.7°F	–56.5°C
Temperature gradient (a)	0.003 566°F/ft	0.0065°C/m

Laws of Gases

BOYLE'S LAW: When the temperature of a given mass of gas remains constant, the volume of the gas varies inversely at its pressure.

$$\frac{V_1}{V_2} = \frac{P_2}{P_1}$$

CHARLES' LAW: When the pressure of a given mass of gas is kept constant, the volume of the gas is approximately proportional to its absolute temperature.

$$\frac{V_1}{V_2} = \frac{T_1}{T_2}$$

GENERAL LAW

$$\frac{P_1 V_1}{T_1} = \frac{P_2 V_2}{T_2}$$

Standard Atmosphere

Altitude, ft	Temperature t		Pressure P			Density	
	°F	°C	inHg	cmHg	kPa	ρ	ρ/ρ_0
0	59.0	15.0	29.920	76.00	101.33	0.002 378	1.0000
1 000	55.4	13.0	38.860	73.30	97.74	0.002 309	0.9710
2 000	51.9	11.0	27.820	70.66	94.22	0.002 242	0.9428
3 000	48.3	9.1	26.810	68.10	90.80	0.002 176	0.9151
4 000	44.7	7.1	25.840	65.63	87.51	0.002 112	0.8881
5 000	41.2	5.1	24.890	63.22	84.29	0.002 049	0.8616
6 000	37.6	3.1	23.980	60.91	81.21	0.001 988	0.8358
7 000	34.0	1.1	23.090	58.65	78.20	0.001 928	0.8106
8 000	30.5	–0.8	22.220	56.44	75.25	0.001 869	0.7859
9 000	26.9	–2.8	21.380	54.31	72.41	0.001 812	0.7619
10 000	23.3	–4.8	20.580	52.27	69.70	0.001 756	0.7384
11 000	19.8	–6.8	19.790	50.27	67.02	0.001 702	0.7154
12 000	16.2	–8.8	19.030	48.34	64.45	0.001 648	0.6931
13 000	12.6	–10.8	18.290	46.46	61.94	0.001 596	0.6712
14 000	9.1	–12.7	17.570	44.63	59.50	0.001 545	0.6499
15 000	5.5	–14.7	16.880	42.88	57.17	0.001 496	0.6291

Standard Atmosphere *(Continued)*

Altitude, ft	Temperature t °F	Temperature t °C	Pressure P inHg	Pressure P cmHg	Pressure P kPa	Density ρ	Density ρ/ρ_0
16 000	1.9	−16.7	16.210	41.47	54.90	0.001 448	0.6088
17 000	−1.6	−18.7	15.560	39.52	52.70	0.001 401	0.5891
18 000	−5.2	−20.7	14.940	37.95	50.60	0.001 355	0.5698
19 000	−8.8	−22.6	14.330	36.40	48.87	0.001 311	0.5509
20 000	−12.3	−24.6	13.750	34.93	46.57	0.001 267	0.5327
21 000	−15.9	−26.6	13.180	33.48	44.64	0.001 225	0.5148
22 000	−19.5	−28.6	12.630	32.08	42.77	0.001 183	0.4974
23 000	−23.0	−30.6	12.100	30.73	40.98	0.001 143	0.4805
24 000	−26.6	−32.5	11.590	29.44	39.25	0.001 103	0.4640
25 000	−30.2	−34.5	11.100	28.19	37.59	0.001 065	0.4480
26 000	−33.7	−36.5	10.620	26.97	35.97	0.001 028	0.4323
27 000	−37.3	−38.5	10.160	25.81	34.41	0.000 992	0.4171
28 000	−40.9	−40.5	9.720	24.69	32.92	0.000 957	0.4023
29 000	−44.4	−42.5	9.293	23.60	31.47	0.000 922	0.3879
30 000	−48.0	−44.4	8.880	22.56	30.07	0.000 889	0.3740
31 000	−51.6	−46.4	8.483	21.55	28.73	0.000 857	0.3603
32 000	−55.1	−48.4	8.101	20.58	27.44	0.000 826	0.3472
33 000	−58.7	−50.4	7.732	19.64	26.19	0.000 795	0.3343
34 000	−62.2	−52.4	7.377	18.74	24.98	0.000 765	0.3218
35 000	−65.8	−54.3	7.036	17.87	23.83	0.000 736	0.3098
36 000	−69.4	−56.3	6.708	17.04	22.72	0.000 704	0.2962
37 000	−69.7	−56.5	6.395	16.24	21.62	0.000 671	0.2824
38 000	−69.7	−56.5	6.096	15.48	20.65	0.000 640	0.2692
39 000	−69.7	−56.5	5.812	14.76	19.68	0.000 610	0.2566
40 000	−69.7	−56.5	5.541	14.07	18.77	0.000 582	0.2447
41 000	−69.7	−56.5	5.283	13.42	17.89	0.000 554	0.2332
42 000	−69.7	−56.5	5.036	12.79	17.06	0.000 529	0.2224
43 000	−69.7	−56.5	4.802	12.20	16.26	0.000 504	0.2120
44 000	−69.7	−56.5	4.578	11.63	15.50	0.000 481	0.2021
45 000	−69.7	−56.5	4.364	11.08	14.78	0.000 459	0.1926
46 000	−69.7	−56.5	4.160	10.57	14.09	0.000 437	0.1837
47 000	−69.7	−56.5,	3.966	10.07	13.43	0.000 417	0.1751
48 000	−69.7	−56.5	3.781	9.60	12.81	0.000 397	0.1669
49 000	−69.7	−56.5	3.604	9.15	12.21	0.000 379	0.1591
50 000	−69.7	−56.5	3.436	8.73	11.64	0.000 361	0.1517

Greek Alphabet

Name	Capital	Lowercase	Use	Name	Capital	Lowercase	Use
Alpha	A	α	Angles	Xi	Ξ	ξ	
Beta	B	β		Omicron	O	o	
Gamma	Γ	γ	Ratio of specific heats	Pi	Π	π	Ratio of circumference to diameter (3.1416)
Delta	Δ	δ	Relative absolute pressure				
Epsilon	E	ε	Expansion ratio, surface emissivity	Rho	P	ρ	Mass density
				Sigma	Σ	σ	Capital: sign of summation; lowercase: relative density
Zeta	Z	ζ					
Eta	H	η	Coefficient of kinematic viscosity, efficiency				
				Tau	T	τ	
Theta	Θ	θ		Upsilon	Υ	υ	Kinematic viscosity
Iota	I	ι		Phi	Φ	φ	Capital: relative viscosity; lowercase: phase
Kappa	K	κ					
Lambda	Λ	λ	Capital: wing sweep angle; lowercase: taper rate, wavelength	Chi	X	χ	
				Psi	Ψ	ψ	
				Omega	Ω	ω	Capital: ohms; lowercase: specific weight
Mu	M	μ	Absolute viscosity				
Nu	N	ν					

Conversion Factors

Multiply	By	To Obtain
Atmospheres (atm)	76.0	cmHg at 0°C
	29.92	inHg at 0°C
	33.90	ftH$_2$O at 4°C
	1.033	kg/cm^2
	14.696	psi
	101.33	kPa
	2116.0	psf
	1.0133	bar, hectopieze
Bars (bar)	75.01	cmHg at 0°C
	29.53	inHg at 0°C
	14.50	psi
	100.0	kPa
Barns (b)	10^{-24}	cm^2 (nuclear cross section)
British thermal units (Btu)	778.26	ft-1b
	3.930×10^{-4}	hp-h
	2.931×10^{-4}	kW-h
	2.520×10^{-1}	kg-cal
	1.076×10^{2}	kg-m
	1055	J
Btu/s	1055	W
Centimeters (cm)	0.3937	in
	3.281×10^{-2}	ft
cmHg	5.354	inH$_2$O at 4°C
	4.460×10^{-1}	ftH$_2$O at 4°C
	1.934×10^{-1}	psi
	27.85	psf
	135.95	kg/m^2
	1.333	kPa
cm/s	3.281×10^{-2}	ft/s
	2.237×10^{-2}	mph
	3.60×10^{-2}	km/h
cm^2	1.550×10^{-1}	in^2
	1.076×10^{-3}	ft^2
	10^{4}	m
cm^3	10^{-3}	L
	6.102×10^{-2}	in^3
	2.642×10^{-4}	U.S. gal
centipoise (cP)	6.72×10^{-4}	lb/s-ft
	3.60	kg/h-m
circular mils (cmil)	7.854×10^{-7}	in^2
	5.067×10^{-4}	mm^2
	7.854×10^{-1}	mil^2
curies (Ci)	3.7×10^{10}	disintegrations/s
degrees (arc) (°)	1.745×10^{-2}	rad
dynes (dyn)	1.020×10^{-3}	g
	2.248×10^{-6}	lb
	10^{5}	N
	7.233×10^{-5}	poundals
electron volts (eV)	1.602×10^{-12}	ergs
ergs (*or* dyn-cm)	9.478×10^{-11}	Btu
	6.2×10^{11}	eV
	7.376×10^{-8}	ft-lb
	1.020×10^{-3}	g-cm
	10^{-7}	Joule
	2.388×10^{-11}	kg-cal
feet (ft)	3.048×10^{-1}	m
	3.333×10^{-1}	yd
	1.894×10^{-4}	mi
	1.646×10^{-4}	nm
ft^2	929.0	cm^2
	144.0	in^2
	9.294×10^{-2}	m^2
	1.111×10^{-1}	yd^2
	2.296×10^{-5}	acres
ft^3	2.832×10^{4}	cm^3
	1728	in^3
	3.704×10^{-2}	yd^3
	7.481	U.S. gal
	28.32	L
	2.83×10^{-2}	m^3
ft^3/min	4.719×10^{-1}	L/s
	2.832×10^{-2}	m^3/min
ft^{3}H$_2$O	62.428	lb
ftH$_2$O at 4°C	2.950×10^{-2}	atm
	4.335×10^{-1}	psi
	62.43	psf
	3.048×10^{2}	kg/m^2
	2.999	kPa
	8.826×10^{-1}	inHg at 0°C
	2.240	cmHg at 0°C
ft/min	1.136×10^{-2}	mph
	1.829×10^{-2}	km/h
	5.080×10^{-1}	cm/s
ft/s	6.818×10^{-1}	mph
	1.097	km/h
	30.48	cm/s
	5.925×10^{-1}	kn
ft-lb	1.383×10^{-1}	kg-m
	1.356	N-m *or* J
	1.285×10^{-3}	Btu
	3.776×10^{-7}	kW-h
ft-lb/min	3.030×10^{-5}	hp
	4.06×10^{-5}	kW
ft-lb/s	1.818×10^{-3}	hp
	1.356×10^{-3}	kW
fluidounce (fluid oz)	8	drams
	29.6	cm^3
gallon (gal), Imperial	277.4	in^3
	1.201	U.S. gal
	4.546	L
gal, U.S., dry	268.8	in^3
	1.556×10^{-1}	ft^3
	1.164	U.S. gal, liquid
	4.405	L
gal, U.S., liquid	231.0	in^3
	1.337×10^{-1}	ft^3
	3.785	L
	8.327×10^{-1}	imperial gal
	128.0	fluid oz
gigagram (Gg)	10^{6}	kg
grains	6.480×10^{-2}	g
grams (g)	15.43	grains
	3.527×10^{-2}	oz avdp
	2.205×10^{-3}	lb avdp
	1000	mg
	10^{-3}	kg
	980.67	dyn
g-cal	3.969×10^{-3}	Btu
g of U^{235} fissioned	23 000	kW-h heat generated
g/cm	0.1	kg/m
	6.721×10^{-2}	lb/ft
	5.601×10^{-3}	lb/in
g/cm^3	1000	kg/m^3
	62.43	lb/ft^3
hectare	10^{4}	m^2
	2.471	acre
hectopieze	29.53	inHg
	75.01	cmHg
horsepower (hp)	33 000	ft-lb/min
	550	ft-lb/s
	76.04	m-kg/s
	1.014	metric hp

Multiply	By	To Obtain
	7.457×10^{-1}	kW
	745.7	W
	7.068×10^{-1}	Btu-s
hp, metric	75.0	m-kg/s
	9.863×10^{-1}	hp
	7.355×10^{-1}	kW
	6.971×10^{-1}	Btu-s
hp-h	2.545×10^{3}	Btu
	1.98×10^{6}	ft-lb
	2.737×10^{5}	m-kg
inch (in)	2.54	cm
	83.33×10^{-3}	ft
inHg at 0°C	40.66	inAcBr$_4$
	3.342×10^{-2}	atm
	13.60	inH$_2$O at 4°C
	1.133	ftH$_2$O
	4.912×10^{-1}	psi
	70.73	psf
	3.386	kPa
	3.453×10^{2}	kg/m^2
inH$_2$O at 4°C	2.99	AcBr$_4$
	7.355×10^{-2}	inHg at 0°C
	1.868×10^{-1}	cmHg at 0°C
	3.613×10^{-2}	psi
	2.49×10^{-1}	kPa
	25.40	kg/m^2
in^2	6.452	cm^2
	6.94×10^{-3}	ft^2
	6.452×10^{-4}	m^2
in^3	16.39	cm^3
	1.639×10^{-2}	L
	4.329×10^{-3}	U.S. gal
	1.639×10^{-5}	m^3
	1.732×10^{-2}	qt
joules (J)	9.480×10^{-4}	Btu
	7.375×10^{-1}	ft-lb
	2.389×10^{-4}	kg-cal
	1.020×10^{-1}	kg-m
	2.778×10^{-4}	Watt-h
	3.725×10^{-7}	hp-h
	10^{7}	ergs
kilograms (kg)	2.205	lb
	9.808	N
	35.27	oz
	10^{3}	g
kg-cal	3.9685	Btu
	3087	ft-lb
	4.269×10^{2}	kg-m
	4.1859×10^{3}	J
kg-m	7.233	ft-lb
	9.809	J
kg/m^3	62.43×10^{-3}	lb/ft^3
	10^{-3}	g/cm^3
kg/cm^2	14.22	psi
	2.048×10^{3}	psf
	28.96	inHg at 0°C
	32.8	ftH$_2$O at 4°C
	98.077	kPa
kilometers (km)	3.281×10^{3}	ft
	6.214×10^{-1}	mi
	5.400×10^{-1}	nmi
	10^{5}	cm
km/h	9.113×10^{-1}	ft/s
	5.396×10^{-1}	kn
	6.214×10^{-1}	mph
	2.778×10^{-1}	m/s
kPa	1000	N/m^2
	14.503×10^{-2}	psi
kilowatts (kW)	9.480×10^{-1}	Btu/s

Multiply	By	To Obtain
	7.376×10^{2}	ft-lb/s
	1.341	hp
	2.839×10^{-1}	kg-cal/s
kW-h heat generated	4.35×10^{-5}	g U^{235} fissioned
knots (kn)	1.0	nmi/h
	1.688	ft/s
	1.151	mph
	1.852	km/h
	5.148×10^{-1}	m/s
liters (L)	10^{2}	cm^3
	61.03	in^3
	3.532×10^{-2}	ft^3
	2.642×10^{-1}	U.S. gal
	2.200×10^{-1}	imperial gal
	1.057	qt
megagrams (Mg)	10^{3}	kg
megapascals (MPa)	10^{3}	kPa
meters (m)	39.37	in
	3.281	ft
	1.094	yd
	6.214×10^{-4}	mi
m/s	3.281	ft/s
	2.237	mph
	3.600	km/h
m^2	10.76	ft^2
	1.196	yd^2
m^3	35.31	ft^3
	264.17	gal (U.S.)
	61023	in^3
	1.308	yd^3
microamperes (μA)	6.24×10^{12}	unit charges/s
microns (μm)	3.937×10^{5}	in
	10^{-6}	m
miles (mi)	5280	ft
	1.609	km
	8.690×10^{-1}	nmi
mph	1.467	ft/s
	4.470×10^{-1}	m/s
	1.609	km/h
	8.690×10^{-1}	kn
millibars (mbar)	2.953×10^{-2}	inHg at 0°C
	100.0×10^{-3}	kPa
nautical miles (nmi)	6076.1	ft
	1.852	km
	1.151	mi
	1852	m
newtons (N)	10^{5}	dyn
	2.248×10^{-1}	lb
ounces (oz) avdp	6.250×10^{-2}	lb avdp
	28.35	g
	4.375×10^{2}	grains
oz, fluid	29.57	cm^3
	1.805	in^3
pascals (Pa)	1.0	N/m^2
	14.503×10^{-5}	psi
	29.5247×10^{-5}	inHg at 0°C
	10^{-1}	bar
	100	mbar
pounds (lb)	453.6	g
	7000	grains
	16.0	oz avdp
	4.448	N
	3.108×10^{-2}	slugs
lb/ft^3	16.02	kg/m^3
lb/in^3	1728	lb/ft^3
	27.68	gm/cm^3

Multiply	By	To Obtain	Multiply	By	To Obtain
psi	2.036	inHg at 0°C	tons	2×10^3	lb
	2.307	ftH$_2$O at 4°C		907.2	kg
	6.805×10^{-2}	atm	tons, metric	10^3	kg
	7.031×10^2	kg/m^2		2.205×10^3	lb
	6.895	kPa	unit charges/s	1.6×10^{-13}	µA
radians (rad)	57.30	°(arc)	watts (W)	9.481×10^{-4}	Btu/s
rad/s	57.30	°/s		1.340×10^{-3}	hp
	15.92×10^{-2}	rev/s	yards (yd)	3.0	ft
	9.549	rpm		36.0	in
revolutions (rev)	6.283	rad		9.144×10^{-1}	m
rpm	1.047×10^{-1}	rad/s	yd^2	9.0	ft^2
slugs	14.59	kg		1.296×10^3	in^2
	32.18	lb		8.361×10^{-1}	m^2
stokes	10^{-4}	m^2/s	yd^3	27.0	ft^3
stones	14	lb		2.022×10^2	gal (U.S.)
	6.35	kg		7.646×10^{-1}	m^3

Trigonometric Functions: Degrees, Radians

Deg	Rad	Sin	Cos	Tan	Cot	Sec	Csc		
0.0	0.0000	0.000 00	1.0000	0.000 00	—	1.000	—	1.5708	90.0
.1	0.0017	0.001 75	1.0000	0.001 75	573.0	1.000	573.0	1.5691	.9
.2	0.0035	0.003 49	1.0000	0.003 49	286.5	1.000	286.5	1.5673	.8
.3	0.0052	0.005 24	1.0000	0.005 24	191.0	1.000	191.0	1.5656	.7
.4	0.0070	0.006 98	1.0000	0.006 98	143.2	1.000	143.2	1.5638	.6
.5	0.0087	0.008 73	1.0000	0.008 73	114.6	1.000	114.6	1.5621	.5
.6	0.0105	0.010 47	0.9999	0.010 47	95.49	1.000	95.49	1.5603	.4
.7	0.0122	0.012 22	0.9999	0.012 22	81.85	1.000	81.85	1.5586	.3
.8	0.0139	0.013 96	0.9999	0.013 96	71.62	1.000	71.62	1.5568	.2
.9	0.0157	0.015 71	0.9999	0.015 71	63.66	1.000	63.67	1.5551	.1
1.0	0.0175	0.017 45	0.9998	0.017 46	57.29	1.000	57.30	1.5533	89.0
.1	0.0192	0.019 20	0.9998	0.019 20	52.08	1.000	52.09	1.5516	.9
.2	0.0209	0.020 94	0.9998	0.020 95	47.74	1.000	47.75	1.5499	.8
.3	0.0227	0.022 69	0.9997	0.022 69	44.07	1.000	44.08	1.5481	.7
.4	0.0244	0.024 43	0.9997	0.024 44	40.92	1.000	40.93	1.5464	.6
.5	0.0262	0.026 18	0.9997	0.026 19	38.19	1.000	38.20	1.5446	.5
.6	0.0279	0.027 92	0.9996	0.027 93	35.80	1.000	35.82	1.5429	.4
.7	0.0297	0.029 67	0.9996	0.029 68	33.69	1.000	33.71	1.5411	.3
.8	0.0314	0.031 41	0.9995	0.031 43	31.82	1.000	31.84	1.5394	.2
.9	0.0332	0.033 16	0.9995	0.033 17	30.14	1.001	30.16	1.5376	.1
2.0	0.0349	0.034 90	0.9994	0.034 92	28.64	1.001	28.65	1.5359	88.0
.1	0.0367	0.036 64	0.9993	0.036 67	27.27	1.001	27.29	1.5341	.9
.2	0.0384	0.038 39	0.9993	0.038 42	26.03	1.001	26.05	1.5324	.8
.3	0.0401	0.040 13	0.9992	0.040 16	24.90	1.001	24.92	1.5307	.7
.4	0.0419	0.041 88	0.9991	0.041 91	23.86	1.001	23.88	1.5289	.6
.5	0.0436	0.043 62	0.9990	0.043 66	22.90	1.001	22.93	1.5272	.5
.6	0.0454	0.045 36	0.9990	0.045 41	22.02	1.001	22.04	1.5254	.4
.7	0.0471	0.047 11	0.9989	0.047 16	21.20	1.001	21.23	1.5237	.3
.8	0.0489	0.048 85	0.9988	0.048 91	20.45	1.001	20.47	1.5219	.2
.9	0.0506	0.050 59	0.9987	0.050 66	19.74	1.001	19.77	1.5202	.1
		Cos	Sin	Cot	Tan	Csc	Sec	Rad	Deg

Deg	Rad	Sin	Cos	Tan	Cot	Sec	Csc		
3.0	0.0524	0.052 34	0.9986	0.052 41	19.08	1.001	19.11	1.5184	87.0
.1	0.0541	0.054 08	0.9985	0.054 16	18.46	1.001	18.49	1.5167	.9
.2	0.0559	0.055 82	0.9984	0.055 91	17.89	1.002	17.91	1.5149	.8
.3	0.0576	0.057 56	0.9983	0.057 66	17.34	1.002	17.37	1.5132	.7
.4	0.0593	0.059 31	0.9982	0.059 41	16.83	1.002	16.86	1.5115	.6
.5	0.0611	0.061 05	0.9981	0.061 16	16.35	1.002	16.38	1.5097	.5
.6	0.0628	0.062 79	0.9980	0.062 91	15.90	1.002	15.93	1.5080	.4
.7	0.0645	0.064 53	0.9979	0.064 67	15.46	1.002	15.50	1.5062	.3
.8	0.0663	0.066 27	0.9978	0.066 42	15.06	1.002	15.09	1.5045	.2
.9	0.0681	0.068 02	0.9977	0.068 17	14.67	1.002	14.70	1.5027	.1
4.0	0.0698	0.069 76	0.9976	0.069 93	14.30	1.002	14.34	1.5010	86.0
.1	0.0716	0.071 50	0.9974	0.071 68	13.95	1.003	13.99	1.4992	.9
.2	0.0733	0.073 24	0.9973	0.073 44	13.62	1.003	13.65	1.4975	.8
.3	0.0750	0.074 98	0.9972	0.075 19	13.30	1.003	13.34	1.4957	.7
.4	0.0768	0.076 72	0.9971	0.076 95	13.00	1.003	13.03	1.4940	.6
.5	0.0785	0.078 46	0.9969	0.078 70	12.71	1.003	12.75	1.4923	.5
.6	0.0803	0.080 20	0.9968	0.080 46	12.43	1.003	12.47	1.4905	.4
.7	0.0820	0.081 94	0.9966	0.082 21	12.16	1.003	12.20	1.4888	.3
.8	0.0838	0.083 68	0.9965	0.083 97	11.91	1.004	11.95	1.4870	.2
.9	0.0855	0.085 42	0.9963	0.085 73	11.66	1.004	11.71	1.4853	.1
5.0	0.0873	0.087 16	0.9962	0.087 49	11.43	1.004	11.47	1.4835	85.0
.1	0.0890	0.088 89	0.9960	0.089 25	11.20	1.004	11.25	1.4818	.9
.2	0.0908	0.090 63	0.9959	0.091 01	10.99	1.004	11.03	1.4800	.8
.3	0.0925	0.092 37	0.9957	0.092 77	10.78	1.004	10.83	1.4783	.7
.4	0.0942	0.094 11	0.9956	0.094 53	10.58	1.004	10.63	1.4765	.6
.5	0.0960	0.095 85	0.9954	0.096 29	10.39	1.005	10.43	1.4748	.5
.6	0.0977	0.097 58	0.9952	0.098 05	10.20	1.005	10.25	1.4731	.4
.7	0.0995	0.099 32	0.9951	0.099 81	10.02	1.005	10.07	1.4713	.3
.8	0.1012	0.101 1	0.9949	0.101 6	9.845	1.005	9.896	1.4696	.2
.9	0.1030	0.102 8	0.9947	0.103 3	9.677	1.005	9.728	1.4678	.1
6.0	0.1047	0.104 5	0.9945	0.105 1	9.514	1.006	9.567	1.4661	84.0
.1	0.1065	0.106 3	0.9943	0.106 9	9.357	1.006	9.411	1.4643	.9
.2	0.1082	0.108 0	0.9942	0.108 6	9.205	1.006	9.259	1.4626	.8
.3	0.1100	0.109 7	0.9940	0.110 4	9.058	1.006	9.113	1.4608	.7
.4	0.1117	0.111 5	0.9938	0.112 2	8.915	1.006	8.971	1.4591	.6
.5	0.1134	0.113 2	0.9936	0.113 9	8.777	1.006	8.834	1.4574	.5
.6	0.1152	0.114 9	0.9934	0.115 7	8.643	1.007	8.700	1.4556	.4
.7	0.1169	0.116 7	0.9932	0.117 5	8.513	1.007	8.571	1.4539	.3
.8	0.1187	0.118 4	0.9930	0.119 2	8.386	1.007	8.446	1.4521	.2
.9	0.1204	0.120 1	0.9928	0.121 0	8.264	1.007	8.324	1.4504	.1
7.0	0.1222	0.121 9	0.9925	0.122 8	8.144	1.008	8.206	1.4486	83.0
.1	0.1239	0.123 6	0.9923	0.124 6	8.028	1.008	8.091	1.4469	.9
.2	0.1257	0.125 3	0.9921	0.126 3	7.916	1.008	7.979	1.4451	.8
.3	0.1274	0.127 1	0.9919	0.128 1	7.806	1.008	7.870	1.4434	.7
.4	0.1292	0.128 8	0.9917	0.129 9	7.700	1.008	7.764	1.4416	.6
.5	0.1309	0.130 5	0.9914	0.131 7	7.596	1.009	7.661	1.4399	.5
.6	0.1326	0.132 3	0.9912	0.133 4	7.495	1.009	7.561	1.4382	.4
.7	0.1344	0.134 0	0.9910	0.135 2	7.396	1.009	7.463	1.4364	.3
.8	0.1361	0.135 7	0.9907	0.137 0	7.300	1.009	7.368	1.4347	.2
.9	0.1379	0.137 4	0.9905	0.138 8	7.207	1.010	7.276	1.4329	.1
8.0	0.1396	0.139 2	0.9903	0.140 5	7.115	1.010	7.185	1.4312	82.0
.1	0.1414	0.140 9	0.9900	0.142 3	7.026	1.010	7.097	1.4294	.9
.2	0.1431	0.142 6	0.9898	0.144 1	6.940	1.010	7.011	1.4277	.8
.3	0.1449	0.144 4	0.9895	0.145 9	6.855	1.011	6.927	1.4259	.7
.4	0.1466	0.146 1	0.9893	0.147 7	6.772	1.011	6.845	1.4242	.6
.5	0.1484	0.147 8	0.9890	0.149 5	6.691	1.011	6.765	1.4224	.5
.6	0.1501	0.149 5	0.9888	0.151 2	6.612	1.011	6.687	1.4207	.4
.7	0.1518	0.151 3	0.9885	0.153 0	6.535	1.012	6.611	1.4190	.3
.8	0.1536	0.153 0	0.9882	0.154 8	6.460	1.012	6.537	1.4172	.2
.9	0.1553	0.154 7	0.9880	0.156 6	6.386	1.012	6.464	1.4155	.1
		Cos	Sin	Cot	Tan	Csc	Sec	Rad	Deg

Deg	Rad	Sin	Cos	Tan	Cot	Sec	Csc		
9.0	0.1571	0.156 4	0.9877	0.158 4	6.314	1.012	6.392	1.4137	81.0
.1	0.1588	0.158 2	0.9874	0.160 2	6.243	1.013	6.323	1.4120	.9
.2	0.1606	0.159 9	0.9871	0.162 0	6.174	1.013	6.255	1.4102	.8
.3	0.1623	0.161 6	0.9869	0.163 8	6.107	1.013	6.188	1.4085	.7
.4	0.1641	0.163 3	0.9866	0.165 5	6.041	1.014	6.123	1.4067	.6
.5	0.1658	0.165 0	0.9863	0.167 3	5.976	1.014	6.059	1.4050	.5
.6	0.1676	0.166 8	0.9860	0.169 1	5.912	1.014	5.996	1.4032	.4
.7	0.1693	0.168 5	0.9857	0.170 9	5.850	1.015	5.935	1.4015	.3
.8	0.1710	0.170 2	0.9854	0.172 7	5.789	1.015	5.875	1.3998	.2
.9	0.1728	0.171 9	0.9851	0.174 5	5.730	1.015	5.816	1.3980	.1
10.0	0.1745	0.173 6	0.9848	0.176 3	5.671	1.015	5.759	1.3963	80.0
.1	0.1763	0.175 4	0.9845	0.178 1	5.614	1.016	5.702	1.3945	.9
.2	0.1780	0.177 1	0.9842	0.179 9	5.558	1.016	5.647	1.3928	.8
.3	0.1798	0.178 8	0.9839	0.181 7	5.503	1.016	5.593	1.3910	.7
.4	0.1815	0.180 5	0.9836	0.183 5	5.449	1.017	5.540	1.3893	.6
.5	0.1833	0.182 2	0.9833	0.185 3	5.396	1.017	5.487	1.3875	.5
.6	0.1850	0.184 0	0.9829	0.187 1	5.343	1.017	5.436	1.3858	.4
.7	0.1868	0.185 7	0.9826	0.189 0	5.292	1.018	5.386	1.3840	.3
.8	0.1885	0.187 4	0.9823	0.190 8	5.242	1.018	5.337	1.3823	.2
.9	0.1902	0.189 1	0.9820	0.192 6	5.193	1.018	5.288	1.3806	.1
11.0	0.1920	0.190 8	0.9816	0.194 4	5.145	1.019	5.241	1.3788	79.0
.1	0.1937	0.192 5	0.9813	0.196 2	5.097	1.019	5.194	1.3771	.9
.2	0.1955	0.194 2	0.9810	0.198 0	5.050	1.019	5.148	1.3753	.8
.3	0.1972	0.195 9	0.9806	0.199 8	5.005	1.020	5.103	1.3736	.7
.4	0.1990	0.197 7	0.9803	0.201 6	4.959	1.020	5.059	1.3718	.6
.5	0.2007	0.199 4	0.9799	0.203 5	4.915	1.020	5.016	1.3701	.5
.6	0.2025	0.201 1	0.9796	0.205 3	4.872	1.021	4.973	1.3683	.4
.7	0.2042	0.202 8	0.9792	0.207 1	4.829	1.021	4.931	1.3666	.3
.8	0.2059	0.204 5	0.9789	0.208 9	4.787	1.022	4.890	1.3648	.2
.9	0.2077	0.206 2	0.9785	0.210 7	4.745	1.022	4.850	1.3641	.1
12.0	0.2094	0.207 9	0.9781	0.212 6	4.705	1.022	4.810	1.3614	78.0
.1	0.2112	0.209 6	0.9778	0.214 4	4.665	1.023	4.771	1.3596	.9
.2	0.2129	0.211 3	0.9774	0.216 2	4.625	1.023	4.732	1.3579	.8
.3	0.2147	0.213 0	0.9770	0.218 0	4.586	1.023	4.694	1.3561	.7
.4	0.2164	0.214 7	0.9767	0.219 9	4.548	1.024	4.657	1.3544	.6
.5	0.2182	0.216 4	0.9763	0.221 7	4.511	1.024	4.620	1.3526	.5
.6	0.2199	0.218 1	0.9759	0.223 5	4.474	1.025	4.584	1.3509	.4
.7	0.2217	0.219 8	0.9755	0.225 4	4.437	1.025	4.549	1.3491	.3
.8	0.2234	0.221 5	0.9751	0.227 2	4.402	1.025	4.514	1.3474	.2
.9	0.2251	0.223 3	0.9748	0.229 0	4.366	1.026	4.479	1.3456	.1
13.0	0.2269	0.225 0	0.9744	0.230 9	4.331	1.026	4.445	1.3439	77.0
.1	0.2286	0.226 7	0.9740	0.232 7	4.297	1.027	4.412	1.3422	.9
.2	0.2304	0.228 4	0.9736	0.234 5	4.264	1.027	4.379	1.3404	.8
.3	0.2321	0.230 0	0.9732	0.236 4	4.230	1.028	4.347	1.3387	.7
.4	0.2339	0.231 7	0.9728	0.238 2	4.198	1.028	4.315	1.3369	.6
.5	0.2356	0.233 4	0.9724	0.240 1	4.165	1.028	4.284	1.3352	.5
.6	0.2374	0.235 1	0.9720	0.241 9	4.134	1.029	4.253	1.3334	.4
.7	0.2391	0.236 8	0.9715	0.243 8	4.102	1.029	4.222	1.3317	.3
.8	0.2409	0.238 5	0.9711	0.245 6	4.071	1.030	4.192	1.3299	.2
.9	0.2426	0.240 2	0.9707	0.247 5	4.041	1.030	4.163	1.3282	.1
14.0	0.2443	0.241 9	0.9703	0.249 3	4.011	1.031	4.134	1.3265	76.0
.1	0.2461	0.243 6	0.9699	0.251 2	3.981	1.031	4.105	1.3247	.9
.2	0.2478	0.245 3	0.9694	0.253 0	3.952	1.032	4.077	1.3230	.8
.3	0.2496	0.247 0	0.9690	0.254 9	3.923	1.032	4.049	1.3212	.7
.4	0.2513	0.248 7	0.9686	0.256 8	3.895	1.032	4.021	1.3195	.6
.5	0.2531	0.250 4	0.9681	0.258 6	3.867	1.033	3.994	1.3177	.5
.6	0.2548	0.252 1	0.9677	0.260 5	3.839	1.033	3.967	1.3160	.4
.7	0.2566	0.253 8	0.9673	0.262 3	3.812	1.034	3.941	1.3142	.3
.8	0.2583	0.255 4	0.9668	0.264 2	3.785	1.034	3.915	1.3125	.2
.9	0.2600	0.257 1	0.9664	0.266 1	3.758	1.035	3.889	1.3107	.1
		Cos	Sin	Cot	Tan	Csc	Sec	Rad	Deg

Deg	Rad	Sin	Cos	Tan	Cot	Sec	Csc		
15.0	0.2618	0.258 8	0.9659	0.267 9	3.732	1.035	3.864	1.3090	75.0
.1	0.2635	0.260 5	0.9655	0.269 8	3.706	1.036	3.839	1.3073	.9
.2	0.2653	0.262 2	0.9650	0.271 7	3.681	1.036	3.814	1.3055	.8
.3	0.2670	0.263 9	0.9646	0.273 6	3.655	1.037	3.790	1.3038	.7
.4	0.2688	0.265 6	0.9641	0.275 4	3.630	1.037	3.766	1.3020	.6
.5	0.2705	0.267 2	0.9636	0.277 3	3.606	1.038	3.742	1.3003	.5
.6	0.2723	0.268 9	0.9632	0.279 2	3.582	1.038	3.719	1.2985	.4
.7	0.2740	0.270 6	0.9627	0.281 1	3.558	1.039	3.695	1.2968	.3
.8	0.2758	0.272 3	0.9622	0.283 0	3.534	1.039	3.673	1.2950	.2
.9	0.2775	0.274 0	0.9617	0.284 9	3.511	1.040	3.650	1.2933	.1
16.0	0.2793	0.275 6	0.9613	0.286 7	3.487	1.040	3.628	1.2915	74.0
.1	0.2810	0.277 3	0.9608	0.288 6	3.465	1.041	3.606	1.2898	.9
.2	0.2827	0.279 0	0.9603	0.290 5	3.442	1.041	3.584	1.2881	.8
.3	0.2845	0.280 7	0.9598	0.292 4	3.420	1.042	3.563	1.2863	.7
.4	0.2862	0.282 3	0.9593	0.294 3	3.398	1.042	3.542	1.2846	.6
.5	0.2880	0.284 0	0.9588	0.296 2	3.376	1.043	3.521	1.2828	.5
.6	0.2897	0.285 7	0.9583	0.298 1	3.354	1.043	3.500	1.2811	.4
.7	0.2915	0.287 4	0.9578	0.300 0	3.333	1.044	3.480	1.2793	.3
.8	0.2932	0.289 0	0.9573	0.301 9	3.312	1.045	3.460	1.2776	.2
.9	0.2950	0.290 7	0.9568	0.303 8	3.291	1.045	3.440	1.2758	.1
17.0	0.2967	0.292 4	0.9563	0.305 7	3.271	1.046	3.420	1.2741	73.0
.1	0.2985	0.294 0	0.9558	0.307 6	3.251	1.046	3.401	1.2723	.9
.2	0.3002	0.295 7	0.9553	0.309 6	3.230	1.047	3.382	1.2706	.8
.3	0.3019	0.297 4	0.9548	0.311 5	3.211	1.047	3.363	1.2689	.7
.4	0.3037	0.299 0	0.9542	0.313 4	3.191	1.048	3.344	1.2671	.6
.5	0.3054	0.300 7	0.9537	0.315 3	3.172	1.049	3.326	1.2654	.5
.6	0.3072	0.302 4	0.9532	0.317 2	3.152	1.049	3.307	1.2636	.4
.7	0.3089	0.304 0	0.9527	0.319 1	3.133	1.050	3.289	1.2619	.3
.8	0.3107	0.305 7	0.9521	0.321 1	3.115	1.050	3.271	1.2601	.2
.9	0.3124	0.307 4	0.9516	0.323 0	3.096	1.051	3.254	1.2584	.1
18.0	0.3142	0.309 0	0.9511	0.324 9	3.078	1.051	3.236	1.2566	72.0
.1	0.3159	0.310 7	0.9505	0.326 9	3.060	1.052	3.219	1.2549	.9
.2	0.3177	0.312 3	0.9500	0.328 8	3.042	1.053	3.202	1.2531	.8
.3	0.3194	0.314 0	0.9494	0.330 7	3.024	1.053	3.185	1.2514	.7
.4	0.3211	0.315 6	0.9489	0.332 7	3.006	1.054	3.168	1.2497	.6
.5	0.3229	0.317 3	0.9483	0.334 6	2.989	1.054	3.152	1.2479	.5
.6	0.3246	0.319 0	0.9478	0.336 5	2.971	1.055	3.135	1.2462	.4
.7	0.3264	0.320 6	0.9472	0.338 5	2.954	1.056	3.119	1.2444	.3
.8	0.3281	0.322 3	0.9466	0.340 4	2.937	1.056	3.103	1.2427	.2
.9	0.3299	0.323 9	0.9461	0.342 4	2.921	1.057	3.087	1.2409	.1
19.0	0.3316	0.325 6	0.9455	0.344 3	2.904	1.058	3.072	1.2392	71.0
.1	0.3334	0.327 2	0.9449	0.346 3	2.888	1.058	3.056	1.2374	.9
.2	0.3351	0.328 9	0.9444	0.348 2	2.872	1.059	3.041	1.2357	.8
.3	0.3368	0.330 5	0.9438	0.350 2	2.856	1.060	3.026	1.2339	.7
.4	0.3386	0.332 2	0.9432	0.352 2	2.840	1.060	3.011	1.2322	.6
.5	0.3403	0.333 8	0.9426	0.354 1	2.824	1.061	2.996	1.2305	.5
.6	0.3421	0.335 5	0.9421	0.356 1	2.808	1.062	2.981	1.2287	.4
.7	0.3438	0.337 1	0.9415	0.358 1	2.793	1.062	2.967	1.2270	.3
.8	0.3456	0.338 7	0.9409	0.360 0	2.778	1.063	2.952	1.2252	.2
.9	0.3473	0.340 4	0.9403	0.362 0	2.762	1.064	2.938	1.2235	.1
20.0	0.3491	0.342 0	0.9397	0.364 0	2.747	1.064	2.924	1.2217	70.0
.1	0.3508	0.343 7	0.9391	0.365 9	2.733	1.065	2.910	1.2200	.9
.2	0.3526	0.345 3	0.9385	0.367 9	2.718	1.066	2.896	1.2182	.8
.3	0.3543	0.346 9	0.9379	0.369 9	2.703	1.066	2.882	1.2165	.7
.4	0.3560	0.348 6	0.9373	0.371 9	2.689	1.067	2.869	1.2147	.6
.5	0.3578	0.350 2	0.9367	0.373 9	2.675	1.068	2.855	1.2130	.5
.6	0.3595	0.351 8	0.9361	0.375 9	2.660	1.068	2.842	1.2113	.4
.7	0.3613	0.353 5	0.9354	0.377 9	2.646	1.069	2.829	1.2095	.3
.8	0.3630	0.355 1	0.9348	0.379 9	2.633	1.070	2.816	1.2078	.2
.9	0.3648	0.356 7	0.9342	0.381 9	2.619	1.070	2.803	1.2060	.1
		Cos	Sin	Cot	Tan	Csc	Sec	Rad	Deg

Deg	Rad	Sin	Cos	Tan	Cot	Sec	Csc		
21.0	0.3665	0.358 4	0.9336	0.383 9	2.605	1.071	2.790	1.2043	69.0
.1	0.3683	0.360 0	0.9330	0.385 9	2.592	1.072	2.778	1.2025	.9
.2	0.3700	0.361 6	0.9323	0.387 9	2.578	1.073	2.765	1.2008	.8
.3	0.3718	0.363 3	0.9317	0.389 9	2.565	1.073	2.753	1.1990	.7
.4	0.3735	0.364 9	0.9311	0.391 9	2.552	1.074	2.741	1.1973	.6
.5	0.3752	0.366 5	0.9304	0.393 9	2.539	1.075	2.729	1.1956	.5
.6	0.3770	0.368 1	0.9298	0.395 9	2.526	1.076	2.716	1.1938	.4
.7	0.3787	0.369 7	0.9291	0.397 9	2.513	1.076	2.705	1.1921	.3
.8	0.3805	0.371 4	0.9285	0.400 0	2.500	1.077	2.693	1.1903	.2
.9	0.3822	0.373 0	0.9278	0.402 0	2.488	1.078	2.681	1.1886	.1
22.0	0.3840	0.374 6	0.9272	0.404 0	2.475	1.079	2.669	1.1868	68.0
.1	0.3857	0.376 2	0.9265	0.406 1	2.463	1.079	2.658	1.1851	.9
.2	0.3875	0.377 8	0.9259	0.408 1	2.450	1.080	2.647	1.1833	.8
.3	0.3892	0.379 5	0.9252	0.410 1	2.438	1.081	2.635	1.1816	.7
.4	0.3910	0.381 1	0.9245	0.412 2	2.426	1.082	2.624	1.1798	.6
.5	0.3927	0.382 7	0.9239	0.414 2	2.414	1.082	2.613	1.1781	.5
.6	0.3944	0.384 3	0.9232	0.416 3	2.402	1.083	2.602	1.1764	.4
.7	0.3962	0.385 9	0.9225	0.418 3	2.391	1.084	2.591	1.1746	.3
.8	0.3979	0.387 5	0.9219	0.420 4	2.379	1.085	2.581	1.1729	.2
.9	0.3997	0.389 1	0.9212	0.422 4	2.367	1.086	2.570	1.1711	.1
23.0	0.4014	0.390 7	0.9205	0.424 5	2.356	1.086	2.559	1.1694	67.0
.1	0.4032	0.392 3	0.9198	0.426 5	2.344	1.087	2.549	1.1676	.9
.2	0.4049	0.393 9	0.9191	0.428 6	2.333	1.088	2.538	1.1659	.8
.3	0.4067	0.395 5	0.9184	0.430 7	2.322	1.089	2.528	1.1641	.7
.4	0.4084	0.397 1	0.9178	0.432 7	2.311	1.090	2.518	1.1624	.6
.5	0.4102	0.398 7	0.9171	0.434 8	2.300	1.090	2.508	1.1606	.5
.6	0.4119	0.400 3	0.9164	0.436 9	2.289	1.091	2.498	1.1589	.4
.7	0.4136	0.401 9	0.9157	0.439 0	2.278	1.092	2.488	1.1572	.3
.8	0.4154	0.403 5	0.9150	0.441 1	2.267	1.093	2.478	1.1554	.2
.9	0.4171	0.405 1	0.9143	0.443 1	2.257	1.094	2.468	1.1537	.1
24.0	0.4189	0.406 7	0.9135	0.445 2	2.246	1.095	2.459	1.1519	66.0
.1	0.4206	0.408 3	0.9128	0.447 3	2.236	1.095	2.449	1.1502	.9
.2	0.4224	0.409 9	0.9121	0.449 4	2.225	1.096	2.439	1.1484	.8
.3	0.4241	0.411 5	0.9114	0.451 5	2.215	1.097	2.430	1.1467	.7
.4	0.4259	0.413 1	0.9107	0.453 6	2.204	1.098	2.421	1.1449	.6
.5	0.4276	0.414 7	0.9100	0.455 7	2.194	1.099	2.411	1.1432	.5
.6	0.4294	0.416 3	0.9092	0.457 8	2.184	1.100	2.402	1.1414	.4
.7	0.4311	0.417 9	0.9085	0.459 9	2.174	1.101	2.393	1.1397	.3
.8	0.4328	0.419 5	0.9078	0.462 1	2.164	1.102	2.384	1.1379	.2
.9	0.4346	0.421 0	0.9070	0.464 2	2.154	1.102	2.375	1.1362	.1
25.0	0.4363	0.422 6	0.9063	0.466 3	2.145	1.103	2.366	1.1345	65.0
.1	0.4381	0.424 2	0.9056	0.468 4	2.135	1.104	2.357	1.1327	.9
.2	0.4398	0.425 8	0.9048	0.470 6	2.125	1.105	2.349	1.1310	.8
.3	0.4416	0.427 4	0.9041	0.472 7	2.116	1.106	2.340	1.1292	.7
.4	0.4433	0.428 9	0.9033	0.474 8	2.106	1.107	2.331	1.1275	.6
.5	0.4451	0.430 5	0.9026	0.477 0	2.097	1.108	2.323	1.1257	.5
.6	0.4468	0.432 1	0.9018	0.479 1	2.087	1.109	2.314	1.1240	.4
.7	0.4485	0.433 7	0.9011	0.481 3	2.078	1.110	2.306	1.1222	.3
.8	0.4503	0.435 2	0.9003	0.483 4	2.069	1.111	2.298	1.1205	.2
.9	0.4520	0.436 8	0.8996	0.485 6	2.059	1.112	2.289	1.1188	.1
26.0	0.4538	0.438 4	0.8988	0.487 7	2.050	1.113	2.281	1.1170	64.0
.1	0.4555	0.439 9	0.8980	0.489 9	2.041	1.114	2.273	1.1153	.9
.2	0.4573	0.441 5	0.8973	0.492 1	2.032	1.115	2.265	1.1135	.8
.3	0.4590	0.443 1	0.8965	0.494 2	2.023	1.115	2.257	1.1118	.7
.4	0.4608	0.444 6	0.8957	0.496 4	2.014	1.116	2.249	1.1100	.6
.5	0.4625	0.446 2	0.8949	0.498 6	2.006	1.117	2.241	1.1083	.5
.6	0.4643	0.447 8	0.8942	0.500 8	1.997	1.118	2.233	1.1065	.4
.7	0.4660	0.449 3	0.8934	0.502 9	1.988	1.119	2.226	1.1048	.3
.8	0.4677	0.450 9	0.8926	0.505 1	1.980	1.120	2.218	1.1030	.2
.9	0.4695	0.452 4	0.8918	0.507 3	1.971	1.121	2.210	1.1013	.1
		Cos	Sin	Cot	Tan	Csc	Sec	Rad	Deg

Deg	Rad	Sin	Cos	Tan	Cot	Sec	Csc		
27.0	0.4712	0.454 0	0.8910	0.509 5	1.963	1.122	2.203	1.0996	63.0
.1	0.4730	0.455 5	0.8902	0.511 7	1.954	1.123	2.195	1.0978	.9
.2	0.4747	0.457 1	0.8894	0.513 9	1.946	1.124	2.188	1.0961	.8
.3	0.4765	0.458 6	0.8886	0.516 1	1.937	1.125	2.180	1.0943	.7
.4	0.4782	0.460 2	0.8878	0.518 4	1.929	1.126	2.173	1.0926	.6
.5	0.4800	0.461 7	0.8870	0.520 6	1.921	1.127	2.166	1.0908	.5
.6	0.4817	0.463 3	0.8862	0.522 8	1.913	1.128	2.158	1.0891	.4
.7	0.4835	0.464 8	0.8854	0.525 0	1.905	1.129	2.151	1.0873	.3
.8	0.4852	0.466 4	0.8846	0.527 2	1.897	1.130	2.144	1.0856	.2
.9	0.4869	0.467 9	0.8838	0.529 5	1.889	1.132	2.137	1.0838	.1
28.0	0.4887	0.469 5	0.8829	0.531 7	1.881	1.133	2.130	1.0821	62.0
.1	0.4904	0.471 0	0.8821	0.534 0	1.873	1.134	2.123	1.0804	.9
.2	0.4922	0.472 6	0.8813	0.536 2	1.865	1.135	2.116	1.0786	.8
.3	0.4939	0.474 1	0.8805	0.538 4	1.857	1.136	2.109	1.0769	.7
.4	0.4957	0.475 6	0.8796	0.540 7	1.849	1.137	2.103	1.0751	.6
.5	0.4974	0.477 2	0.8788	0.543 0	1.842	1.138	2.096	1.0734	.5
.6	0.4992	0.478 7	0.8780	0.545 2	1.834	1.139	2.089	1.0716	.4
.7	0.5009	0.480 2	0.8771	0.547 5	1.827	1.140	2.082	1.0699	.3
.8	0.5027	0.481 8	0.8763	0.549 8	1.819	1.141	2.076	1.0681	.2
.9	0.5044	0.483 3	0.8755	0.552 0	1.811	1.142	2.069	1.0664	.1
29.0	0.5061	0.484 8	0.8746	0.554 3	1.804	1.143	2.063	1.0647	61.0
.1	0.5079	0.486 3	0.8738	0.556 6	1.797	1.144	2.056	1.0629	.9
.2	0.5096	0.487 9	0.8729	0.558 9	1.789	1.146	2.050	1.0612	.8
.3	0.5114	0.489 4	0.8721	0.561 2	1.782	1.147	2.043	1.0594	.7
.4	0.5131	0.490 9	0.8712	0.563 5	1.775	1.148	2.037	1.0577	.6
.5	0.5149	0.492 4	0.8704	0.565 8	1.767	1.149	2.031	1.0559	.5
.6	0.5166	0.493 9	0.8695	0.568 1	1.760	1.150	2.025	1.0542	.4
.7	0.5184	0.495 5	0.8686	0.570 4	1.753	1.151	2.018	1.0524	.3
.8	0.5201	0.497 0	0.8678	0.572 7	1.746	1.152	2.012	1.0507	.2
.9	0.5219	0.498 5	0.8669	0.575 0	1.739	1.154	2.006	1.0489	.1
30.0	0.5236	0.500 0	0.8660	0.577 4	1.732	1.155	2.000	1.0472	60.0
.1	0.5253	0.501 5	0.8652	0.579 7	1.725	1.156	1.994	1.0455	.9
.2	0.5271	0.503 0	0.8643	0.582 0	1.718	1.157	1.988	1.0437	.8
.3	0.5288	0.504 5	0.8634	0.584 4	1.711	1.158	1.982	1.0420	.7
.4	0.5306	0.506 0	0.8625	0.586 7	1.704	1.159	1.976	1.0402	.6
.5	0.5323	0.507 5	0.8616	0.589 0	1.698	1.161	1.970	1.0385	.5
.6	0.5341	0.509 0	0.8607	0.591 4	1.691	1.162	1.964	1.0367	.4
.7	0.5358	0.510 5	0.8599	0.593 8	1.684	1.163	1.959	1.0350	.3
.8	0.5376	0.512 0	0.8590	0.596 1	1.678	1.164	1.953	1.0332	.2
.9	0.5393	0.513 5	0.8581	0.598 5	1.671	1.165	1.947	1.0315	.1
31.0	0.5411	0.515 0	0.8572	0.600 9	1.664	1.167	1.942	1.0297	59.0
.1	0.5428	0.516 5	0.8563	0.603 2	1.658	1.168	1.936	1.0280	.9
.2	0.5445	0.518 0	0.8554	0.605 6	1.651	1.169	1.930	1.0263	.8
.3	0.5463	0.519 5	0.8545	0.608 0	1.645	1.170	1.925	1.0245	.7
.4	0.5480	0.521 0	0.8536	0.610 4	1.638	1.172	1.919	1.0228	.6
.5	0.5498	0.522 5	0.8526	0.612 8	1.632	1.173	1.914	1.0210	.5
.6	0.5515	0.524 0	0.8517	0.615 2	1.625	1.174	1.908	1.0193	.4
.7	0.5533	0.525 5	0.8508	0.617 6	1.619	1.175	1.903	1.0175	.3
.8	0.5550	0.527 0	0.8499	0.620 0	1.613	1.177	1.898	1.0158	.2
.9	0.5568	0.528 4	0.8490	0.622 4	1.607	1.178	1.892	1.0140	.1
32.0	0.5585	0.529 9	0.8480	0.624 9	1.600	1.179	1.887	1.0123	58.0
.1	0.5603	0.531 4	0.8471	0.627 3	1.594	1.180	1.882	1.0105	.9
.2	0.5620	0.532 9	0.8462	0.629 7	1.588	1.182	1.877	1.0088	.8
.3	0.5637	0.534 4	0.8453	0.632 2	1.582	1.183	1.871	1.0071	.7
.4	0.5655	0.535 8	0.8443	0.634 6	1.576	1.184	1.866	1.0053	.6
.5	0.5672	0.537 3	0.8434	0.637 1	1.570	1.186	1.861	1.0036	.5
.6	0.5690	0.538 8	0.8425	0.639 5	1.564	1.187	1.856	1.0018	.4
.7	0.5707	0.540 2	0.8415	0.642 0	1.558	1.188	1.851	1.0000	.3
.8	0.5725	0.541 7	0.8406	0.644 5	1.552	1.190	1.846	0.9983	.2
.9	0.5742	0.543 2	0.8396	0.646 9	1.546	1.191	1.841	0.9966	.1
		Cos	Sin	Cot	Tan	Csc	Sec	Rad	Deg

Deg	Rad	Sin	Cos	Tan	Cot	Sec	Csc		
33.0	0.5760	0.544 6	0.8387	0.649 4	1.540	1.192	1.836	0.9948	57.0
.1	0.5777	0.546 1	0.8377	0.651 9	1.534	1.194	1.831	0.9931	.9
.2	0.5794	0.547 6	0.8368	0.654 4	1.528	1.195	1.826	0.9913	.8
.3	0.5812	0.549 0	0.8358	0.656 9	1.522	1.196	1.821	0.9896	.7
.4	0.5829	0.550 5	0.8348	0.659 4	1.517	1.817	1.817	0.9879	.6
.5	0.5847	0.551 9	0.8339	0.661 9	1.511	1.199	1.812	0.9861	.5
.6	0.5864	0.553 4	0.8329	0.664 4	1.505	1.201	1.807	0.9844	.4
.7	0.5882	0.554 8	0.8320	0.666 9	1.499	1.202	1.802	0.9826	.3
.8	0.5889	0.556 3	0.8310	0.669 4	1.494	1.203	1.798	0.9809	.2
.9	0.5917	0.557 7	0.8300	0.672 0	1.488	1.205	1.793	0.9791	.1
34.0	0.5934	0.559 2	0.8290	0.674 5	1.483	1.206	1.788	0.9774	56.0
.1	0.5952	0.560 6	0.8281	0.677 1	1.477	1.208	1.784	0.9756	.9
.2	0.5969	0.562 1	0.8271	0.679 6	1.471	1.209	1.779	0.9739	.8
.3	0.5986	0.563 5	0.8261	0.682 2	1.466	1.211	1.775	0.9721	.7
.4	0.6004	0.565 0	0.8251	0.684 7	1.460	1.212	1.770	0.9704	.6
.5	0.6021	0.566 4	0.8241	0.687 3	1.455	1.213	1.766	0.9687	.5
.6	0.6039	0.567 8	0.8231	0.689 9	1.450	1.215	1.761	0.9669	.4
.7	0.6056	0.569 3	0.8221	0.692 4	1.444	1.216	1.757	0.9652	.3
.8	0.6074	0.570 7	0.8211	0.695 0	1.439	1.218	1.752	0.9634	.2
.9	0.6091	0.572 1	0.8202	0.697 6	1.433	1.219	1.748	0.9617	.1
35.0	0.6109	0.573 6	0.8192	0.700 2	1.428	1.221	1.743	0.9599	55.0
.1	0.6126	0.575 0	0.8181	0.702 8	1.423	1.222	1.739	0.9852	.9
.2	0.6144	0.576 4	0.8171	0.705 4	1.418	1.224	1.735	0.9564	.8
.3	0.6161	0.577 9	0.8161	0.708 0	1.412	1.225	1.731	0.9547	.7
.4	0.6178	0.579 3	0.8151	0.710 7	1.407	1.227	1.726	0.9530	.6
.5	0.6196	0.580 7	0.8141	0.713 3	1.402	1.228	1.722	0.9512	.5
.6	0.6213	0.582 1	0.8131	0.715 9	1.397	1.230	1.718	0.9495	.4
.7	0.6231	0.583 5	0.8121	0.718 6	1.392	1.231	1.714	0.9477	.3
.8	0.6248	0.585 0	0.8111	0.721 2	1.387	1.233	1.710	0.9460	.2
.9	0.6266	0.586 4	0.8100	0.723 9	1.381	1.235	1.705	0.9442	.1
36.0	0.6283	0.587 8	0.8090	0.726 5	1.376	1.236	1.701	0.9425	54.0
.1	0.6301	0.589 2	0.8080	0.729 2	1.371	1.238	1.697	0.9407	.9
.2	0.6318	0.590 6	0.8070	0.731 9	1.366	1.239	1.693	0.9390	.8
.3	0.6336	0.592 0	0.8059	0.734 6	1.361	1.241	1.689	0.9372	.7
.4	0.6353	0.593 4	0.8049	0.737 3	1.356	1.242	1.685	0.9355	.6
.5	0.6370	0.594 8	0.8039	0.740 0	1.351	1.244	1.681	0.9338	.5
.6	0.6388	0.596 2	0.8028	0.742 7	1.347	1.246	1.677	0.9320	.4
.7	0.6405	0.597 6	0.8018	0.745 4	1.342	1.247	1.673	0.9303	.3
.8	0.6423	0.599 0	0.8007	0.748 1	1.337	1.249	1.669	0.9285	.2
.9	0.6440	0.600 4	0.7997	0.750 8	1.332	1.250	1.666	0.9269	.1
37.0	0.6458	0.601 8	0.7986	0.753 6	1.327	1.252	1.662	0.9250	53.0
.1	0.6475	0.603 2	0.7976	0.756 3	1.322	1.254	1.658	0.9233	.9
.2	0.6493	0.604 6	0.7965	0.759 0	1.317	1.255	1.654	0.9215	.8
.3	0.6510	0.606 0	0.7955	0.761 8	1.313	1.257	1.650	0.9198	.7
.4	0.6528	0.607 4	0.7944	0.764 6	1.308	1.259	1.646	0.9180	.6
.5	0.6545	0.608 8	0.7934	0.767 3	1.303	1.260	1.643	0.9163	.5
.6	0.6562	0.610 1	0.7923	0.770 1	1.299	1.262	1.639	0.9146	.4
.7	0.6580	0.611 5	0.7912	0.772 9	1.294	1.264	1.635	0.9128	.3
.8	0.6597	0.612 9	0.7902	0.775 7	1.289	1.266	1.632	0.9111	.2
.9	0.6615	0.614 3	0.7891	0.778 5	1.285	1.267	1.628	0.9093	.1
38.0	0.6632	0.615 7	0.7880	0.781 3	1.280	1.269	1.624	0.9076	52.0
.1	0.6650	0.617 0	0.7869	0.784 1	1.275	1.271	1.621	0.9058	.9
.2	0.6667	0.618 4	0.7859	0.786 9	1.271	1.272	1.617	0.9041	.8
.3	0.6685	0.619 8	0.7848	0.789 8	1.266	1.274	1.613	0.9023	.7
.4	0.6702	0.621 1	0.7837	0.792 6	1.262	1.276	1.610	0.9006	.6
.5	0.6720	0.622 5	0.7826	0.795 4	1.257	1.278	1.606	0.8988	.5
.6	0.6737	0.623 9	0.7815	0.798 3	1.253	1.280	1.603	0.8971	.4
.7	0.6754	0.625 2	0.7804	0.810 2	1.248	1.281	1.599	0.8954	.3
.8	0.6772	0.626 6	0.7793	0.804 0	1.244	1.283	1.596	0.8936	.2
.9	0.6789	0.628 0	0.7782	0.806 9	1.239	1.285	1.592	0.8919	.1
		Cos	**Sin**	**Cot**	**Tan**	**Csc**	**Sec**	**Rad**	**Deg**

Deg	Rad	Sin	Cos	Tan	Cot	Sec	Csc		
39.0	0.6807	0.629 3	0.7771	0.809 8	1.235	1.287	1.589	0.8901	51.0
.1	0.6824	0.630 7	0.7760	0.812 9	1.230	1.289	1.586	0.8884	.9
.2	0.6842	0.632 0	0.7749	0.815 6	1.226	1.290	1.582	0.8866	.8
.3	0.6859	0.633 4	0.7738	0.818 5	1.222	1.292	1.579	0.8849	.7
.4	0.6877	0.634 7	0.7727	0.821 4	1.217	1.294	1.575	0.8831	.6
.5	0.6894	0.636 1	0.7716	0.824 3	1.213	1.296	1.572	0.8814	.5
.6	0.6912	0.637 4	0.7705	0.827 3	1.209	1.298	1.569	0.8796	.4
.7	0.6929	0.638 8	0.7694	0.830 2	1.205	1.300	1.566	0.8779	.3
.8	0.6946	0.640 1	0.7683	0.833 2	1.200	1.302	1.562	0.8762	.2
.9	0.6964	0.641 4	0.7672	0.836 1	1.196	1.304	1.559	0.8744	.1
40.0	0.6981	0.642 8	0.7660	0.839 1	1.192	1.305	1.556	0.8727	50.0
.1	0.6999	0.644 1	0.7649	0.842 1	1.188	1.307	1.552	0.8709	.9
.2	0.7016	0.645 5	0.7639	0.845 1	1.183	1.309	1.549	0.8691	.8
.3	0.7034	0.646 8	0.7627	0.848 1	1.179	1.311	1.546	0.8674	.7
.4	0.7051	0.648 1	0.7615	0.851 1	1.175	1.313	1.543	0.8656	.6
.5	0.7069	0.649 4	0.7604	0.854 1	1.171	1.315	1.540	0.8639	.5
.6	0.7086	0.650 8	0.7593	0.857 1	1.167	1.317	1.537	0.8622	.4
.7	0.7103	0.652 1	0.7581	0.860 1	1.163	1.319	1.534	0.8604	.3
.8	0.7121	0.653 4	0.7570	0.863 2	1.159	1.321	1.530	0.8587	.2
.9	0.7138	0.654 7	0.7559	0.866 2	1.154	1.323	1.527	0.8570	.1
41.0	0.7156	0.656 1	0.7547	0.869 3	1.150	1.325	1.524	0.8552	49.0
.1	0.7173	0.657 4	0.7536	0.872 4	1.146	1.327	1.521	0.8535	.9
.2	0.7191	0.658 7	0.7524	0.875 4	1.142	1.329	1.518	0.8517	.8
.3	0.7208	0.660 0	0.7513	0.878 5	1.138	1.331	1.515	0.8500	.7
.4	0.7226	0.661 3	0.7501	0.881 6	1.134	1.333	1.512	0.8482	.6
.5	0.7243	0.662 6	0.7490	0.884 7	1.130	1.335	1.509	0.8465	.5
.6	0.7261	0.663 9	0.7478	0.887 8	1.126	1.337	1.506	0.8447	.4
.7	0.7278	0.665 2	0.7466	0.891 0	1.122	1.339	1.503	0.8430	.3
.8	0.7295	0.666 5	0.7455	0.894 1	1.118	1.341	1.500	0.8412	.2
.9	0.7313	0.667 8	0.7443	0.897 2	1.115	1.344	1.497	0.8395	.1
42.0	0.7330	0.669 1	0.7431	0.900 4	1.111	1.346	1.494	0.8378	48.0
.1	0.7348	0.670 4	0.7420	0.903 6	1.107	1.348	1.492	0.8360	.9
.2	0.7365	0.671 7	0.7408	0.906 7	1.103	1.350	1.489	0.8343	.8
.3	0.7383	0.673 0	0.7396	0.909 9	1.099	1.352	1.486	0.8325	.7
.4	0.7400	0.674 3	0.7385	0.913 1	1.095	1.354	1.483	0.8308	.6
.5	0.7418	0.675 6	0.7373	0.916 3	1.091	1.356	1.480	0.8290	.5
.6	0.7435	0.676 9	0.7361	0.919 5	1.087	1.359	1.477	0.8273	.4
.7	0.7453	0.678 2	0.7349	0.922 8	1.084	1.361	1.475	0.8255	.3
.8	0.7470	0.679 4	0.7337	0.926 0	1.080	1.363	1.472	0.8238	.2
.9	0.7487	0.680 7	0.7325	0.929 3	1.076	1.365	1.469	0.8221	.1
43.0	0.7505	0.682 0	0.7314	0.932 5	1.072	1.367	1.466	0.8203	47.0
.1	0.7522	0.683 3	0.7301	0.935 8	1.069	1.370	1.464	0.8186	.9
.2	0.7540	0.684 5	0.7290	0.939 1	1.065	1.372	1.461	0.8168	.8
.3	0.7558	0.685 8	0.7278	0.942 4	1.061	1.374	1.458	0.8151	.7
.4	0.7575	0.687 1	0.7266	0.945 7	1.057	1.376	1.455	0.8133	.6
.5	0.7592	0.688 4	0.7254	0.949 0	1.054	1.379	1.453	0.8116	.5
.6	0.7610	0.689 6	0.7242	0.952 3	1.050	1.381	1.450	0.8098	.4
.7	0.7627	0.690 9	0.7230	0.955 6	1.046	1.383	1.447	0.8081	.3
.8	0.7645	0.692 1	0.7218	0.959 0	1.043	1.386	1.445	0.8063	.2
.9	0.7662	0.693 4	0.7206	0.962 3	1.039	1.388	1.442	0.8046	.1
44.0	0.7679	0.694 7	0.7193	0.965 7	1.036	1.390	1.440	0.8029	46.0
.1	0.7697	0.695 9	0.7181	0.969 1	1.032	1.393	1.437	0.8011	.9
.2	0.7714	0.697 2	0.7169	0.972 5	1.028	1.395	1.434	0.7994	.8
.3	0.7732	0.698 4	0.7157	0.975 9	1.025	1.397	1.432	0.7976	.7
.4	0.7749	0.699 7	0.7145	0.979 3	1.021	1.400	1.429	0.7959	.6
.5	0.7767	0.700 9	0.7133	0.982 7	1.018	1.402	1.427	0.7941	.5
.6	0.7784	0.702 2	0.7120	0.986 1	1.014	1.404	1.424	0.7924	.4
.7	0.7802	0.703 4	0.7108	0.989 6	1.011	1.407	1.422	0.7907	.3
.8	0.7819	0.704 6	0.7096	0.993 0	1.007	1.409	1.419	0.7889	.2
.9	0.7837	0.705 9	0.7083	0.996 5	1.003	1.412	1.417	0.7871	.1
45.0	0.7854	0.707 1	0.7071	1.000 0	1.000	1.414	1.414	0.7854	45.0
		Cos	Sin	Cot	Tan	Csc	Sec	Rad	Deg

ABBREVIATIONS, SYMBOLS, AND ACRONYMS

The abbreviations, symbols, and acronyms listed here are many of those likely to be encountered by aviation technicians and others involved in the operation and maintenance of aircraft. Additional abbreviations are constantly being issued by various agencies; therefore, technicians will find many others in use during the course of their activities in aviation.

A	Amperes, area
AA	The Aluminum Association
AAIP	Approved aircraft inspection program
ABC	After bottom center (piston)
ABDC	After bottom dead center
AC	Air Corps; advisory circular
ac	Alternating current
AD	Airworthiness directive
ADF	Automatic direction finder
ADMA	Aviation Distributors and Manufacturers Association
AF	Air Force
AIAA	American Institute of Aeronautics and Astronautics
AISI	American Iron and Steel Institute
AMS	Aeronauticals Materials Specification
AN	Air Force—Navy
AND	Air Force—Navy Design
A & P	Airframe and Powerplant
APU	Auxiliary power unit
AR	Aspect ratio
AS	Aeronautical Standard
ASTM	American Society for Testing Materials
ATA	Air Transport Association of America
ATC	After top center
ATDC	After top dead center
avdp	Avoirdupois pound
b	Wing span
BBC	Before bottom center
BBDC	Before bottom dead center
BC	Bottom center
BDC	Bottom dead center
BF	Buoyant force
bhp	Brake horsepower
BITE	Built-in test equipment
bmep	Brake mean effective pressure
bsfc	Brake specific fuel consumption
BTC	Before top center
Btu	British thermal unit
c	Average wing chord
C	Celsius; center; centigrade
CAB	Civil Aeronautics Board
CAS	Calibrated airspeed
CAT	Carburetor air temperature
C_D	Coefficient of drag
C_L	Coefficient of lift
C_M	Coefficient of pitching moment
C-D	Converging-diverging nozzle (jet engine)
CDP	Compressor discharge pressure
CDT	Compressor discharge temperature
CG	Center of gravity
CHT	Cylinder head temperature
CIP	Compressor inlet pressure
CIT	Compressor inlet temperature
₵	Center line
CP	Center of pressure
CPR	Compressor discharge (pressure) ratio
CRS	Certified repair station

c_T	Tip chord
CSD	Constant-speed drive
c_s	Root chord
D	Drag
dB	Decibel
dm	Decimeter
DME	Distance-measuring equipment compatible with TACAN
dyn	Dyne
E	Electromotive force (voltage)
EAS	Equivalent airspeed
EC	Exhaust valve closes
EGT	Exhaust gas temperature
EPR	Engine pressure ratio
eshp	Equivalent shaft horsepower
EVC	Engine vane control
EW	Empty weight
EWCG	Empty weight center of gravity
F	Fahrenheit; force; thrust
FAA	Federal Aviation Administration
FAR	Federal Aviation Regulation
FE	Flight environment
FED	Federal
Fg	Gross thrust
FM	Fan marker for ILS; frequency modulation
F	Net thrust
F	Ram drag of engine airflow
FSDO	Flight standards district office (FAA)
g	Gram
GADO	General aviation district office (FAA)
GAMA	General Aviation Manufacturers Association
GAW	General-Aircraft-Wing
GS	Glide slope of ILS
Hz	Hertz
Hg	Mercury
IA	Inspection authorization (FAA)
I	Electric current (amperage)
ICAO	International Civil Aviation Organization
IC	Intake valve closes
IFR	Instrument flight rules
IGV	Inlet guide vanes
ILS	Instrument landing system
IO	Intake valve opens
IPC	Illustrated parts catalog
K	Kelvin
Kn	Knot
kPa	Kilopascals
L	Lift
L/D	Lift/drag ratio
LDA	Localizer-type directional aid
LE	Leading edge
LF	Low frequency
LOC	Localizer of ILS: location
LORAN	Long-range navigation equipment
M	Moment
M	Mach number
MA	Mechanical advantage
MAC	Mean aerodynamic chord
MAP	Manifold pressure (absolute)
M or D	Malfunction or defect report
MEL	Minimum equipment list
MEK	Methyl-ethyl-ketone

METO	Maximum except takeoff (power)
MIL	Military
MM	Middle marker for ILS
MS	Material Standard; Military Standard
MSL	Mean sea level
MWCL	Main-wheel center line
N	Newton
NACA	National Advisory Committee for Aeronautics (superseded)
NAF	Naval Aircraft Factory
NAS	National Aerospace Standard; Navy Aircraft Standard
NASA	National Aeronautics and Space Administration
NDB	Nondirectional beacon (ADF)
NPRM	Notice of proposed rule making
NTC	Negative torque control
NTS	Negative torque signal
OAT	Outside air temperature
OBAWS	On-board aircraft weighing system
OGV	Outlet guide vanes
OM	Outer marker for ILS
P	Pressure
PALS	Precision aircraft landing system
PMA	Parts manufacturing authorization
PVA	Polyvinyl alcohol
PVC	Polyvinyl chloride
p	Density
psi	Per square inch
Q	Dynamic pressure
R	Rankine; Reynolds number
RBN	Radio beacon
rpm	Revolutions per minute
S	Airfoil area
SAE	Society of Automotive Engineers
sfc	Specific fuel consumption
SPEC	Specification
STC	Supplemental type certificate
STD	Standard
Σ	The sum of
T	Temperature
TACAN	Ultra-high frequency tactical air navigation aid
TAS	True airspeed
TAFI	Turn-around fault isolation
TC	Top center; type certificate
TCDS	Type certificate data sheet
TDC	Top dead center
TET	Turbine exhaust temperature
T/F	Turbofan
TIP	Turbine inlet pressure
TIT	Turbine inlet temperature
T/J	Turbojet
TOP	Turbine outlet pressure
TOT	Turbine outlet temperature
T/S	Turboshaft
tsfc	Thrust specific fuel consumption
TSO	Technical standard order (FAA)
TSS	Thrust sensitive signal
UHF	Ultra-high frequency
V	Velocity; volts; volume
V_A	Design maneuvering speed (aircraft operation)
V_B	Design speed for maximum gust intensity

V_C	Design cruising speed
V_D	Design diving speed
V_{DF}/M_{DF}	Demonstrated flight diving speed
V_F	Design flap speed
V_{FC}/M_{FC}	Maximum speed for stability characteristics
V_{FE}	Maximum flap extended speed
VFR	Visual flight rules
V_H	Maximum speed in level flight with maximum continuous power
VHF	Very high frequency
VIGV	Variable inlet guide vanes
V_{LE}	Maximum landing gear extended speed
V_{LO}	Maximum landing gear operating speed
V_{LOF}	Lift-off speed
V_{MC}	Minimum control speed with critical engine inoperative
V_{MO}/M_{MO}	Maximum operating limit speed
V_{NE}	Never exceed speed
VOR	Very-high-frequency omnirange station
VORTAC	Collocated VOR and TACAN
V_R	Rotation speed (aircraft takeoff)
V_S	Stalling speed or minimum steady flight speed at which the airplane is controllable
V_{SO}	Stalling speed or the minimum steady flight speed in the landing configuration
V_{SI}	Stalling speed or the minimum steady flight speed obtained in a specific configuration
VTOL	Vertical takeoff and landing
VV	Variable viscosity
V_X	Speed for best angle of climb
V_Y	Speed for best rate of climb
V_1	Critical engine failure speed
V_2	Takeoff safety speed
W	Weight

U.S. DEPARTMENT OF TRANSPORTATION

FAA AERONAUTICAL CENTER

The following offices at the Mike Monroney Aeronautical Center (Post Office Box 25082, Oklahoma City, Oklahoma 73125; 405/680-3296) should be contacted for the services indicated:

Aircraft Registration Branch (AVN-450) (405/680-3131)—Registering aircraft ($5 application fee); recording aircraft titles and security documents ($5 fee for recording); requesting duplicate of lost aircraft registration certificates ($2 each duplicate); requesting special N number ($10); request to reserve, change, or reassign an N number ($10 fee); request for dealer's registration ($10 fee, $2 for each duplicate). Address communication to Post Office Box 25504, Oklahoma City, Oklahoma 73125.

Airman Certification Branch (AVN-460) (405/680-3261)—Notification of change in permanent mailing address (must be given in writing within 30 days) or change of name (must furnish legal proof of change); requesting duplicate of lost pilot certificate (must furnish full name, date of birth, place of birth, certificate number and/or Social Security number, and $2 fee).

Aeromedical Certification Branch (405/680-4821, 680-4827)—Requesting duplicate of lost medical certificate ($2 fee); requesting information on medical certification problems, if not being handled by regional flight surgeon. Address communications to: Post Office Box 26080, Oklahoma City, Oklahoma 73126.

FEDERAL AVIATION ADMINISTRATION REGIONAL OFFICES

Alaskan Region—Anchorage
Governing Alaska and Aleutian Islands
Frank L. Cunningham, Administrator; 222 West Seventh Avenue,
Anchorage, Alaska 99513; 907/271-5645

Central Region—Kansas City
Governing Iowa, Kansas, Missouri, Nebraska
Paul K. Bohr, Director; 601 East 12th Street, Federal Building,
Kansas City, Missouri 64106; 816/426-5626

Eastern Region—New York
*Governing Delaware, District of Columbia, Maryland, New
Jersey, New York, Pennsylvania, Virginia, West Virginia*
Daniel J. Peterson, Administrator; John F. Kennedy International
Airport. Fitzgerald Federal Building, Jamaica, New York 11430;
718/917-1005

Europe, Africa, and Middle East Region—Brussels, Belgium
Governing Europe, Africa, and the Middle East
Benjamin Demps, Jr., Director; c/o American Embassy—FAA;
APO New York 09667; (32-2) 513.38.30 extension 2701

Great Lakes Region—Chicago
*Governing Illinois, Indiana, Minnesota, Michigan, Ohio,
Wisconsin, North and South Dakota*
Timothy P. Forté, Administrator; O'Hare Lake Office Center, 2300
East Devon Avenue, Des Plaines, Illinois 60018; 312/694-7000

New England Region—Boston
*Governing Connecticut, Maine, Massachusetts, New Hampshire,
Rhode Island, Vermont*
Arlene B. Feldman, Administrator; 12 New England Executive
Park, Burlington, Massachusetts 01803; 617/273-7244

Northwest Mountain Region—Seattle
*Governing Colorado, Idaho, Montana, Oregon, Utah,
Washington, Wyoming*
Frederick M. Issac, Administrator; 17900 Pacific Highway South,
C-68966, Seattle, Washington 98168; 206/431-2001

Southern Region—Atlanta
*Governing Alabama, Florida, Georgia, Kentucky, Mississippi,
North Carolina, South Carolina, Tennessee, Puerto Rico,
U.S. Virgin Islands*
Garland P. Castleberry, Administrator; Post Office Box 20636,
Atlanta, Georgia 30320; 404/763-7222

Southwest Region—Fort Worth
Governing Arkansas, Louisiana, New Mexico, Oklahoma, Texas
Don P. Watson, Administrator; 4400 Blue Mound Road, Fort
Worth, Texas 76193; 817/624-5000

Western-Pacific Region—Los Angeles
*Governing Arizona, California, Nevada, Hawaii, Pacific Ocean,
and Asia*
Jerold M. Chavkin, Administrator; Post Office Box 92007, World-
way Postal Center, Los Angeles, California 90009; 213/297-1427

FAA DISTRICT OFFICES

Address correspondence to the General Aviation District Office
(GADO), Flight Standards District Office (FSDO), or Flight Stan-
dards Field Office (FSFO) (as appropriate), Federal Aviation
Administration.

Alaskan Region

Anchorage, Alaska—(FSDO), 4510 West International Airport
Road, Suite 216, 99502-1088; 907/243-1902
Fairbanks, Alaska—(FSDO), 6348 Old Airport Way, 99701;
907/474-0276

Juneau, Alaska—(FSDO), 1910 Alex Holden Way, Suite A,
99801; 907/789-0231

Central Region

Des Moines, Iowa—(FSDO), 3021 Army Post Road, 50321;
515/284-4094
Wichita, Kansas—(FSDO), FAA Building, Room 103, 1801 Air-
port Road, 67209; 316/946-4462
Kansas City, Missouri—(FSDO), Kansas City International Air-
port, 525 Mexico City Avenue, 64153; 816/243-3800
St. Louis, Missouri—(FSDO-62), FAA Building, 10801 Pear
Tree Lane, Suite 200, St. Ann, 63074; 314/429-1006
Lincoln, Nebraska—(FSDO), General Aviation Building, Lin-
coln Municipal Airport, 68524; 402/437-5485

Eastern Region

Washington, D.C.—(FSDO), GT Building, Post Office Box
17325, Dulles International Airport, Washington, D.C. 20041;
703/557-5360
Baltimore, Maryland—(GADO), Baltimore-Washington Interna-
tional Airport, North Administration Building, Elm Road, 21240;
301/859-5780
Teterboro, New Jersey—(GADO), 150 Fred Wehren Drive,
07608; 201/288-1745
Albany, New York—(GADO), Albany County Airport, Old Page
Building, 12211; 518/869-8482
Farmingdale, New York—(GADO), Administrative Building,
Gate Number 1, Republic Airport, 11735; 516/694-5530
Rochester, New York—(GADO), Rochester-Monroe County
Airport, 1295 Scottsville Road, 14624; 716/235-3438
Valley Stream, New York—(FSDO-15), 181 South Franklin Av-
enue, 4th Floor, 11581; 718/917-1848
Allentown, Pennsylvania—(GADO), Allentown-Bethlehem-
Easton Airport, RAS Aviation Center Building, 18103;
215/264-2888
New Cumberland, Pennsylvania—(GADO), Capital City Air-
port, Room 201, Administration Building, 17070; 717/782-4528
Philadelphia, Pennsylvania—(GADO), Northeast Philadelphia
Airport, 19114; 215/597-9708
Philadelphia, Pennsylvania—(GADO), Philadelphia Internation-
al Airport, 19153; 215/596-0673
Pittsburgh, Pennsylvania—(GADO), Allegheny County Airport,
Administration Building, Room 213, West Mifflin, 15122;
412/462-5507
Richmond, Virginia—(GADO), Richmond International/Byrd
Field, Executive Terminal Building, 2nd Floor, Sandston, 23150;
804/222-7494
Charleston, West Virginia—(GADO), Yeager Airport, 301
Eagle Mountain Road, Room 144, 25311; 304/343-4689

Great Lakes Region

Chicago, Illinois—(GADO), DuPage Airport, West Chicago,
60185; 312/377-4500
Springfield, Illinois—(GADO), 3 North Airport Drive, Capitol
Airport, 62708; 217/492-4238
Indianapolis, Indiana—(GADO), Indianapolis International Air-
port, FAA Building 1, 46241; 317/247-2491
South Bend, Indiana—(GADO), Michiana Regional Airport,
1843 Commerce Drive, 46628; 219/236-8480
Detroit, Michigan—(FSDO), Willow Run Airport-East Side,
8800 Beck Road, Belleville, 48111; 313/485-2550
Grand Rapids, Michigan—(FSDO), 5500 44th Street, S.E., Kent
County Airport, 49508; 616/456-2427
Minneapolis, Minnesota—(GADO), Minneapolis-St. Paul Inter-
national Airport, 6201 34th Avenue, South, Room 201, 55450;
612/725-3341
Fargo, North Dakota—(GADO), Hector International, Room
216, Administration Building, 58102; 701/232-8949
Cincinnati, Ohio—(GADO), Lunken Airport Executive Building,
Ground Floor, 4242 Airport Road, 45226; 513/533-8110

Cleveland, Ohio—(GADO), Cleveland-Hopkins International Airport, 21046 Brookpark Road, 44135; 216/267-0220
Columbus, Ohio—(GADO), East 17th Avenue, Port Columbus International Airport, 43219; 614/469-7476
Rapid City, South Dakota—(FSDO), Rapid City Regional Airport, RR 2, Box 4750, 57701; 605/393-1359
Milwaukee, Wisconsin—(GADO), General Mitchell Field, 53207; 414/747-5531

New England Region

Windsor Lock, Connecticut—(FSDO), Bradley International Airport, 1st Floor, Building 85-214, 06096-1009; 203/654-1000
Portland, Maine—(GADO), Portland International Jetport, 2A1 McKay Avenue, 04102; 207/774-4484
Bedford, Massachusetts—(FSDO), 2nd Floor, Civil Terminal Building, L. G. Hanscom Field, 01730; 617/274-7130

Northwest Mountain Region

Aurora, Colorado—(FSDO-60), 10455 East 25th Avenue, Suite 202, 80010; 303/286-5400
Boise, Idaho—(GADO), 3975 Rickenbacker Street, 83705; 208/334-1238
Billings, Montana—(GADO), Room 216, Administration Building, Billings Logan International Airport, 59101; 406/245-6270, 245-6279
Helena, Montana—(GADO), Room 3, FAA Building, Helena Airport, 59601; 406/449-5270
Eugene, Oregon—(GADO), Mahlon Sweet Airport, 90606 Greenhill Road, 97402; 503/688-9721
Portland, Oregon—(GADO), Portland-Hillsboro Airport, 3355 N.E. Cornell Road, Hillsboro, Oregon 97123; 503/326-2104
Salt Lake City, Utah—(GADO), 116 North 2400 West, Room 103, 84116; 801/524-4247, 524-4248
Seattle, Washington—(FSDO), Boeing Field-King County International, 7300 Perimeter Road, South, 98108; 206/431-1363
Spokane, Washington—(FSFO), Felts Field, 6105 East Rutter, Post Office Box 11649, 99206; 509/353-2434
Casper, Wyoming—(GADO), Natrona County International Airport, 1187 Fuller Street, 82601; 307/234-8959

Southern Region

Birmingham, Alabama—(FSDO), Municipal Airport, 6500 43rd Avenue, North, 35206; 205/731-1557
Clearwater, Florida—(GADO), St. Petersburg-Clearwater Airport, Terminal Building, West Wing, 33520; 813/531-1434
Fort Lauderdale, Florida—(FSDO), Fort Lauderdale Airport, 286 S.W. 34th Street, 33315; 305/527-0884
Jacksonville, Florida—(GADO), Craig Municipal Airport, FAA Building, 32211; 904/641-7311
Miami, Florida—(FSDO), Miami International Airport, Post Office Box 592015, 33159; 305/526-2562
Orlando, Florida—(FSDO), 9355 Tradeport Drive, 32827-5397; 407/648-6840
Atlanta, Georgia—(FSDO), Fulton County Airport-Brown Field, 1680 Phoenix Parkway, 2nd Floor, College Park, 30349; 404/994-5276
Louisville, Kentucky—(FSDO), Bowman Field, 40205; 502/582-6116
Jackson, Mississippi—(FSDO), Jackson International Airport, 120 North Hangar Drive, Suite 6, 39208; 601/960-4633
Charlotte, North Carolina—(FSDO), Charlotte-Douglas International Airport, 5318 Morris Field Drive, 28208; 704/392-3214
Raleigh, North Carolina—(GADO), Raleigh-Durham Airport, Route 1, Box 486A, Morrisville, 27560; 919/755-4240
Winston-Salem, North Carolina—(FSDO), 8025 North Point Boulevard, Room 250, 27106; 919/761-3147
San Juan, Puerto Rico—(FSDO), Luis Munoz Marin International Airport, Room 203A, Santurce, 00914; 809/791-5050
Columbia, South Carolina—(FSDO), Columbia Metropolitan

Airport, 2819 Aviation Way, West Columbia, 39169; 803/765-5931
Memphis, Tennessee—(FSDO), 3385 Airways Boulevard, Suite 115, 38116; 901/521-3820
Nashville, Tennessee—(FSDO), 322 Knapp Boulevard, Room 101, Nashville Metropolitan Airport, 37217; 615/251-5661

Southwest Region

Little Rock, Arkansas—(GADO), FAA/NWS Building, Room 201, Adams Field, 72202; 800/632-9566, 501/372-3437
Baton Rouge, Louisiana—(GADO), Ryan Field, 9191 Plank Road, 70811; 800/821-1960, 504/356-5701
Albuquerque, New Mexico—(GADO), 1601 Randolph Road, S.E., Suite 200N, 87106; 505/247-0156
Oklahoma City, Oklahoma—(GADO), Suite 601, FAA Building, 1300 South Meridian, Wiley Post Airport, Bethany, 73108; 405/231-4196
Dallas, Texas—(GADO), 8032 Air Freight Lane, Love Field, 75235; 800/442-3394, 214/357-0142
Fort Worth, Texas—(FSDO), Alliance Airport, 14400 Perimeter Road, 76177; 817/491-5000
Houston, Texas—(GADO), Hobby Airport, 8800 Paul B. Koonce Drive, Room 152, 77061; 713/643-6504
Lubbock, Texas—(GADO), Route 3, Box 51, 79401; 800/858-4115, 806/762-0335 (New Mexico), 800/692-4115 (Texas)
San Antonio, Texas—(GADO), International Airport, 10100 Reunion Place, Suite 201, 78216; 800/292-2023, 512/341-4371

Western-Pacific Region

Scottsdale, Arizona—(GADO), Scottsdale Municipal Airport, 15041 North Airport Drive, 85260; 602/640-2561
Fresno, California—(FSDO-4), Fresno Air Terminal, 4955 East Anderson Avenue, Number 110, 93727-1573; 209/487-5306
Long Beach, California—(FSDO), Long Beach Airport, 2815 East Spring Street, 90801; 213/426-7134
Los Angeles, California—(FSDO), Los Angeles International Airport, 5885 West Imperial Highway, 90045; 213/215-2150
Oakland, California—(FSDO), Oakland International Airport Building, L-105, Earhart Road, 94614, 415/273-7155
Riverside, California—(GADO), Riverside Municipal Airport, 6961 Flight Road, 92504; 714/351-6701
Sacramento, California—(FSDO-25), 6650 Belleau Wood Lane, 95822; 916/551-1721
San Diego, California—(GADO), Montgomery Field Airport, 8665 Gibbs Drive, Suite 110, 92123; 619/557-5281
San Jose, California—(GADO), San Jose Municipal Airport, 1387 Airport Boulevard, 95110; 408/291-7681
Van Nuys, California—(FSDO), 16501 Sherman Way, Suite 330, 91406; 818/904-6291
Honolulu, Hawaii—(FSDO), 90 Nakolo Place, Room 215, 96819; 808/836-0615
Las Vegas, Nevada—(FSDO-6), 241 East Reno Avenue, Suite 200, 89119; 702/388-6482
Reno, Nevada—(GADO), Cannon International Airport, 601 South Rock Boulevard, D-102, 89502; 702/784-5321

FAA INTERNATIONAL OFFICES

North America
Montreal, Canada—U.S. Member ICAO Air Navigation Commission. Address: 1000 Sherbrooke Street, Room 753, West Montreal, Quebec, Canada. Telephone 514/285-8304

Europe, Africa, and Middle East Headquarters
FAA, Brussels, Belgium—Director. APO Address: c/o American Embassy, APO New York, New York 09667-1011. Location: 15 Rue de la Loi, B-1040 Brussels, Belgium. International Air Mail Address: FAA, c/o American Embassy, 27 Boulevard du Regent, 1000 Brussels, Belgium. Telephone (32) 2 513-3830, extension 2236.

Europe, Africa, Middle East Regional Offices

Amman, Jordan—Chief Civil Aviation Assistance Group. APO Address: FAA, c/o American Embassy, APO New York, New York 09892. Telephone (962) 6 644 371, extension 276 and 277.

Berlin, Germany—U.S. Administration for Aeronautics. APO Address: EU-9, U.S. Mission, Berlin, Air Post Office Box 5500, APO New York, New York 09742. Telephone (49) 30 690-9666.

Frankfurt, Germany—FAA Representative. APO Address: c/o American Consulate General, Flight Inspection Field Office, APO New York, New York 09213. International Air Mail Address: FAA, Flight Inspection Field Office, c/o American Consulate General, Post Fach 10 01 34, D-6000 Frankfurt/Main 1, Federal Republic of Germany. Telephone (49) 69 705-211.

Paris, France—FAA Representative. APO Address: c/o American Embassy, APO New York, New York 09777. Location and International Air Mail Address: FAA, c/o American Embassy, Two Avenue Gabriel, 75382 Paris Cedex 08, France. Telephone (33) 4 296-1202, extension 2901, 2931; (33) 4 261-8075.

London, England—FAA Representative. FPO Address: c/o American Embassy, Box 41, FPO New York, New York 09509. Location and International Air Mail Address: FAA, c/o American Embassy, Grosvenor Square, London WIAE 1AE, Great Britain. Telephone (44) 7 1 499-9000, extension 2410 or 2440; (44) 7 1 408-8094.

Madrid, Spain—Chief Civil Aviation Assistance Group/CASLO. APO Address: c/o American Embassy, Box 38, APO New York, New York 09285-0001. Telephone (34) 1 320-5392, extension 2426. International address: c/o American Embassy, Serrano 75, Box 38, Madrid 28002, Spain.

Muscat, Oman—Chief Civil Aviation Assistance Group. International Air Mail Address: c/o American Embassy Muscat, Department of State Mail Room, Washington, D.C. 20520-6220 (less than 2 lbs) or 20521-6220 (more than 2 lbs). Telephone (968) 519390, Embassy. FAA International CCAAG. Post Office Box 50202, Madinat, Qaboos, Sultanate of Oman.

Rome, Italy—FAA Representative. APO Address: c/o American Embassy, APO New York, New York 09794-0007. Location and International Air Mail Address: FAA, c/o American Embassy Annex, Via V Veneto 119, 00187 Rome, Italy. Telephone (39) 6 4674-2298, 4674-2347, 4674-1 (extensions 2298 and 2347).

Dakar, Senegal—FAA Representative. Air Mail Address: FAA, Dakar, Department of State, Washington, D.C. 20520. International Air Mail: FAA, c/o American Embassy, Boite Postale 49 Dakar Senegal. Telephone (22) 1 236753, 234296 Embassy, extension 2180; (22) 1 216753.

Southern Headquarters

(Responsible for South America, Caribbean Sea, and Central America, excluding Mexico)

FAA Southern Region—Director. Address: Post Office Box 20636, Atlanta, Georgia, USA, 30320. Location: 3400 Norman Berry Drive, East Point, Georgia 30344. Telephone 404/763-7222.

Southern Regional Office

Rio de Janeiro, Brazil—FAA Representative. APO Address: c/o U.S. Consulate General, APO Miami, Florida 34030. Telephone 5521-240-9853.

Southwest Headquarters

(Responsible for Mexico)

FAA Southwest Region—Director. Address: Post Office Box 1689, Fort Worth, Texas 76193. Location: 4400 Blue Mound Road, Fort Worth, Texas 76193. Telephone 817/624-5000.

Western-Pacific Region

Finegayan, Guam—Resident Director/Sector Manager. Location and International Air Mail Address: Route 008, Finegayan, Guam 96912. Telephone 671/355-5026.

Pago Pago, Samoa—Resident Director. Location and International Air Mail Address: Post Office Box 8, Pago Pago, American Samoa 96799. Telephone 684/639-9485.

Tokyo, Japan—FAA Representative. APO Address: East Asian IFO, c/o American Embassy, APO San Francisco 96503. Telephone Tokyo 583-7141, extension 7625; Manager, Flight Inspection Field Office. APO Address: Building 1376, APO San Francisco 96328. Telephone AUT 225-5682.

NATIONAL TRANSPORTATION SAFETY BOARD

800 Independence Avenue, S.W.
Washington, D.C. 20594
202/382-6600

NTSB FIELD OFFICES

Anchorage, Alaska—222 West 7th Avenue, Room 142, Box 11, 99513; 907/271-5001

Atlanta, Georgia—1720 Peachtree Street, N.W., Suite 321, 30309; 404/347-7385

Chicago, Illinois—31 West 775 North Avenue, 60185; 708/377-8177

Denver, Colorado—4760 Oakland Street, Suite 500, 80239; 303/361-0600

Fort Worth, Texas—1200 Copeland Road, Suite 300, Arlington, Texas 76011; 817/885-6800

Kansas City, Missouri—8725 Rosehill, Suite 109, Lenexa, Kansas 66215; 913/492-1196

Los Angeles, California—Federal Building, 15000 Aviation Boulevard, Post Office Box 6117, Lawndale, California 90261; 213/297-1041

Miami, Florida—8405 N.W. 53rd Street, Suite B-103, 33166; 305/536-5057

Parsippany, New Jersey—2001 Route 46, Suite 203, 07054; 201/334-6420

Seattle, Washington—19518 Pacific Highway, South, Room 201, Seattle, Washington 98188; 206/764-3782

Washington, D.C.—Bureau of Field Operations, 800 Independence Avenue, S.W., Washington, D.C. 20594; 202/382-6820

FEDERAL COMMUNICATIONS COMMISSION

1919 M Street, N.W.
Washington, D.C. 20554
Consumer Assistance and Small Business Division, 202/632-7000

FEDERAL AVIATION REGULATIONS

AIRCRAFT NATIONALITY MARKS

AN	Nicaragua	OO	Belgium
AP	Pakistan	OY	Denmark
B	China (Nationalist)	PH	Netherlands
CB, CP	Bolivia	PI	Philippine Republic
CC	Chile	PJ	Antilles
CF	Canada	PK	Indonesia
CN	Morocco	PP, PT	Brazil
CS, CR	Portugal	PZ	Surinam
CU	Cuba	SE	Sweden
CX	Uruguay	SN	Sudan
D	Germany	SP	Poland
EC	Spain	SU	Egypt
EI, EJ	Ireland	SX	Greece
EA	Liberia	TC	Turkey
EP	Iran	TF	Iceland
ET	Ethiopia	TG	Guatemala
F	France	TI	Costa Rica
G	Great Britain	VH	Australia
HA	Hungary	VP, VQ, VR	Great Britain
HB	Switzerland		Colonies & Protectorates
HC	Ecuador	VT	India
HH	Haiti	XA, XB, XC	Mexico
HI	Dominican Republic	XH	Honduras
HK	Columbia	XV	Vietnam
HL	Korea	XZ, Xy	Burma
HP	Panama	YA	Afghanistan
HS	Thailand	YE	Yemen
HZ	Saudi Arabia	YJ	New Hebrides
I	Italy	YR	Romania
JA	Japan	YS	El Salvador
JY	Jordan	YV	Venezuela
LN	Norway	ZA	Albania
LV, LQ	Argentina	ZK, ZL, ZM	New Zealand
LX	Luxembourg	ZP	Paraguay
LZ	Bulgaria	S, ZT, ZU	Union of South Africa
MR	Monaco	4R	Sri Lanka (Ceylon)
N	United States	4X	Israel
OB	Peru	5A	Libya
OD	Lebanon	9G	Ghana
OE	Austria		
OH	Finland		

Glossary

The following list of words and terms represents those that are often encountered in the field of aviation. Additional terms are defined in the text; however, there are still other words or terms that technicians hear or see and for which they may want a definition. For a more complete list of definitions, technicians should consult a good aviation dictionary.

A & P mechanic. An aircraft mechanic who is certificated by the Federal Aviation Administration. A & P certification indicates that the mechanic is qualified to perform certain maintenance and inspection functions on aircraft that are licensed in the United States.

Absolute altitude. The actual altitude above the surface of the earth, either land or water. Also means *terrain clearance*.

Adhesive. A substance used to bond two materials together by chemical means.

Adiabatic. Denotes a change in volume or pressure without a gain or loss of heat.

Aerodynamics. The science relating to the effects of air or other gases in motion.

Aileron. A moveable control surface attached to the trailing edge of a wing to control an airplane in roll.

Airfoil. Any surface that is designed to obtain a useful reaction, or lift, from air as it passes over its surface. Wings, control surfaces, propeller blades, and helicopter rotors are all examples of airfoils.

Airfoil profile. The outline of an airfoil section.

Airfoil section. The cross-sectional shape of an airfoil, viewed as if it were cut through in a fore-and-aft plane.

Airworthiness. The state or quality of an aircraft or of an aircraft component which will enable safe performance according to specifications.

Airworthiness directive. A directive issued by the FAA requiring that certain inspections and/or repairs be performed on specific makes and models of aircraft, engines, propellers, rotors, or appliances and setting forth time limits for such operations.

Alloy. A solid solution consisting of two or more metallic constituents. The alloy usually contains one predominant metal to which are added small amounts of other metals to improve strength and heat resistance.

Alloy steel. A steel that contains metallic elements other than those found in carbon steels.

Ambient. Surrounding. Ambient conditions are those conditions existing in the surrounding area.

Angle of attack. The acute angle between the chord line of a wing and the direction that the air strikes the wing (relative wind).

Angle of incidence. The acute angle formed by the chord of the airfoil and the longitudinal axis of the aircraft. In most helicopters, this angle continually changes while the helicopter is in horizontal flight; each blade varies its angle of incidence in making a complete cycle of rotation.

Arm (weight and balance). The horizontal distance in inches between a reference datum line and the center of gravity of an object. If the object is behind the datum, the arm is positive, and if it is ahead of the datum, the arm is negative.

Articulated rotor. A rotor system whose individual or collective main rotor blades are free to flap, feather, and drag individually or collectively.

Aspect ratio. The ratio of the span of the wing (length) to the mean chord of the wing (width). The wing span squared, divided by area.

Atmospheric pressure. The pressure that is exerted on the surface of the earth by the air that surrounds the earth. Under standard conditions, at sea level, the atmospheric pressure is 14.69 pounds per square inch.

Autogiro. A rotary-wing aircraft whose rotor is turned throughout its flight by air forces resulting from the motion of the craft through the air.

Autorotation. The process of producing lift with airfoils that rotate freely and are not engine-driven. When a helicopter enters into autorotation, the flow of air is upward through the main rotor system, rather than downward, as is the case when the rotors are engine-driven.

Bearing (direction). The horizontal direction of an object or point, usually measured clockwise from a reference line or direction through 360°.

Bearing (mechanical). A part of a machine that supports a journal, pivot, or pin that rotates, oscillates, or slides.

Bernoulli's principle. A principle of physics that explains the relationship that exists between kinetic and potential energy as it is found in a fluid that is moving at a subsonic speed in a smooth (laminar) flow. Bernoulli's principle states that if the total energy in the stream of fluid is not changed, any increase in the kinetic energy in the fluid (the velocity of the fluid), will cause a corresponding decrease in the potential energy of the fluid (its pressure).

Blind rivet. A special rivet designed to be used where only one side is accessible and it is not possible to use a bucking bar for riveting.

Bond. An attachment of one material to another or the finish to a metal or fabric.

Bonded structure. A structure whose parts are joined together by chemical rather than mechanical methods.

Bonding agent. An adhesive used to bond structure parts together.

Brazing. A method of joining two pieces of metal by wetting their surface with a molten alloy of copper, zinc, or tin.

Bulkhead. A wall, partition, or similar member or structure in an airplane or missile fuselage at right angles to the longitudinal axis of the body and serving to strengthen, divide, or help give shape to the body.

Cabane. A pyramidal arrangement of struts used to support a wing above the fuselage of an airplane or to provide a point of attachment for the inner ends of half axles in some types of landing gear.

Calibrated airspeed. An airspeed value derived when corrections have been applied to an indicated airspeed to compensate for installation errors, instrument errors, errors in the pitot-static system, and aircraft attitude errors.

Camber. The amount of curve of an airfoil section. The curve on the top of the airfoil is called upper camber, and the curve on the bottom of the airfoil is called lower camber.

Canard. An aircraft or aircraft configuration having its horizontal stabilizing and control surfaces in front of the wing or wings.

Cantilever. A beam or member supported at or near one end only, without external bracing.

Catalyst. A material which is used to bring about a change but does not actually enter into the change itself.

Cathedral. The downward angle of a wing as measured from horizontal on the lateral axis. The outer end of the lower surface of the wing is lower than the root end.

Center of gravity. The point within an aircraft on which, for balance purposes, the total force due to gravity is concentrated.

Center of pressure. The point between the leading and trailing edges of an airfoil at which the aerodynamic forces acting on the airfoil are considered to be concentrated.

Centrifugal force. A force created in a rotating system that is exerted in a direction away from the axis of rotation. More specifically, on a helicopter with a rotating rotor system, centrifugal force tends to pull the blades off the rotor head.

Centripetal force. The force within a body that opposes centrifugal force as a body rotates or spins. Centripetal

force acts inside a piece of rotating machinery and it tries to pull the rotating object in toward the center of its rotation.

Chord. The reference line from which the upper and lower contours of an airfoil are measured. A straight line directly across an airfoil from the leading edge to the trailing edge.

Cohesion. The act or process of holding tightly together.

Collective pitch change. A means of simultaneously increasing or decreasing the pitch of the main rotor blades. The pitch change is equal on all the main rotor blades.

Compression rib. A rib, more strongly built than others, designed to resist compression forces.

Conduction. The transfer of energy through a conductor by means of molecular activity without any external motion.

Coning. The upward flexing of the rotor blades resulting from the vectorially combined effects of centrifugal force and lift.

Convection. The process by which heat is transferred by movement.

Crack. A partial separation of material usually caused by excessive internal stresses resulting from overloading, vibration fatigue, temperature changes, or defective assemblies.

Critical angle of attack. The highest angle of attack at which the air will flow over the airfoil in a smooth flow. At angles of attack that are greater than the critical angle of attack, the air will burble (it will flow in a disturbed pattern), and lift will be lost. The critical angle of attack is sometimes called the stalling angle of attack.

Cyclic-pitch change. A mechanical means employed to change the pitch of the main rotor blades when the rotor system makes a cycle of rotation. The angular change of opposing rotor blades is equal and opposite.

Dash numbers. Numbers following, and separated from, a part number by a dash. The number usually used in identifying either the components or size of the part.

Datum. An imaginary vertical plane from which all horizontal measurements are taken.

Declination. The angular distance of a body from the celestial equator, measured along the hour circle passing through the body and named north or south according to the direction of the body from the celestial equator.

Density. The weight per unit volume of any substance.

Density altitude. The pressure altitude corrected for free air temperature.

Dihedral. The upward angle of a wing as measured from horizontal on the lateral axis. The outer end of the lower surface of the wing is higher than the root end.

Directional stability. Stability of an aircraft with respect to the vertical axis, that is, the tendency of an aircraft not to turn unless induced to do so with rudder control.

Dissymmetry of lift. The unequal lift across a rotor disk that occurs in horizontal flight as a result of the difference in velocity of the air passing over the advancing half of the disk area and the velocity of the air passing over the retreating half of the disk area.

Drag. The force that tends to resist an airfoil's passage through the air. Drag is parallel to the relative wind and varies as the square of the velocity.

Drift. The lateral divergence or movement of the flight path of a flying vehicle from the direction of its heading, measured between the heading and track, usually caused by a crosswind.

Eddy. A region of undirected or swirling flow, as in the flow of air about or behind a body; a vortex.

Effective span. The span of an airfoil less corrections for tip loss.

Elevator. A horizontal, hinged control surface, attached to the trailing edge of a horizontal stabilizer of an airplane, designed to control the airplane about the lateral axis.

Elevon. A control surface that functions as both an elevator and an aileron.

Empennage. The assembly of stabilizing and control surfaces at the tail of an aircraft.

Equivalent airspeed. A calibrated airspeed corrected for the effect of compression of air in the pitot system.

Fairing. A part of a structure having a smooth, streamlined outline, used to cover a nonstreamlined object or to smooth a junction.

Fastener. A device such as a rivet or bolt used to fasten two objects together.

Fatigue strength. The measured resistance of a body to failure caused by repeated applications of stress.

Federal Aviation Regulations. Regulations that have been established by the Federal Aviation Administration. These regulations are the rules that govern the operation of aircraft, airways, and aviation personnel.

Ferrous. The term describing metal that is derived from an iron base.

Fill. The direction across the width of a fabric.

Fill threads. Threads running across the width of a piece of fabric.

Fin. A common term for the vertical stabilizer.

Flange. The formed or widened portion at the edges of an I beam or U-channel-shaped part or around a lightening hole. Flanges are formed to add stiffness.

Flapping. The movement of a blade in a direction parallel to the axis of rotation. Flapping results from the dissymmetry of lift that exists between the advancing and retreating halves of the disk area during horizontal flight. Flapping is also induced when the tip-path plane is tilted (application of cyclic control).

Flutter. A vibration or oscillation of definite period set up in an aileron, wing, or other surface of aerodynamic forces and maintained by the aerodynamic forces and by the elastic inertial forces of the object itself.

G force. An accelerating force.

Grip range. The difference between the maximum and minimum thickness of material that may be joined by a fastener.

Ground speed. The speed of an aircraft relative to the earth's surface.

Gust. A sudden and brief change of wind speed or direction.

Gyroscopic precession. A characteristic of all rotating bodies. Such bodies will be uniformly displaced 90° in the direction of rotation from where a force is applied.

Heading. The horizontal direction in which a craft points as it flies through the air, usually expressed as an angle measured clockwise from north to the longitudinal axis of the craft.

Heat sink. In thermodynamic theory, a means by which heat is stored or dissipated in or transferred from the system under consideration. The thick metal shield on the nose cone of a rocket is often designed as a heat sink to absorb heat generated by air friction.

Helicopter. A rotorcraft that, for its horizontal motion, depends principally on its engine-driven rotors.

Honeycomb. A hexagonal cellular material made of thin metal, paper, or plastic used as core material for sandwich structure. Named after the honeybee's honeycomb because of its appearance.

Hypersonic flow. A flow at very high supersonic speed.

IFR. The abbreviation of *instrument flight rules*. An airplane flying IFR is flying according to instrument flight rules.

ILS. The abbreviation for *instrument landing system*.

Impact pressure. That pressure of a moving fluid brought to rest which is in excess of the pressure the fluid possesses when it does not flow. Also, it is the dynamic pressure at lower flow speeds.

Indicated airspeed. The airspeed measurement shown by an airspeed indicator.

Induced drag. That part of the drag of an airfoil caused by the lift, that is, the change in the direction of the airflow.

Inspection authorization. An authorization that is issued by the Federal Aviation Administration to certain licensed and experienced A & P mechanics that allows them to perform annual and progressive inspections. This authorization allows them to approve aircraft for return to service after certain types of major repairs and alterations.

Interference drag. Drag due to the interference of the airflow around aircraft components close to one another.

Knot. A rate of speed equivalent to 1 nmi/h (6076.1033 ft/h) [1852 m/h].

Laminated. Composed of thin layers of material firmly bonded or united together.

Laminated structure. A structure of an aircraft made up of layers of material bonded together to form complex shapes.

Lapse rate. The rate of change of temperature, pressure, or some other meteorological phenomenon with altitude, usually the rate of decrease of temperature with increased height.

Lateral stability. The tendency of an aircraft to resist rolling.

Lead/lag (hunting). The angular motion of the blade about the lag axis. This type of motion results in a speeding up or slowing down of the blade in the plane of rotation.

Load. The ratio of a specified load to the total weight of the aircraft. Load factors are expressed in terms of G units.

Longitudinal stability. The stability of an aircraft with respect to pitching motions.

Mach number. A ratio of the true airspeed of the aircraft divided by the speed of sound in the air through which the aircraft is flying at the time.

Manometer. A gage for measuring the pressure of gases or vapors having a sensing device consisting of a column of liquid in a glass tube.

Mass. The quantity of matter in a body as measured in its relation to inertia.

Mean aerodynamic chord. The chord of an imaginary rectangular airfoil that has the same aerodynamic force vectors as the actual airfoil.

MIL spec. Standard specifications for materials and parts to ensure compliance with quality and performance standards. The specifications were originally for military purposes; hence the abbreviation MIL.

Negative dihedral. A downward inclination of a wing or other surface.

Parasite drag. The drag produced by air flowing over surfaces not involved in producing lift.

Pitot-static system. A system of ports, receptacles and associated tubing to capture total and static pressure and route it to the pitot-static instruments. Pitot-static instruments normally include the airspeed indicator, altimeter, and vertical speed indicator.

Positive g. A force acting on a body undergoing positive acceleration.

Power loading. The ratio of the gross weight of an airplane to its power.

Pressure altitude. Altitude above standard sea level—29.92 inHg [101.31 kPa]—measured with a barometric altimeter.

Primary flight controls. The aircraft flight controls that cause the aircraft to rotate about its three axes. In an airplane, the primary controls are the ailerons, the elevators, and the rudder.

Rigid rotor. A helicopter main rotor that is not hinged at the hub to permit independent movement of each blade. The blades have freedom of movement only along the longitudinal axis to permit the changing of pitch.

Rotary-wing aircraft (rotor craft). A type of aircraft that is supported in the air wholly or in part by wings or blades rotating about a vertical axis.

Rudder. A vertical control surface hinged to the tail post, aft of the vertical stabilizer, designed to apply yawing moments to the airplane.

SAE number Any of a series of numbers established as standards by the Society of Automotive Engineers for grading materials, components, and other products.

Sandwich construction. A type of construction in which two sheets, sides, or plates are separated by a core of stiffening material, such as honeycomb or balsa wood.

Semirigid rotor. A two-bladed rotor system whose blades are fixed to a hub that is free to teeter on the mast. The blades also have a means to feather.

Slug. A unit of mass having a value of approximately 32.175 lb [14.594 kg] under standard conditions of gravity.

Speed of sound. The speed of propagation of sound waves. At sea level, in the standard atmosphere, the speed of sound is 761 mph or 661 kn.

Stabilator. A horizontal all-movable tail surface.

Stability. The property of an aircraft or other body to resist displacement and, if displaced, to develop forces that will tend to restore the original condition (straight and level flight).

Stress raiser. A scratch, groove, rivet hole, forgoing defect, or other structural discontinuity causing the concentration of stress.

Thermal. A rising current of warm air.

Thrust. A reaction force measured in pounds.

Tip-path plane. The imaginary circular surface formed by a plane passed through the average tip path of the rotor blades.

Translational lift. The additional lift obtained when leaving the ground cushion and entering horizontal flight.

Transonic speed. The speed of a body relative to the surrounding fluid at which the flow is in some places subsonic and other places supersonic; usually from Mach 0.8 to 1.2.

Type certificate data sheet. The official specifications of an aircraft, engine, or propeller that are issued by the Federal Aviation Administration.

Velocity. A vector quantity equal to speed in a given direction.

Vertical stabilizer. A vertical airfoil fixed approximately parallel to the plane of symmetry of an airplane; also called the fin.

Viscosity. The resistance of a fluid to flow.

Washin. A permanent warp or twist given a wing such that some specified angle of attack is greater at the tip than at the root.

Washout. The opposite of washin; that is, a wing warp or twist in which the tip angle is less than the root angle.

Wing loading. The ratio of the total gross weight of the aircraft divided by the total wing area. Wing loading is expressed in pounds per square feet.

Yaw. The movement of an aircraft about the vertical axis.

Zero-lift angle of attack. The angle of attack at which no lift is created.

Index